VOLUME ONE HUNDRED AND THIRTY

ADVANCES IN PROTEIN CHEMISTRY AND STRUCTURAL BIOLOGY

Protein Design and Structure

VOLUME ONE HUNDRED AND THIRTY

ADVANCES IN PROTEIN CHEMISTRY AND STRUCTURAL BIOLOGY

Protein Design and Structure

Edited by

ROSSEN DONEV

MicroPharm Ltd,
United Kingdom

ELSEVIER

ACADEMIC PRESS
An imprint of Elsevier

Academic Press is an imprint of Elsevier
50 Hampshire Street, 5th Floor, Cambridge, MA 02139, United States
525 B Street, Suite 1650, San Diego, CA 92101, United States
The Boulevard, Langford Lane, Kidlington, Oxford OX5 1GB, United Kingdom
125 London Wall, London, EC2Y 5AS, United Kingdom

First edition 2022

Notices

Knowledge and best practice in this field are constantly changing. As new research and experience broaden our understanding, changes in research methods, professional practices, or medical treatment may become necessary.

Practitioners and researchers must always rely on their own experience and knowledge in evaluating and using any information, methods, compounds, or experiments described herein. In using such information or methods they should be mindful of their own safety and the safety of others, including parties for whom they have a professional responsibility.

To the fullest extent of the law, neither the Publisher nor the authors, contributors, or editors, assume any liability for any injury and/or damage to persons or property as a matter of products liability, negligence or otherwise, or from any use or operation of any methods, products, instructions, or ideas contained in the material herein.

ISBN: 978-0-323-99229-9
ISSN: 1876-1623

For information on all Academic Press publications
visit our website at https://www.elsevier.com/books-and-journals

Publisher: Zoe Kruze
Acquisitions Editor: Leticia Lima
Developmental Editor: Naiza Ermin Mendoza
Production Project Manager: Abdulla Sait
Cover Designer: Victoria Pearson

Typeset by STRAIVE, India

Contents

Contributors

Mustfa Alkhanani
Emergency Service Department, College of Applied Sciences, Al Maarefa University, Riyadh, Saudi Arabia

Abdulaziz S. Alothaim
Department of Biology, College of Science in Zulfi, Majmaah University, Majmaah, Saudi Arabia

Kamakshi Balsekar
Laboratory of Integrative Genomics, Department of Integrative Biology, School of Bio Sciences and Technology, Vellore Institute of Technology, Vellore, Tamil Nadu, India

R. Bithia
Laboratory of Integrative Genomics, Department of Integrative Biology, School of Bio Sciences and Technology, Vellore Institute of Technology, Vellore, Tamil Nadu, India

Aparna Boral
Department of Life Sciences, Presidency University, Kolkata, West Bengal, India

Ana Camara-Artigas
Departamento de Química Física, Universidad de Almería, Campus de Excelencia Internacional Agroalimentario ceiA3 y CIAMBITAL, Almeria, Spain

Francisco Castillo
Fundación MEDINA, Centro de Excelencia en Investigación de Medicamentos Innovadores en Andalucía, Parque Tecnológico Ciencias de la Salud, Granada, Spain

Lei Chen
Division of Cardiovascular Diseases, Cardiovascular Research Institute, University of Kansas Medical Center, Kansas City, KS, United States

Buddhadeb Dawn
Division of Cardiovascular Diseases, Cardiovascular Research Institute, University of Kansas Medical Center, Kansas City, KS, United States

Maria Di Bernardo
Consiglio Nazionale delle Ricerche, Istituto per la Ricerca e l'Innovazione Biomedica (IRIB), Palermo, Italy

C. George Priya Doss
Laboratory of Integrative Genomics, Department of Integrative Biology, School of Bio Sciences and Technology, Vellore Institute of Technology, Vellore, Tamil Nadu, India

Mohammad ElGamacy
University Hospital Tübingen, Division of Translational Oncology; Max Planck Institute for Biology, Tübingen, Germany

Harold Elias
Division of Cardiovascular Diseases, Cardiovascular Research Institute, University of Kansas Medical Center, Kansas City, KS, United States

Nikita Jain
Laboratory of Integrative Genomics, Department of Integrative Biology, School of Bio Sciences and Technology, Vellore Institute of Technology, Vellore, Tamil Nadu, India

Ritika Kabra
National Centre for Cell Science, NCCS Complex, SP Pune University Campus, Pune, India

Balu Kamaraj
Department of Neuroscience Technology, College of Applied Medical Sciences, Imam Abdulrahman Bin Faisal University, Jubail, Saudi Arabia

Tim J. Kamerzell
Department of Pharmaceutical Chemistry, The University of Kansas, Lawrence; Division of Internal Medicine, HCA MidWest Health; Applied AI Technologies, LLC, Overland Park; Division of Cardiovascular Diseases, Cardiovascular Research Institute, University of Kansas Medical Center, Kansas City, KS; Cardiovascular Division, Brigham and Women's Hospital and Harvard Medical School, Boston, MA, United States

Madhurima Khamaru
Department of Life Sciences, Presidency University, Kolkata, West Bengal, India

Irene Luque
Departamento de Química Física e Instituto de Biotecnología, Facultad de Ciencias, Universidad de Granada, Granada, Spain

Calum MacRae
Cardiovascular Division, Brigham and Women's Hospital and Harvard Medical School, Boston, MA, United States

Jose C. Martinez
Departamento de Química Física e Instituto de Biotecnología, Facultad de Ciencias, Universidad de Granada, Granada, Spain

C. Russell Middaugh
Department of Pharmaceutical Chemistry, The University of Kansas, Lawrence, KS, United States

Brittney Mikell
Cardiovascular Division, Brigham and Women's Hospital and Harvard Medical School, Boston, MA, United States

Devrani Mitra
Department of Life Sciences, Presidency University, Kolkata, West Bengal, India

Javier Murciano-Calles
Departamento de Química Física e Instituto de Biotecnología, Facultad de Ciencias, Universidad de Granada, Granada, Spain

Florencio Pazos
Computational Systems Biology Group, National Center for Biotechnology (CNB-CSIC), Madrid, Spain

Annalisa Pinsino
Consiglio Nazionale delle Ricerche, Istituto di Farmacologia Traslazionale (IFT); Consiglio Nazionale delle Ricerche, Istituto per la Ricerca e l'Innovazione Biomedica (IRIB), Palermo, Italy

V. Anu Preethi
School of Computer Science and Engineering, Vellore Institute of Technology, Vellore, Tamil Nadu, India

Vibin Ramakrishnan
Department of Biosciences and Bioengineering, Indian Institute of Technology Guwahati, Guwahati, Assam, India

Ondipilliraja Rudhra
Computer Aided Drug Design and Molecular Modelling Lab, Department of Bioinformatics, Science Block, Alagappa University, Karaikudi, Tamil Nadu, India

Javier Ruiz-Sanz
Departamento de Química Física e Instituto de Biotecnología, Facultad de Ciencias, Universidad de Granada, Granada, Spain

Sajitha Sasidharan
Department of Biosciences and Bioengineering, Indian Institute of Technology Guwahati, Guwahati, Assam, India

Chandrabose Selvaraj
Computer Aided Drug Design and Molecular Modelling Lab, Department of Bioinformatics, Science Block, Alagappa University, Karaikudi, Tamil Nadu, India

Mariem Sidenna
Department of Biomedical Sciences, College of Health and Sciences, QU Health, Qatar University, Doha, Qatar

Sanjeev Kumar Singh
Computer Aided Drug Design and Molecular Modelling Lab, Department of Bioinformatics, Science Block, Alagappa University, Karaikudi, Tamil Nadu, India

Shailza Singh
National Centre for Cell Science, NCCS Complex, SP Pune University Campus, Pune, India

R. Siva
Laboratory of Integrative Genomics, Department of Integrative Biology, School of Bio Sciences and Technology, Vellore Institute of Technology, Vellore, Tamil Nadu, India

Nikolai N. Sluchanko
A.N. Bach Institute of Biochemistry, Federal Research Center "Fundamentals of Biotechnology" of the Russian Academy of Sciences, Moscow, Russian Federation

Baviri Sowmya
Laboratory of Integrative Genomics, Department of Integrative Biology, School of Bio Sciences and Technology, Vellore Institute of Technology, Vellore, Tamil Nadu, India

Sekhar Talluri
Department of Biotechnology, GITAM, Visakhapatnam, India

D. Thirumal Kumar
Laboratory of Integrative Genomics, Department of Integrative Biology, School of Bio Sciences and Technology, Vellore Institute of Technology, Vellore; Meenakshi Academy of Higher Education and Research (Deemed to be University), Chennai, Tamil Nadu, India

S. Udhaya Kumar
Laboratory of Integrative Genomics, Department of Integrative Biology, School of Bio Sciences and Technology, Vellore Institute of Technology, Vellore, Tamil Nadu, India

Rinku Polachirakkal Varghese
Laboratory of Integrative Genomics, Department of Integrative Biology, School of Bio Sciences and Technology, Vellore Institute of Technology, Vellore, Tamil Nadu, India

Hatem Zayed
Department of Biomedical Sciences, College of Health and Sciences, QU Health, Qatar University, Doha, Qatar

CHAPTER ONE

Algorithms for protein design

Sekhar Talluri*

Department of Biotechnology, GITAM, Visakhapatnam, India

*Corresponding author: e-mail address: stalluri.gitam@gmail.com

Contents

Abstract

Computational Protein Design has the potential to contribute to major advances in enzyme technology, vaccine design, receptor-ligand engineering, biomaterials, nanosensors, and synthetic biology. Although Protein Design is a challenging problem, proteins can be designed by experts in Protein Design, as well as by non-experts whose primary interests are in the applications of Protein Design. The increased accessibility of Protein Design technology is attributable to the accumulated knowledge and experience with Protein Design as well as to the availability of software and online resources. The objective of this review is to serve as a guide to the relevant literature with a focus on the novel methods and algorithms that have been developed or applied for Protein Design, and to assist in the selection of algorithms for Protein Design. Novel algorithms and models that have been introduced to utilize the enormous amount of experimental data and novel computational hardware have the potential for producing substantial increases in the accuracy, reliability and range of applications of designed proteins.

Advances in Protein Chemistry and Structural Biology, Volume 130
ISSN 1876-1623
https://doi.org/10.1016/bs.apcsb.2022.01.003

1. Introduction

The objective of protein design is to compute the optimal sequence of amino acids for a specified task (Traoré, Allouche, André, Schiex, & Barbe, 2017). In some cases, the task may involve specification of the fold or the shape of a protein (Kuhlman, 2019). The objective may specify protein function, thermal stability, solubility, sequence length, combinations of the above, or other properties (Korendovych & DeGrado, 2020). The activity, shape, stability and other properties that may be part of the objective depend upon the amino acid sequence of the designed protein. The amino acid sequence of a protein determines the statistical weights of the accessible conformational states, and the set of accessible conformational states determine its physical, chemical and biological properties (Michael & Simonson, 2021). Force fields, such as those that were initially developed for protein structure prediction can be utilized for the prediction of the conformational energies and these conformational energies can be utilized for computation of the statistical weights of the conformers for a specified amino acid sequence (Simonson et al., 2013). The structure (conformation) of a protein determines its properties. A variety of algorithms and functions have been developed for using the structure of a protein for computation of other properties of interest, such as thermal stability, ligand binding and enzymatic activity (Lucas & Kortemme, 2020; Talluri, 2011a). Therefore, an objective function can be constructed to relate the properties of interest and the sequence of amino acids (Zhou et al., 2005). Such objective functions are also referred to as target functions or scoring functions (Pan & Kortemme, 2021). Target functions for protein design should predict the energy of a specified conformation of a specified protein sequence with sufficient accuracy to accurately model its stability (Huang, Pearce, & Zhang, 2020). In addition, the target function should be simple enough to enable sampling or enumeration of the vast space of conformations and sequences (Vorobieva, 2021).

Protein Design may be stated as the problem of computing the amino acid sequence that optimizes the objective function value (Hallen & Donald, 2019). There is no analytical solution for this inverse problem. Most of the current protein design algorithms generate amino acid sequences and then evaluate the target function for the selected amino acid sequence. And the objective function has to be optimized by searching through the space of amino acid sequences. Alternative approaches based on training of neural networks are also very promising (Pearce & Zhang, 2021).

The earliest attempts at protein design, that were based on the use of hydrophobic and hydrophilic building blocks, produced molten globules. Such problems were addressed by computational redesign of the packing of core residues (Korendovych & DeGrado, 2020). The earliest successful applications of protein design involved the redesign of a sequence of amino acids, in a fragment, usually in a loop or a set of residues at the active site. For example, PROTDES was utilized for the design of MHC-1 binding peptides and to change the specificity of the λ cI repressor protein (Suárez, Tortosa, & Jaramillo, 2008). Subsequent applications of protein design utilized a fixed backbone to serve as a scaffold for attachment of amino acid side chains. Later designs are based on a flexible backbone and are powerful enough to generate novel folds (Huang, Boyken, & Baker, 2016). The progress in Computational Protein Design has been reviewed earlier (Kuhlman & Bradley, 2019; Lippow & Tidor, 2007; Woolfson, 2021).

There are two basic approaches for Protein Design, namely template based Protein Design and *de novo* Protein Design (Kuhlman & Bradley, 2019). Template based Protein Design utilizes the available structure of a protein to construct a scaffold, and generates amino acid sequences with optimum fit to the scaffold. Whereas, *de novo* Protein Design does not utilize any structural template, and it may be utilized for the design of proteins with novel folds.

2. Applications of protein design

Different algorithms are likely to be optimum for different classes of problems. An overview of the problems that have been addressed by Protein Design provides the necessary contextual information regarding the different types of algorithms that can be used to address such problems. Protein design has been applied for a wide variety of applications, such as the design of novel protein folds (Kuhlman et al., 2003), nanobiosensors (Yang et al., 2021), enzymes (Löffler, Schmitz, Hupfeld, Sterner, & Merkl, 2017), polypeptide inhibitors (Reynolds et al., 2008), biological therapeutics (Talluri, 2019), antibodies (Sesterhenn et al., 2020) and vaccines (Ong, Huang, Pearce, Zhang, & He, 2021). In addition, Protein Design methodology is applicable for the improvement of specific properties of selected proteins, such as thermal stability (Barakat, Barakat, & Love, 2010; Talluri, 2011b), enzymatic activity (Liao et al., 2007), or biological compatibility (Choi, Verma, Griswold, & Bailey-Kellogg, 2017; Salvat et al., 2014).

3. Design goals

Proteins are capable of adopting a wide variety of different shapes and they may be used for many different applications (Dawson, Rhys, & Woolfson, 2019). Therefore, the list of constraints that need to be satisfied depends upon the application for which the protein is being designed. The goal may specify the design of a protein with a sequence of amino acids that satisfies one or more requirements, such as the following:

Shape of the protein, *e.g.*, globular, rod shaped, *etc.*

Length of the amino acid sequence which may be important for economy of production

Physical Stability in solution within a specified range of pH, temperature, solvent and ionic strength

Chemical stability in solution within the range of operating conditions, such as resistance to autoproteolysis and deamidation

Resistance to proteolytic cleavage in solution

Physical and Chemical stability in solution in the presence of chemical additives such as urea

Solubility within the range of operating conditions

Functional efficiency changes within a range of operating conditions

Functional specificity for catalysis within a range of operating conditions

Specificity of interaction with inorganic ions, metabolites and biomolecules

Tailored antigenicity such as absence of certain antigenic sites

Solubility in the cytosol in the denatured state

Transportability by the cellular machinery

Stability in the lyophilized state

Reconstitutability of structure and function from lyophilized powder

Subset of standard amino acids

Non-standard amino acids

Range of conformations that are sampled at equilibrium for specified operating conditions

Reversibility of functional changes in response to changes in environment

Changes in function triggered by environmental stimuli

- Conformational changes upon binding of ligands for controlled activity
- Conformational changes upon change of environment, such as pH, temperature, *etc.*
- Conformational changes upon chemical modification such as phosphorylation for controlled activity
- Conformational changes induced by light

Design goals may be simple and precise, such as Protein Design for thermal stability (Jiang, Farid, Pistor, & Farid, 2000; Talluri, 2011a). Other problems, such as Design of a ligand activated switch, may specify requirements for multiple stable conformations with maximization of differences in ligand binding affinity and maximization of differences in functional activity upon binding to a specified ligand (Karimi & Shen, 2018; Kuhlman, 2019; Leaver-Fay, Jacak, Stranges, & Kuhlman, 2011).

4. Target function design

Conceptually, protein design can be regarded as a problem involving the optimization of a target function, if the target function is designed to quantitatively evaluate the relationship between the design goals and a specific amino acid sequence (Grigoryan et al., 2006; Jiang et al., 2000). Formulation of the protein design problem as an optimization problem provides an overview of the requirements and challenges involved in protein design.

A target function, an objective function or a scoring scheme can be used to predict the properties of interest for a specified amino acid sequence (Jiang et al., 2000; Kraemer-Pecore, Wollacott, & Desjarlais, 2001). Protein design algorithms may utilize a hierarchy of scoring functions. A scoring function, usually based on estimation of energy, can be used to predict the most probable conformation of the amino acid sequence (Huang & Bystroff, 2013; Huang et al., 2020; Norn et al., 2021). And the predicted structure/conformation can be utilized to predict the property of interest, such as stability, enzymatic activity or binding energy (Lippow & Tidor, 2007; Michael & Simonson, 2021; Talluri, 2011a). A wide range of scoring functions have been developed for both of these tasks (Li, Yang, Zhan, Dai, & Zhou, 2013; Pan & Kortemme, 2021).

A generalized scoring function for protein design can expressed in the following manner.

$$\text{Score} = \text{Max}_{\{\text{All possible sequences}\}} \Sigma_i \, W_i \, \mathit{PredProp}_i(\mathit{PredConf}(\text{Sequence}))$$

The summation is over all properties of interest.

PredConf(Sequence) is a function that predicts the energies or probabilities of all relevant conformations for a specific sequence of amino acids. Its output is a list of the relevant conformations and their energies or probabilities. The number of potential conformations is infinite, therefore the list is a truncated list of conformations that are deemed to be "relevant" for the problem of interest. This may be a single conformation of lowest energy, or a list of the most probable conformers. Prediction of the conformation of a protein sequence is the classic "Protein Structure Prediction" problem.

$PredProp_i$ predicts the value of the ith property of interest, such as the conformational entropy, enzymatic activity, thermal stability or ligand binding energy for a specified conformation or a set of conformations. The objective of many protein design applications is to obtain a protein with a single stable conformation. Such as design goal can be encoded by a *PredProp* term based on the conformational entropy or by using an explicit function that returns a high value if the conformational ensemble has only one conformation with significant statistical weight.

W_i is the weight associated with the property i in the design goal. The weight may be positive for desirable properties (positive design) or negative for undesirable properties (negative design), or a combination of both. For example, Enzyme design may require a trade-off between stability and activity. Such trade-offs can be explicitly specified by adjusting the relative weights for the *PredProp* functions that evaluate stability and activity.

Prediction of the energy of a specified conformation enables one to predict the stability of the designed protein, specified by its amino acid sequence (Huang et al., 2020; Jiang et al., 2000). The energy of a particular conformation of a particular amino acid sequence can be predicted by using quantum mechanics, molecular mechanics, knowledge based or hybrid functions (Boas & Harbury, 2007). Molecular mechanics and knowledge based functions are used in most applications of protein design due to the lower computational demands of these methods compared to quantum mechanics based methods. Most of the successful estimates of energy utilize separate terms for single residue, pairwise interactions and global properties (Huang & Bystroff, 2013). Each of these terms represents several different types of physical interactions. For example, single residue terms are used to account for the dependence of energy on bond lengths, bond angles and torsion angles. Pairwise terms are used to account for electrostatic energy, van der Waal's interactions and hydrogen bonds. And global terms account for solvent accessible surface area, internal cavities and local conformational entropy. Although such empirical and knowledge based functions have been under development for a long time for the purpose of protein structure prediction, there is a renewed interest to optimize the relative weights and assess the correlation between the different terms that contribute to the conformational energy, for protein design applications (Das & Baker, 2008; Huang & Bystroff, 2013). In addition, there is one important difference between the force fields utilized in Molecular

Mechanics studies for protein structure prediction and the functions used for evaluating energy in protein design studies. The functional form and parameters for the empirical energy functions utilized in Molecular Mechanics are optimized to evaluate the relative energies of different conformations of the same amino acid sequence, whereas in protein design studies it is necessary to compare the function value for different amino acid sequences (Dawson et al., 2019). Conformational landscape optimization results in fewer alternative minima for sequence design (Norn et al., 2021).

Although the objective of most protein design problems involves finding a sequence of amino acids that will adopt a single conformation or a narrow range of closely related conformations, certain protein design problems, require two or more conformations. For example, the design of proteins whose function can be controlled by an external switch, such as light (Kuhlman, 2019). Specialized target functions have been developed for Multi-state protein design to enable optimization of an ensemble of states, with specification for positive and negative selection of states (Karimi & Shen, 2018).

Knowledge based target functions are also being developed for directly encoding the relationships between amino acid sequences and the properties of interest (Mackenzie, Zhou, & Grigoryan, 2016). The development of knowledge based target functions, as well as physics based scoring functions, has benefited enormously from the structural data available in databases such as PDB as well as the extensive knowledge obtained from other biophysical methods for characterization of native and mutant proteins (Li et al., 2013).

Pareto optimal search is an alternative approach for design that does not involve a weighted summation of terms of interest in the objective function. The EpiSweep algorithm, takes into consideration the competing goals of function and immunogenicity, by seeking pareto optimal solutions (Choi et al., 2017; Parker, Choi, Griswold, & Bailey-Kellogg, 2013).

5. Algorithms for protein design

An analytical solution for Protein Design is not available. Algorithms that are applicable for protein design utilize a simplified biophysical representation of a protein, or a stochastic approach, or an ensemble generation approach, or an iterative or evolutionary approach, or a combination of these methods. A trade-off exists between the utilization of simplified representations that enable an exhaustive search and the utilization of accurate representations with intractable search space. These extremes can be avoided by algorithms that utilize intermediate level representations that are simple enough to guarantee a solution with available resources and are accurate enough to be of practical use for protein design (Gainza, Nisonoff, & Donald, 2016). Alternatively, a hierarchy of representations of increasing complexity can be utilized by algorithms that attempt to provide solutions for accurate representations by optimizing the approximate solutions provided by lower level representations. A variety of iterative and evolutionary algorithms are available for optimization of feasible solutions. Many algorithms that were developed earlier with the primary objective of protein structure prediction or protein engineering are also applicable to problems of protein design with minor modifications (Jumper et al., 2021; Kuhlman & Bradley, 2019). Some of these algorithms are also included in this review, but the algorithms that have been *deployed* for the solution of protein design problems are emphasized in this review. Algorithms with provable guarantees of accuracy for a specific model are termed as "provable algorithms" (Hallen & Donald, 2019).

A canonical algorithm for protein design involves five steps: (i) Generate a new sequence of amino acids. (ii) Predict the set of accessible conformations and their statistical weights. (iii) For each conformation, predict the property of interest (*e.g.*, binding energy, enzyme activity, *etc.*). (iv) Calculate the objective function value for the generated sequence by using the predicted values of the properties for each conformation and their respective statistical weights. (v) Select the sequence having the highest value of the objective function. Stop, if the objective function value meets the specifications of the problem, or if all sequences of interest have been investigated, or if resource limitations are encountered. Otherwise, go to step (i). Accurate prediction of protein conformation and stability for a specified sequence of amino acids is a challenging and as yet unsolved problem that has been under investigation for several decades. Similarly, accurate prediction of the binding energy or enzymatic activity of a protein is still a challenging task, even if accurate structural information is available regarding its dominant conformation. In spite of such difficulties, many proteins have been designed successfully, the success rate is increasing rapidly, and software for protein design is now available for deployment by non-experts.

Numerous methods and algorithms have been developed for Computational Protein Design. Algorithms and methods that are similar to each other have been grouped together in the following sections, for convenience. However, many protein design algorithms may be assigned to more than one of the categories listed below. For example, an algorithm for global minimization of conformational energy may also be described as an algorithm for searching the conformational space. Furthermore, many graph search algorithms that were initially developed for discretized search space have been extended for neighborhood search with continuous functions.

5.1 Algorithms for discretized search space

A variety of simplified biophysical representations with discretized search space have been developed. Discretization of search space enables the utilization of enumerative or search algorithms that are guaranteed to locate the optimal sequence for a specified biophysical model, or to find a sequence within specified tolerance from the optimal sequence, within a finite time, which can be estimated before application of the algorithm (Gainza et al., 2016).

5.1.1 Enumeration

An exhaustive search over all possible sequences for identification of the optimal sequence is theoretically feasible, if the search is restricted to discretely sampled conformations. However, the search space is extremely large, and the computed search-time is not practically accessible with currently available computational resources even for moderate size problems. For example, for a protein having 100 amino acids, and restriction of the discretely sample conformations to one of two states, α or β, the minimum number of function evaluations would be $(20^{100}) \times (2^{100}) = (1.27 \times 10^{130}) \times (1.27 \times 10^{30}) = 1.61 \times 10^{160}$. If these computations were to be performed on a computer capable of 10^{15} function evaluations per second, then the time requirement would be $\sim 10^{137}$ years. This time limit, though finite, is clearly impractical as it is greater than the estimated age of our universe.

However, enumeration may be practical for the design of a short subsequence of a larger protein, or for a simplified representation of a protein such as the lattice model.

5.1.1.1 Side chain conformations and rotamer enumeration

Complete enumeration is feasible for short segments of proteins for simplified representations. For example, the backbone conformation could be fixed and the conformational sampling of side chains may be limited to rotamers. Rotamers are a discrete set of statistically favored conformations of amino acid side chains (Ponder & Richards, 1987). All possible conformations for an amino acid can be enumerated for a fixed backbone conformation, if the conformations are sampled with fixed increments of torsion angles (Dunbrack & Karplus, 1994). If the conformation of the polypeptide backbone for a short segment of length 6–8 residues is fixed, then the side chain conformations can be exhaustively enumerated, if the conformations are generated by using fixed increments of torsion angles or by selection from a rotamer library.

5.1.1.2 Lattice models

Simplified representations of proteins can be utilized to create models whose properties can be investigated by enumeration. Lattice models are simplified representations that permit exhaustive enumeration for sequences of length up to about 30. Sequences with only two type of residues (hydrophobic and polar) have been optimized on a two dimensional lattice structure (Yue & Dill, 1993). Simulations on lattice proteins were used to identify criteria for fast folding, such as Δ_{10}, the energy gap between the global energy minimum native state E_0 and the next highest energy E_1 (Šali, Shakhnovich, & Karplus, 1994). Exhaustive enumeration and the computational tractability of lattice models was utilized for investigating questions regarding the designability of protein structures (Buchler & Goldstein, 2000) and to explore the relationship between the designability of protein structures and statistical thermodynamic descriptors such as Δ_{10} and foldability. Monte Carlo simulations with lattice models have been utilized for the design of β-barrel proteins (Kolinski, Galazka, & Skolnick, 1995).

5.1.2 Graph algorithms

Design problems involving protein-protein interactions or protein-ligand interactions require optimization of the hydrogen bond network. A graph representation of potential hydrogen bond partners enables enumeration of all possible side chain hydrogen bond networks for a specified backbone structure (Boyken, 2016; Maguire, Boyken, Baker, & Kuhlman, 2018).

5.1.2.1 Tree search

Simplified biophysical representations that enable the formulation of the Protein Design problem as a search tree have been developed. The problem of searching through all possible combinations of rotamers to find the conformation that has the lowest energy for a fixed backbone can be formulated as a tree search problem. Each node of the tree represents a residue in the protein chain, and the branches are the rotamers for the selected residue.

Each terminal node (leaf-node) of the tree represents one conformation of the complete protein. The set of all nodes that are at a distance, d, from the root node, represent all possible conformations for a fragment of length d. The depth of the tree would be equal to the length of the polypeptide chain. If there are m rotamers per residue and the number of residues in the polypeptide chain is l. Then the number of possible conformations are l^m. Tree search algorithms can be used for searching the space of protein sequences (Dahiyat & Mayo, 1996; Hallen & Donald, 2019).

5.1.2.1.1 The dead end elimination algorithm (DEE) The Dead-End-Elimination algorithm (DEE) is used for pruning of the rotamers that cannot be present in any low-energy conformation of the protein (Desmet, Maeyer, Hazes, & Lasters, 1992). The DEE algorithm compares pairs of rotamers of a given residue. A rotamer, h, cannot be present in any low-energy conformation, if *every* conformation containing the rotamer h is higher in energy than a conformation obtained by replacement of rotamer h with the rotamer l. This can be ascertained in time linear in the number of residues for an energy function that can be written in the following form, because it is sufficient to compare the residue specific conformational energies of rotamers h and l, and the pairwise interactions of h and l with rotamers of the remaining residues.

$$E(\mathbf{r}) = \Sigma_i E(i_\mathbf{r}) + \Sigma_{i<j} E(i_\mathbf{r}, j_\mathbf{r})$$

The Dead-End-Elimination algorithm was applied for multi-state protein design (Yanover, Fromer, & Shifman, 2007).

Tree search algorithms can be modified to incorporate continuous flexibility to model the lower energy conformations in the vicinity of the enumerated discrete conformations (Hallen & Donald, 2017). iMinDEE algorithm is a provable algorithm for removal of continuous rotamers that will not contribute to the optimal solution (Gainza, Roberts, & Donald, 2012). The DEEPer algorithm extends this approach to models with flexible sidechains and backbones (Hallen, Keedy, & Donald, 2013).

5.1.2.1.2 The branch-and-bound algorithm The branch-and-bound algorithm can be used to identify the lowest energy conformation without exhaustive search of all possible conformations, if the energy function is defined is such a way that the energy of the conformation of any fragment of the protein can never exceed the energy of the complete protein formed by inclusion of that fragment in the same conformation. This requires that addition of residues to a growing polypeptide chain should *always* increase the energy (or remain unchanged) if the conformation of the previous residues is fixed. In addition, the branch-and-bound algorithm requires a systematic method that is capable of generating a tree that represents all possible conformations of the protein, as well as sub-trees that represent the conformations of fragments of the protein. The branch-and-bound algorithm first evaluates the energy of one conformation of the complete protein. Then the branch-and-bound algorithm starts a systematic traversal starting from the root node. At each node, it evaluates the conformational energy for the tree/subtree, that represents the full protein or a fragment of the protein. If the energy is less than the energy of the full protein conformation found in the search up to that point, it continues the search by moving to a daughter node of the same branch. However, if the energy at any node is greater than the energy of the best conformation identified up to that point, it does not explore the set of nodes that are branching from that point, because addition of residues would only increase the conformational energy. This enables the branch-and-bound algorithm to eliminate search of all conformations of fragments whose extension would give rise to energy higher than the best one found in the search up to that point. Since the branch-and-bound is efficient at eliminating the requirement to search in large hypervolumes of the conformational subspace that could not possibly be the location of the optimal conformation, the search will be focused on those areas of the conformational subspace that are likely to be the location of the optimal conformation.

The branch-and-bound algorithm for tree search is a depth-first search algorithm. The branch-and-bound algorithm is the basis for several recently proposed protein design algorithms. It has been shown that methods that are based on the branch-and-bound algorithm are effective for finding solutions to a simple biophysical model, involving a rigid polypeptide backbone, pairwise interactions energies and finite number of rotamers per residue. A variety of suggestions for calculating the bounds have been proposed and evaluated. BroMAP uses DEE and Maximum-A-Posteriori estimation (MAP) for reduction of problem size within each node (Hong, Lippow, Tidor, & Lozano-Pérez, 2009).

iCFN (interconnected Cost Function Networks) is a weighted constraint satisfaction based exact algorithm for multi-state protein design (Karimi & Shen, 2018). iCFN uses a branch and bound algorithm for searching a tree structure composed of sequences, substates and conformations.

5.1.2.1.3 The A* algorithm The A* algorithm in Protein Design is generally used after application of Dead-End-Elimination algorithm

(DEE) (Leach & Lemon, 1998). After pruning of rotamers with DEE, A* can be used to enumerate the remaining conformations in the order of increasing energy (Gainza et al., 2013). The non-terminal nodes of the tree represent the conformations of the fragments of the protein and the terminal nodes represent the conformations of the complete protein. The branches of each node represent the rotamers for the last residue of the fragment. The energy function has to define in such a manner the energy of any conformation of any fragment cannot decrease upon addition of the next residue in the protein sequence to the fragment. The A* algorithm uses a guided tree-search. The guidance is provided by the use of a heuristic function. For each node, x, that is explored, an f-score is calculated. The f-score is the sum of the energy for the conformation of the fragment represented by the node x, g(x), and a heuristic function that estimates the minimum energy for any protein conformation incorporating the fragment in the conformation represented by the node, x, h(x).

$$f(x) = g(x) + h(x)$$

The A* algorithm evaluates the f-scores for each node reachable from the current node and then moves to the node with the lowest f-score. The process is repeated until the lowest f-score node is a leaf node. The resulting leaf node will be guaranteed to represent the lowest energy conformation. The next lowest conformation can also be found by continuing the process from the node with the next lowest f-score. Therefore, conformations can be enumerated in the order of lowest energy, if desired. This can be useful for ensemble based methods and for multi-state design.

The A*-CFN search algorithm has been implemented with a Cost Function Network (CFN) based heuristic lower bound (Traoré et al., 2016). Use of Existential Directed Arc Consistency (EDAC) provided a stronger lower bound for the implementation of the A* algorithm and a substantial increase in its efficiency, compared to the use of Directed Arc Consistency bounds used in earlier implementations of the DEE/A* algorithm.

The COMETS algorithm, uses an A* search algorithm with a "tree within a tree" data structure for encoding the Protein Design problem (Hallen & Donald, 2016). The nodes of the main tree represent partial sequences, and the function value at each node of the main tree is evaluated by a tree search in conformational space by using an inner tree. COMETS was proposed for solving multi-state protein design problems.

The Tree search algorithms can also be combined with a local model of continuous flexibility (Hallen, Jou, & Donald, 2017). Continuous flexibility around discrete, low energy conformations can be modeled and incorporated into a tree-search algorithm such as A* to compute bounds on the energies of continuous regions around discrete conformations.

The BBK* algorithm for binding affinities, uses multisequence bounds to avoid computations for a subset of sequences (Ojewole, Jou, Fowler, & Donald, 2018). Hence, BBK* algorithm run time is sublinear in the number of sequences.

5.1.2.1.4 Dynamic programming Dynamic programming algorithms compute optimum values of the scoring function by building up from optimum solutions of partial problem. BWM* (Branch Width Minimization) is a provable dynamic programming algorithm for enumeration of a gap-free ensemble of conformations in order of increasing energy (Jou, Jain, Georgiev, & Donald, 2016). The performance of BWM* can be predicted in polynomial time before running the algorithm.

5.2 Algorithms for optimization

A wide variety of optimization methods, applicable to proteins, have been developed for protein structure prediction, and some of these can be utilized, with minor modifications, for protein design problems. Gradient based optimization methods are a component of the workflow of most protein design algorithms. The BFGS algorithm, L-BFGS algorithm and the Polak-Ribiere/Fletcher-Reeves algorithms for search along conjugate directions are the most common gradient based optimizers used in Protein design workflows. Numerical gradients are often utilized for optimization of functions whose analytical gradients cannot be easily computed, or in those cases where calculation of numerical gradient is computationally more efficient than calculation of the analytical gradient. Gradient based methods are very efficient for local optimization. However, gradient based methods are usually unsuitable for global optimization problems because the complexity of the functions of interest in protein design produces manifolds with multiple local minima, and the gradient based methods get trapped in these local minima. In addition, gradient based methods have several limitations, such as inapplicability to problems involving discrete variables, such as changes of amino acid sequence. However, gradient based energy minimization has been successfully incorporated into Rosetta's Iterative relaxation/design algorithm by using an iterative optimization process involving alternation between rotamer-sequence discrete optimization and gradient based optimization of backbone and side chain torsion angles (Huang et al., 2016).

5.2.1 Linear programming

Cluster expansion can be used to convert structure based energies into functions of sequence, by implicit treatment of protein conformational energy (Zhou et al., 2005). Cluster expansion was used to model backbone flexibility in template based protein design (Apgar, Hahn, Grigoryan, & Keating, 2009). A cluster expansion method, implemented in CLEVER, maps the functional force field to a linear function that is dependent only on the sequence. The linearized function was then optimized using integer linear programming solvers (Negron & Keating, 2013).

Prune Local Unrealistic Geometries (PLUG) is a linear programming based method for pruning of conformations based on lower bounds for interatomic distances (Hallen, 2019).

5.2.2 Backrub

Although most optimization methods that are currently used in protein design are general optimization methods that were developed earlier in other fields, the backrub is a specialized perturbation method that was developed for protein structure optimization. Observation of high resolution protein structures revealed the presence of a correlation between backbone and side-chain conformational changes (Davis, Arendall, Richardson, & Richardson, 2006). A change of a side-chain or a conformational change of the side-chain is accompanied by a local deformation of the protein backbone. A side-chain conformational change at residue i is accompanied by a local deformation of the backbone that involves a rotation around the axis defined by the $C^{\alpha}_{i-1} \ldots C^{\alpha}_{i+1}$ vector and compensating changes in the backbone torsion angles between the backbone atoms $C^{\alpha}_{i-1} \ldots C^{\alpha}_{i+1}$. An algorithm for dead end elimination (DEE) of rotamers with limited local backbone flexibility involving backrub motion has been developed for protein design applications (Georgiev, Keedy, Richardson, Richardson, & Donald, 2008).

5.2.3 Backpropagation

Backpropagation of gradients is one of the standard methods used for training of Neural Networks that have been utilized for Protein Design. Direct optimization of all possible amino acid sequences and all possible structures has been demonstrated by backpropagation of gradients in the transform-restrained Rosetta (trRosetta) structure prediction network (Norn et al., 2021).

5.2.4 Genetic algorithms

Genetic algorithms may be used to optimize a target function that encodes information regarding the design goals (Jones, 1994). Genetic algorithms were designed to simulate the process of natural evolution. Genetic algorithms optimize populations of sequences. An initial population set can be generated randomly, or derived from a profile. The algorithm alters the sequences in the population based on their fitness. The fitness function may be the same as the target function or a function that is related to the target function. For example, the fitness function may be based on the predicted conformational energy or the predicted binding constant. The fitness of each member of the population is evaluated. A selection operator is utilized to select sequences that will contribute to the next generation. One or more copies of the selected sequences are added to the new set, based considerations such as fitness and diversity. The selected sequences are then operated upon by mutation or hybridization operators to generate diversity. The new set of sequences are then evaluated for their fitness. The process of selection, reproduction and mutation is repeated until sequences with the desired level of fitness are obtained, or until a fixed number of cycles are completed.

Elfin is a genetic algorithm for the design of proteins of specified shape (Yeh, Brunette, Baker, McIntosh-Smith, & Parmeggiani, 2018). It utilizes structural building blocks derived from repeat proteins for building large non-symmetric structures of specified shapes. Proteins with more than 1000 amino acids can be built by this method.

5.2.5 Mean field optimization

Mean field optimization provides an alternative to the combinatorial explosion of states for exploring sequence space or conformational space. For each site, the probability of alternate choices is listed. For example, this could be the probability of a rotamer in a specific location of the fold of interest. The initial set of probabilities may be random or derived from a profile. The mean field energy is calculated based on the relative contributions of alternatives at all sites. The mean field energy is optimized by considering the probabilities as variables. The advantage of this method is that for a problem involving N sites, with m potential alternatives for each site. The number of variables that need optimization is $m \times N$, whereas if all the specific configurations are to be enumerated, m^N function evaluations would be required. Unfortunately, this method can be trapped in a local minimum—a problem that is common to many other optimization methods.

A self-consistent mean field approach with a semi-empirical physical potential was used for design of sequences for templates consisting of fixed backbones (Mach & Koehl, 2013). A hidden Markov model based profile was constructed for the sequences designed for a template.

5.3 Parametric modeling and model fitting

5.3.1 Parametric modeling

Parametric modeling uses a mathematical relationship between the design variables and a small number of parameters to model the structural variability of interest in a selected fold (Crick, 1953). For example, parametric models of the coiled-coil may be utilized to design backbone conformation of proteins with specified values of supercoil radius and supercoil twist. Parametric design has been utilized for design of α-helical bundles, transmembrane proteins and α-helical barrels (Huang et al., 2014; Lu, 2018; Thomson, 2014).

5.3.2 Constraint satisfaction

Loop closure algorithms can be used to fill gaps between segments of a scaffold.

For small loops, the Go and Scheraga loop closure algorithm can enumerate all backbone conformations for specified end distances and end orientations (Go & Scheraga, 1970). The finite list of rotamers for all potential amino acid side chains can be identified for these backbone conformations. The Cyclic Coordinate Descent is a robotics algorithm for loop closure (Canutescu & Dunbrack, 2003). The KIC algorithm, is another robotics inspired algorithm that uses a kinematic closure algorithm to generate all possible analytical solutions by using a set of three pivot residues and then computes a library of torsion values for the remaining torsion angles for loops of arbitrary sizes (Mandell, Coutsias, & Kortemme, 2009). The KIC algorithm has been implemented in RosettaRemodel (Huang et al., 2011).

5.3.3 Threading or fold prediction

Threading is the problem of identifying the fold that has the best match to a specified protein sequence. This method has been utilized extensively for protein structure prediction (Bazzoli, Tettamanzi, & Zhang, 2011). The method can be included as a part of a computational protein design algorithm, to identify the set of sequences that are compatible with a specified protein fold (Brylinski, 2013). The space of known native protein folds is limited and small.

5.4 Stochastic and ensemble based algorithms

5.4.1 Stochastic algorithms

Stochastic algorithms are efficient for calculating the integrals of multivariable functions and provide estimates of error based on the extent of sampling. These methods have the advantage that the extent of sampling can be adjusted based on the required tolerance for error. And the error can be reduced to any desired value subject to availability of sufficient computational resources for sampling. Therefore, such methods are provably correct to any desired finite level of precision in a finite time, and provably correct with an error value that approaches zero for ergodic sampling. Such methods have been utilized extensively for estimates of free energy and other thermodynamic properties in studies of Protein Folding, Protein Engineering and Protein Design (Panel, Villa, Opuu, Mignon, & Simonson, 2021).

The first attempt at automation of Protein Design utilized a Monte Carlo Simulated Annealing algorithm for optimization of an empirical function simultaneously in sequence and structure space (Hellinga & Richards, 1994).

The Markov Chain Monte Carlo algorithm is used in Rosetta Design (Das & Baker, 2008). Rosetta's Packer tool uses Simulated Annealing for exploring rotamer space. And, although it is not guaranteed to locate the global minimum, it is one of the most efficient algorithms for finding high quality solutions near the optimum.

5.4.1.1 Evolutionary algorithms

Certain evolutionary algorithms, such as EvoDesign, utilize information deduced from homology to generate sequences that have an increased probability of high quality structure (Mitra, Shultis, & Zhang, 2013). EvoDesign uses a protein structure scaffold as input. Proteins of similar folds are identified from PDB based on structural similarity. An evolutionary profile of the proteins having folds that match the design goals is then constructed by using structural alignments of PDB homologs as input. The evolutionary profile is then used as a generator of decoy sequences. The secondary structure, solvent accessibility and backbone torsion angles of the decoy sequences are then predicted by using a Neural Network trained on a PSI-BLAST Position Specific Scoring Matrix (PSSM). The EvoDesign force field contains four evolutionary terms and an optional term for energy computed from a physics based force field (FoldX). The evolutionary terms in the EvoDesign force field are based on the similarity of the sequence, secondary structure, solvent accessibility and torsion angles between the decoy and the target profile. EvoDesign uses a Metropolis Monte Carlo simulation for search of sequence space guided by the *Z*-score of the decoy energy. Simulations used for evaluation of EvoDesign demonstrated that EvoDesign without the optional physics based force field term is five times faster, with a small decrease in accuracy.

5.4.2 Ensemble based methods

Ensemble based methods are generally used to obtain accurate free energies required for protein design. Free energy based methods are not restricted to the pairwise interaction models which are utilized by almost all the deterministic provable finite resource limit algorithms. Free energy calculations can include non-linear terms attributable to electrostatics and solvation. Simulations were utilized to estimate the absolute free energy of a macromolecular state and also to construct a sequence based energy function for protein design (Grigoryan, 2013).

The use of a fixed backbone and rigid rotamers may result in the incorrect rejection of an optimal sequence. Such a problem can be overcome by the use of multistate design, which uses an ensemble of backbone conformations to simulate backbone flexibility (Davey & Chica, 2017).

5.4.2.1 Protein design using electrostatics

Mutations that alter long range electrostatic interactions can cause significant changes in intermolecular association with minor changes in the structure of the individual molecules (Ronald, Kieslich, & Morikis, 2011). Perturbation based methods that evaluate Poisson-Boltzmann free energy and Columbic electrostatics have been developed for the design of mutant proteins with desirable association properties (Ronald et al., 2011).

5.4.3 Distance Geometry

The Distance Geometry method has been utilized primarily for protein structure determination from NMR spectroscopic data (Havel & Wuthrich, 1985). However, it may also be utilized for computational protein design (Talluri, 2017; Zhu & Braun, 1999). The distance geometry algorithm is very efficient for those problems where the problem definition can be formulated as a set of pairwise distance bounds. The specified distance bounds are utilized in the distance geometry algorithm, to generate self-consistent pairwise distances between all pairs of atoms, using a distribution function. The distribution function can be derived from frequencies of residue pairs in high resolution structures of proteins (Zhu & Braun, 1999). The distribution function can incorporate information obtained from any relevant biomolecular force field (Talluri, 2017). Subsequently, embedding is used to project data from a higher level space to three dimensional space for generation of the Cartesian coordinates of biopolymers (Aszodi, Gradwell, & Taylor, 1995; Havel, 1990). The resulting conformation is then subjected to conventional structure refinement procedures (Wagner et al., 1987). The resulting conformation can be assessed for satisfaction of protein design goals using a suitable scoring function. This methodology can be utilized to obtain a sampling of an ensemble of conformations for specified distribution functions.

Self-correcting Distance Geometry (SECODG) calculations using distance constraints derived from low energy pairs of beta-strands correctly identified the native pairing of strands in the beta-sheet of bovine pancreatic trypsin inhibitor (BPTI) (Zhu & Braun, 1999).

5.4.4 Quantum Annealing algorithm

A Quantum Computer can simultaneously represent 2^B different configurations by using a word consisting of B qubits, if each qubit can simultaneously represent two states (Feynman, 1986). This is called a superposition of states. Quantum Annealing is a Quantum Algorithm that can shift the probability distribution of the superposed quantum states in a manner such that the state representing the global minimum is overwhelmingly probable upon measurement (Kadowaki & Nishimori, 1998). The basis of the Quantum Annealing algorithm is that a quantum-mechanical system in the ground state remains in the ground state during a reversible adiabatic change. Therefore, the global minimum of an optimization problem can be obtained by adiabatically evolving a quantum mechanical system from an easy to prepare ground state, represented by a Hamiltonian Hs, to a state that is described by the Hamiltonian Ht that is a representation of the classical objective function to be minimized. If Ht is equivalent to the Hamiltonian for a transverse Ising model, then the transition between the states can be controlled by the strength of the transverse field (Kadowaki & Nishimori, 1998). QPacker is a Quantum Design algorithm based on a direct mapping of the classical Rosetta Packer algorithm to a quantum computer's architecture. Qpacker uses Quantum Annealing and a rotamer level pairwise decomposable energy function to explore the rotamer conformational space for a peptide (Mulligan et al., 2020). The parameters for the energy function were the same as those in the classical Rosetta energy function used in Packer.

5.5 Neural networks

Artificial Intelligence and Machine Learning methods, such as Neural Networks with Deep Learning, have made a substantial impact on Protein Design (LeCun, Bengio, & Hinton, 2015; Pearce & Zhang, 2021). A neural network can be specified by a list of nodes that represent neurons, the strengths of the connections between pairs of nodes and the transfer function used by each neuron to map its input to output. A neuron can have multiple inputs and multiple outputs. And neurons are generally arranged in layers. A variety of representations have been utilized for Protein Design, such as 2D convolution of contact maps, 3D convolution of voxelized representations, language models and graphs (Gao, Mahajan, Sulam, & Gray, 2020; Maguire, Grattarola, Mulligan, Klyshko, & Melo, 2021). In Feed Forward Neural Networks (FFNN), the flow of information always occurs unidirectionally from one layer to another. In Recurrent Neural Networks (RNN) the requirement of unidirectional flow of information is not applicable.

5.5.1 Deep learning

Deep learning neural networks have a large number of layers of neurons and this enables extraction and analysis of high level features from the input data (Gao et al., 2020). The profiles of protein sequences compatible with a specified structure were predicted by using a neural network with fragment based and energy based profiles (Li, Yang, Faraggi, Zhan, & Zhou, 2014). A recurrent neural network was utilized for peptide design (Müller, Hiss, & Schneider, 2018). A Deep Learning Neural Network was used to predict the identity of an amino acid based on the identity and position of neighboring residues (Wang, Cao, Zhang, & Qi, 2018). The features of a target residue and its closest neighbor, based on C^{α}-C^{α} distance, were used as input for a residue probability network consisting of an input layer, several hidden layers and a softmax layer as output layer with 20-dimensional output. The 20 dimensions were used to represent the probabilities of 20 possible amino acid types for the target residue. The same parameters are shared by residue probability networks involving 15 closest neighbors. The outputs of the residue probability networks are then multiplied by weights obtained by using the same inputs that are used for computing outputs from the residue probability networks. The weighted outputs are used as inputs for fully connected layers, and the final output layer is a softmax layer with 20-dimensional output representing the probabilities of the 20 possible amino acid types for the target residue.

A Deep Convolutional Neural Network (ProDCoNN) was formulated and trained to predict the residue type based on the 3D structural environment around the C^{α} atom of an amino acid residue (Zhang et al., 2020). The Neural Network consisting of nine convolutional layers was designed to learn structural information of different types and at different scales, such as, bond lengths, bond angles, torsional angles and secondary structure. The input for the 3D Deep Convolutional Neural Network was a gridded box with atomic coordinates and types around a residue.

5.5.2 Graph convolutional neural network

The information between the interaction between rotamers can be coded in the edge attributes of a graph convolutional neural network. The efficiency of quantum computers with a limited number of connected qubits can be utilized for protein design by using such a graph convolutional neural network (Maguire et al., 2021).

5.5.3 Variational AutoEncoders

Deep Learning techniques can be utilized for the learning of complex distributions that encode protein sequence, structural and functional features (Greener, Moffat, & Jones, 2018). Such learned distributions can then be utilized for generative applications. The Variational AutoEncoder (VAE) is a semi-supervised learning method that utilizes an inference mechanism and a generative mechanism. The inference mechanism can be utilized to generate a compact representation of the data. Such a representation can be utilized by the generative mechanism to generate a synthetic sample of sequences that are similar to those in the input data. The generated sequences can be evaluated for the presence and quality of desired features, based on a scoring scheme, and then the selected generated sequences can be used as input for the next iteration (Greener et al., 2018).

Grammar Variational AutoEncoders that utilize a generative grammar for molecular structure representation have been developed (Kusner, Paige, & Hernández-Lobato, 2017). Conditional Variational AutoEncoders (CVAE) can be used to train deep generative models that generate a sample of sequences that are conditioned by a specified grammar. Conditional variational autoencoders (CVAE) have been utilized for the addition of metal binding sites to proteins and also for the design of proteins with a novel topology (Greener et al., 2018). A context free grammar (CFG) for representation of protein 3D structure has been defined and utilized in the application of CVAE for the design of novel folds (Greener et al., 2018).

5.5.4 Generative adversarial networks (GANs)

Generative Adversarial Networks (GANs) consist of a generative module and a discriminative module that are trained simultaneously. The discriminative module is trained to distinguish real from fake instances, whereas the generative module is trained to generate fake instances that are similar to real instances. During the training phase the generative module attempts to generate sequences that can fool the discriminative module, while the discriminative module attempts to improve its ability to predict real sequences from those that were generated by the generative module (Lin, Lin, & Lane, 2020).

GANs have been applied for design of peptides and proteins (Lin et al., 2020). A Long Short Term Memory Generative Adversarial Network (LSTM-GAN) was used to generate novel protein topologies (Sabban & Markovsky, 2020). A Feedback GAN was used to generate antimicrobial peptides (Gupta & Zou, 2019). The feedback loop of the Feedback GAN consists of a generative network module for producing amino acid sequences and a differentiable neural network analyzer for scoring the sequences. Deep Convolutional Generative Adversarial Networks (DCGANs) that utilize deep convolutional networks for implementation of the generative as well as the discriminative functions have also been utilized for protein design (Anand & Huang, 2018; Bian, Wang, Jun, & Xie, 2019; Rossetto & Zhou, 2019).

Guided Conditional Wasserstein Generative Adversarial Networks (gcWGANs) were utilized for generating novel protein folds (Karimi, Zhu, Cao, & Shen, 2020). Conditional GANs utilize conditional variables, such as class labels, to augment the training of the generative as well as the discriminative modules. gcWGAN utilizes a conditional Wasserstein GAN to generate sequences and an oracle to predict the fold of the generated sequence. The predicted fold serves as a feedback for the generative network module. The Wasserstein GAN uses a distance metric that permits comparison of distributions that have no overlap. The introduction of this metric alleviates problems related to instabilities in the training of the GANs.

5.6 Hybrid algorithms

Most protein design algorithms involve a combination of systematic search, statistical sampling and optimization.

5.6.1 Tertiary structural motifs

The scaffold of interest is divided into spatially adjacent residue positions and then a Protein Structure Database is searched for residues that are in similar structural environment—these are the Tertiary Structural Motifs (TERMS) (Mackenzie et al., 2016). The sequence preferences from the Protein Structure Database are then utilized to construct a knowledge based energy function. The sequence of interest is then obtained by optimization of this knowledge based energy function.

5.7 Multilevel design

Multilevel design methods utilize representations of structures with increasing level of detail. For example, a protein may be modeled as an ellipsoid of revolution, and the same protein can be modeled at a greater level of detail by representing each domain or subunit by an ellipsoid of revolution (Monticelli, 2008). Other levels of details may utilize simplified representations for each of the amino acids or small groups of atoms (united atoms), or more detailed representations at the level of individual atoms or subatomic representations for description of electron cloud polarization.

Novel models for representation of the protein structure may be utilized to reduce the number of independent variables. In addition to the models based on discretization of search space, models utilizing united atoms and connected ellipsoids of revolution have been proposed for reducing the number of independent variables. Such models have been demonstrated to provide substantial reduction in the number of independent degrees of freedom without a substantial reduction in the estimation of the conformation dependent energy of proteins (Maisuradze, Senet, Czaplewski, Liwo, & Scheraga, 2010). Such representations have the potential for reducing the search space significantly and they can be utilized with many of the conformational search and optimization algorithms developed for full-atom representations.

A coarse-grained potential energy function for the representation of protein backbone conformation at the α-carbon level has been developed and validated with loops (MacDonald, Kelley, & Freemont, 2013). The coarse-grained potential energy function includes terms for soft steric repulsion, pseudo hydrogen bond, pseudo Cα-Cα bond, pseudo Cα-Cα-Cα angle and pseudo Cα-Cα-Cα-Cα torsion angle. Such low level representations can be utilized for rapid low resolution sampling of the backbone required for efficient incorporation of geometrical constraints in protein backbone design (MacDonald, Maksimiak, Sadowski, & Taylor, 2010). A reconstruction algorithm for generating the full backbone model from the low resolution representation has been developed (Moore, Kelley, Barber, Murray, & MacDonald, 2013). Coarse-grained representations of the protein backbone and the reconstruction algorithm have been utilized for fragment free *de novo* protein backbone design (MacDonald et al., 2016).

Tunable course graining of force fields enables model building with a continuous change in the level of resolution (Spiriti & Zuckerman, 2014). Multilevel design methods can switch from low level representations at earlier stages of the design process to representations with increasingly higher level of detail in the later stages of the design process. Such methods are particularly attractive for problems that involve intermolecular interactions as well as for problems where the overall shape of the protein is an important design goal. Multilevel design enables utilization of efficient systematic search algorithms for lower levels of representation and accurate optimization methods for representations with higher level of details (MacDonald & Freemont, 2016). A multiobjective swarm intelligence based algorithm for protein design, Hydra, optimizes solutions for two different energy functions simultaneously (Li, Zhang, Wu, OuYang, & Shen, 2021).

Funneled energy landscapes can be designed by combining low resolution models which disfavor alternative states and high resolution models that create a deep energy minimum for the target structure (Norn et al., 2021).

5.8 Hierarchical design

Hierarchical design may be either top-down or bottom-up. In the top-down approach, a fold is selected or designed and segments that are compatible with the fold are inserted into selected positions (MacDonald et al., 2010). The objective of top down hierarchical design is to increase the efficiency of protein design, by the design of a scaffold followed by generation of a full structural model guided by the scaffold.

In the bottom-up approach, stable fragments are identified and these are used as building blocks for the next stage of assembly (Singh, 2021). This method involves selection of fragments from a database or enumeration for small fragments, followed by function optimization. An exhaustive conformational search is carried out for small fragments. Subsequently, the low energy conformations of small fragments are progressively joined in several stages, that involve structure optimization and growth, to generate the target. The bottom-up approach to *de novo* functional protein design has been illustrated by the design of folds that were designed to accommodate complex structural motifs (Yang et al., 2021).

A database of fragments, obtained from experimental protein structures can be used to bridge sections of a scaffold with gaps. Fragment assembly can also be used for complete assembly of a *de novo* backbone fold (Kuhlman et al., 2003). The fragment library often consists of a database of secondary structural elements extracted from proteins and a database of connectors which are also extracted from proteins. *De novo* folds can be constructed by choosing secondary structural elements from such a library and then connecting pairs of secondary structural elements by using the connector library. This approach of fragment assembly provides a convenient method for the design of *de novo* folds with placement of secondary structural elements at desired locations (Koga et al., 2012; Lin et al., 2015).

The ability to control the scaffold curvature is useful in applications that require the design of proteins with specified intermolecular interactions. The curvature of the scaffold can be controlled by design (Park et al., 2015). Scaffolds with precisely defined curvature can be obtained from repeats by designing a library of self-compatible fragments that can be polymerized to generate scaffolds and a library of fragment specific junction modules for connecting them. LRR repeats were used for the design of a set of self-compatible building block modules. Proteins of variable length and constant curvature can be obtained by insertion of multiple repeats of these building blocks. The connectors were designed to join two non-identical building blocks in such a manner that the curvature of the connected segment depends only on the identity of the adjacent building blocks. Combinations of building blocks and junction modules can be selected to create proteins with defined specified curvature of the scaffold. The application of this method enables the design of proteins with controlled curvature and atomic-level accuracy (Park et al., 2015).

5.9 Protein design with iteration and feedback

5.9.1 Iterative approach

An iterative method for optimization, applicable for protein design problems of small to moderate size has been utilized in PROTDES (Suárez et al., 2008). In this method, one of the positions to be optimized is selected randomly, random rotamers are placed at the selected position and then among the available rotamers, the rotamer that minimizes the free energy of the entire system after local optimization is selected. This step can be followed by molecular dynamics for optimization of backbone flexibility. This process, consisting of selecting a location randomly and identifying the optimal rotamer for that location is repeated for a fixed number of steps. This method has the advantage that it does not rely on a pairwise approximation. Therefore, the rotamer positioning can be implemented by using solvation models such as Accessible Surface Area (ASA) (Ooi, Oobatake, Nemethy, & Scheraga, 1987), Effective energy function (EEF1) (Lazaridis & Karplus, 1999), and Generalized Born using Molecular Volume (GBMV) (Lee, Salsbury, & Brooks, 2002). In addition, the free energy calculation, utilized in the optimization step, can include interactions of the designed protein with other molecules such as DNA. This permits selection of protein sequences based on structural as well as functional requirements.

5.9.2 Design feedback

Algorithms for design feedback are applicable within the Protein Design Automation cycle. The PDA cycle uses design followed by simulation, experimental testing and data analysis (Dahiyat & Mayo, 1996). The objective of the data analysis is to discover correlations between the calculated properties of simulated structures and the experimental observables. The measured properties, such as measured thermal stability of the synthesized proteins may be related to the physical properties of the designed protein by using Quantitative Structure Activity Relationships (QSAR). The QSAR models may be used to identify the property or set of properties that are most highly correlated with the thermal stability or other specified cost function.

6. Software

Several software suites that implement protein design algorithms are available, and many are free and open source. For example, PROTDES, Rosetta3, OSPREY, EvoDesign, FoldX, CCBuilder. PROTDES is a Toolbox based on CHARMM scripts. PROTDES finds the optimum set of amino acids at specified positions in the protein chain, without requiring pairwise approximations (Suárez et al., 2008). Rosetta3 and Rosetta scripts are part of an extensive suite of biomacromodeling tools that includes protein design software (Bender et al., 2016). RosettaRemodel implements a blueprint for flexible protein design based on the use of Rosetta tools and the Rosetta force field (Huang et al., 2011). Interactive protein design software, based on the Rosetta force field, is available in a game format (Koepnick, 2019). Foldit Standalone is a video game derived interactive graphical user interface to Rosetta (Kleffner et al., 2017). OSPREY contains a suite of protein redesign algorithms (Gainza et al., 2013). Toulbar2 is an open source solver used for CFN optimization in computational protein design (Traoré et al., 2013). EvoDesign uses evolutionary profiles to guide the sequence search and it is available on a webserver (Mitra et al., 2013; Pearce, Huang, Setiawan, & Zhang, 2019). FoldX is useful for design of proteins that bind to DNA (Nadra, Serrano, & Alibés, 2011) and to investigate the relationship between missense mutations and disease (Alibés et al., 2010). CCbuilder is a web based application for parametric design of coiled coil proteins (Wood & Woolfson, 2018).

Many of the algorithms described earlier have been optimized to take advantage of parallel computing, distributed computing and/or vector processing (Hallen & Donald, 2019). Substantial increases in the efficiency of some of these algorithms have been obtained by changes in algorithms or in the implementation of the algorithm for specialized hardware such as GPUs or cloud based distributed computing (Pan et al., 2016; Zhou, Donald, & Zeng, 2017).

7. Discussion

Achievement of the goals of protein design requires improvements in search algorithms as well as improvements in force fields. Although the technologies used in the successful solution of related problems such as protein structure prediction are partially applicable to protein design, *de novo* protein design is a far more challenging problem. The most significant recent advancements in protein structure determination utilized knowledge based force fields and machine learning (Jumper et al., 2021). These knowledge based force fields and machine learning methods can be utilized for template based protein design. However, the applicability of these methods to for *de novo* protein design is still in the early stages of development.

The development of scoring functions can be improved significantly by the use of deep learning and datasets that contain both positive as well as negative controls (Pearce & Zhang, 2021). Unfortunately, there is a paucity of easily accessible knowledge regarding negative controls because most of the literature on protein engineering and protein design only documents positive achievements (Li et al., 2013). Identification of the sources of failed design can be difficult because of the enormous number of different ways that a design may fail. Design failure may occur because of inadequate stability of the desired conformation, or because of the existence of alternate conformations with higher stability, or because of high/low conformational entropy associated with the desired conformation, or because of increased stability of the unfolded state, or because of kinetic barriers to folding, or because of undesirable intermolecular interactions such as aggregation, or because of enhanced chemical instability such as deamidation, or because of expression related problems such as problems with transcription, post-translational processing, proteolysis or export.

Simplified biophysical models enable the deployment of efficient search algorithms (Gainza et al., 2016). Graph Algorithms such as Dead-End-Elimination or A* provide efficient solutions for discretized models with force-fields that are limited to singleton or pairwise terms (Hallen & Donald, 2019). However, the incorporation of terms for elimination of voids or solvation terms, that depend upon the solvent accessible surface area, may require computation of global properties that cannot be easily incorporated into such algorithms. A trade-off exists between the advantages of model simplification which enables utilization of provable algorithms, and the efficiency of stochastic algorithms which are applicable to a wider range of models. Novel algorithms, possibly based on novel biophysical representations are required for addressing these problems. In addition, problems that require consideration of multiple conformations and conformational dynamics could also benefit from algorithms based on novel biophysical representations.

The efficiency of protein design algorithms for more realistic biophysical models can be improved by more efficient utilization of the partial or approximate solutions provided by design algorithms that are based on simplified biophysical models.

8. Conclusion

Protein Design is a very challenging problem, because it is NP-hard for protein models that are realistic enough to be of practical utility (Pierce & Winfree, 2002). Proteins have to be designed in such a manner that the lowest energy conformations have the desired properties such as the specified fold, and the desired conformations are kinetically accessible from different regions of the conformational space. In addition, there should be an absence of undesirable conformations with energies that are close to the global minimum.

Substantial progress has been made in the application of Protein Design by using the cyclical Design, Build, Test and Learn (DBT) paradigm (Dahiyat & Mayo, 1996). The current state of progress in Protein Design necessitates validation of the design by experimental synthesis followed by characterization of all the relevant properties. For design problems related to stabilization, the number of designs that require testing can be less than

10 with a single DBT cycle, with the state of the art protein design methodology. However, design problems related to enzyme specificity and molecular interaction require either more design cycles or the testing of a larger number of designs, or a combination of both.

The successful design of a protein requires the development of a scoring function that can accurately predict the desired properties of a specified amino acid sequence, such as its 3D structure and enzyme activity (Conway & DiMaio, 2016; Li et al., 2013). And a search algorithm for optimizing the scoring function (Gainza et al., 2016). Both of these aspects have been the object of extensive studies. A wide variety of algorithms that are relevant for Protein Design have been reviewed in this article. Some of these algorithms have been demonstrated to be effective and can be implemented with moderate computational resources. A knowledge of the strengths and limitations of the available protein design algorithms is necessary for the selection of the methods that are best suited for the problem of interest.

Most of the successful applications of protein design, to-date, have utilized combinations of enumerative, stochastic, iterative, evolutionary and search algorithms (Huang et al., 2016). More capable hardware, better algorithms, and more accurate scoring functions will continue to improve the efficiency, accuracy and scope of Protein Design. In addition, novel approaches such as quantum computing algorithms or novel biophysical representations may lead to improvements that may result in a substantial increase in the specificity, size and complexity of problems that can be addressed by Protein Design (Maguire et al., 2021; Mulligan et al., 2020).

Acknowledgments

Sincere apologies for not including or for very brief coverage of some outstanding contributions to protein design due to limitations of space in this review.

References

Alibés, A., Nadra, A. D., De Masi, F., Bulyk, M. L., Serrano, L., & Stricher, F. (2010). Using protein design algorithms to understand the molecular basis of disease caused by protein-DNA interactions: The Pax6 example. *Nucleic Acids Research*, *38*, 7422–7431.

Anand, N., & Huang, P. (2018). Generative modeling for protein structures. In *Proceedings of the advances in neural information processing systems, Montreal, QC, Canada, 3–8 December 2018*.

Apgar, J. R., Hahn, S., Grigoryan, G., & Keating, A. E. (2009). Cluster expansion models for flexible-backbone protein energetics. *Journal of Computational Chemistry*, *30*, 2402–2413.

Aszodi, A., Gradwell, M. J., & Taylor, W. R. (1995). Global fold determination from a small number of distance restraints. *Journal of Molecular Biology*, *251*(2), 308–326.

Barakat, N. H., Barakat, N. H., & Love, J. J. (2010). Combined use of experimental and computational screens to characterize protein stability. *Protein Engineering, Design & Selection: PEDS, 23*, 799–807.

Bazzoli, A., Tettamanzi, A. G. B., & Zhang, Y. (2011). Computational protein design and large-scale assessment by I-TASSER structure assembly simulations. *Journal of Molecular Biology, 407*(5), 764–776. https://doi.org/10.1016/j.jmb.2011.02.017.

Bender, B. J., et al. (2016). Protocols for molecular modeling with Rosetta3 and RosettaScripts. *Biochemistry, 55*, 4748–4763.

Bian, Y., Wang, J., Jun, J., & Xie, X.-Q. (2019). Deep convolutional generative adversarial network (dcGAN) models for screening and design of small molecules targeting cannabinoid receptors. *Molecular Pharmaceutics, 16*, 4451–4460.

Boas, F. E., & Harbury, P. B. (2007). Potential energy functions for protein design. *Current Opinion in Structural Biology, 17*, 199–204.

Boyken, S. E. (2016). De novo design of protein homo-oligomers with modular hydrogen-bond network-mediated specificity. *Science, 352*, 680–687.

Brylinski, M. (2013). eVolver: An optimization engine for evolving protein sequences to stabilize the respective structures. *BMC Research Notes, 6*, 303.

Buchler, N. E. G., & Goldstein, R. A. (2000). Surveying determinants of protein structure designability across different energy models and amino-acid alphabets: A consensus. *The Journal of Chemical Physics, 112*, 2533.

Canutescu, A. A., & Dunbrack, R. L. (2003). Cyclic coordinate descent: A robotics algorithm for protein loop closure. *Protein Science, 12*, 963–972.

Choi, Y., Verma, D., Griswold, K. E., & Bailey-Kellogg, C. (2017). EpiSweep: Computationally driven reengineering of therapeutic proteins to reduce immunogenicity while maintaining function. *Methods in Molecular Biology (Clifton, N.J.), 1529*, 375–398.

Conway, P., & DiMaio, F. (2016). Improving hybrid statistical and physical force fields through local structure enumeration. *Protein Science: A Publication of the Protein Society, 25*, 1525–1534.

Crick, F. H. C. (1953). The Fourier transform of a coiled-coil. *Acta Crystallographica, 6*, 685–689.

Dahiyat, B. I., & Mayo, S. L. (1996). Protein design automation. *Protein Science, 5*, 895–903.

Das, R., & Baker, D. (2008). Macromolecular modeling with Rosetta. *Annual Review of Biochemistry*, 77(1), 363–382.

Davey, J. A., & Chica, R. A. (2017). Multistate computational protein design with backbone ensembles. *Methods in Molecular Biology (Clifton, N.J.), 1529*, 161–179.

Davis, I. W., Arendall, W. B., Richardson, D. C., & Richardson, J. S. (2006). The backrub motion: How protein backbone shrugs when a sidechain dances. *Structure, 14*, 265–274. https://doi.org/10.1016/j.str.2005.10.007.

Dawson, W. M., Rhys, G. G., & Woolfson, D. N. (2019). Towards functional de novo designed proteins. *Current Opinion in Chemical Biology, 52*, 102–111.

Desmet, J., Maeyer, M., Hazes, B., & Lasters, I. (1992). The dead-end elimination theorem and its use in protein sidechain positioning. *Nature, 356*, 539–542.

Dunbrack, R. L., & Karplus, M. (1994). Conformational analysis of the backbone-dependent rotamer preferences of protein sidechains. *Nature Structural Biology, 1*(5), 334–340.

Feynman, R. P. (1986). Quantum mechanical computers. *Foundations of Physics, 16*, 507–531.

Gainza, P., Nisonoff, H. M., & Donald, B. R. (2016). Algorithms for protein design. *Current Opinion in Structural Biology, 39*, 16–26.

Gainza, P., Roberts, K., & Donald, B. (2012). Protein design using continuous rotamers. *PLoS Computational Biology, 8*(1), 1–15, e1002335.

Gainza, P., et al. (2013). OSPREY: Protein design with ensembles, flexibility, and provable algorithms. *Methods in Enzymology, 523*(23422427), 87–107.

Gao, W., Mahajan, S. P., Sulam, J., & Gray, J. J. (2020). Deep learning in protein structural modeling and design. *Patterns*, *1*(9), 100142.
Georgiev, I., Keedy, D., Richardson, J. S., Richardson, D. C., & Donald, B. R. (2008). Algorithm for backrub motions in protein design. *Bioinformatics (Oxford, England)*, *24*, i196–i204.
Go, N., & Scheraga, H. A. (1970). Ring closure and local conformational deformations of chain molecules. *Macromolecules*, *3*(2), 178–187.
Greener, J. G., Moffat, L., & Jones, D. T. (2018). Design of metalloproteins and novel protein folds using variational autoencoders. *Scientific Reports*, *8*, 16189.
Grigoryan, G. (2013). Absolute free energies of biomolecules from unperturbed ensembles. *Journal of Computational Chemistry*, *34*, 2726–2741.
Grigoryan, G., Zhou, F., Lustig, S. R., Ceder, G., Morgan, D., & Keating, A. E. (2006). Ultra-fast evaluation of protein energies directly from sequence. *PLoS Computational Biology*, *2*(6), e63. https://doi.org/10.1371/journal.pcbi.0020063.
Hallen, M. A. (2019). PLUG (Pruning of Local Unrealistic Geometries) removes restrictions on biophysical modeling for protein design. *Proteins*, *87*(1), 62–73. https://doi.org/10.1002/prot.25623.
Gupta, A., & Zou, J. (2019). Feedback GAN for DNA optimizes protein functions. *Nature Machine Intelligence*, *1*(2), 105–111.
Hallen, M., & Donald, B. (2016). COMETS (constrained optimization of multistate energies by TreeSearch): A provable and efficient protein design algorithm to optimize binding affinity and specificity with respect to sequence. *Journal of Computational Biology*, *23*(5), 311–321.
Hallen, M. A., & Donald, B. R. (2017). CATS (Coordinates of Atoms by Taylor Series): Protein design with backbone flexibility in all locally feasible directions. *Bioinformatics (Oxford, England)*, *33*(14), i5–i12. https://doi.org/10.1093/bioinformatics/btx277.
Hallen, M. A., & Donald, B. R. (2019). Protein design by provable algorithms. *Communications of the ACM*, *62*, 76–84.
Hallen, M. A., Jou, J. D., & Donald, B. R. (2017). LUTE (local unpruned tuple expansion): Accurate continuously flexible protein design with general energy functions and rigid rotamer-like efficiency. *Journal of Computational Biology: A Journal of Computational Molecular Cell Biology*, *24*(6), 536–546. https://doi.org/10.1089/cmb.2016.0136.
Hallen, M., Keedy, D., & Donald, B. (2013). Dead-end elimination with perturbations (DEEPer): A provable protein design algorithm with continuous sidechain and backbone flexibility. *Proteins: Structure, Function, and Bioinformatics*, *81*(1), 18–39.
Havel, T. F. (1990). The sampling properties of some distance geometry algorithms applied to unconstrained polypeptide chains: A study of 1830 independently computed conformations. *Biopolymers*, *29*(12 − 13), 1565–1585.
Havel, T. F., & Wuthrich, K. (1985). An evaluation of the combined use of nuclear magnetic resonance and distance geometry for the determination of protein conformations in solution. *Journal of Molecular Biology*, *182*(2), 281–294.
Hellinga, H. W., & Richards, F. M. (1994). Optimal sequence selection in proteins of known structure by simulated evolution. *Proceedings of the National Academy of Sciences of the United States of America*, *91*(13), 5803–5807.
Hong, E.-J., Lippow, S. M., Tidor, B., & Lozano-Pérez, T. (2009). Rotamer optimization for protein design through MAP estimation and problem-size reduction. *Journal of Computational Chemistry*, *30*, 1923–1945.
Huang, P.-S., Ban, Y.-E. A., Richter, F., Andre, I., Vernon, R., Schief, W. R., et al. (2011). RosettaRemodel: A generalized framework for flexible backbone protein design. *PLoS One*, *6*(8), e24109. https://doi.org/10.1371/journal.pone.0024109.
Huang, P., Boyken, S. E., & Baker, D. (2016). The coming of age of de novo protein design. *Nature*, *537*, 320–327.

Huang, Y.-M., & Bystroff, C. (2013). Expanded explorations into the optimization of an energy function for protein design. *IEEE/ACM Transactions on Computational Biology and Bioinformatics*, *10*(5), 1176–1187. https://doi.org/10.1109/TCBB.2013.113.
Huang, P.-S., Oberdorfer, G., Xu, C., Pei, X. Y., Nannenga, B. L., Rogers, J. M., et al. (2014). High thermodynamic stability of parametrically designed helical bundles. *Science*, *346*, 481–485. https://doi.org/10.1126/science.1257481.
Huang, X., Pearce, R., & Zhang, Y. (2020). EvoEF2: Accurate and fast energy function for computational protein design. *Bioinformatics (Oxford, England)*, *36*(4), 1135–1142. https://doi.org/10.1093/bioinformatics/btz740.
Jiang, X., Farid, H., Pistor, E., & Farid, R. S. (2000). A new approach to the design of uniquely folded thermally stable proteins. *Protein Science: A Publication of the Protein Society*, *9*, 403–416.
Jones, D. T. (1994). De novo protein design using pairwise potentials and a genetic algorithm. *Protein Science*, *3*, 567–574.
Jou, J. D., Jain, S., Georgiev, I. S., & Donald, B. R. (2016). BWM*: A novel, provable, ensemble-based dynamic programming algorithm for sparse approximations of computational protein design. *Journal of Computational Biology: A Journal of Computational Molecular Cell Biology*, *23*(6), 413–424. https://doi.org/10.1089/cmb.2015.0194.
Jumper, J., Evans, R., Pritzel, A., Green, T., Figurnov, M., et al. (2021). Highly accurate protein structure prediction with AlphaFold. *Nature*, *596*(7873), 583–589.
Kadowaki, T., & Nishimori, H. (1998). Quantum annealing in the transverse Ising model. *Physical Review E*, *58*, 5355–5363.
Karimi, M., & Shen, Y. (2018). iCFN: An efficient exact algorithm for multistate protein design. *Bioinformatics (Oxford, England)*, *34*, i811–i820.
Karimi, M., Zhu, S., Cao, Y., & Shen, Y. (2020). De novo protein design for novel folds using guided conditional Wasserstein generative adversarial networks. *Journal of Chemical Information and Modeling*, *60*(12), 5667–5681.
Kleffner, R., Flatten, J., Leaver-Fay, A., Baker, D., Siegel, J. B., Khatib, F., et al. (2017). Foldit standalone: A video game-derived protein structure manipulation interface using Rosetta. *Bioinformatics (Oxford, England)*, *33*, 2765–2767.
Koepnick, B. (2019). De novo protein design by citizen scientists. *Nature*, *570*(776), 390–394.
Koga, N., Tatsumi-Koga, R., Liu, G., Xiao, R., Acton, T. B., Montelione, G. T., et al. (2012). Principles for designing ideal protein structures. *Nature*, *491*, 222–227. https://doi.org/10.1038/nature11600.
Kolinski, A., Galazka, W., & Skolnick, J. (1995). Computer design of idealized β-motifs. *The Journal of Chemical Physics*, *103*, 10286. https://doi.org/10.1063/1.469930.
Korendovych, I. V., & DeGrado, W. F. (2020). De novo protein design, a retrospective. *Quarterly Reviews of Biophysics*, *53*, e3.
Kraemer-Pecore, C. M., Wollacott, A. M., & Desjarlais, J. R. (2001). Computational protein design. *Current Opinion in Chemical Biology*, *5*, 690–695.
Kuhlman, B. (2019). Designing protein structures and complexes with the molecular modeling program Rosetta. *The Journal of Biological Chemistry*, *294*(50), 19436–19443. https://doi.org/10.1074/jbc.aw119.008144.
Kuhlman, B., & Bradley, P. (2019). Advances in protein structure prediction and design. *Nature Reviews. Molecular Cell Biology*, *20*, 681–697.
Kuhlman, B., Dantas, G., Ireton, G. C., Varani, G., Stoddard, B. L., & Baker, D. (2003). Design of a novel globular protein fold with atomic-level accuracy. *Science*, *302*, 1364–1368. https://doi.org/10.1126/science.1089427.
Kusner, M. J., Paige, B., & Hernández-Lobato, J. M. (2017). Grammar variational autoencoder. In *70. Proceedings of the 34th international conference on machine learning* (pp. 1945–1954).

Lazaridis, T., & Karplus, M. (1999). Effective energy function for proteins in solution. *Proteins*, *35*(3), 133–152.

Leach, A. R., & Lemon, A. P. (1998). Exploring the conformational space of protein side chains using dead-end elimination and the A* algorithm. *Proteins*, *33*, 227–239.

Leaver-Fay, A., Jacak, R., Stranges, P. B., & Kuhlman, B. (2011). A generic program for multistate protein design. *PLoS One*, *6*(7), 1–17. e20937.

LeCun, Y., Bengio, Y., & Hinton, G. (2015). Deep learning. *Nature*, *521*, 436–444.

Lee, M. S., Salsbury, F. R., & Brooks, C. L. (2002). Novel generalized Born methods. *Journal of Computational Chemistry*, *116*, 10606–10614.

Li, Z., Yang, Y., Faraggi, E., Zhan, J., & Zhou, Y. (2014). Direct prediction of profiles of sequences compatible with a protein structure by neural networks with fragment-based local and energy-based nonlocal profiles. *Proteins*, *82*, 2565–2573.

Li, Z., Yang, Y., Zhan, J., Dai, L., & Zhou, Y. (2013). Energy functions in de novo protein design: Current challenges and future prospects. *Annual Review of Biophysics*, *42*(1), 315–335.

Li, R.-X., Zhang, N.-N., Wu, B., OuYang, B., & Shen, H.-B. (2021). Multiobjective heuristic algorithm for *de novo* protein design in a quantified continuous sequence space. *Computational and Structural Biotechnology Journal*, *19*, 2575–2587. https://doi.org/10.1016/j.csbj.2021.04.046.

Liao, J., Warmuth, M. K., Govindarajan, S., Ness, J. E., Wang, R. P., Gustafsson, C., et al. (2007). Engineering proteinase K using machine learning and synthetic genes. *BMC Biotechnology*, 7, 16.

Lin, Y.-R., Koga, N., Tatsumi-Koga, R., Liu, G., Clouser, A., Montelione, G. T., et al. (2015). Control over overall shape and size in de novo designed proteins. *Proceedings of the National Academy of Sciences of the United States of America*, *112*, E5478–E5485. https://doi.org/10.1073/pnas.1509508112.

Lin, E., Lin, C.-H., & Lane, H.-Y. (2020). Relevant applications of generative adversarial networks in drug design and discovery: Molecular de novo design, dimensionality reduction, and de novo peptide and protein design. *Molecules (Basel, Switzerland)*, *25*, 3250.

Lippow, S. M., & Tidor, B. (2007). Progress in computational protein design. *Current Opinion in Biotechnology*, *18*(4), 305–311. https://doi.org/10.1016/j.copbio.2007.04.009.

Löffler, P., Schmitz, S., Hupfeld, E., Sterner, R., & Merkl, R. (2017). Rosetta:MSF: A modular framework for multi-state computational protein design. *PLoS Computational Biology*, *13*, e1005600.

Lu, P. (2018). Accurate computational design of multipass transmembrane proteins. *Science*, *359*, 1042–1046.

Lucas, J. E., & Kortemme, T. (2020). New computational protein design methods for de novo small molecule binding sites. *PLoS Computational Biology*, *16*, e1008178.

MacDonald, J. T., & Freemont, P. S. (2016). Computational protein design with backbone plasticity. *Biochemical Society Transactions*, *44*, 1523–1529. https://doi.org/10.1042/BST20160155.

MacDonald, J. T., Kelley, L. A., & Freemont, P. S. (2013). Validating a coarse-grained potential energy function through protein loop modelling. *PLoS One*, *8*, e65770. https://doi.org/10.1371/journal.pone.0065770.

MacDonald, J. T., Maksimiak, K., Sadowski, M. I., & Taylor, W. R. (2010). De novo backbone scaffolds for protein design. *Proteins: Structure, Function, and Bioinformatics*, *78*(5), 1311–1325. https://doi.org/10.1002/prot.22651.

MacDonald, J. T., et al. (2016). Synthetic beta-solenoid proteins with the fragment-free computational design of a beta-hairpin extension. *Proceedings of the National Academy of Sciences of the United States of America*, *113*(37), 10346–10351. https://doi.org/10.1073/pnas.1525308113.

Mach, P., & Koehl, P. (2013). Capturing protein sequence-structure specificity using computational sequence design. *Proteins, 81*, 1556–1570.

Mackenzie, C. O., Zhou, J., & Grigoryan, G. (2016). Tertiary alphabet for the observable protein structural universe. *Proceedings of the National Academy of Sciences of the United States of America, 113*, E7438–E7447.

Maguire, J. B., Boyken, S. E., Baker, D., & Kuhlman, B. (2018). Rapid sampling of hydrogen bond networks for computational protein design. *Journal of Chemical Theory and Computation, 2018*, 2751–2760.

Maguire, J. B., Grattarola, D., Mulligan, V. K., Klyshko, E., & Melo, H. (2021). XENet: Using a new graph convolution to accelerate the timeline for protein design on quantum computers. *PLoS Computational Biology, 17*, e1009037.

Maisuradze, G. G., Senet, P., Czaplewski, C., Liwo, A., & Scheraga, H. A. (2010). Investigation of protein folding by coarse-grained molecular dynamics with the UNRES force field. *The Journal of Physical Chemistry. A, 114*, 4471–4485.

Mandell, D. J., Coutsias, E. A., & Kortemme, T. (2009). Sub-angstrom accuracy in protein loop reconstruction by robotics-inspired conformational sampling. *Nature Methods, 6*, 551–552.

Michael, E., & Simonson, T. (2021). How much can physics do for protein design? *Current Opinion in Structural Biology, 72*, 46–54. https://doi.org/10.1016/j.sbi.2021.07.011.

Mitra, P., Shultis, D., & Zhang, Y. (2013). EvoDesign: De novo protein design based on structural and evolutionary profiles. *Nucleic Acids Research, 41*(23671331), W273–W280.

Monticelli, L. (2008). The MARTINI coarse-grained force field: Extension to proteins. *Journal of Chemical Theory and Computation, 4*, 819–834.

Moore, B. L., Kelley, L. A., Barber, J., Murray, J. W., & MacDonald, J. T. (2013). High-quality protein backbone reconstruction from alpha carbons using Gaussian mixture models. *Journal of Computational Chemistry, 34*, 1881–1889. https://doi.org/10.1002/jcc.23330.

Müller, A. T., Hiss, J. A., & Schneider, G. (2018). Recurrent neural network model for constructive peptide design. *Journal of Chemical Information and Modeling, 58*, 472–479.

Mulligan, V. K., et al. (2020). Designing peptides on a quantum computer. *Bioarxiv*. https://doi.org/10.1101/752485.

Nadra, A. D., Serrano, L., & Alibés, A. (2011). DNA-binding specificity prediction with FoldX. *Methods in Enzymology, 498*, 3–18.

Negron, C., & Keating, A. E. (2013). Multistate protein design using CLEVER and CLASSY. *Methods in Enzymology, 523*, 171–190.

Norn, C., et al. (2021). Protein sequence design by conformational landscape optimization. *Proceedings of the National Academy of Sciences of the United States of America, 118*(11). https://doi.org/10.1073/pnas.2017228118.

Ojewole, A. A., Jou, J. D., Fowler, V. G., & Donald, B. R. (2018). BBK* (branch and bound over K*): A provable and efficient ensemble-based protein design algorithm to optimize stability and binding affinity over large sequence spaces. *Journal of Computational Biology: A Journal of Computational Molecular Cell Biology, 25*, 726–739.

Ong, E., Huang, X., Pearce, R., Zhang, Y., & He, Y. (2021). Computational design of SARS-CoV-2 spike glycoproteins to increase immunogenicity by T cell epitope engineering. *Computational and Structural Biotechnology Journal, 19*, 518–529.

Ooi, T., Oobatake, M., Nemethy, G., & Scheraga, H. (1987). Accessible surface areas as a measure of the thermodynamics parameters of hydration of peptides. *Proceedings of the National Academy of Sciences of the United States of America, 84*, 3086–3090.

Pan, Y., Dong, Y., Zhou, J., Hallen, M., Donald, B. R., Zeng, J., et al. (2016). cOSPREY: A cloud-based distributed algorithm for large-scale computational protein design. *Journal of Computational Biology: A Journal of Computational Molecular Cell Biology, 23*, 737–749.

Pan, X., & Kortemme, T. (2021). Recent advances in de novo protein design: Principles, methods, and applications. *The Journal of Biological Chemistry*, *296*, 100558. https://doi.org/10.1016/j.jbc.2021.100558.

Panel, N., Villa, F., Opuu, V., Mignon, D., & Simonson, T. (2021). Computational design of PDZ-peptide binding. *Methods in Molecular Biology (Clifton, N.J.)*, *2256*, 237–255. https://doi.org/10.1007/978-1-0716-1166-1_14.

Park, K., Shen, B. W., Parmeggiani, F., Huang, P.-S., Stoddard, B. L., & Baker, D. (2015). Control of repeat protein curvature by computational protein design. *Nature Structural & Molecular Biology*, *22*(2), 167–174.

Parker, A., Choi, Y., Griswold, K., & Bailey-Kellogg, C. (2013). Structure-guided deimmunization of therapeutic proteins. *Journal of Computational Biology*, *20*(2), 152–165.

Pearce, R., Huang, X., Setiawan, D., & Zhang, Y. (2019). EvoDesign: Designing protein-protein binding interactions using evolutionary interface profiles in conjunction with an optimized physical energy function. *Journal of Molecular Biology*, *431*(13), 2467–2476. https://doi.org/10.1016/j.jmb.2019.02.028.

Pearce, R., & Zhang, Y. (2021). Deep learning techniques have significantly impacted protein structure prediction and protein design. *Current Opinion in Structural Biology*, *68*, 194–207.

Pierce, N. A., & Winfree, E. (2002). Protein design is NP-hard. *Protein Engineering*, *15*(10), 779–782. https://doi.org/10.1093/protein/15.10.779.

Ponder, J. W., & Richards, F. M. (1987). Tertiary templates for proteins—Use of packing criteria in the enumeration of allowed sequences for different structural classes. *Journal of Molecular Biology*, *193*, 775–791.

Reynolds, K. A., et al. (2008). Computational redesign of the SHV-1 beta-lactamase/beta-lactamase inhibitor protein interface. *Journal of Molecular Biology*, *382*, 1265–1275.

Ronald, D. G. J., Kieslich, C. A., & Morikis, D. (2011). Electrostatic clustering and free energy calculations provide a foundation for protein design and optimization. *Annals of Biomedical Engineering*, *39*(4), 1252–1263. https://doi.org/10.1007/s10439-010-0226-9.

Rossetto, A., & Zhou, W. (2019). GANDALF: A prototype of a GAN-based peptide design method. In *Proceedings of the 10th ACM international conference on bioinformatics, computational biology and health informatics, Niagara Falls, NY, USA, 7–10 September 2019* (pp. 61–66).

Sabban, S., & Markovsky, M. (2020). RamaNet: Computational de novo helical protein backbone design using a long short-term memory generative adversarial neural network. *F1000Research*, *9*, 298.

Šali, A., Shakhnovich, E. I., & Karplus, M. J. (1994). How does a protein fold? *Nature*, *369*(6477), 248–251.

Salvat, R. S., Parker, A. S., Guilliams, A., Choi, Y., Bailey-Kellogg, C., & Griswold, K. E. (2014). Computationally driven deletion of broadly distributed T cell epitopes in a biotherapeutic candidate. *Cellular and Molecular Life Sciences: CMLS*, *71*, 4869–4880.

Sesterhenn, F., et al. (2020). De novo protein design enables the precise induction of RSV-neutralizing antibodies. *Science (New York, N.Y.)*, *368*, 6492.

Simonson, T., et al. (2013). Computational protein design: The Proteus software and selected applications. *Journal of Computational Chemistry*, *34*(28), 2472–2484. https://doi.org/10.1002/jcc.23418.

Singh, A. (2021). Bottom-up de novo protein design. *Nature Methods*, *18*, 233.

Spiriti, J., & Zuckerman, D. M. (2014). Tunable coarse graining for Monte Carlo simulations of proteins via smoothed energy tables: Direct and exchange simulations. *Journal of Chemical Theory and Computation*, *10*(11), 5161–5177. https://doi.org/10.1021/ct500622z.

Suárez, M., Tortosa, P., & Jaramillo, A. (2008). PROTDES: CHARMM toolbox for computational protein design. *Systems and Synthetic Biology*, *2*, 105–113.

Talluri, S. (2011a). PLS for prediction of thermal stability of protein mutants. *Journal of Advanced Bioinformatics Applications and Research*, *2*(2), 155–160.
Talluri, S. (2011b). Advances in engineering of proteins for thermal stability. *International Journal of Advanced Biotechnology and Research*, *2*(1), 190–200.
Talluri, S. (2017). Distance geometry with improved sampling of conformational space for biomolecular structure determination by NMR spectroscopy and protein design. *Research Journal of Life Sciences Bioinformatics Pharmaceutical and Chemical Sciences*, *3*(2), 84–91. https://doi.org/10.26479/2017.0302.08.
Talluri, S. (2019). Computational protein design of bacteriocins based on structural scaffold of aureocin A53. *International Journal of Bioinformatics Research and Applications*, *15*(2), 129–143. https://doi.org/10.1504/IJBRA.2019.099575.
Thomson, A. R. (2014). Computational design of water-soluble α-helical barrels. *Science*, *346*, 485–488.
Traoré, S., Allouche, D., André, I., de Givry, S., Katsirelos, G., Schiex, T., et al. (2013). A new framework for computational protein design through cost function network optimization. *Bioinformatics (Oxford, England)*, *29*(17), 2129–2136. https://doi.org/10.1093/bioinformatics/btt374.
Traoré, S., Allouche, D., André, I., Schiex, T., & Barbe, S. (2017). Deterministic search methods for computational protein design. *Methods in Molecular Biology (Clifton, N.J.)*, *1529*, 107–123.
Traoré, S., Roberts, K. E., Allouche, D., Donald, B. R., André, I., Schiex, T., et al. (2016). Fast search algorithms for computational protein design. *Journal of Computational Chemistry*, *37*(12), 1048–1058. https://doi.org/10.1002/jcc.24290.
Vorobieva, A. A. (2021). Principles and methods in computational membrane protein design. *Journal of Molecular Biology*, *433*(20), 167154. https://doi.org/10.1016/j.jmb.2021.167154.
Wagner, G., Braun, W., Havel, T. F., Schaumann, T., Go, N., & Wuthrich, K. (1987). Protein structures in solution by nuclear magnetic resonance and distance geometry. The polypeptide fold of the basic pancreatic trypsin inhibitor determined using two different algorithms, DISGEO and DISMAN. *Journal of Molecular Biology*, *196*(3), 611–639.
Wang, J., Cao, H., Zhang, J. Z. H., & Qi, Y. (2018). Computational protein design with deep learning neural networks. *Scientific Reports*, *8*, 6349.
Wood, C. W., & Woolfson, D. N. (2018). CCBuilder 2.0: Powerful and accessible coiled-coil modeling. *Protein Science: A Publication of the Protein Society*, *27*, 103–111.
Woolfson, D. N. (2021). A brief history of De novo protein design: Minimal, rational, and computational. *Journal of Molecular Biology*, *433*(20), 167160. https://doi.org/10.1016/j.jmb.2021.167160.
Yang, C., et al. (2021). Bottom-up de novo design of functional proteins with complex structural features. *Nature Chemical Biology*, *17*, 492–500.
Yanover, C., Fromer, M., & Shifman, J. (2007). Dead end elimination for multistate protein design. *Journal of Computational Chemistry*, *28*(13), 2122–2129.
Yeh, C.-T., Brunette, T., Baker, D., McIntosh-Smith, S., & Parmeggiani, F. (2018). Elfin: An algorithm for the computational design of custom three-dimensional structures from modular repeat protein building blocks. *Journal of Structural Biology*, *201*(2), 100–107. https://doi.org/10.1016/j.jsb.2017.09.001.
Yue, K. Z., & Dill, K. A. (1993). Designing representations of protein conformations. *Abstracts of Papers of the American Chemical Society*, *206*, 115.
Zhang, Y., Chen, Y., Wang, C., Lo, C.-C., Liu, X., Wu, W., et al. (2020). ProDCoNN: Protein design using a convolutional neural network. *Proteins*, *88*(7), 819–829. https://doi.org/10.1002/prot.25868.
Zhou, Y., Donald, B. R., & Zeng, J. (2017). Parallel computational protein design. *Methods in Molecular Biology (Clifton, N.J.)*, *1529*, 265–277.

Zhou, F., Grigoryan, G., Lustig, S. R., Keating, A. E., Ceder, G., & Morgan, D. (2005). Coarse-graining protein energetics in sequence variables. *Physical Review Letters*, *95*, 148103.

Zhu, H., & Braun, W. (1999). Sequence specificity, statistical potentials, and three-dimensional structure prediction with self-correcting distance geometry calculations of beta-sheet formation in proteins. *Protein Science*, *8*(2), 326–342.

CHAPTER TWO

Computational prediction of protein functional sites—Applications in biotechnology and biomedicine

Florencio Pazos*
Computational Systems Biology Group, National Center for Biotechnology (CNB-CSIC), Madrid, Spain
*Corresponding author: e-mail address: pazos@cnb.csic.es

Contents

Abstract

There are many computational approaches for predicting protein functional sites based on different sequence and structural features. These methods are essential to cope with the sequence deluge that is filling databases with uncharacterized protein sequences. They complement the more expensive and time-consuming experimental approaches by pointing them to possible candidate positions. In many cases they are jointly used to characterize the functional sites in proteins of biotechnological and biomedical interest and eventually modify them for different purposes. There is a clear trend towards approaches based on machine learning and those using structural information, due to the recent developments in these areas. Nevertheless, "classic" methods based on sequence and evolutionary features are still playing an important role as these features are strongly related to functionality. In this review, the main approaches for predicting general functional sites in a protein are discussed, with a focus on sequence-based approaches.

Advances in Protein Chemistry and Structural Biology, Volume 130
ISSN 1876-1623
https://doi.org/10.1016/bs.apcsb.2021.12.001

1. Introduction

Thanks to the advances in sequencing technologies (van Dijk, Auger, Jaszczyszyn, & Thermes, 2014) today is relatively inexpensive and fast to obtain the amino-acid sequence of the whole protein repertory of a given organism, by sequencing its genome. Metagenomics projects (Segata et al., 2012) are also increasing the number of sequenced proteins and protein fragments. Nevertheless, the functional characterization of these massive amounts of data is not that easy, and experimental approaches for inferring functional features cannot cope with that deluge of sequence information. Consequently, most proteins for which we know the amino acid sequence are not functionally characterized. Less than 10% of proteins in the well curated UniProt database (The UniProt, 2017) have associated functional information (Rauer, Sen, Waman, Abbasian, & Orengo, 2021), a figure than is even lower for other less annotated protein sequence databases. And the functional characterization at the residue level (i.e. residues involved in some aspect of the protein's function) is one order of magnitude lower (~1%).

"Protein function" is a complex, multidimensional, and difficult to define phenomenon, in spite of the fact that everyone has an intuitive idea of what it means. This translates into the definition of "functional site", although a functional definition could be that it comprises any residue whose mutation alters the normal biological functioning of a protein without substantially changing its three-dimensional (3D) structure. This includes not only the active/catalytic sites of enzymes, but also the binding sites for small molecules, nucleic acids or other proteins, post-translational modification sites, etc.

Knowing which residues of a protein are functional is important not only for understanding the protein's function at the atomic level, but also to devise ways for modifying that function in our benefit, for biotechnological/biomedical purposes. The experimental characterization of active sites is challenging as it mainly involves mutating all the protein residues. For this reason, a plethora of computational methods have been developed that complement these experimental approaches by predicting a set of possible functional sites, consequently restricting the number of required experiments. There are many examples illustrating this interplay between computational and experimental methods, where in-silico approaches produce a set of candidate residues that are subsequently mutated to confirm their involvement in function or, more interestingly, to change specificity or design new functions (see some examples in (Chagoyen, Garcia-Martin, & Pazos, 2016). As another timely example, the in-silico prediction of functional sites is also providing important information at the molecular level on current diseases such as COVID19 (Pontes, Ruiz-Serra, Lepore, & Valencia, 2021) or Ebola (Pappalardo et al., 2016).

This review tries to provide an overview of the current landscape of methods for the in-silico prediction of general protein functional sites from different sequence and structural features. By "general" I mean methods that try to recognize important residues irrespective of their particular functional role. There is a plethora of specific prediction methods focused on particular types of functional sites (e.g. catalytic sites, glycosylation sites, DNA-binding sites, ...) that are not covered here. This review also pays special attention to the sequence-based methods, as sequence information is much more abundant than structural data, and sequence features, especially those based on evolution, are still among the most informative for determining a site's functional importance.

2. Sequence-based approaches

The number of known protein amino-acid sequences deposited in public databases grows exponentially, and this number is much higher than, for example, that of known 3D structures. It has been estimated that sequence data are around 300-fold larger than structural data (Rauer et al., 2021). For this reason, most methodologies for the in-silico prediction of protein functional sites use sequence information (Pietrosemoli, Lopez, Segura-Cabrera, & Pazos, 2012).

2.1 Based on sequence similarity

The most obvious way to detect functional sites in a sequence of interest is to carry out plain sequence searches against databases of known functional/active sites, such as *PROSITE* (Sigrist et al., 2002) or the *Catalytic Site Atlas* (Ribeiro et al., 2018). A simple *BLAST* search (Altschul et al., 1997) against any sequence database where the proteins contain functional annotations at the residue level, such as Uniprot (The UniProt, 2017) could also serve for this purpose (Fig. 1).

Although some interesting attempts have been made for predicting functional sites in single sequences (e.g. Ofran & Rost, 2007), most methods use, in way or another, evolutionary information in the form of multiple sequence

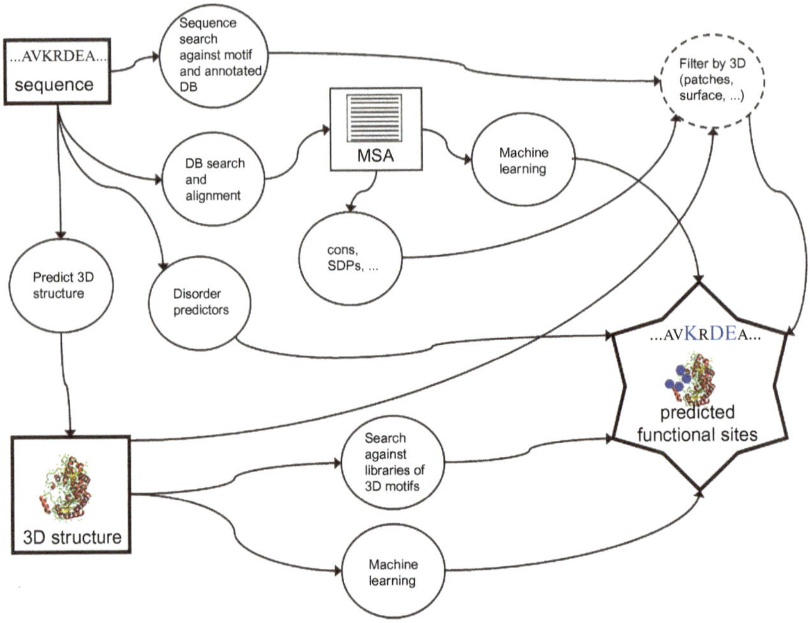

Fig. 1 Workflow showing the main strategies for predicting functional sites starting from sequence and/or structure information. Only the prediction of "general" functional sites is covered, not the prediction of specific types of functional sites. Circles represent processes. The dotted circle represents the optional step of filtering a set of predictions with structural information, if available.

alignments (MSAs) of homologous sequences (Pazos & Sanchez-Pulido, 2014). Homologous proteins, those arisen from a common ancestor, share a similar overall 3D structure (Chothia & Lesk, 1986) and many functional features (Devos & Valencia, 2000), both inherited from that ancestor. Consequently, MSAs of homologous proteins are rich sources of evolutionary, functional and structural information, as their positions represent the amino-acids allowed by the evolution at particular (homologous) sites. The patterns of residue variation at particular positions can be seen as landmarks left by the evolutionary process constrained by structural and functional requirements (Juan, Pazos, & Valencia, 2013) (Fig. 2).

Multiple sequence alignments can be easily generated by looking for homologous via a sequence search, and subsequently aligning them. Many web servers for performing sequence search incorporate the option to automatically align the found homologs. There are also databases with pre-compiled families of homologs and their associated MSAs, such as Pfam (Finn et al., 2014).

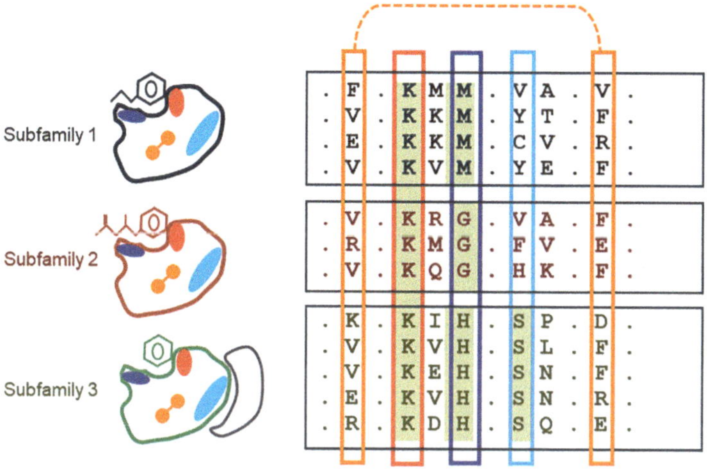

Fig. 2 Position mutational patterns related to structure and function extracted from protein multiple sequence alignments. The figure shows a MSA with 12 homologous proteins (hence with the same overall 3D structure and sharing a common function) where three functional subfamilies with different specificities can be defined (illustrated by the bound substrate). Fully conserved positions (red) tend to map in the active site of the proteins where the common function of the family resides, as well as in the structural core (not shown here for simplicity). Specificity-determining positions (SDPs) (light and dark blue) tend to map in the regions responsible for the functional specificity, in this case the binding sites for the variable part of the substrate and the protein binding site specific to subfamily 3. Correlated mutations (orange) tend to point to pairs of positions close in the 3D structure of the protein.

2.2 Conservation

The most obvious pattern of residue variation related to structure and function in MSAs is, indeed, non-variation (i.e. full conservation). A protein site where the evolution did not allow any amino-acid change in any of the homologous proteins has to be important for the structure or function of the family. Historically, this was the first variation pattern related to functionality, as it might be evident even in families with a limited number of known sequences (Zuckerkandl & Pauling, 1965). Although at first sight it could look trivial to detect conserved positions in MSAs, issues such as non-perfect conservation, conservation of subtle chemical properties, etc. makes it a non-trivial issue, and hence there are many methods for this purpose (Valdar, 2002). In general, the best methods are those that take into account the family's phylogeny in order to correct spurious conservation caused by very close sequences, for example (Pupko, Bell, Mayrose, Glaser, & Ben-Tal, 2002). If available, structural information can be incorporated to any of these methodologies. The most obvious way to combine conservation with structural information is to filter the set of conserved sites and keep those with the structural features attributed to functional sites, such as being at the surface (Ashkenazy, Erez, Martz, Pupko, & Ben-Tal, 2010) and forming patches. This allows, for example, to discard conservation due to structural reasons, such as that happening at structural cores.

In spite of being one of the oldest proxies for functionality, sequence conservation is still the strongest indicator of functional importance: for example, when combined with other features in predictors, conservation is shown to have the highest contribution determining the functionality of a site (Rauer et al., 2021).

2.3 Correlated mutations

Another informative pattern in MSAs is that involving pairs of positions with a correlated variation pattern. That is, changes in one of the positions are not independent but related to the equivalent changes in the other (Fig. 2). In general, this behavior indicates some type of relationship or linkage between the two corresponding sites, what in many cases has been related to closeness of these sites in the 3D structure, hence placing this phenomenon at the basis of a plethora of methods for predicting 3D contacts using only sequence information (Juan et al., 2013). The first methodologies achieved low performances predicting contacts from MSAs using correlated mutations, far from those needed to fold a protein or to get topological models. Nevertheless, in recent years a new wave of methods emerged with enough accuracy for generating reliable 3D models from the constraints imposed by the contacts predicted from correlated mutations (i.e. Hopf et al., 2012; Jones, Buchan, Cozzetto, & Pontil, 2012; Morcos et al., 2011). In one hand, this was achieved thanks to the accumulation of sequences in databases, which results in larger MSAs with the thousands of sequences these methodologies require in order to work. In the other hand, these methods include sophisticated approaches for disentangling the correlation signals due to 3D contacts from those due to other factors. Indeed, one of these factors is functionality: in many cases the sequence distribution of functional sites (see next point) makes them show up as "correlating" with other positions in the MSA (Chakrabarti & Panchenko, 2009), even if they are not in contact. So, this relationship between correlated mutations and functional sites, even if being a problem for contact prediction, can be exploited for the detection of functional regions (i.e. da Fonseca, & de Oliveira, & Bleicher, 2019; Kuipers et al., 2009; McMurrough, Dickson, Thibert, Gloor, & Edgell, 2014; Teppa, Wilkins, Nielsen and Buslje, 2012).

2.4 Specificity-determining positions

A subtler pattern of conservation in MSAs is the "subfamily-specific conservation." Those positions are conserved but only for particular subfamilies within the alignment: they are variable in other subfamilies or also conserved but with a different amino-acid (Fig. 2). Their conservation indicates functional importance, while the difference between subfamilies indicates that this importance is subfamily-specific and not shared by all homologous proteins in the MSA. If the partition of the MSA into different subfamilies is due to functional reasons, i.e. slightly different functional specificities within the common function shared by all members of the MSA, these subfamily-specific conservation patterns point to "specificity-determining positions" (SDPs). For example, a family of enzymes catalyzing the same chemical transformation (e.g. dehydrogenation), could be divided into different subfamilies according with the particular substrate where that transformation is applied (e.g. malate vs lactate dehydrogenases). That substrate specificity will be reflected in the amino-acid patterns of particular positions involved in binding the substrate: they will tend to be conserved due to their importance in binding, but with a different amino-acid in each subgroup to accommodate its particular substrate. On the other hand, positions involved in the common function of the family (dehydrogenation) would tend to be fully conserved. Consequently, SDPs complement fully conserved positions as predictors of functionality, as they are markers of different functional aspects (Fig. 2).

This complementarity between SDPs and conserved positions in detecting functional sites is important from a biotechnological point of view due to the differences between the types of sites they detect. In many cases, the detection of functional sites is followed by experimental mutation of these sites. Mutation of fully conserved position usually leads to the inactivation of the protein, and hence its main utility is confirming their involvement in function. Nevertheless, mutating SDPs it is possible to swap specificity or even design new specificities, what is more interesting for protein design and biotechnology (Chagoyen et al., 2016).

Many approaches for detecting SDPs in MSAs have been developed. They differ in their methodological basis as well as on whether, apart from the MSA itself, they require an additional input (generally some sort of functional classification of the proteins) (Fig. 3). In the first case, the methods are termed "unsupervised," and they infer the functional classification (i.e. functional subfamilies) from the sequence relationships implicit in the MSA. On the contrary, in "supervised" methods the functional classification is external to the MSA, and hence it has to be provided as additional input (Pazos, Rausell, & Valencia, 2006). Most methods belong to the first class as in most cases the protein classification implicit in the MSA (i.e. phylogeny) recapitulates the functional classification, in agreement with a scenario of "divergent evolution". But in some cases, or for some particular functional features of the many that are "pushing" the evolution of a family, there is a "function/phylogeny disagreement" and supervised approaches are required (e.g. (Bradley & Beltrao, 2019; Creixell et al., 2015; Huang et al., 2015; Pazos et al., 2006)). Most approaches have both, an unsupervised and

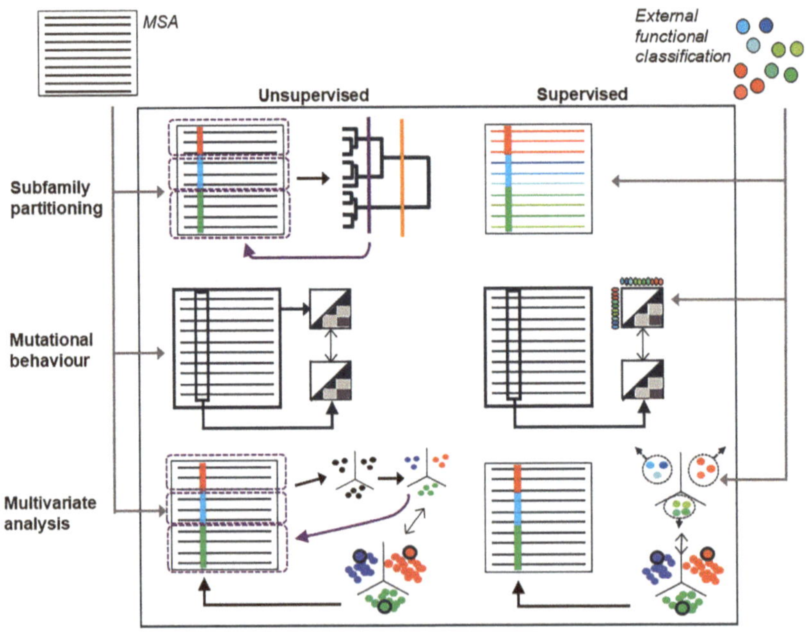

Fig. 3 Schemas and classification of the methods for detecting specificity-determining residues in multiple sequence alignments. The three main classes of approaches are illustrated for a MSA with three subfamilies (red, blue and green) and a SDP position. The main families of methods can be classified according with two axes: the methodology itself (rows); and depending on whether the method uses the MSA as the only input, inferring the subfamily classification from it ("unsupervised," left column), or apart from the MSA it requires an external functional classification of the proteins as additional input ("supervised," right column).

a supervised version, as for generating the second one only has to "bypass" the particular way used by the methodology for inferring the functional classification from the MSA, and incorporate an external functional classification instead (Fig. 3).

2.4.1 Approaches bases on subfamily partitioning

The first and simplest family of approaches comprises the "subfamily partitioning approaches." These methods split the set of sequences of the MSA in a number of disjoint groups and then infer the SDPs from them using different approaches. Note that once the subgroups are split, detecting the SDPs becomes similar to detecting conserved positions (previous point). One of the first methods to detect SDPs was the *Evolutionary Trace* (ET) method (Lichtarge, Bourne, & Cohen, 1996). ET takes as input a MSA and the phylogenetic tree of its proteins (that can be generated from the same MSA). By going along the levels of that tree from the root to the leaves, the method explores successive hierarchical partitions of the MSA, from very general to more specific subfamilies, and look for the conserved positions showing up at each partition (Fig. 3). ET ranks the positions of the MSA according with the partition where they become conserved: fully conserved positions appear at partition 0 (the whole MSA, root of the tree), in partition 1 the positions that are differentially conserved in the two main subfamilies within the MSA show up, and so on. From its initial implementation, ET was improved in different ways, for example incorporating an entropy-based criteria for evaluating the subfamily-specific conservation (Mihalek, Res, & Lichtarge, 2004). Indeed, many "subfamily partitioning approaches" use an entropy-based criterion for quantifying the "SDPeness" of a position, as the mutual information between its amino acid pattern and the pattern defined by a given subfamily classification. They differ on the way they look for an "optimal" subfamily classification from where to extract SDPs. One possibility is to explore different partitions (for example following the phylogenetic tree, as in *ET*), and look for the "optimal" one according with some criterion, such as that delivering the largest number of SDPs (Hannenhalli & Russell, 2000). Another possibility is to split the MSA into orthologous groups, as these tend to be functionally homogenous (Mirny & Gelfand, 2002). It is also possible, using heuristic approaches, to explore a very large number of possible subfamily groupings (not only those coherent with a given phylogenetic tree) and report that which maximizes some criterion, together with its associated SDPs. This is the strategy followed, for example, by *CEO* (Reva, Antipin, & Sander, 2007).

Turning any of these methods into supervised is as simple as imposing a given subfamily partition externally, bypassing its partition-generation stage and leaving only the SDP detection part (Fig. 3).

2.4.2 Approaches based on mutational behavior

What characterizes this family of approaches is that they do not use an explicit partition of the MSA into a given number of disjoint subfamilies, but the whole set of protein pairwise similarities is taken into account (Fig. 3). In this context, a possible strategy for locating SDPs is to look for MSA positions whose pattern of change "mutational behavior" resembles that of the whole MSA, represented by that matrix of protein pair-wise similarities. That is the expected behavior of a SDP position (Fig. 2), as similar amino acids correspond to globally similar proteins (same subfamily), and the other way around. The pattern of change of a position is represented by a matrix containing the similarities for all pairs of amino acids at that position, extracted from a standard substitution matrix. The positions with matrices most similar to that of the whole alignment (Fig. 3) are selected as the predicted SDPs (del Sol Mesa, Pazos, & Valencia, 2003; Pazos et al., 2006). Variations of this general strategy are, for example to compare the protein similarity matrix with that derived from an alignment window of consecutive positions, instead of a single position (La, Sutch, & Livesay, 2005), or derived for a set of positions that form a patch in the 3D structure, if available (Landgraf, Xenarios, & Eisenberg, 2001).

Using these strategies in "supervised" mode is also straightforward. The matrix of protein pair-wise similarities, where the phylogenetic and functional relationships are implicit, has to be substituted by an equivalent matrix representing a set of generic "functional similarities" (Fig. 3) (Pazos et al., 2006). This can be, for example, a matrix with the structural similarities of the ligands the proteins bind (Pazos et al., 2006), or a matrix representing similarities of interaction contexts in order to look for SDPs pointing to regions responsible for interaction specificity (Pitarch, Ranea, & Pazos, 2021).

2.4.3 Approaches based on multivariate analysis

A whole family of approaches are based on a vectorial representation of the proteins within the MSA followed by some dimensionality reduction step to find a lower dimensional space where the vector clusters correspond to the protein subfamilies (Fig. 2). The ability of these methods to detect SDPs resides in performing a similar vectorial treatment for the individual residues, which generates an equivalent space where the residue vectors co-locate with those of the subfamilies they are SDPs for (Fig. 3). The *SequenceSpace* program was the first in implementing this approach, using "principal component analysis" (PCA) as the dimensionality reduction step (Casari, 1995). In spite of its success in detecting SDPs, a drawback of that program was that it required a manual interactive inspection of the vectorial spaces in order to locate the subfamilies and their corresponding SDPs what, among other things, precluded its large scale application. A more recent representative of this type of methods is *S3Det* (Rausell, Juan, Pazos, & Valencia, 2010), which uses "multiple correspondence analysis" (MCA) for dimensionality reduction and incorporates automatic procedures for locating the clusters of proteins (subfamilies) and their corresponding SDPs.

An advantage of these approaches is that, concomitantly with the SDPs, they also report subfamily partitions (clusters) detected with an approach alternative to those used for generating phylogenetic trees.

Using these approaches in supervised mode consists basically in "forcing" the placement of the protein vectors according with the subfamily partition we want to impose: i.e. clustering together the proteins within the same subfamily and apart from the others (Fig. 3).

All these methods can be combined with 3D structure information if available. As with conservation, the most straightforward approach is to filter the set of SDPs and keep those preferentially at the surface and clustering in patches (e.g. Neuwald, Aravind, & Altschul, 2018).

2.5 Disordered regions

Proteins lacking a fixed 3D structure in their native state (either globally or in particular regions) are widespread and involved in core biological processes (Bondos, Dunker, & Uversky, 2021). Intrinsically-disordered protein regions (IDRs) are frequently involved in interactions with other proteins or small molecules, as they undergo a disorder to order transition upon binding (Mészáros, Simon, & Dosztányi, 2009). This makes these interactions mediated by IDRs especially suitable for certain types of functional interactions, such as transient interactions with multiple partners. For this reason, detecting possible IDRs in a sequence of interest is important not only for generating a structural model, but to detect functional regions involved in binding as well. There are different methods for detecting IDRs, based on training (Ward, Sodhi, McGuffin, Buxton, & Jones, 2004) or in physical principles (Erdős & Dosztányi, 2020). Moreover, due to this importance in binding, disorder is being incorporated in specific predictors of binding sites, e.g. (Zhao, Peng, & Yang, 2018).

3. Structure-based approaches

As commented in the previous sections, in principle any sequence-based prediction of functional sites can be combined with structural information, if available. But besides that, there are many methods which specifically use structure information as input. These methods have some advantages respect to those based on sequences alone, especially for some types of functional sites, as structure is "closer" to function than sequence, but they have the obvious drawback that structural information is more scarce than sequence data. Moreover, in most cases, leaving apart structural genomics projects, structural determination is a later step in the workflow for characterizing a protein, and mostly follows its functional characterization.

The most obvious way to detect functional/active sites in 3D structures is to look for structurally similar motifs in functionally characterized proteins. There are many methods for doing that which use different libraries of motifs of different types, e.g. (de Melo-Minardi, Bastard, & Artiguenave, 2010; Moraes, Pappa, Pires, & Izidoro, 2017; Torrance, Bartlett, Porter, & Thornton, 2005; Toti, Viet Hung, Tortosa, Brandi, & Polticelli, 2018; Xu et al., 2018). Since some of these methods or the database of sites they use are focused on specific types of functional sites, it might be interesting to use meta-predictors that combine many of them in order to have a broader picture of the possible functional sites of a given 3D structure, e.g. (Wu, Peng, Zhang, & Yang, 2018).

In many cases, binding and active sites are energetically unstable. Among other things, this is due to the need to place together energetically unfavorable combinations of residues due to functional requirements (for example two positive residues in a catalytic site), or to maintain unstable regions that become stabilized upon binding. Luque and Freire related the uneven distribution of the stabilization energy in proteins with binding sites and cooperativity effects (Luque & Freire, 2000). It has also been found a relationship between functionally important residues and those contributing negatively to the electrostatic stabilization energy (Elcock, 2001).

Some methods use a network representation of a protein's structure, where the nodes are the residues and the edges represent physical contacts between residues, and apply different graph-theoretical approaches to locate important nodes in that network, as a way to predict the importance of the corresponding residues. For example, residue importance has been related to the "closeness" (Amitai et al., 2004) and "betweeness" (Del Sol & O'Meara, 2005) network parameters of the corresponding nodes in the network representation.

With the recent revolution in the prediction of protein 3D structure represented by the machine learning-based *AlphaFold* system (Jumper et al., 2021) we are approaching a situation where we could have reliable structural models for almost any protein sequence. In that situation, it might be possible to apply these structure-based methods for predicting functional sites to any protein after generating its reliable structural model. For that it is essential that these structural models are not only globally correct, as they have been shown to be, but that their details at the functional sites are accurate as well, something that is still under investigation (Mullard, 2021). As commented previously, it could be said that active sites "do not follow the same rules" as the rest of the protein (e.g. they are energetically unstable) and consequently they always were problematic for structure prediction methods. In any case, it is clear that this turning point in protein 3D structure prediction will revolutionize the detection of functional sites as well.

4. Machine learning

Recent advances in machine learning (e.g. "deep learning"), coupled to the accumulation of the biological information these data-hungry algorithms require, is revolutionizing many aspects of structural biology and bioinformatics, as illustrated by the *AlphaFold* system for predicting protein 3D structure with unprecedented accuracy (Jumper et al., 2021). The prediction of functional sites is not an exception, and the recent years were characterized by an increase number of methodologies using that approach, a tendency that for sure will continue in the future.

As with any machine learning (ML) approach, these methods require a training set of examples (functional and non-functional sites, in this case) represented by a set of features (e.g. sequence and structure-derived features). From this dataset, the system "learns" how to distinguish the two classes based on these features. Once trained, when an unseen example is presented to the system (given in terms of these features), it generates a prediction for it.

There are ML systems for predicting functional sites based on sequence features (e.g. (Cui, Dong, Hong, & Wang, 2019; Lee, Keum, & Nam, 2019)), on structure (e.g. (Jiménez, Doerr, Martínez-Rosell, Rose, & De Fabritiis, 2017; Pu, Govindaraj, Lemoine, Wu, & Brylinski, 2019)), and on both (Das, Scholes, Sen, & Orengo, 2021).

A ML system is as good as the training set it is based on. It has to be large (specially for the newer deep-learning approaches), curated/reliable, and well balanced (similar number of examples in the two classes: functional and non-functional). In the case of the prediction of functional/binding sites, obtaining good datasets is problematic as there is not a univocal definition of "functional site" (see Introduction). In most cases binding sites (small molecule, nucleic acid and protein binding sites) are used as proxies for functional sites as most protein "functions" involve binding in one way or another, and reliable and abundant binding information can be obtained from structural databases. Moreover, specific predictors can be generated for different types of functional sites (e.g. binding, catalytic, ...) just by using specific training sets. In this regard, initiatives for generating curated annotations of structural databases at the residue level, such as *PDBe-KB* (PDBe-KB consortium, 2020) are important for the future development of ML predictors of binding sites. Balancing the training set is also important as there are far more examples of non-binding than binding sites.

5. Conclusions and future trends

Due to the pace at which raw protein sequences accumulate in public databases, computational methods for predicting protein function and functional sites are required to complement the experimental determination of functional features.

The current landscape of approaches for predicting protein functional sites is mature enough, both in terms of performance and usability, for these approaches to become part of the standard toolbox of molecular biologists. Today, these approaches are routinely applied, many times in combination with experimental approaches, to decipher the functional sites in families of proteins of interest and eventually modify them with biomedical or biotechnological purposes (Chagoyen et al., 2016).

Nevertheless, new methodologies appear every day and the existing approaches are continuously improving. The continuous growth of the sequence databases and the consequent enlargement of the known protein functional space will also prompt new developments in the field. It was also expected this deluge of sequence data to improve existing approaches, for example by enriching the multiple sequence alignments many of them use. Nevertheless, this issue is far from being clear (Garrido-Martín & Pazos, 2018). It looks like the mere accumulation of sequences does not improve some of these methods as they are now, and maybe more directed sequencing efforts should be considered, e.g. to complete underrepresented protein families.

It is clear that the future of the field, as in many other bioinformatics areas, will be dominated by machine learning methods trained with these growing sources of information.

Another foreseeable trend will be a shift of the methodologies towards structure-based approaches. Although structural information is still orders of magnitude scarcer than sequence information, this will slowly change thanks to improvements in techniques for structure determination at high resolution (e.g. Cryo-EM (Assaiya, Burada, Dhingra, & Kumar, 2021)) as well as the new groundbreaking approaches for the reliable prediction of protein structures (e.g. AlphaFold (Jumper et al., 2021)).

In spite of these trends, "classical" methods based on sequence conservation patterns, whose foundations date decades ago, are still very powerful as these sequence landmarks are tightly associated to functional constraints.

As new methods continue to be developed, an important issue is to create initiatives to objectively benchmark these approaches based on common rules and testing sets, similar to those already running, for example, for protein structure prediction methods: e.g. CASP (Kryshtafovych, Schwede, Topf, Fidelis, & Moult, 2019). This is especially important for the machine learning systems, which could be highly biased to the training sets used during development, and hence their benchmark in external validation sets is crucial. Although some steps have been carried out in this direction of benchmarking methods (e.g. (Chakraborty & Chakrabarti, 2015; Zhang & Kurgan, 2018)) they are still limited in the number of methods evaluated and the datasets of functional sites used for benchmarking. These initiatives will provide users with a clear picture of the current landscape of available methodologies and their relative performances and ranges of applicability.

Acknowledgments

This work was partially supported by project PID2019-108096RB-C22 from the Spanish Ministry for Science and Innovation with European Regional Development Fund.

References

Altschul, S. F., Madden, T. L., Schaffer, A. A., Zhang, J., Zhang, Z., Miller, W., et al. (1997). Gapped BLAST and PSI-BLAST: A new generation of protein database search programs. *Nucleic Acids Research*, *25*, 3389–3402.

Amitai, G., Shemesh, A., Sitbon, E., Shklar, M., Netanely, D., Venger, I., et al. (2004). Network analysis of protein structures identifies functional residues. *Journal of Molecular Biology*, *344*(4), 1135–1146.

Ashkenazy, H., Erez, E., Martz, E., Pupko, T., & Ben-Tal, N. (2010). ConSurf 2010: Calculating evolutionary conservation in sequence and structure of proteins and nucleic acids. *Nucleic Acids Research*, *38*, W529–W533.

Assaiya, A., Burada, A. P., Dhingra, S., & Kumar, J. (2021). An overview of the recent advances in cryo-electron microscopy for life sciences. *Emerging Topics in Life Sciences*, *5*(1), 151–168. https://doi.org/10.1042/ETLS20200295.

Bondos, S. E., Dunker, A. K., & Uversky, V. N. (2021). On the roles of intrinsically disordered proteins and regions in cell communication and signaling. *Cell Communication and Signaling: CCS*, *19*(1), 88. https://doi.org/10.1186/s12964-021-00774-3.

Bradley, D., & Beltrao, P. (2019). Evolution of protein kinase substrate recognition at the active site. *PLoS Biology*, *17*(6), e3000341. https://doi.org/10.1371/journal.pbio.3000341.

Casari, G. (1995). A method to predict functional residues in proteins. *Nature Structural Biology*, *2*, 171–178.

Chagoyen, M., Garcia-Martin, J. A., & Pazos, F. (2016). Practical analysis of specificity-determining residues in protein families. *Briefings in Bioinformatics*, *17*(2), 255–261. https://doi.org/10.1093/bib/bbv045.

Chakrabarti, S., & Panchenko, A. R. (2009). Coevolution in defining the functional specificity. *Proteins*, *75*(1), 231–240.

Chakraborty, A., & Chakrabarti, S. (2015). A survey on prediction of specificity-determining sites in proteins. *Briefings in Bioinformatics*, *16*(1), 71–88.

Chothia, C., & Lesk, A. M. (1986). The relation between the divergence of sequence and structure in proteins. *The EMBO Journal*, *5*, 823–826.

Creixell, P., Palmeri, A., Miller, C. J., Lou, H. J., Santini, C. C., Nielsen, M., et al. (2015). Unmasking determinants of specificity in the human kinome. *Cell*, *163*(1), 187–201. https://doi.org/10.1016/j.cell.2015.08.057.

Cui, Y., Dong, Q., Hong, D., & Wang, X. (2019). Predicting protein-ligand binding residues with deep convolutional neural networks. *BMC Bioinformatics*, *20*(1), 93. https://doi.org/10.1186/s12859-019-2672-1.

da Fonseca, N. J., Afonso, M. Q. L., Jr., de Oliveira, L. C., & Bleicher, L. (2019). A new method bridging graph theory and residue co-evolutionary networks for specificity determinant positions detection. *Bioinformatics*, *35*(9), 1478–1485. https://doi.org/10.1093/bioinformatics/bty846.

Das, S., Scholes, H. M., Sen, N., & Orengo, C. (2021). CATH functional families predict functional sites in proteins. *Bioinformatics*, *37*(8), 1099–1106. https://doi.org/10.1093/bioinformatics/btaa937.

de Melo-Minardi, R. C., Bastard, K., & Artiguenave, F. (2010). Identification of subfamily-specific sites based on active sites modeling and clustering. *Bioinformatics*, *26*(24), 3075–3082.

del Sol Mesa, A., Pazos, F., & Valencia, A. (2003). Automatic methods for predicting functionally important residues. *Journal of Molecular Biology*, *326*(4), 1289–1302.

Del Sol, A., & O'Meara, P. (2005). Small-world network approach to identify key residues in protein-protein interaction. *Proteins*, *58*(3), 672–682.

Devos, D., & Valencia, A. (2000). Practical limits of function prediction. *Proteins*, *41*, 98–107.

Elcock, A. (2001). Prediction of functionally important residues based solely on the computed energetics of protein structure. *Journal of Molecular Biology*, *312*, 885–896.

Erdős, G., & Dosztányi, Z. (2020). Analyzing protein disorder with IUPred2A. *Current Protocols in Bioinformatics*, *70*(1), e99. https://doi.org/10.1002/cpbi.99.

Finn, R. D., Bateman, A., Clements, J., Coggill, P., Eberhardt, R. Y., Eddy, S. R., et al. (2014). Pfam: The protein families database. *Nucleic Acids Research*, *42*(Database issue), D222–D230.

Garrido-Martín, D., & Pazos, F. (2018). Effect of the sequence data deluge on the performance of methods for detecting protein functional residues. *BMC Bioinformatics*, *19*(1), 67. https://doi.org/10.1186/s12859-018-2084-7.

Hannenhalli, S. S., & Russell, R. B. (2000). Analysis and prediction of functional sub-types from protein sequence alignments. *Journal of Molecular Biology*, *303*, 61–76.

Hopf, T. A., Colwell, L. J., Sheridan, R., Rost, B., Sander, C., & Marks, D. S. (2012). Three-dimensional structures of membrane proteins from genomic sequencing. *Cell*, *149*(7), 1607–1621. https://doi.org/10.1016/j.cell.2012.04.012.

Huang, H., Pandya, C., Liu, C., Al-Obaidi, N. F., Wang, M., Zheng, L., et al. (2015). Panoramic view of a superfamily of phosphatases through substrate profiling. *Proceedings of the National Academy of Sciences of the United States of America, 112*(16), E1974–E1983. https://doi.org/10.1073/pnas.1423570112.

Jiménez, J., Doerr, S., Martínez-Rosell, G., Rose, A. S., & De Fabritiis, G. (2017). DeepSite: Protein-binding site predictor using 3D-convolutional neural networks. *Bioinformatics, 33*(19), 3036–3042. https://doi.org/10.1093/bioinformatics/btx350.

Jones, D. T., Buchan, D. W. A., Cozzetto, D., & Pontil, M. (2012). PSICOV: Precise structural contact prediction using sparse inverse covariance estimation on large multiple sequence alignments. *Bioinformatics, 28*, 184–190.

Juan, D., Pazos, F., & Valencia, A. (2013). Emerging methods in protein co-evolution. *Nature Reviews. Genetics, 14*(4), 249–261.

Jumper, J., Evans, R., Pritzel, A., Green, T., Figurnov, M., Ronneberger, O., et al. (2021). Highly accurate protein structure prediction with AlphaFold. *Nature, 596*(7873), 583–589. https://doi.org/10.1038/s41586-021-03819-2.

Kryshtafovych, A., Schwede, T., Topf, M., Fidelis, K., & Moult, J. (2019). Critical assessment of methods of protein structure prediction (CASP)—Round XIII. *Proteins: Structure, Function, and Bioinformatics, 87*(12), 1011–1020. https://doi.org/10.1002/prot.25823.

Kuipers, R. K., Joosten, H. J., Verwiel, E., Paans, S., Akerboom, J., van der Oost, J., et al. (2009). Correlated mutation analyses on super-family alignments reveal functionally important residues. *Proteins, 76*(3), 608–616.

La, D., Sutch, B., & Livesay, D. R. (2005). Predicting protein functional sites with phylogenetic motifs. *Proteins, 58*(2), 309–320.

Landgraf, R., Xenarios, I., & Eisenberg, D. (2001). Three-dimensional cluster analysis identifies interfaces and functional residue clusters in proteins. *Journal of Molecular Biology, 307*, 1487–1502.

Lee, I., Keum, J., & Nam, H. (2019). DeepConv-DTI: Prediction of drug-target interactions via deep learning with convolution on protein sequences. *PLoS Computational Biology, 15*(6), e1007129. https://doi.org/10.1371/journal.pcbi.1007129.

Lichtarge, O., Bourne, H. R., & Cohen, F. E. (1996). An evolutionary trace method defines binding surfaces common to protein families. *Journal of Molecular Biology, 257*, 342–358.

Luque, I., & Freire, E. (2000). Structural stability of binding sites: Consequences for binding affinity and allosteric effects. *Proteins, S4*, 63–71.

McMurrough, T. A., Dickson, R. J., Thibert, S. M., Gloor, G. B., & Edgell, D. R. (2014). Control of catalytic efficiency by a coevolving network of catalytic and noncatalytic residues. *Proceedings of the National Academy of Sciences of the United States of America, 111*(23), E2376–E2383.

Mészáros, B., Simon, I., & Dosztányi, Z. (2009). Prediction of protein binding regions in disordered proteins. *PLoS Computational Biology, 5*(5), e1000376. https://doi.org/10.1371/journal.pcbi.1000376.

Mihalek, I., Res, I., & Lichtarge, O. (2004). A family of evolution-entropy hybrid methods for ranking protein residues by importance. *Journal of Molecular Biology, 336*(5), 1265–1282.

Mirny, L. A., & Gelfand, M. S. (2002). Using orthologous and paralogous proteins to identify specificity-determining residues in bacterial transcription factors. *Journal of Molecular Biology, 321*, 7–20.

Moraes, J. P. A., Pappa, G. L., Pires, D. E. V., & Izidoro, S. C. (2017). GASS-WEB: A web server for identifying enzyme active sites based on genetic algorithms. *Nucleic Acids Research, 45*(W1), W315–W319. https://doi.org/10.1093/nar/gkx337.

Morcos, F., Pagnani, A., Lunt, B., Bertolino, A., Marks, D. S., Sander, C., et al. (2011). Direct-coupling analysis of residue coevolution captures native contacts across many protein families. *Proceedings of the National Academy of Sciences of the United States of America, 108*(49), E1293–E1301.

Mullard, A. (2021). What does AlphaFold mean for drug discovery? *Nature Reviews Drug Discovery*, *20*(10), 725–727. https://doi.org/10.1038/d41573-021-00161-0.

Neuwald, A. F., Aravind, L., & Altschul, S. F. (2018). Inferring joint sequence-structural determinants of protein functional specificity. *eLife*, *7*, e29880. https://doi.org/10.7554/eLife.29880.

Ofran, Y., & Rost, B. (2007). ISIS: Interaction sites identified from sequence. *Bioinformatics*, *23*(2), e13–e16.

Pappalardo, M., Juliá, M., Howard, M. J., Rossman, J. S., Michaelis, M., & Wass, M. N. (2016). Conserved differences in protein sequence determine the human pathogenicity of ebolaviruses. *Scientific Reports*, *6*, 23743. https://doi.org/10.1038/srep23743.

Pazos, F., Rausell, A., & Valencia, A. (2006). Phylogeny-independent detection of functional residues. *Bioinformatics*, *22*(12), 1440–1448.

Pazos, F., & Sanchez-Pulido, L. (2014). *Protein Superfamilies. En ELS.* John Wiley & Sons, Ltd. https://doi.org/10.1002/9780470015902.a0025587).

PDBe-KB consortium. (2020). PDBe-KB: A community-driven resource for structural and functional annotations. *Nucleic Acids Research*, *48*(D1), D344–D353. https://doi.org/10.1093/nar/gkz853.

Pietrosemoli, N., Lopez, D., Segura-Cabrera, A., & Pazos, F. (2012). Computational prediction of important regions in protein sequences. *IEEE Signal Processing Magazine*, *29*(6), 143–147.

Pitarch, B., Ranea, J. A. G., & Pazos, F. (2021). Protein residues determining interaction specificity in paralogous families. *Bioinformatics*, *37*(8), 1076–1082. https://doi.org/10.1093/bioinformatics/btaa934.

Pontes, C., Ruiz-Serra, V., Lepore, R., & Valencia, A. (2021). Unraveling the molecular basis of host cell receptor usage in SARS-CoV-2 and other human pathogenic β-CoVs. *Computational and Structural Biotechnology Journal*, *19*, 759–766. https://doi.org/10.1016/j.csbj.2021.01.006.

Pu, L., Govindaraj, R. G., Lemoine, J. M., Wu, H.-C., & Brylinski, M. (2019). DeepDrug3D: Classification of ligand-binding pockets in proteins with a convolutional neural network. *PLoS Computational Biology*, *15*(2), e1006718. https://doi.org/10.1371/journal.pcbi.1006718.

Pupko, T., Bell, R. E., Mayrose, I., Glaser, F., & Ben-Tal, N. (2002). Rate4Site: An algorithmic tool for the identification of functional regions in proteins by surface mapping of evolutionary determinants within their homologues. *Bioinformatics*, *18*(S1), S71–S77.

Rauer, C., Sen, N., Waman, V. P., Abbasian, M., & Orengo, C. A. (2021). Computational approaches to predict protein functional families and functional sites. *Current Opinion in Structural Biology*, *70*, 108–122. https://doi.org/10.1016/j.sbi.2021.05.012.

Rausell, A., Juan, D., Pazos, F., & Valencia, A. (2010). Protein interactions and ligand binding: From protein subfamilies to functional specificity. *Proceedings of the National Academy of Sciences of the United States of America*, *107*(5), 1995–2000.

Reva, B., Antipin, Y., & Sander, C. (2007). Determinants of protein function revealed by combinatorial entropy optimization. *Genome Biology*, *8*(11), R232.

Ribeiro, A. J. M., Holliday, G. L., Furnham, N., Tyzack, J. D., Ferris, K., & Thornton, J. M. (2018). Mechanism and catalytic site atlas (M-CSA): A database of enzyme reaction mechanisms and active sites. *Nucleic Acids Research*, *46*(D1), D618–D623. https://doi.org/10.1093/nar/gkx1012.

Segata, N., Waldron, L., Ballarini, A., Narasimhan, V., Jousson, O., & Huttenhower, C. (2012). Metagenomic microbial community profiling using unique clade-specific marker genes. *Nature Methods*, *9*(8), 811–814. https://doi.org/10.1038/nmeth.2066.

Sigrist, C. J., Cerutti, L., Hulo, N., Gattiker, A., Falquet, L., Pagni, M., et al. (2002). PROSITE: A documented database using patterns and profiles as motif descriptors. *Briefings in Bioinformatics*, *3*(3), 265–274.

Teppa, E., Wilkins, A. D., Nielsen, M., & Buslje, C. M. (2012). Disentangling evolutionary signals: Conservation, specificity determining positions and coevolution. Implication for catalytic residue prediction. *BMC Bioinformatics*, *13*, 235.

The UniProt, C. (2017). UniProt: The universal protein knowledgebase. *Nucleic Acids Research*, *45*(D1), D158–D169. https://doi.org/10.1093/nar/gkw1099.

Torrance, J. W., Bartlett, G. J., Porter, C. T., & Thornton, J. M. (2005). Using a library of structural templates to recognise catalytic sites and explore their evolution in homologous families. *Journal of Molecular Biology*, *347*(3), 565–581.

Toti, D., Viet Hung, L., Tortosa, V., Brandi, V., & Polticelli, F. (2018). LIBRA-WA: A web application for ligand binding site detection and protein function recognition. *Bioinformatics*, *34*(5), 878–880. https://doi.org/10.1093/bioinformatics/btx715.

Valdar, W. S. (2002). Scoring residue conservation. *Proteins*, *48*(2), 227–241.

van Dijk, E. L., Auger, H., Jaszczyszyn, Y., & Thermes, C. (2014). Ten years of next-generation sequencing technology. *Trends in Genetics*, *30*(9), 418–426.

Ward, J. J., Sodhi, J. S., McGuffin, L. J., Buxton, B. F., & Jones, D. T. (2004). Prediction and functional analysis of native disorder in proteins from the three kingdoms of life. *Journal of Molecular Biology*, *337*(3), 635–645.

Wu, Q., Peng, Z., Zhang, Y., & Yang, J. (2018). COACH-D: Improved protein–ligand binding sites prediction with refined ligand-binding poses through molecular docking. *Nucleic Acids Research*, *46*(W1), W438–W442. https://doi.org/10.1093/nar/gky439.

Xu, Y., Wang, S., Hu, Q., Gao, S., Ma, X., Zhang, W., et al. (2018). CavityPlus: A web server for protein cavity detection with pharmacophore modelling, allosteric site identification and covalent ligand binding ability prediction. *Nucleic Acids Research*, *46*(W1), W374–W379. https://doi.org/10.1093/nar/gky380.

Zhang, J., & Kurgan, L. (2018). Review and comparative assessment of sequence-based predictors of protein-binding residues. *Briefings in Bioinformatics*, *19*(5), 821–837. https://doi.org/10.1093/bib/bbx022.

Zhao, Z., Peng, Z., & Yang, J. (2018). Improving sequence-based prediction of protein–peptide binding residues by introducing intrinsic disorder and a consensus method. *Journal of Chemical Information and Modeling*, *58*(7), 1459–1468. https://doi.org/10.1021/acs.jcim.8b00019.

Zuckerkandl, E., & Pauling, L. (1965). Molecules as documents of evolutionary history. *Journal of Theoretical Biology*, *8*, 357–366.

CHAPTER THREE

Structure and chemistry of enzymatic active sites that play a role in the switch and conformation mechanism

Chandrabose Selvaraj[a,*], Ondipilliraja Rudhra[a], Abdulaziz S. Alothaim[b], Mustfa Alkhanani[c], and Sanjeev Kumar Singh[a,*]

[a]Computer Aided Drug Design and Molecular Modelling Lab, Department of Bioinformatics, Science Block, Alagappa University, Karaikudi, Tamil Nadu, India
[b]Department of Biology, College of Science in Zulfi, Majmaah University, Majmaah, Saudi Arabia
[c]Emergency Service Department, College of Applied Sciences, Al Maarefa University, Riyadh, Saudi Arabia
*Corresponding authors: e-mail address: selnikraj@bioclues.org; sksingh@alagappauniversity.ac.in

Contents

Advances in Protein Chemistry and Structural Biology, Volume 130
ISSN 1876-1623
https://doi.org/10.1016/bs.apcsb.2022.02.002

Abstract

Enzymes, which are biological molecules, are constructed from polypeptide chains, and these molecules are activated through reaction mechanisms. It is the role of enzymes to speed up chemical reactions that are used to build or break down cell structures. Activation energy is reduced by the enzymes' selective binding of substrates in a protected environment. In enzyme tertiary structures, the active sites are commonly situated in a "cleft," which necessitates the diffusion of substrates and products. The amino acid residues of the active site may be far apart in the primary structure owing to the folding required for tertiary structure. Due to their critical role in substrate binding and attraction, changes in amino acid structure at or near the enzyme's active site usually alter enzyme activity. At the enzyme's active site, or where the chemical reactions occur, the substrate is bound. Enzyme substrates are the primary targets of the enzyme's active site, which is designed to assist in the chemical reaction. This chapter elucidates the summary of structure and chemistry of enzymes, their active site features, charges and role of water in the structures to clarify the biochemistry of the enzymes in the depth of atomic features.

1. Introduction

X-ray crystallography has made it possible to examine enzyme structures at the molecular level of detail and recent advantage in computational power and software algorithms has made to understand the atom level insights of the enzymes (Srivastava, Nagai, Srivastava, Miyashita, & Tama, 2018; Trabuco et al., 2011). These computational features provide detail information on (a) protein movements in the dynamic motion, especially in the loop regions and side chains; (b) understanding the interactions between two chemical substances like enzymes, cofactor, ligands, drugs, metal and water molecules; (c) reaction mechanisms with reactant to product stage with detailed intermediates and (d) energy details of the reactions, complex formations and other biological formations (Selvaraj, Sakkiah, Tong, & Hong, 2018). The interactions of amino acid side chains with each other and with the environment determine the structure of proteins (Luscombe, Laskowski, & Thornton, 2001). The secondary structure is the product of interactions between the peptide groups that link the primary structure (Krieger & Kececioglu, 2020). The proteins have a linear order of amino acids called as primary and segmented as secondary and tertiary structures that fold or coil the amino acids into a 3D shape due to hydrogen bonds (Jiang, Han, & Wu, 2013). There are several proteins in which quaternary structures are created when many polypeptide chains come together to form one complex (Bozic, Doles, Gradisar, & Jerala, 2013). A wide array of proteins, including structural and enzyme proteins, regulatory proteins and transport proteins, are involved in many biological processes (Guo, Sirkis, & Schekman, 2014). Translation (mRNA to protein) results in a polypeptide that serves a specific function in the cell and function of those proteins determined based on its structure and sequences (Buccitelli & Selbach, 2020). Structural proteins aid in the construction and maintenance of cellular forms and biological forms. There are other proteins that are capable of carrying out biochemical activities, such as those enzymes involved in cellular respiration or biosynthesis. Proteins that regulate transcription or translation by binding to DNA or RNA are known as regulatory proteins (Yokoyama et al., 2006). In addition to nucleic acids, regulatory proteins frequently bind to other smaller signal molecules. Ensuring the cell's functions can be understood by allowing molecules like carbohydrates to enter and exit through the mode of transport proteins (Chen, Cheung, Feng, Tanner, & Frommer, 2015). Besides this, the excretion metabolism is aided by these transport proteins in the plasma membrane, which remove waste from the cell. Some proteins, despite not being part of the plasma membrane, yet play a function in the transport mechanism. Red blood cells (RBC) contain hemoglobin protein, which carries oxygen throughout the human body (Reid & Mohandas, 2004).

2. Proteins as drug targets

Disease-causing proteins are the most common targets for therapeutic drugs because of their importance in disease mechanisms as exemplified by proteins. Some of the drugs are known to work on ribosomes, DNA (Minor groove and Major groove) or other pathogenic targets (Nakamoto & Osada, 2013). These proteins must be present in the diseased state and have binding sites for the medication within their structure. Enzymes and receptor proteins have repeatedly emerged as the most popular therapeutic targets in studies attempting to estimate the number of licensed medications that act on this class of target proteins (Quan, Wang, Chu, & Zhang, 2018). In terms of receptor proteins, the most prevalent targets are G protein-coupled receptors, which comprise the majority of membrane receptors. Approximately, 800 GPCRs are encoded in the human genome, and they may respond to a wide range of signals ranging in size from photons to tiny proteins. The study of disease molecular pathways relies on the identification of potential drug target proteins as a critical stage in drug discovery (Hauser, Attwood, Rask-Andersen, Schioth, & Gloriam, 2017), and Fig. 1 shows the representation of proteins as drug targets. Because most proteins interact with one another, researchers are looking at the human protein–protein interaction

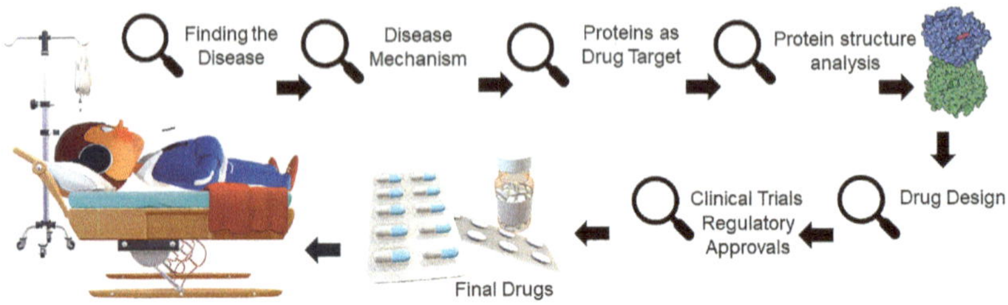

Fig. 1 Graphical representation of using the proteins as drug targets for disease mechanism.

network to find potential target proteins. From "one drug, one target, one disease," the discovery and development of new drugs has evolved into "one drug, many targets, many diseases." One of the newest areas of research in the pharmaceutical industry involves the use of existing drugs in new ways, such as drug repurposing, in order to avoid costly mistakes during the early stages of drug development and preclinical trials, as well as lower attrition rates during the final stages of clinical trials. Drug repurposing has profited greatly from the rapid development of bioinformatics and cheminformatics tools and omics technology in the biomedical arena (Pushpakom et al., 2019).

3. Protein conformations

The 3D structure of a protein restricts its function in the stable environment, but its dynamic motion help to define it. Protein functional conformational variations can aid a wide range of biological processes in cells. Protein framework to be effective, it must be able to adapt to changing conditions and evolve under external limitations (Fuglebakk, Tiwari, & Reuter, 2015). In today's world of computational biology, proteomics is one of the most essential and relevant topics, raising an abundance of intriguing and thought-provoking issues. Research into protein dynamics and function, as well as new insights into protein folding, are all hot topics in bioinformatics and medicine (Stetz et al., 2019). Recent Insights into G protein activation afforded by recently resolved structures of GPCRs and G protein complexes are critical for understanding G protein signaling. Real-time binding and dissociation of GPCR-G proteins in the cell take place in a highly dynamic environment (Latorraca, Venkatakrishnan, & Dror, 2017). Recently, the bacterial protein sortase A (SrtA) is shown with the binding site loop fluctuate in forming the open to closed lid

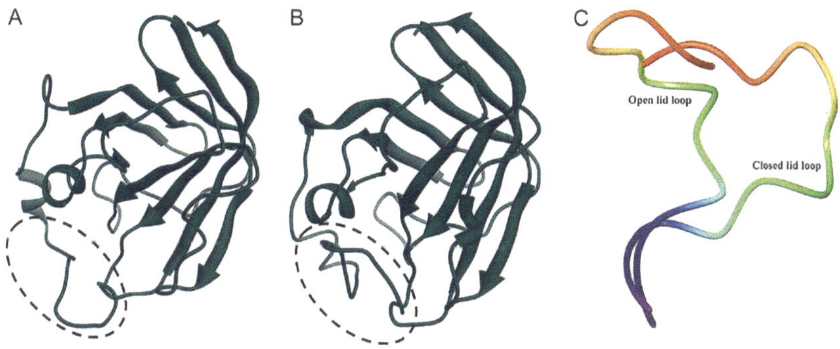

Fig. 2 Protein conformation of open (A) closed (B) conformation in SrtA in *Bacillus anthracis* and the (C) indicates the tail conformational changes.

conformation as gated mechanism to explore the inhibitory potential of reported compounds and through this study, the authors have shown the structural changes in the active sites (Fig. 2) have done an impact in the inhibitory profiles of drug candidates (Selvaraj et al., 2021).

4. Enzymes in reaction mechanism

Enzymes, which are biological molecules, are made up of chains of polypeptides and triggered by reaction mechanisms. Enzymes play an important part in the construction and destruction of cellular structures by speeding up chemical processes (Cooney, 2017). The enzyme–substrate complex refers to the relationship between an enzyme and its substrates. Reaction mechanisms can be inferred from the study of reaction kinetics. It is the electrostatic and hydrophobic forces that draw substrates to the active site, which are known as noncovalent bonds since they are physical rather than chemical bonds (Delaune & Alsayouri, 2021). A ternary complex has two substrates and one enzyme; a binary complex has only one substrate and one enzyme. The active site of an enzyme attracts substrates, catalyzes the chemical reaction that produces products, and then permits the products to dissociate (Zhu, Bilgin, & Snyder, 2003). The active site of an enzyme may bind molecules that are strikingly similar to the enzyme's substrate, but these molecules are unable to undergo reaction. In order to prevent the actual substrate from attaching to the active site, these molecules cover the active site. Competition between the inhibitor and substrate for the active site explains why this suppression of enzyme activity is of a competitive character (Prothiwa & Bottcher, 2020).

5. Enzymes active site

All of the amino acid residues that make up the enzyme's active site are placed in specific order. Depending on the amino acids around them, these residues might be positively or negatively charged (Tsou, 1998). The active site's amino acid composition and 3D location have a substantial impact on the enzyme's structure and chemical properties (Silman & Sussman, 2008). The substrates are drawn to the active site by non-covalent connections formed by hydrogen bonds, ionic bonds, electrostatic forces, and hydrophobic forces (Fu, Li, Wang, Luan, & Dai, 2021). These active site amino acid residues also create a unique chemical environment for the enzyme's proper binding and operation, and they determine whether a specific substrate binds or not. Due to their critical role in substrate binding and attraction, changes in amino acid structure at or near the enzyme's active site usually alter enzyme activity as shown in the Fig. 3 (Jiang, You, & Zhang, 2019). At the enzyme's active site, or where the chemical reactions occur, the substrate is bound. Enzyme substrates are the primary targets of the enzyme's active site, which is designed to assist in the chemical reaction. As an enzyme catalyst, it lowers the activation energy of biological processes and converts the substrate to products at physiological temperatures. The temperature of the environment, the pH, the location of the enzyme in the body, and the presence of other substances all influence the activity of

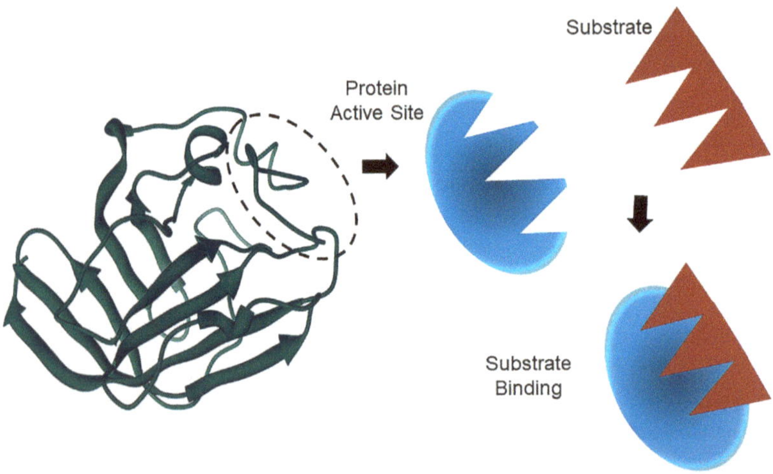

Fig. 3 Representation of substrate binding to the active site of an enzyme molecule.

enzymes. The rate of enzyme activity varies according to the enzyme's surroundings. Enzyme activity refers to the rate at which the enzyme performs its function. When an enzyme binds to its substrate, a complex is formed and at the end of the reaction, the enzyme will always return to its original state (Blat, 2010). The catalytic activity of an enzyme is dependent on the presence of a cofactor, a non-protein component found in many enzymes. Enzymes can only perform at their maximum efficiency or endurance, if the biochemical process is supported by cofactors, which are typically small organic compounds or metal ions. The enzyme active site is where the substrate normally binds, but for some other substances, the enzyme binds with the enzyme at a different location called the allosteric site (Fuller, Wilson, Eberhart, & Alexandrova, 2019). In the allosteric site, molecules can either activate or inhibit enzyme activity, or even shut it down completely.

The cofactor molecules or ligand molecules alter the enzyme's conformation or shape by binding to the allosteric site. Molecules that turn off enzymes, by binding to these sites are called allosteric inhibitors and activating molecules are called Allosteric activators. Allosteric inhibitors bind to the enzyme's allosteric site, which serves as the enzyme's "off switch." An enzyme's "on switch," the Allosteric activators bind to the enzyme at its allosteric site. To produce a reaction product, enzymes chemically modify the bound substrate to produce a new molecule. Enzymes are proteins having this property: they can bind substrate at their active site. In the same way that ligands attach to proteins, substrates bind to enzymes. Substrates, on the other hand, undergo an enzyme-induced chemical change and are transformed to products when they bind to enzymes (Tsou, 1998). Every chemical reaction needs an initial energy input before the reaction can take place. The reaction will proceed rapidly and readily if the initial energy demand (also known as the activation energy or energy barrier) is low. The reaction will take longer to complete if the activation energy is high (Bigley & Raushel, 2013).

6. Metal and ion oriented active site

Retroviral integrases are enzymes that integrate viral DNA into the host genome (INs). INs from retroviruses have two known metal binding domains. The zinc finger motif in the N-terminal domain has been found to bind Zn^{2+}, whereas a trio of acidic amino acids in the central catalytic core domain bind Mn^{2+} or Mg^{2+}, the metal cofactors essential for enzymatic activity (Reddy & Singh, 2014). "Processing" and "joining" are two distinct phases of the integration reaction. Endo nucleolytic cleavage is known as "processing," while polynucleotide transfer is known as "joining." Earlier findings revealed that $Mn^{2+} > Mg^{2+}$ is the preferred metal for in vitro activity of avian sarcoma virus INs, Ca^{2+}, Zn^{2+}, and Cd^{2+} can also bind at the active site of the catalytic domain, according to recent study. Two zinc and cadmium cations are also bound in the active site, with all three active site triad residues (Asp64, Asp121, and Glu157) contributing to their coordination. These findings point to a two-metal mechanism for retroviral integrase catalysis. The researchers further showed that Zn^{2+} may be used as a cofactor for Endo nucleolytic processes catalyzed by avian sarcoma virus IN's full-length protein, a variant lacking the N-terminal domain, or the isolated catalytic domain (Reddy, Singh, & Singh, 2014). In the presence of Zn^{2+}, polynucleotidyl transferase activities are significantly reduced or undetectable. Despite the fact that the processing and joining phases of integrase use the same mechanism and active site triad, their metal preferences separate them (Jacques, Coincon, & Sygusch, 2018). HEME halo-peroxidases are biological enzymes that combine H^2O^2 and halides on a HEME core to produce hypohalide, which then interacts with a substrate through halide transfer. The reaction pathway starting with an iron (III)-hydrogen peroxide-HEME complex was investigated using model complexes of the active site of HEME halo-peroxidase (Brown, Yates, Bielecki, Olczak, & Smalley, 2018). The research shows that the conversion of H_2O_2 and halide on a HEME core is quick, and only a few intermediates in the reaction pathway have a long enough lifespan to be trapped and characterized using experimental techniques.

7. Substrate binding models

The (i) Lock and Key model and (ii) Induced fit model are the two important models have been proposed to illustrate the binding process between substrate and enzyme. Apart from this model, the conformational selection model also has been proposed, and here we will discuss these three models in detail.

7.1 Lock and key model

Fisher's working with glycolytic enzymes and their ability to distinguish between different sugar stereoisomers was the inspiration for this Lock and Key theory. A model known as Fisher's theory, or the "lock and key" model, is one of two that describe the interaction between enzymes and their substrates (Du et al., 2016). The Lock and Key model are based on the idea that enzymes and substrates have complementary shapes. According to this theory, the enzyme's active site can accommodate a substrate of any shape or size but should match to the partner molecule. It is based on this theory that both enzymes and substrates have a set of characteristics that allow them to fit together. The enzyme acts as a "lock," accepting only one or a few substrates, and the substrates serve as the "keys." For this reason, enzymes can fit into a substrate just like a key can fit into the lock of its own substrate. In explaining enzyme specificity, this theory proposes that enzyme activity and reaction ground state are complementary (Abdelsattar, Mansour, & Aboul-Ela, 2021). The substrate's geometric structure and orientation are compatible with enzyme activity, but the shapes of each do not influence each other. Due to the predetermined shape of both enzyme and substrate, the substrate remains stable. These Lock and Key model apply for small molecule interactions to macromolecular interactions, especially at the binding interface of protein–protein interactions, protein surfaces are found to be complementary to each other in shape (Selvaraj, Omer, Singh, & Singh, 2015). This geometric descriptor is called shape complementarity. Proteins' convex and concave surfaces are complementary in general, and this can be defined as a match for the protein–protein interactions. Even though the lock-and-key model can explain a lot of interactions, it has the major flaw of ignoring the bound substrate that is transformed into the transition state (Beltran-Villegas, Colon-Melendez, Solomon, & Larson, 2016).

7.2 Induced fit model

Until Daniel Koshland induced fit hypothesis was proposed, the theory of Lock and Key model was widely accepted. The "conformational plasticity" of enzymes is key to the induced-fit hypothesis (Xu & Lill, 2013). The induced fit model state that, during the ligand-free and ligand-bound states of an enzyme, the active site conformations are different. An enzyme and its substrate may be likened to putting your hand in a glove, with the enzyme acting as the glove and the substrate as the hand, according to the hypothesis. Three important points are proposed by the Induced fit model:

1. The precise orientation of the catalytic group is necessary for enzyme activity.
2. The substrate may alter the three-dimensional arrangement of the active site residues.
3. Non-substrates cannot perform any action because substrate-binding is required to bring about the necessary changes in the enzyme active site.

The major flaw in the theory of induced fit has to do with an Enzyme–substrate complex that is not favorable thermodynamically. Even though the substrate is bound to enzyme, the energetically favorable conformational change will confirm the suitable substrate binding pose. Recent studies on crystal structure-based analysis have clarify the existence of conformational state, that are favored by several energy factors.

7.3 Conformational selection model

The conformational selection model was born as a result of a change to the induced fit modeling framework. This concept does not require the enzyme to undergo a conformational change as a result of the binding event. However, in this situation, ligand binding occurs directly to the enzyme via a mechanism analogous to the MWC (Monod–Wyman–Changeux) model of cooperation (Vogt & Di Cera, 2013). It is common for the molecules to change their conformation after binding, making it easier for them to bind again. An enzyme's inactive "active site," its active "active site" (the active conformation), and the enzyme's binding to the ligand are all kept in balance in this way (Boehr, Nussinov, & Wright, 2009).

8. Understanding allosteric site and active site

Unlike the rest of the molecule, an enzyme's active site is more susceptible to denaturation. The active site orientation can be disrupted if an inhibitor binds somewhere other than the active site. The ability to alter the enzyme's activity is made possible by its flexibility nature through conformational changes (Daura, 2019). Furthermore, the enzyme can be triggered by some agonist or antagonist molecules that cause the active site to function in the protein. Some of these molecules, which are extremely similar to an enzyme's substrate, may bind to the active site but fail to undergo reaction. Molecules such as this encapsulate the active site and prevent a substrate from binding to it, as shown in the Fig. 4. This inhibition of enzyme activity is competitive because the inhibitor molecule competes with the substrate for the active site. For example, the molecule namely Sulfanilamide is having identical with a substrate called p-aminobenzoic acid, which is playing functional role in folic acid metabolism and here Sulfanilamide bind to it and prevents from reacting (Huang, Nussinov, & Zhang, 2017). The Non-Competitive mechanism, on the other hand, occurs when a molecule binds to the binding site and does not interfere with the binding of the substrate. In this case, the inhibitor molecule prevents the

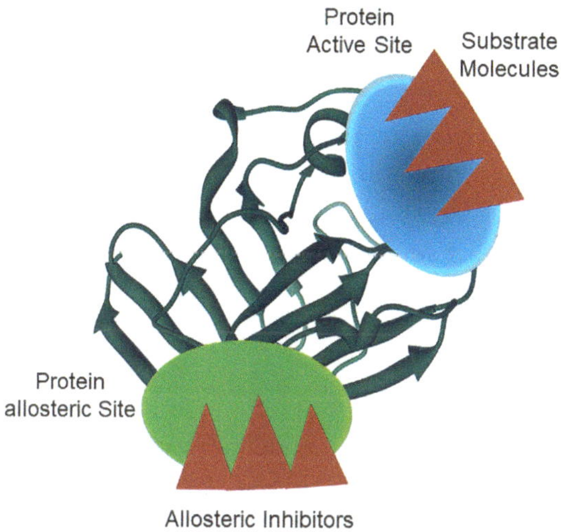

Fig. 4 Protein showing the active site and allosteric site for the inhibitors.

catalytic group from reacting by drawing the binding group away from the substrate. Since it is not the active site, the allosteric site is known as a region where the inhibitor can interact with enzymes. Here, the catalysis is slowed by the inhibitor, but the substrate can still connect to the active site (Lu, Shen, & Zhang, 2019). Allosteric sites are those that can bind both inhibitors and activators, thus they are referred to as such. The enzyme's structure can be activated or inhibited by allosteric sites, which act as regulatory sites. But in terms of active site, certain molecules function as inhibitors and they restrict or block enzyme's function but interacting or binding inside the active site space. In this, the factors like Temperature and pH can have an impact on the enzyme's active site and function. An enzyme's ability to perform its role as a catalyst is reduced when the temperature and pH of the environment change rapidly (Changeux, 2013).

9. Role of charges in active site

The enzyme's active site is a pocket or groove on its surface where a substrate, or reactant, is introduced. Chemical groups are used to cover the active site in a way that is complementary to the substrate's structure and charge. In order to understand enzyme activity and function, it has been proposed that complementary shapes and charges between enzyme and substrate are essential (Wallner et al., 2012). Charged groups in protein active sites interact electrostatically with substrates, transition states, and products.

The majority of research into charged groups' significance in protein biological activity has focused on these groups, and the mechanisms of their action are frequently known qualitatively rather than statistically (Hiller et al., 2020). A charged group in active site has an important role in protein function and its activity. For example, Development of protein–protein and protein–nucleic acid complexes; interaction of receptors with charged ligands; binding of substrates by enzymes and catalysis of processes; selective ion transport in protein channels; protein folding and stability; protein denaturation at high and low pH levels; protein solubilization and precipitation; protein integration into amyloid fibers (Fuller et al., 2019). Researchers have focused on mutations in the protein's active site using site-directed mutagenesis as a primary tool for studying charged groups in protein biochemistry. Protein charge ladders were created as an additional experimental strategy to studying charged groups in proteins, and they are a complement to site-directed mutagenesis (Harris & Turner, 2002). Charge contributions to protein stability and interactions with ligands are clarified using protein charge ladders. Charge ladder studies give a different sort of data than site-directed mutagenesis studies in that they allow for the assessment of average contributions of a certain type of charged group. In most circumstances, shape and charge are interrelated, thus changing the shape of a molecule also alters the charge distribution. By transferring a hydrogen ion from one carbon to another, the enzyme ketosteroid isomerase (KSI) rearranges the bonds inside its substrate, a multiring steroid molecule. The creation of two weak, transient hydrogen bonds between KSI and an oxygen atom on the substrate will occur in this process. This oxygen becomes partly negatively charged when the substrate deforms toward the transition state, and the hydrogen bonds become stronger. Other compounds that suit the active site, such as a phenolate anion, can be bound by KSI (Pollack, 2004). This chemical has an oxygen atom in the same place as the steroid oxygen, but the oxygen in phenolate is negatively charged, simulating the steroid's transition state. Protease 3 (Pr3) was first identified as an elastin-degrading protease with structural and functional similarities to human neutrophil elastase (HNE). Pr3 is kept as an active enzyme within the main granules of HNE alongside HNE and cathepsin G and is selected as a free or membrane-bound protease from activated cells. Despite their structural and functional similarities, the biological roles of human neutrophil protease 3 (Pr3) vary from those of neutrophil elastase. Although both proteases are substantially cationic, the distribution of charged residues in their sequences differs significantly. The substrate binding site in Pr3 is more polar, and certain subsites are more restricted than those in HNE, according to kinetic and structural investigations. The electrostatic surface potentials of Pr3 and HNE substrate binding sites were also differed (Kuki et al., 2018). The discoveries have been exploited to create novel Pr3 substrates with expanded binding sites, which might help explain the unique cleavage of biological protein substrates and reveal new pathophysiologically relevant protein targets.

10. Reported proteins with active site switches

Ras proteins are small GTPase protein molecule, which acts as a molecular switch in many intracellular signaling pathways that control cell growth, proliferation, and differentiation. The transformation between the GDP-bound and the GTP-bound forms was regulated by guanine nucleotide exchange factors and GTPase-activating proteins (Chen et al., 2019). X-ray crystallography and NMR analyses of H-Ras and Rap1A, showed that the exchange of GTP for GDP results in allosteric conformational changes in two adjacent regions, termed switch I (residues 32–38) and switch II (residues 60–75), and enables Ras to complete the downstream signaling through direct interaction with its effectors (Imamura, Matsumoto, Kondo, Kitayama, & Noda, 2003). High-fidelity DNA replication is achieved in part by DNA polymerases evaluating the accuracy of each nucleotide integrated and if a mistake is made, the wrong nucleotide is deleted before further primer extension takes place. Under single turnover conditions, T4 DNA polymerase can remove two erroneous nucleotides, which comprises putative exonuclease-to-polymerase and polymerase-to-exonuclease active site flipping processes (Guo et al., 2017). Many eukaryotic cell processes are governed by cytosolic calcium [Ca^{2+}] immobilization (Sunarso, Toita, Tsuru, & Ishikawa, 2016). Phospholipase Cβ (PLCβ) enzymes are activated by G protein βγ subunits from activated Gαi-Gβγ heterotrimers. The presence of active Gαq is required for the proper functioning of Gαi-Gβγ-PLCβ-Ca^{2+} signaling module. Thus, Pfeil EM, reported that Gαq acted as a master switch for controlling the Gβγ-Mediated Calcium Mobilization by Gi-coupled GPCRs (Rebres et al., 2011). The single-stranded DNA (ssDNA) cytosine deaminase APOBEC3B (A3B) plays a vital role in innate immunity against HIV, HBV, HCV virus, and transposon, but it is also a primary endogenous source of cancer.

11. Open and closed conformation in active sites

Coxsackievirus A16 is a member of the Picornaviridae family and is a primary cause of hand, foot, and mouth disease, which primarily affects children (Mao et al., 2014). There are currently no vaccinations or antiviral treatments available to treat this virus. The enterovirus 2A protease is a non-structural protein that has the potential to self-cleave as well as cleave the eukaryotic translation initiation factor 4G (Visser et al., 2019). Crystal structure analysis of coxsackievirus A16 2A protease showed that structure has an open conformation, with its active site accessible, substrate binding, and cleavage activity. Conformational conversion triggered by two "switcher" residues Glu88 and Tyr89 located within the bll2-cII loop. This switch was responsible for conformational conversion and substrate specificity of 2A protease from coxsackievirus A16 (Sun et al., 2013). These new findings should aid in the development of effective 2A protease inhibitors in the future. COVID-19, the viral infection triggered by the SARS-CoV-2 virus, is becoming a global pandemic, with hundreds of thousands of people dying every day. The spike protein is used by these viruses to enter host cells. Cryo-electron microscopy revealed the structure of SARS-spike CoV-2's protein. The solved structures showed that the spike proteins can be found in two conformations: closed (pre-fusion) and open (post-fusion) conformation. The spike protein undergoes structural modifications to bind with the ACE2 receptor on the host cell (Ke et al., 2020). Streptavidin (derived from the bacteria Streptomyces avidinni) is a tetrameric protein with 159 residues per chain, each of which may bind one biotin molecule noncovalently. Biotin attaches to the open end of each streptavidin subunit's barrel, and an eight-residue surface loop (45–52) (Ser, Ala, Val, Gly, Asn, Ala, Glu, Ser) folds to cap the barrel. In the streptavidin–biotin complex, the eight-residue surface loop 45–52 (Ser, Ala, Val, Gly, Asn, Ala, Glu, Ser) of the homotetrameric protein streptavidin possesses a "closed" conformation, with one of the greatest binding affinities (G − 18 kcal mol). This loop, which is unbound streptavidin has an "open" conformation, has been recognized as a key element in streptavidin's biotin affinity (General, Dragomirova, & Meirovitch, 2011). Lipases (triacylglycerol ester hydrolases, EC 3.1.1.3) are one of the most significant industrial enzymes. Detergents, food, bioenergy, flavors, medicines, and enantiopure esters and amino acid derivatives used in fine chemicals and agrochemicals are all examples of where these enzymes are employed (Yao et al., 2021). These enzymes work at the lipid–water interface. The existence of a moveable subdomain lid or flap positioned above the active site is a significant characteristic of many lipases. The active site is shielded from the environment and inaccessible to substrates while the lid is closed, therefore, the lipase remains inactive. Substrates can reach the active sites of lipases in an open conformation and be transformed. To put it another way, only in "open conformation" lipases have catalytic activity (Khan et al., 2017). The structure of lipase from *Thermomyces lanuginosus*, for example, has been determined in both its closed and open conformations. Given lipases' relatively poor catalytic activity in predominantly aqueous environments, it's reasonable to suppose that the "closed" conformation predominates under these conditions. In more hydrophobic (organic) media or the presence of organic–aqueous interphase, however, the "open" form is likely to be the dominant structure, which is consistent with lipases' typically greater catalytic activity. "Interfacial activation" refers to the enhanced activity of lipases in the presence of apolar–aqueous interphases (Khan et al., 2017). The HIV protease is a key therapeutic target for HIV/AIDS treatment, and its structure and activity have been studied massively. This enzyme is required for viral maturation because it processes certain cleavage sites in the Gag and Gag-Pol precursor polyproteins, allowing them to mature. The active site is located in a central cavity capped by two flexible flap sections in these 99 amino acids aspartic protease, which operates as a homodimer (Lv, Chu, & Wang, 2015). This dimer has two conformations: closed and open, which are important in substrate binding and release. Investigation of an HIV-1 protease data set comprising 552 dimer structures showed the largest number of HIV protease structures appeared in closed conformations and some protease sequences are conversely located in specific parts of the conformational landscape of this protein (Palese, 2017).

12. Conformational changes in active sites

The structural changes in the enzyme active site have a significant impact on its activity. The active site undergoes a conformational shift when the substrates bind. When compared to water-based catalysis, ligand-gated conformational changes are one of the hallmarks of enzyme catalysis' extraordinary efficiency. Pan et al. proposed in their studies that changes in protease active site conformation caused by substrates have an impact on subsequent substrate reactions (Pan et al., 2010). According to comparative investigations of conformation and activity changes in different enzymes following denaturation, inactivation generally occurs before apparent global structural changes. As a result, active site flexibility would be required for the complete expression of enzyme activity. It has recently been demonstrated that conformational change occurs during enzyme catalysis, particularly in the active site, and that the active site structure loosens up during the activation of many enzymes (Zou, 2001). Ras family small GTPases exist in two interexchange conformations, "inactive" state and "active" state, in their GTP-bound forms. Shima et al. elucidated the mechanism behind the Ras protein state transition from active to inactive state using the X-ray crystal structure analysis of H-Ras combined with 5-(β,γ-imido) triphosphate (GppNHp) (Shima et al., 2010). Identification of definite crystal structures of H-RasT35S-GppNHp revealed guanine nucleotide-mediated interaction and its rearrangement by a nucleotide positional change were responsible for GTP-bound Ras protein state transition. These results showed the conformational changes differentiating the two states and gave new insights into the molecular basis for the state transition of Ras protein (Shima et al., 2010). Spike glycoprotein on SARS-CoV-2 virus binds to ACE2 receptors on the host cell membrane and starts its entry and replication. The S protein could be in a receptor inaccessible (closed) or accessible (open) state depending on the down and up positions of its receptor-binding domain (RBD). Gur et al. performed all-atom molecular dynamics (MD) simulations of closed and open states of the S protein trimer in the presence of explicit water and ions and found out that MD simulations starting from intermediate conformations between the open and closed states suggested that RBD changes to the up position through a semi-open intermediate, this could lower the free energy barrier between the closed and open states (Gur et al., 2020). Thus, the RBD-ACE2 binding is compatible with the semi-open state, whereas the closed state of the S protein is not, and the formation of the intermediate state is crucial for the host cell recognition (Gur et al., 2020). Zika virus (ZIKV) infection has been reported more in South Central America and the Caribbean. There are no clinically effective antiviral drugs available to treat ZIKV infection. The methyltransferase domain (MTase) of ZIKA nonstructural protein 5 (NS5) can methylate guanine N-7 and ribose 2′-O in the nascent RNA transcripts in a sequential manner to create the m7NGpppA2′Om cap structure and the methylation starts with transferring the methyl group from the donor S-adenosylmethionine (SAM) to the methyl acceptor and then leaves the S-adenosylhomocysteine

(SAH) as a byproduct. This methylation mechanism is required for ZIKA replication and immune evasion, making it a potential therapeutic target. The structure of the ZIKA MTase is remarkably conserved throughout the Flavivirus genus. For the first time in the ZIKA virus, the conformational changes of both SAM and 7-MeGpp binding sites of MTase during catalysis of SAM to SAH have been reported (Zhou et al., 2017). Hepatitis delta virus (HDV) is a human disease and the satellite RNA of the hepatitis B virus. It uses a self-cleaving catalytic RNA motif to manage multimeric intermediates in its genome's double-rolling circular replication. Local conformational changes in the catalytic core of the transacting hepatitis delta virus ribozyme accompany efficient catalysis (Cunha, Tavanez, & Gudima, 2015). The two-component dengue virus NS2B-NS3 protease (DEN NS2B-NS3pro) is a well-known therapeutic target, but creating inhibitors is problematic due to the lack of a crystal structure of the protease in its fully active state (Wu et al., 2015). de la Cruz et al. (2011) analyzed the fold of DEN NS2B-NS3pro in solution with and without bound inhibitor by nuclear magnetic resonance (NMR) spectroscopy and revealed that binding of low molecular weight inhibitors promoted large conformational changes in the dengue virus NS2B-NS3 protease (de la Cruz et al., 2011). HDV ribozyme is a nuclease enzyme that acts as a catalyst during viral infection, enhancing the chemical process and controlling replication intermediates. The ribozyme catalytic process has been proposed to include universal acid-base substrate instability as well as global and local conformational modifications. In addition to this mechanism, a comparison of the ribozyme's pre- and postcleavage crystal structures revealed a significant structural change in the RNA following cleavage, as well as the removal of a catalytically essential divalent metal ion from the active region, and revealed that RNA structural rearrangements also regulate ribozyme and ribonucleoprotein enzyme reactivity (Ke, Zhou, Ding, Cate, & Doudna, 2004).

13. Role of water in active site

Water molecules are extremely small, containing only two types of atoms, and they play several functions in biological processes. Water molecules' importance in protein–ligand interactions has become clearer over time. Through hydrogen bonding, water molecules aid in the stabilization of the complex between a protein and a ligand. Hydration forces are largely responsible for the packing and stability of 3D protein structures. Waters are typically found at or around the active site of enzymes, implying that they play an important role in active site stability, flexibility, ligand coordination, and residue positioning, as their evolutionary conservation demonstrates. Water molecules have long been known to play an important role in enzymes and proteins. Water, for example, is a fundamental molecular element that can be used as a product or reactant in condensation and hydrolysis reactions. The presence of water molecules at the active site aids in the prediction of optimal binding postures during docking. In chemical processes of reaction mechanism, water molecules can also serve as transition state intermediates. In proteins, water acts as a structural element, connecting them together via hydrogen bonds and preserving and stabilizing the positions of residues and the fold. Water molecules have a major role in the interaction between a ligand and a macromolecule. As a result, the positions of conserved water molecules in the protein binding groove must be determined prior to docking. This can be done by superimposing multiple crystal structures from the Protein Data Bank files with similar binding locations.

In addition, when in vivo assemblies such as enzyme–substrate or protein–inhibitor complexes are formed, the water-exposed surfaces of proteins or ligands change. The inclusion of water molecules dramatically improved the compounds' docking accuracy. As a result, the presence of water molecules on the protein's surface may have an impact on the precision of ligand–protein docking. These water molecules have a low B factor, can make at least three hydrogen bonds, and are commonly found in protein surface grooves. Water molecules were divided into two groups: those that are conserved and are not displaced by any ligands, and those that are displaced by certain ligands. The double decoupling approach with replica exchange thermodynamic integration was used to compute the absolute binding free energies of water molecules, using Monte Carlo simulations, and Bayesian statistics are used to determine the chance that a certain water molecule will be displaced by an adequately designed ligand. X-ray structural analysis was used for the first time to identify catalytic water molecules coexisting with a product peptide in the active site of HIV-1 protease. The significance of water molecules in the stability of ligand binding via strong hydrogen bond interactions was proven by the Radial Distribution Function (RDF) plot of MD simulation, and researchers observed enhanced connections by keeping crystallographic water molecules within 3.0 Å of the ligand. Effective docking simulations improved noticeably when flexible water molecules solvating the ligand's hydrogen bonding groups were included. The concept has been implemented by Molegro Virtual Docker, a docking program (MVD).

The displacement of a specific water molecule has been connected to an increase or decrease in ligand binding affinity with the target protein in numerous investigations. The GIST (Grid Inhomogeneous Solvation Theory) and HSA (Hydration Site Analysis) models show that ligands can increase affinity by displacing energetically unfavorable binding site water molecules. Extensive molecular dynamics modeling studies revealed that conserved water molecules in the active site cavity play a role in the enzyme's substrate identification. The active site water molecules, which act as a template for catalysis reaction. Furthermore, Zhao, Guengerich, Voehler, & Waterman, 2005 findings suggested that Cytochrome P450 158A2 used the substrate hydroxyl groups to stabilize active site water and assisted in the iron-linked dioxygen activation and proton transfer (Zhao et al., 2005). Decaneto et al. (2017) demonstrated that solvent water interactions within the active site were responsible for efficient proton transfer and the enzymatic activity of membrane type I matrix metalloproteinase (Decaneto et al., 2017). On the contrary, analysis of separate water pool (SWPs) molecules' function in the nitrogenase enzyme showed that it did not possess a proton transfer mechanistic role. The pools are most likely non-functional because they are not conserved: they appear to exist to fill gaps in the protein structure. Understanding the structure and function of these water molecules in a protein's active domain should help rational drug design strategies significantly increase their efficacy.

14. Future prospectus

All of the catalytic steps are carried out in a predefined shape that minimizes the production of new conformations while preserving optimal stability of the transition states. Apart from these structure and chemistry of active sites are core important for understanding the enzymatic action and mechanisms. Molecules that are too big to fit in the enzyme's active site and hence are unable to interact with it prevent enzyme reactions from occurring. In these settings, the active site's bond, group, substrate, and optical specificity are all known. Molecular-level insights into the active site, their conformations, and atomic participation have been made possible by to the advent of experimental and computational approaches. There are two points in the reaction pathway where the active site loops must cycle between closed and occluded conformations: immediately following hydride transfer and again following product release. The conformational behavior of these molecules in the active site also has been fully elucidated by molecular level dynamics computations. Overall, this chapter provides the elucidations of active site structure and chemistry, computational advantages and application in studying the enzyme chemistry. Even if there is a lot of information accessible, digging deeper into the chemical and physical features of the active site will lead to a better understanding of the active site and aid to generate novel lead compounds.

Acknowledgment

The authors C.S. and S.K.S. thankfully acknowledge the Tamil Nadu State Council for Higher Education (TANSCHE) for the research grant (Au/S.o. (P&D): TANSCHE Projects: 117/2021). M. Alkhanani would like to express his gratitude to AlMaarefa University, Riyadh, Saudi Arabia, for providing funding (TUMA-2021-53) to do this research.

References

Abdelsattar, A. S., Mansour, Y., & Aboul-Ela, F. (2021). The perturbed free-energy landscape: Linking ligand binding to biomolecular folding. *Chembiochem*, *22*(9), 1499–1516. https://doi.org/10.1002/cbic.202000695.

Beltran-Villegas, D. J., Colon-Melendez, L., Solomon, M. J., & Larson, R. G. (2016). Kinetic modeling and design of colloidal lock and key assembly. *Journal of Colloid and Interface Science*, *463*, 242–257. https://doi.org/10.1016/j.jcis.2015.10.062.

Bigley, A. N., & Raushel, F. M. (2013). Catalytic mechanisms for phosphotriesterases. *Biochimica et Biophysica Acta*, *1834*(1), 443–453. https://doi.org/10.1016/j.bbapap.2012.04.004.

Blat, Y. (2010). Non-competitive inhibition by active site binders. *Chemical Biology & Drug Design*, *75*(6), 535–540. https://doi.org/10.1111/j.1747-0285.2010.00972.x.

Boehr, D. D., Nussinov, R., & Wright, P. E. (2009). The role of dynamic conformational ensembles in biomolecular recognition. *Nature Chemical Biology*, *5*(11), 789–796. https://doi.org/10.1038/nchembio.232.

Bozic, S., Doles, T., Gradisar, H., & Jerala, R. (2013). New designed protein assemblies. *Current Opinion in Chemical Biology*, *17*(6), 940–945. https://doi.org/10.1016/j.cbpa.2013.10.014.

Brown, J. L., Yates, E. A., Bielecki, M., Olczak, T., & Smalley, J. W. (2018). Potential role for Streptococcus gordonii-derived hydrogen peroxide in heme acquisition by Porphyromonas gingivalis. *Molecular Oral Microbiology*, *33*(4), 322–335. https://doi.org/10.1111/omi.12229.

Buccitelli, C., & Selbach, M. (2020). mRNAs, proteins and the emerging principles of gene expression control. *Nature Reviews. Genetics*, *21*(10), 630–644. https://doi.org/10.1038/s41576-020-0258-4.

Changeux, J. P. (2013). 50 years of allosteric interactions: The twists and turns of the models. *Nature Reviews. Molecular Cell Biology*, *14*(12), 819–829. https://doi.org/10.1038/nrm3695.

Chen, L. Q., Cheung, L. S., Feng, L., Tanner, W., & Frommer, W. B. (2015). Transport of sugars. *Annual Review of Biochemistry, 84*, 865–894. https://doi.org/10.1146/annurev-biochem-060614-033904.

Chen, S., Li, F., Xu, D., Hou, K., Fang, W., & Li, Y. (2019). The function of RAS mutation in cancer and advances in its drug research. *Current Pharmaceutical Design, 25*(10), 1105–1114. https://doi.org/10.2174/1381612825666190506122228.

Cooney, M. J. (2017). Kinetic measurements for enzyme immobilization. *Methods in Molecular Biology, 1504*, 215–232. https://doi.org/10.1007/978-1-4939-6499-4_17.

Cunha, C., Tavanez, J. P., & Gudima, S. (2015). Hepatitis delta virus: A fascinating and neglected pathogen. *World Journal of Virology, 4*(4), 313–322. https://doi.org/10.5501/wjv.v4.i4.313.

Daura, X. (2019). Advances in the computational identification of allosteric sites and pathways in proteins. *Advances in Experimental Medicine and Biology, 1163*, 141–169. https://doi.org/10.1007/978-981-13-8719-7_7.

de la Cruz, L., Nguyen, T. H., Ozawa, K., Shin, J., Graham, B., Huber, T., et al. (2011). Binding of low molecular weight inhibitors promotes large conformational changes in the dengue virus NS2B-NS3 protease: Fold analysis by pseudocontact shifts. *Journal of the American Chemical Society, 133*(47), 19205–19215. https://doi.org/10.1021/ja208435s.

Decaneto, E., Vasilevskaya, T., Kutin, Y., Ogata, H., Grossman, M., Sagi, I., et al. (2017). Solvent water interactions within the active site of the membrane type I matrix metalloproteinase. *Physical Chemistry Chemical Physics, 19*(45), 30316–30331. https://doi.org/10.1039/c7cp05572b.

Delaune, K. P., & Alsayouri, K. (2021). *Physiology*. Treasure Island (FL): Noncompetitive Inhibitor. In *StatPearls*.

Du, X., Li, Y., Xia, Y. L., Ai, S. M., Liang, J., Sang, P., et al. (2016). Insights into protein-ligand interactions: Mechanisms, models, and methods. *International Journal of Molecular Sciences, 17*(2). https://doi.org/10.3390/ijms17020144.

Fu, L., Li, J., Wang, G., Luan, Y., & Dai, W. (2021). Adsorption behavior of organic pollutants on microplastics. *Ecotoxicology and Environmental Safety, 217*, 112207. https://doi.org/10.1016/j.ecoenv.2021.112207.

Fuglebakk, E., Tiwari, S. P., & Reuter, N. (2015). Comparing the intrinsic dynamics of multiple protein structures using elastic network models. *Biochimica et Biophysica Acta, 1850*(5), 911–922. https://doi.org/10.1016/j.bbagen.2014.09.021.

Fuller, J., 3rd, Wilson, T. R., Eberhart, M. E., & Alexandrova, A. N. (2019). Charge density in enzyme active site as a descriptor of electrostatic preorganization. *Journal of Chemical Information and Modeling, 59*(5), 2367–2373. https://doi.org/10.1021/acs.jcim.8b00958.

General, I. J., Dragomirova, R., & Meirovitch, H. (2011). Calculation of the absolute free energy of binding and related entropies with the HSMD-TI method: The FKBP12-L8 complex. *Journal of Chemical Theory and Computation, 7*(12), 4196–4207. https://doi.org/10.1021/ct2004897.

Guo, Y., Sirkis, D. W., & Schekman, R. (2014). Protein sorting at the trans-Golgi network. *Annual Review of Cell and Developmental Biology, 30*, 169–206. https://doi.org/10.1146/annurev-cellbio-100913-013012.

Guo, J., Zhang, W., Coker, A. R., Wood, S. P., Cooper, J. B., Ahmad, S., et al. (2017). Structure of the family B DNA polymerase from the hyperthermophilic archaeon Pyrobaculum calidifontis. *Acta Crystallographica Section D: Structural Biology, 73*(Pt 5), 420–427. https://doi.org/10.1107/S2059798317004090.

Gur, M., Taka, E., Yilmaz, S. Z., Kilinc, C., Aktas, U., & Golcuk, M. (2020). Conformational transition of SARS-CoV-2 spike glycoprotein between its closed and open states. *The Journal of Chemical Physics, 153*(7), 075101. https://doi.org/10.1063/5.0011141.

Harris, T. K., & Turner, G. J. (2002). Structural basis of perturbed pKa values of catalytic groups in enzyme active sites. *IUBMB Life*, *53*(2), 85–98. https://doi.org/10.1080/15216540211468.

Hauser, A. S., Attwood, M. M., Rask-Andersen, M., Schioth, H. B., & Gloriam, D. E. (2017). Trends in GPCR drug discovery: New agents, targets and indications. *Nature Reviews. Drug Discovery*, *16*(12), 829–842. https://doi.org/10.1038/nrd.2017.178.

Hiller, D. A., Dunican, B. F., Nallur, S., Li, N. S., Piccirilli, J. A., & Strobel, S. A. (2020). The positively charged active site of the bacterial toxin RelE causes a large shift in the General Base pKa. *Biochemistry*, *59*(17), 1665–1671. https://doi.org/10.1021/acs.biochem.9b01047.

Huang, W., Nussinov, R., & Zhang, J. (2017). Computational tools for allosteric drug discovery: Site identification and focus library design. *Methods in Molecular Biology*, *1529*, 439–446. https://doi.org/10.1007/978-1-4939-6637-0_23.

Imamura, Y., Matsumoto, N., Kondo, S., Kitayama, H., & Noda, M. (2003). Possible involvement of Rap1 and Ras in glutamatergic synaptic transmission. *Neuroreport*, *14*(9), 1203–1207. https://doi.org/10.1097/00001756-200307010-00003.

Jacques, B., Coincon, M., & Sygusch, J. (2018). Active site remodeling during the catalytic cycle in metal-dependent fructose-1,6-bisphosphate aldolases. *The Journal of Biological Chemistry*, *293*(20), 7737–7753. https://doi.org/10.1074/jbc.RA117.001098.

Jiang, F., Han, W., & Wu, Y. D. (2013). The intrinsic conformational features of amino acids from a protein coil library and their applications in force field development. *Physical Chemistry Chemical Physics*, *15*(10), 3413–3428. https://doi.org/10.1039/c2cp43633g.

Jiang, Z., You, Q., & Zhang, X. (2019). Medicinal chemistry of metal chelating fragments in metalloenzyme active sites: A perspective. *European Journal of Medicinal Chemistry*, *165*, 172–197. https://doi.org/10.1016/j.ejmech.2019.01.018.

Ke, Z., Oton, J., Qu, K., Cortese, M., Zila, V., McKeane, L., et al. (2020). Structures and distributions of SARS-CoV-2 spike proteins on intact virions. *Nature*, *588*(7838), 498–502. https://doi.org/10.1038/s41586-020-2665-2.

Ke, A., Zhou, K., Ding, F., Cate, J. H., & Doudna, J. A. (2004). A conformational switch controls hepatitis delta virus ribozyme catalysis. *Nature*, *429*(6988), 201–205. https://doi.org/10.1038/nature02522.

Khan, F. I., Lan, D., Durrani, R., Huan, W., Zhao, Z., & Wang, Y. (2017). The lid domain in lipases: Structural and functional determinant of enzymatic properties. *Frontiers in Bioengineering and Biotechnology*, *5*, 16. https://doi.org/10.3389/fbioe.2017.00016.

Krieger, S., & Kececioglu, J. (2020). Boosting the accuracy of protein secondary structure prediction through nearest neighbor search and method hybridization. *Bioinformatics*, *36*(Suppl_1), i317–i325. https://doi.org/10.1093/bioinformatics/btaa336.

Kuki, M. C., Scapim, C. A., Rossi, E. S., Mangolin, C. A., Amaral Junior, A. T. D., & Pinto, R. J. B. (2018). Genome wide association study for gray leaf spot resistance in tropical maize core. *PLoS One*, *13*(6), e0199539. https://doi.org/10.1371/journal.pone.0199539.

Latorraca, N. R., Venkatakrishnan, A. J., & Dror, R. O. (2017). GPCR dynamics: Structures in motion. *Chemical Reviews*, *117*(1), 139–155. https://doi.org/10.1021/acs.chemrev.6b00177.

Lu, S., Shen, Q., & Zhang, J. (2019). Allosteric methods and their applications: Facilitating the discovery of allosteric drugs and the investigation of allosteric mechanisms. *Accounts of Chemical Research*, *52*(2), 492–500. https://doi.org/10.1021/acs.accounts.8b00570.

Luscombe, N. M., Laskowski, R. A., & Thornton, J. M. (2001). Amino acid-base interactions: A three-dimensional analysis of protein-DNA interactions at an atomic level. *Nucleic Acids Research*, *29*(13), 2860–2874. https://doi.org/10.1093/nar/29.13.2860.

Lv, Z., Chu, Y., & Wang, Y. (2015). HIV protease inhibitors: A review of molecular selectivity and toxicity. *HIV AIDS (Auckland)*, *7*, 95–104. https://doi.org/10.2147/HIV.S79956.

Mao, Q., Wang, Y., Yao, X., Bian, L., Wu, X., Xu, M., et al. (2014). Coxsackievirus A16: epidemiology, diagnosis, and vaccine. *Human Vaccines & Immunotherapeutics*, *10*(2), 360–367. https://doi.org/10.4161/hv.27087.

Nakamoto, H., & Osada, H. (2013). Molecular chaperones as drug targets. *Current Pharmaceutical Design*, *19*(3), 307–308. https://doi.org/10.2174/138161213804143671.

Palese, L. L. (2017). Conformations of the HIV-1 protease: A crystal structure data set analysis. *Biochimica et Biophysica Acta, Proteins and Proteomics*, *1865*(11 Pt A), 1416–1422. https://doi.org/10.1016/j.bbapap.2017.08.009.

Pan, R., Image, I., Zhang, X. J., Image, I. I., Zhang, Z. J., Image, I. I., et al. (2010). Substrate-induced changes in protease active site conformation impact on subsequent reactions with substrates. *The Journal of Biological Chemistry*, *285*(30), 22950–22956. https://doi.org/10.1074/jbc.M110.103549.

Pollack, R. M. (2004). Enzymatic mechanisms for catalysis of enolization: Ketosteroid isomerase. *Bioorganic Chemistry*, *32*(5), 341–353. https://doi.org/10.1016/j.bioorg.2004.06.005.

Prothiwa, M., & Bottcher, T. (2020). Competitive profiling for enzyme inhibitors using chemical probes. *Methods in Enzymology*, *633*, 49–69. https://doi.org/10.1016/bs.mie.2019.10.031.

Pushpakom, S., Iorio, F., Eyers, P. A., Escott, K. J., Hopper, S., Wells, A., et al. (2019). Drug repurposing: Progress, challenges and recommendations. *Nature Reviews. Drug Discovery*, *18*(1), 41–58. https://doi.org/10.1038/nrd.2018.168.

Quan, Y., Wang, Z. Y., Chu, X. Y., & Zhang, H. Y. (2018). Evolutionary and genetic features of drug targets. *Medicinal Research Reviews*, *38*(5), 1536–1549. https://doi.org/10.1002/med.21487.

Rebres, R. A., Roach, T. I., Fraser, I. D., Philip, F., Moon, C., Lin, K. M., et al. (2011). Synergistic Ca2+ responses by G{alpha}i- and G{alpha}q-coupled G-protein-coupled receptors require a single PLC{beta} isoform that is sensitive to both G{beta}{gamma} and G{alpha}q. *The Journal of Biological Chemistry*, *286*(2), 942–951. https://doi.org/10.1074/jbc.M110.198200.

Reddy, K. K., & Singh, S. K. (2014). Combined ligand and structure-based approaches on HIV-1 integrase strand transfer inhibitors. *Chemico-Biological Interactions*, *218*, 71–81. https://doi.org/10.1016/j.cbi.2014.04.011.

Reddy, K. K., Singh, P., & Singh, S. K. (2014). Blocking the interaction between HIV-1 integrase and human LEDGF/p75: Mutational studies, virtual screening and molecular dynamics simulations. *Molecular BioSystems*, *10*(3), 526–536. https://doi.org/10.1039/c3mb70418a.

Reid, M. E., & Mohandas, N. (2004). Red blood cell blood group antigens: Structure and function. *Seminars in Hematology*, *41*(2), 93–117. https://doi.org/10.1053/j.seminhematol.2004.01.001.

Selvaraj, C., Omer, A., Singh, P., & Singh, S. K. (2015). Molecular insights of protein contour recognition with ligand pharmacophoric sites through combinatorial library design and MD simulation in validating HTLV-1 PR inhibitors. *Molecular BioSystems*, *11*(1), 178–189. https://doi.org/10.1039/c4mb00486h.

Selvaraj, C., Sakkiah, S., Tong, W., & Hong, H. (2018). Molecular dynamics simulations and applications in computational toxicology and nanotoxicology. *Food and Chemical Toxicology*, *112*, 495–506. https://doi.org/10.1016/j.fct.2017.08.028.

Selvaraj, C., Selvaraj, G., Mohamed Ismail, R., Vijayakumar, R., Baazeem, A., Wei, D. Q., et al. (2021). Interrogation of bacillus anthracis SrtA active site loop forming open/close lid conformations through extensive MD simulations for understanding binding selectivity of SrtA inhibitors. *Saudi Journal of Biological Sciences*, *28*(7), 3650–3659. https://doi.org/10.1016/j.sjbs.2021.05.009.

Shima, F., Ijiri, Y., Muraoka, S., Liao, J., Ye, M., Araki, M., et al. (2010). Structural basis for conformational dynamics of GTP-bound Ras protein. *The Journal of Biological Chemistry*, *285*(29), 22696–22705. https://doi.org/10.1074/jbc.M110.125161.

Silman, I., & Sussman, J. L. (2008). Acetylcholinesterase: How is structure related to function? *Chemico-Biological Interactions*, *175*(1–3), 3–10. https://doi.org/10.1016/j.cbi.2008.05.035.

Srivastava, A., Nagai, T., Srivastava, A., Miyashita, O., & Tama, F. (2018). Role of computational methods in going beyond X-ray crystallography to explore protein structure and dynamics. *International Journal of Molecular Sciences*, *19*(11). https://doi.org/10.3390/ijms19113401.

Stetz, M. A., Caro, J. A., Kotaru, S., Yao, X., Marques, B. S., Valentine, K. G., et al. (2019). Characterization of internal protein dynamics and conformational entropy by NMR relaxation. *Methods in Enzymology*, *615*, 237–284. https://doi.org/10.1016/bs.mie.2018.09.010.

Sun, Y., Wang, X., Yuan, S., Dang, M., Li, X., Zhang, X. C., et al. (2013). An open conformation determined by a structural switch for 2A protease from coxsackievirus A16. *Protein & Cell*, *4*(10), 782–792. https://doi.org/10.1007/s13238-013-3914-z.

Sunarso, Toita, R., Tsuru, K., & Ishikawa, K. (2016). Immobilization of calcium and phosphate ions improves the osteoconductivity of titanium implants. *Materials Science & Engineering. C, Materials for Biological Applications*, *68*, 291–298. https://doi.org/10.1016/j.msec.2016.05.090.

Trabuco, L. G., Schreiner, E., Gumbart, J., Hsin, J., Villa, E., & Schulten, K. (2011). Applications of the molecular dynamics flexible fitting method. *Journal of Structural Biology*, *173*(3), 420–427. https://doi.org/10.1016/j.jsb.2010.09.024.

Tsou, C. L. (1998). Active site flexibility in enzyme catalysis. *Annals of the New York Academy of Sciences*, *864*, 1–8. https://doi.org/10.1111/j.1749-6632.1998.tb10282.x.

Visser, L. J., Langereis, M. A., Rabouw, H. H., Wahedi, M., Muntjewerff, E. M., de Groot, R. J., et al. (2019). Essential role of enterovirus 2A protease in counteracting stress granule formation and the induction of type I interferon. *Journal of Virology*, *93*-(10). https://doi.org/10.1128/JVI.00222-19.

Vogt, A. D., & Di Cera, E. (2013). Conformational selection is a dominant mechanism of ligand binding. *Biochemistry*, *52*(34), 5723–5729. https://doi.org/10.1021/bi400929b.

Wallner, S., Winkler, A., Riedl, S., Dully, C., Horvath, S., Gruber, K., et al. (2012). Catalytic and structural role of a conserved active site histidine in berberine bridge enzyme. *Biochemistry*, *51*(31), 6139–6147. https://doi.org/10.1021/bi300411n.

Wu, H., Bock, S., Snitko, M., Berger, T., Weidner, T., Holloway, S., et al. (2015). Novel dengue virus NS2B/NS3 protease inhibitors. *Antimicrobial Agents and Chemotherapy*, *59*(2), 1100–1109. https://doi.org/10.1128/AAC.03543-14.

Xu, M., & Lill, M. A. (2013). Induced fit docking, and the use of QM/MM methods in docking. *Drug Discovery Today: Technologies*, *10*(3), e411–e418. https://doi.org/10.1016/j.ddtec.2013.02.003.

Yao, W., Liu, K., Liu, H., Jiang, Y., Wang, R., Wang, W., et al. (2021). A valuable product of microbial cell factories: Microbial lipase. *Frontiers in Microbiology*, *12*, 743377. https://doi.org/10.3389/fmicb.2021.743377.

Yokoyama, K., Ishijima, S. A., Clowney, L., Koike, H., Aramaki, H., Tanaka, C., et al. (2006). Feast/famine regulatory proteins (FFRPs): Escherichia coli Lrp, AsnC and related archaeal transcription factors. *FEMS Microbiology Reviews*, *30*(1), 89–108. https://doi.org/10.1111/j.1574-6976.2005.00005.x.

Zhao, B., Guengerich, F. P., Voehler, M., & Waterman, M. R. (2005). Role of active site water molecules and substrate hydroxyl groups in oxygen activation by cytochrome P450 158A2: A new mechanism of proton transfer. *The Journal of Biological Chemistry*, *280*(51), 42188–42197. https://doi.org/10.1074/jbc.M509220200.

Zhou, H., Wang, F., Wang, H., Chen, C., Zhang, T., Han, X., et al. (2017). The conformational changes of Zika virus methyltransferase upon converting SAM to SAH. *Oncotarget*, *8*(9), 14830–14834. https://doi.org/10.18632/oncotarget.14780.
Zhu, H., Bilgin, M., & Snyder, M. (2003). Proteomics. *Annual Review of Biochemistry*, *72*, 783–812. https://doi.org/10.1146/annurev.biochem.72.121801.161511.
Zou, C. L. (2001). Conformational flexibility of enzyme active sites. *Sheng Li Ke Xue Jin Zhan*, *32*(1), 7–12 (Retrieved from https://www.ncbi.nlm.nih.gov/pubmed/12545769).

CHAPTER FOUR

Accelerating therapeutic protein design

Mohammad ElGamacy[a,b,*]
[a]University Hospital Tübingen, Division of Translational Oncology, Tübingen, Germany
[b]Max Planck Institute for Biology, Tübingen, Germany
*Corresponding author: e-mail address: mohammad.elgamacy@tuebingen.mpg.de

Contents

Abstract

Protein structures provide for defined microenvironments that can support complex pharmacological functions, otherwise unachievable by small molecules. The advent of therapeutic proteins has thus greatly broadened the range of manageable disorders. Leveraging the knowledge and recent advances in *de novo* protein design methods has the prospect of revolutionizing how protein drugs are discovered and developed. This review lays out the main challenges facing therapeutic proteins discovery and development, and how present and future advancements of protein design can accelerate the protein drug pipelines.

Advances in Protein Chemistry and Structural Biology, Volume 130
ISSN 1876-1623
https://doi.org/10.1016/bs.apcsb.2022.01.004

1. Introduction

Proteins are polypeptides that mostly exist in a structured form that is driven by intra-molecular and molecule-solvent interactions. This capacity to fold into unique shapes enables proteins to provide large, diverse, chiral physico-chemical environments that can be either dynamically restrained or capable of undergoing defined modes of motion; what is otherwise unachievable by small molecules. This positions proteins to play complex pharmacological roles, and has already broadened the scope of manageable diseases (Leader et al., 2008). The recent advancements in computer hardware, biophysical models, as well as the growth in protein sequence and structural data, have laid sufficient fundamental and technological framework to design therapeutic proteins *de novo*. This rational process not only promises a more directed and efficient means to generate therapeutic molecules, but opens prospects for the design of novel pharmacological function, on-demand.

This chapter discusses the different aspects of therapeutic protein design, including approaches, challenges and prospects. The first section gives an overview of the advantages and challenges of deploying proteins as pharmacological agents. The second section covers the basic notions and advantages of computational protein design. The third section provides a rundown of some remarkable examples of *de novo* designed therapeutic proteins. The fourth and fifth sections summarize the standing challenges in the field and provide perspectives to tackling some of them.

1.1 Proteins as sophisticated therapeutics

The accessible surfaces of proteins generally provide large and constrained chemical environments. In relation to binding other targets, these large interfaces enable protein drugs to interact with high specificity to convex and hydrophilic target surfaces. In contrast, small molecules tend to mostly bind buried, concave, and hydrophobic patches of small surface areas. Furthermore, protein binders also eclipse the paradigm of one-drug, one-target, and can be multivalent and multispecific. Protein also are very dynamic molecules, that can conditionally undergo conformational changes as a result of a physical or a chemical trigger. Such dynamic nature allows protein drugs to serve even more complex roles such as enzyme replacement therapies or as pharmacological agents that can be switched or regulated *in vivo*. Protein drugs have also shown to be generally safer and require shorter clinical development times compared to small molecule therapeutics (Kinch, 2015), and as genetically-encodable molecules can also be re-engineered as gene therapies (Anguela & High, 2019).

1.2 Challenges in protein drug discovery and development

Several fundamental challenges, however, stand in the face of broader utility of therapeutic proteins. So far, the development of therapeutic proteins has been based on either making recombinant forms of native proteins, or on engineering standardized scaffolds to serve the novel function of binding a specific target. In the case of recombinant protein therapeutics, such as therapeutic cytokines or enzyme replacement therapies, the protein drugs inherit most of the characteristics of their parent natural templates with minor modifications, if any. This leads to several pharmacological, production and formulation challenges (Frokjaer & Otzen, 2005). On the other hand, in the case of binders made from a standard template, such as monoclonal antibodies, the development relies on directed evolution through either *in vitro* screening or *in vivo* immunization processes (Brekke & Sandlie, 2003). These directed evolution processes require specialized experimental setups, iterative steps of binding and specificity screens, and are not epitope-targeted.

Furthermore, past the development stage, peptides larger than about 50 amino acids are not amenable to chemical synthesis, necessitating their production through biosynthesis in living cells (Demain & Vaishnav, 2009). This complicates the production of protein drugs as they must be cleanly purified from among thousands of cellular proteins present in their expression system. The protein properties can thus greatly influence, positively or negatively, the complexity and cost of these production and purification processes. These requirements are particularly pronounced for post-transitionally modified proteins, a particularly demanding class (Sethuraman & Stadheim, 2006). Unlike small molecules, proteins are primary suspects of immune surveillance, and may elicit adaptive immune responses that can neutralize their efficacy or even lead to fatal consequences (Jawa et al., 2013).

The main challenges facing therapeutic proteins can be broadly attributed to: (i) stability-related, (ii) physicochemical, as well as (iii) clinical aspects.

1.2.1 Structural stability aspects

The observation that most natural proteins are marginally stable could be largely explained by the functional constraints and the evolutionary forces at the molecular level. Many functions require the protein to undergo dynamic exchange between at least two conformational states, in order to support roles such as transport, catalysis, switching, and signaling. Therefore, a marginally stabilized structure has better accessibility to its alternate conformational states. Moreover, the selective pressure exerted over natural proteins imposes a functionality threshold, this permits a functionally-neutral mutational drift, which is mostly not stabilizing (Martin & Vila, 2020; Taverna & Goldstein, 2002). This marginal stability may also fall into a bigger choreographed picture of post-translational regulation, where growing evidence points to proteolysis as an important physiological negative feedback mechanism in living cells, even though it comes at a high energy cost (Wolf & Menssen, 2018). Conversely, many congenital monogenic diseases simply emerge as a result of destabilizing mutations that lead to reduced expression or activity of the mutated protein (Yue et al., 2005).

From a pharmacological perspective though, the stability of a protein drug represents a critical parameter that influences its production, formulation, shelf-life, as well as pharmacokinetics (PK) and pharmacodynamics (PD) *in vivo* (Bansal et al., 2021; Böttger et al., 2017; Schuster et al., 2020). The thermal stabilization of the folded, active form of a protein drug has been shown to improve its recombinant expression yield, proteolytic resistance, and efficacy (Asial et al., 2013; Kazlauskas, 2018; McConnell et al., 2014). Furthermore, given the trade-off between stability and function, highly stable proteins are more evolvable; i.e., more amenable to functional engineering (Bloom et al., 2006).

A nascent polypeptide chain ideally folds into its functional conformation at a certain rate (kinetic stability) and free energy cost (thermodynamic stability). Owing to its complexity, however, such polymer can still partition to alternate (mostly undesirable) states such as misfolded, partially folded, unfolded, oligomeric, amyloid, or aggregated states. This partitioning is subject to the protein's folding kinetics, folding thermodynamics, and its solvation free energy (Dobson, 2003). Different designable structural features can strongly influence these three parameters; namely, a protein's size, topology, and sequence. For instance, the size of a protein domain affects its folding rate (Ivankov et al., 2003), as well as its free energy, as the configurational entropy lost upon folding grows faster than the favorable enthalpy change with increasing length of the polypeptide chain (Xu & Nussinov, 1998). A protein's topology was also shown to affect its folding rate and free energy (Dinner & Karplus, 2001). Lastly, optimizing the sequence of a protein to either strengthen the intramolecular contacts or increase its surface hydrophilicity can improve its folding or solvation free energies, respectively.

1.2.2 Physicochemical aspects

Protein drugs face additional challenges, relating to their composition and biophysical features, that can be overcome by design. For instance, the best heterologous expression system for a recombinant protein is decided according to the nature of the protein, particularly, its requirement for post-translational modifications (PTMs) or chaperone-assisted folding (Andersen & Krummen, 2002). PTMs are particularly ubiquitous in mammalian proteins and represent challenges toward their recombinant expression, especially glycosylation and disulfide bridges. Even though native proteins may functionally rely on glycan moieties for binding or intracellular targeting, most of the natural or engineered protein glycosylation serves to improve the protein's structural stability and solubility. On the downside, glycoproteins demand the use of costly eukaryotic expression systems, and entrain micro- and macro-heterogeneities of the glycan nature and glycosylation sites, respectively, further obfuscating quality control and regulatory processes (Solá & Griebenow, 2010). Disulfide bonds also serve the structural role of cross-linking, and thus stabilizing the native structure, but more than two disulfide linkages may undergo reshuffling, resulting in misfolded or aggregated species *in vitro* and *in vivo* (Liu et al., 2008; Wypych et al., 2008).

Enhancing solubility is another example challenge that not just depends on the polarity of surface-exposed atoms in a protein, but on a multitude of parameters such as the amyloidogenic propensity (Fernandez-Escamilla et al., 2004), conformational stability (Costanzo et al., 2014), distribution of surface charges (O'Brien et al., 2016), colloidal properties (Valente et al., 2005), among others. Poor solubility of protein drugs can have a broad range of impacts; from production, storage, and adsorption issues, to parenteral administration issues (e.g., dilute infusions) (Hawe & Frieß, 2007) and potential off-target interactions (Rego et al., 2021), and even aggregation-enhanced immunogenicity (Moussa et al., 2016).

1.2.3 Clinical aspects

Five key challenges stand in the way of broader clinical utility for protein drugs. First, given the established inverse relationship between the molecular weight (or more accurately the hydrodynamic radius) of a protein therapeutic and its glomerular filtration rate, low and medium molecular weight proteins (i.e., roughly $<50\,kDa$) undergo fast renal clearance (Brewer, 1951; Carone, 1978). This constitutes a particular barrier to the development of unmodified small or miniprotein therapeutics. Second, large proteins, and particularly antibodies and their conjugates (i.e., roughly $>150\,kDa$), suffer poor bioavailability and tissue distribution properties. This problem is especially pronounced in solid tumors, that are generally characterized by poor vascularization and blood perfusion (Minchinton & Tannock, 2006; Pallasch et al., 2014). Third, the development of an immune response against a protein drug can result in neutralizing its activity, development of immune cross-reactivity to an endogenous homologue, or even life-threatening anaphylaxis (Sauna et al., 2018). This constitutes a clear immunogenicity risk associated with protein drug development, particularly, exogenous proteins with novel sequences. Fourth, protein drugs suffer poor oral bioavailability, which has to do with the dual problem of harsh pH and proteolytic conditions, and the inefficient transport across the gastrointestinal epithelium (Renukuntla et al., 2013). Fifth, protein drugs can only tackle extracellular targets, as generalizable approaches for their intracellular delivery, let alone cell- or compartment-specific delivery, are still lacking (Singh et al., 2019).

All these aspects are intricately linked to sequence and structure of the protein molecule, which in principle can be rationalized and factored into the design process.

2. Computational design of protein drugs

Computational protein design techniques provide for an innovative approach whereby the designer can mount a function over a constellation of possible sequences and structures, resulting in functional molecules with tailored properties. The design process can guide minimal mutagenesis or create fully novel sequences (Dahiyat & Mayo, 1997), design can also control molecular shapes not just recapitulating classical folds, but even delivering novel folds at atomic accuracy (Woolfson, 2021). This section starts by introducing what protein design is, and then enumerates the advantages over traditional protein engineering approaches.

2.1 Elements of protein design

In its simplest incarnation, the protein design problem is a sequence search in a fixed-backbone context, whereby the goal is to find a sequence that would achieve the most negative free energy change upon folding into a predefined backbone structure (i.e., blueprint). In this framework, the conformations and identities of the amino acid side chains (i.e., rotamers) are emplaced at the designable positions (i.e., sampled), where the interaction with their surrounding chemical environment (i.e., protein, solvent, or ligand) is evaluated (i.e., scored). A protein design simulation is simply an intensively iterative process of sampling and scoring routines that aims to minimize the overall energy of a target structure given an optimal sequence.

It is worth noting that despite the impressive successes of computational protein design, it fundamentally cannot evaluate the possible configurational degeneracy of a particular sequence. This is due to the fact that free energy calculation (i.e., scoring) here is a relative measure only pertinent to a single conformational state, and dismisses the energy distribution of all the other conformational states accessible by a single sequence, which is intractable to evaluate. This introduces an element of uncertainty in the evaluation process, even if given an error free scoring function. Nonetheless, recent advances in sampling strategies and scoring functions have greatly improved the design success rates.

2.1.1 Sampling

Molecular mechanics is the most practical modelling method for the design problem, whereby each atom is represented as a particle connected by forces (or constraints) to other atoms in the system. Under this paradigm, the particles move according to Newtonian mechanics toward lower potential energy configurations, and no chemical change to the underlying molecules is assumed. This sampling of moves can be temporal or atemporal. In molecular dynamics (MD) (McCammon et al., 1977) these atomic motions evolve in time, and though it is realistic, it generally requires long sampling intervals to sufficiently cover the distribution of conformations of a single sequence. Larger arbitrary moves however can be introduced via discrete (atemporal) sampling, where moves track successively lower energy values. Discrete sampling methods are better suited for the combinatorial nature of the design problem, where the optimization landscape demands barrier hopping. Such discrete sampling can be of either a deterministic or a stochastic nature, and pick rotamers from a precompiled, statistically representative pool of side chain conformations for each amino acid (i.e., a rotamer library) (Ponder & Richards, 1987).

Deterministic algorithms search rotamer combinations in a reproducible manner, where many of these aim at finding an exact single best solution, a global minimum energy conformation. Dead-end elimination (DEE) algorithm came as one of the earliest exact deterministic algorithms (Desmet et al., 1992). In DEE-related algorithms a decision tree is traversed, in which each depth level represents an amino acid position and each node on that level represents a possible rotamer. As different levels are traversed, sub-trees (i.e., partial sequences) can be pruned from the evaluation process if they inherit a side chain with an energy higher than a threshold that provably excludes the entire branch from containing the global minimum energy conformation. The major drawback of these exact algorithms is that they are too exhaustive to tractably handle more than a few designable residues in a reasonable computing timeframe. To reduce the combinatorial complexity for larger numbers of clustered residues, approaches that rely on building and then simplifying interaction graphs between these residues have been developed, i.e., partitioning them into smaller clusters. Such approaches are more useful in handling full-fledged *de novo* design problems, albeit this comes at the cost of finding an approximate rather than an exact solution (Canutescu et al., 2003).

As learnt from directed evolution, the mutational fitness landscapes of proteins may be sufficiently smooth for continuous downhill energy optimization. Granted that these landscapes are subject to the epistatic and non-additive impact of multiple mutations, which add ruggedness to the landscape, favorable mutations are mostly clustered close to each other in the sequence space. Hence, while some optimization paths can be trapped in local minima, other alternative paths may provide a minimally impeded path toward the vicinity of the optimal sequence (Romero & Arnold, 2009). This organization would permit non-deterministic algorithms to achieve good results given they (i) follow multiple optimization paths, and (ii) contain a basin-hopping mechanism to avoid entrapment in local energy minima. Monte Carlo-based algorithms (Metrop]olis et al., 1953) are a prime example of stochastic methods, especially given the impressive success of the Monte Carlo-based Rosetta design algorithms (Leaver-Fay et al., 2011). These samplers rely on successive random sampling of rotamers, whereby a new rotamer is accepted to replace the old rotamer with a probability of 1 if it lowers the energy of the structure (i.e., a downhill move).

In the cases where the new rotamer results in a higher energy structure, it is accepted with a probability of less than 1, as dictated by the Boltzmann factor[1]: $P_{acc}(s_{old} \rightarrow s_{new}) = e^{\frac{(\epsilon_{old}-\epsilon_{new})}{k_B T}}$. This enables energy barrier hopping, by going through less favorable (i.e., higher) energy states, the ease of which can be tuned by the user-defined temperature (T) metric.

Other examples of stochastic samplers in protein design are genetic algorithms (Liu et al., 2003) or simulated annealing (Lee & Subbiah, 1991). Genetic algorithms emulate the natural processes of conducting iterative cycles of random mutagenesis, recombination, and selection, where the latter step is done via a scoring step. Simulated annealing is another age-old sampling technique used across diverse fields. Simulated annealing also follows a Boltzmann factor-based acceptance criterion, except that the temperature T is gradually reduced along the successive cycles of sampling. This annealing regime aims to gradually reduce the rate of minima-escaping behavior to sample more locally within the minima toward the end of the simulation.

2.1.2 Scoring

All atoms (particles) in a molecular mechanics simulation system are related to each other by either chemical (bonded) or physical (non-bonded) interactions or both. Nonetheless, modelling these interactions is computationally demanding, especially that the design problem entails the evaluation of thousands to millions of mutations and their associated side chain conformations. Thus, interactions with small energy impact are generally ignored in a design context. For instance, non-bonded interactions between atom pairs farther than a defined distance cutoff are not evaluated. Likewise, all chemical bond lengths and angles are considered fixed. In this scenario also, a limited number of discrete side chain conformations are precompiled in a library that represents distributions along the dihedral (i.e., torsional) degrees of freedom, with precomputed internal energies. These rotamer libraries come in a range of different flavors, are mostly compiled from databases of known structures, and can represent correlated or independent backbone (i.e., (ϕ, ψ)) and side chain (i.e., $(\chi_1, \ldots, \chi_n)$) rotameric preferences (Huang et al., 2020).

---

1 Where $P_{acc}(s_{old} \rightarrow s_{new})$ is the acceptance probability of the new rotamer given the old rotamer, and $k_B$is the Boltzmann constant, and ϵ_{old} and ϵ_{new} represent the energies given the old (s_{old}) and new (s_{new}) rotamers, respectively.

Non-bonded interactions are the most important forces driving the folding process as well as the functional features (i.e., intermolecular interactions). The major contributors to these interactions can be of an electromagnetic nature or can result indirectly from the hydrophobic effect. The former class includes interactions between fully or partially charged atoms (i.e., electrostatic forces), or due to transient, short-ranged dipolar interactions (i.e., van der Waals forces). Although the vdW energies can be calculated accurately using an approximate function (i.e., Lennard-Jones potential; LJ), electrostatic interactions are highly dependent on the surrounding chemical environment, representing a challenge to scoring fidelity. Moreover, hydrogen bonding, an important directional interaction that takes place between two electronegative atoms through a hydrogen, is also difficult to evaluate given its complex nature (Weinhold & Klein, 2012). The hydrophobic (solvation) effect constitutes another complex energy term, whereby surface-exposed nonpolar residues can both disrupt water (i.e., solvent) hydrogen bonding networks, and lower the configurational entropy of water molecules near the hydrophobic surface. Evaluating these energy terms constitutes most of the computational cost, as it necessitates evaluating all of the pairwise atom-atom distances or geometric relations, in order to calculate the associated interactions energies (Alford et al., 2017).

Other low-resolution, knowledge-based, and heuristic scoring terms have also been used in design, which can provide a faster, less accurate means for energy calculation. Coarse-graining side chain atoms as a single particle (i.e., *centroid*) can greatly simplify a protein structure and be used to eliminate poor design candidates (*decoys*) at earlier stages (Rohl et al., 2004). Geometric distributions of hydrogen bonds and disulfide bridges in high quality crystal structures were also used in design scoring functions (Alford et al., 2017). RosettaHoles, a geometric descriptor that quantifies underpacking in protein cores, has also been extremely successful in filtering out unrealistically packed decoys (Sheffler & Baker, 2009). More recently, statistical patterns of spatial relationships between amino acids in known structures were used to score the interaction preferences of amino acids instead of evaluating their interaction energies (Zhou et al., 2020).

2.2 Advantages of computational design

From among protein engineering techniques, *in vitro* display technologies stand as the most powerful methods for engineering function or stability in proteins. These methods aim to screen a huge number of protein sequences (i.e., a library) by physically compartmentalizing the nucleic acid sequence (i.e., a genotype) to its corresponding protein, which can fold, bind, or catalyze a reaction (i.e., a phenotype). Typically, this is performed in iterative cycles of *directed evolution*, where the best candidates are isolated and their genotype further diversified and screened for phenotype enhancement. These techniques however remain limited by the maximum library size, and at best can effectively screen fewer than an order of 10^{14} unique sequences per cycle (Amstutz et al., 2001), even with Ribosome display techniques (Plückthun, 2012). This means that fewer than a dozen sites (i.e., 20^{12}) can be randomized simultaneously. Beyond this screening process, the top candidates still have to be individually validated and characterized.

In contrast, protein design rationalizes the discovery process, whereby a *blueprint* is defined and the mutants are sampled according to first principles. This *in silico* process can cover much larger sequence spaces, which can supplement or outright replace *in vitro* display techniques from the protein drug discovery pipeline. Protein design not only provides a better command over the sequence space, but also, over the fold space (Woolfson, 2021); tightly defining the shapes and biophysical properties of the resulting proteins. Moreover, designed binders are epitope-specific, unlike their empirically-discovered counterparts. In addition to its lower time and material costs, the throughput and accuracy of design have been continuously improving given the breakneck speed at which computational infrastructures are advancing. Perhaps the most important reward however, is the role of design in dissecting, analyzing and improving our understanding of different aspects of molecular recognition and protein biophysics through controlled studies.

3. Advances in computational design of therapeutic proteins

Despite the nascency of the protein design field, many achievements have been made in creating proteins with therapeutic potential. These have not only been restricted to creating binders with inhibitory roles, but extended to more complex and diverse functions such as activators, enzymes, or even switchable proteins.

3.1 Inhibitor design

The design of protein-based inhibitors aims at creating reversible binders to block a target epitope, which represents an alternative approach to monoclonal antibody technologies. That challenge of rationalizing binder design, in contrast to what is otherwise an empirical discovery process (i.e., antibody engineering), relates to achieving several physico-chemical merits by the designed protein, in a one-sided interface design process. Firstly, the binder's surface should, ideally, be highly complementary to the target surface, in order to avoid solvent entrapment at the binder:target interface, which entails a low-entropy cost. Secondly, the target's hydrophilic residues that are buried (i.e., desolvated) upon binding have to be enthalpically compensated by polar contacts. Thirdly, a certain ratio of buried polar-polar contacts have to be designed and satisfied to dictate directionality over what otherwise would be a slippery hydrophobic interface with no specific binding orientation. Such buried polar interactions are particularly difficult to accurately score given the available scoring functions. Lastly, to be therapeutically useful, the binder has to encode maximum affinity to the target, while remaining soluble and monomeric in its absence, which are largely opposing objectives. Given these requirements, binder design objectives can vary in difficulty, ranging from redesign of existing binders, to full out *de novo* design.

The simplest and most accessible approach is creating stabilized variants of known binders, whereby the design software would predict individual or combinatorial stabilizing mutations to an existing binder. The introduction of charged surface residues (i.e., supercharging) of a single-chain variable fragment antibody (scFv) could improve its solubility, thermostability, and even its affinity (Miklos et al., 2012). Likewise, such strategy can be used to deploy native receptor domains with binding activity as production-amenable, soluble binders, as was done with the design of a recombinant, antigen-specific T-cell receptor. This, in fusion with a CD3 antibody could direct effector T-cells toward the antigen-presenting cancer cells (Froning et al., 2020).

Another approach that enables structural diversification but still relies on the pre-existing knowledge of a binding epitope is motif grafting (or epitope rescaffolding). In this approach, a continuous or discontinuous epitope known to bind a target is grafted onto a geometrically-accommodating scaffold that can possess completely different topology, shape, or biophysical properties. Several examples of continuous (linear) epitope grafting has been shown. Grafting a complementarity determining region (CDR) loop from a known antibody onto an alternate scaffold, could be designed into a trimeric, high-avidity influenza hemagglutinin binder (Strauch et al., 2017). Blocking the Epstein-Barr virus anti-apoptotic protein (BHRF1; a Bcl-2 homologue) was also achieved by rescaffolding its native binding motif (Bim-BH3) onto different rigid scaffolds (Procko et al., 2014). A very similar strategy was used to create peptides that inhibit the Bcl-2-related anti-apoptotic protein MCL-1 may hold potential as cancer therapeutics (Fletcher et al., 2018).

The most difficult approach is the design of binders against a target epitope *de novo*; in this situation no structural information about the binder is known *a priori*. The low success rate has led designers to either rely on testing a large number of designs—in the order of thousands or tens of thousands, or resort to successive rounds of design-guided mutagenesis and *in vitro* screening. For example, Fleishman *et al.* could design nanomolar binders against the influenza hemagglutinin by the two-step design of isolated-residue docking and scaffold selection, followed by affinity maturation (Fleishman et al., 2011). Alternatively, recent progress with large scale computations and gene synthesis has led to designed libraries of miniproteins that could yield tight, binders with high therapeutic value, against several targets including influenza hemagglutinin, botulinum neurotoxin, and SARS-CoV-2 spike (Cao et al., 2020; Chevalier et al., 2017).

The design of small molecule binding proteins remains largely an unsolved challenge, primarily due to two reasons. One is the inherent inaccuracies of the fast design potentials in evaluating buried polar interactions. The other relates to the lack of high-quality simulation parameters (e.g., partial charges, bond constants, etc.) to cover the diversity of chemotypes of small molecule functional groups, which represent a vast chemical space when compared to the well-studied 20 proteinogenic amino acids. One way to sidestep these difficulties is to graft side chains with known binding geometries to the ligand at hand from previously solved structures of proteins in complex with the same ligand. This strategy was successfully used to design a digoxigenin-binding proteins, which could potentially be used for treating digoxin toxicity or overdosing (Fig. 1A) (Tinberg et al., 2013). Another way is to rely on a statistical heuristic relating an amino acid backbone geometry to a bound functional group of a small molecule ligand, as aggregated from the PDB protein:small molecule complexes. This strategy has led to the direct design of apixaban-binding proteins, with a potential use against apixaban overdosing (Polizzi & DeGrado, 2020).

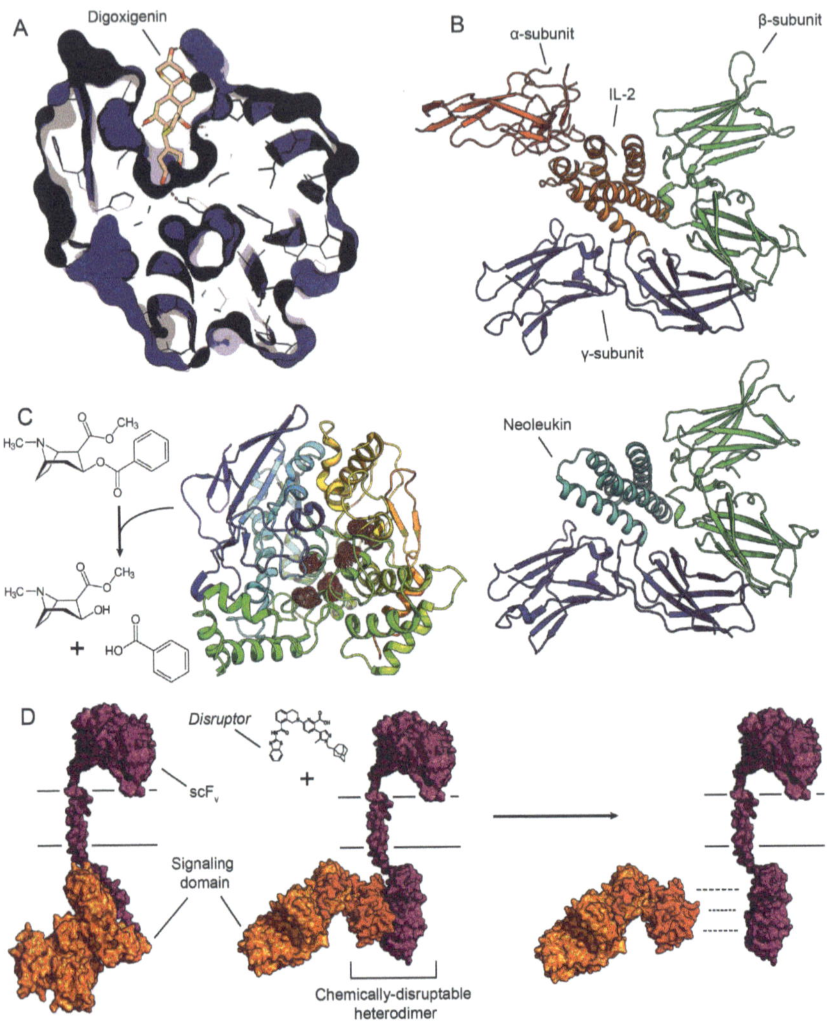

Fig. 1 Examples of *de novo* design proteins with diverse pharmacological roles. (A) A *de novo* designed binding site was built onto the bacterial protein scaffold with unknown function (PDB: 1Z1S). A few rounds of design and *in vitro* screening yielded a nanomolar binder of digoxigenin, that can serve to quench and inhibit the side effects of digoxigenin overdose (dashed red lines designate polar interactions; PDB: 4J9A). (B) The structure of the tetrameric complex between IL-2 and its three receptor subunits (top pane; PDB: 2B5I) is shown in comparison with the designed IL-2 agonist (Neoleukin) that specifically dimerizes the β and γ, but not the α subunit (PDB: 6DG5). (C) Through computational design of a five-point mutant of a butyrylcholinesterase template (mutated residues shown as red spheres; PDB: 1P0P), a novel enzyme capable of degrading cocaine was created, and could rescue cocaine-toxicity *in vivo*. (D) The concept of introducing a chemically disruptable protein-protein interface, could be used to introduce an *off switch* in CAR-T-cells that can be triggered by a small molecule.

3.2 Activator design

To date, the development of antagonists has dominated the field of therapeutic protein engineering. One reason is the broad spectrum of appealing extracellular targets amenable to protein-based inhibition, as evidenced by the continuing boom of therapeutic antibody approvals (Lu et al., 2020). Another reason relates to straightforward *bind-to-quench* approach; occluding a functional epitope to achieve inhibition. Nonetheless, design can leverage more complex structure-function information to encode agonists that mimic physiological effects of native proteins.

Despite the few and rather recent studies, impressive success was realized on this front. Silva *et al.* could design an IL-2 mimic that selectively binds and activates two receptor subunits, and was shown to be a potent activator of T-cells *ex vivo* and *in vivo* (Fig. 1B) (Silva et al., 2019). Using two different design strategies, we could design *de novo* cytokines with granulopoietic function. Through a rescaffolding strategy we could mount the tertiary structural binding epitope from the surface of the granulocyte-colony stimulating factor (G-CSF) onto both *de novo* and natural protein scaffolds. These designs could bind and activate the G-CSF receptor, inducing neutrophil formation from primary human stem cells (Hernandez Alvarez et al., 2020). In another strategy, we could remarkably simplify the G-CSF topology by designing *de novo* loops to reconnect its secondary structural elements into a simpler, yet still active fold, leading to hyperthermostable granulopoietic agonists that were active in zebrafish and mouse models (to be published). Most recently, the grafting and optimization of the binding motif from the programmed cell death-1 ligand 1 (PD-L1) onto designed miniproteins could achieve programmed cell death protein 1 (PD-1) agonists. These were effective at suppressing T-cell activation, leading to potential therapeutics for autoimmunity disorders (Bryan et al., 2021).

Remarkably, agonist design can go beyond exactly mimicking the native ligand, and instead explore alternate configurations by which the target can be modulated. An instructive study by Mohan et al. (2019) has for instance systematically explored a range of geometric configurations by which the erythropoietin receptor can be dimerized, and characterized the detailed impact on the composition and magnitude of downstream secondary messengers. Such a strategy can be used to tune or decompose the biological response (e.g., pro-proliferation vs. pro-differentiation signals) of many related receptors, which has not been done before.

3.3 Enzyme design

De novo design of enzymatic function to catalyze bespoke, non-natural reactions from first principles is still technologically difficult for several reasons. Firstly, enzymes are dynamic effectors with multi-stable states; alternating between substrate binding, one or more intermediate states, and product release. This demands the simultaneous design and sampling of these states, scoring their relative energies, and their rates of interchange. Secondly, enzymes can feature one or more cofactors that can be difficult to accurately parametrize while sampling different chemical environments. Most importantly, applying costly quantum mechanics-based simulation techniques in design-scale calculations is currently not feasible, although it is essential for modelling bond making and breaking. Lastly, evaluating the contribution and impact of solvent molecules rearrangements at each of the conformational or reaction states necessitates explicit solvent representation. Redesign of natural enzymes to alter their catalytic or biophysical properties however represents a more tractable problem, and a great entry point for protein design approaches. Enzyme redesign can thus create a diverse repertoire of therapeutic enzymes, especially given that enzyme replacement therapeutics (ERTs) are well among the most inaccessible and expensive protein therapeutics (Carroll, 2006).

There were few applications of enzyme redesign of therapeutic purposes, with the aim of changing an enzyme's solubility, stability, pH-profile, or substrate specificity. Particularly, redesign of enzymes with altered specificity can introduce new, diverse treatments. For instance, an acid-tolerant endopeptidase from an acidophilic bacterium was successfully redesigned to specifically digest the immunogenic stretches in α-gliadins. α-Gliadins are part of gluten that evoke an autoimmune response in celiac disease patients, thus, these gliadin-digesting designs were made to be acidic-tolerant, proteolytically stable, production amenable, and substrate-specific, to serve as orally administrable therapy (Gordon et al., 2012; Wolf et al., 2015). Steering the specificity of butyryl-cholinesterase away from its native substrate to hydrolyze cocaine has generated a highly active enzyme that could rescue mice from lethal doses of i.p.-administered cocaine (Fig. 1C) (Zheng et al., 2014). Similarly, by redesigning the DNA-recognition site of an endonuclease, it was made to favorably interact with a different base-pair, altering the nuclease cleavage specificity (Ashworth et al., 2006).

Improving the stability and recombinant expression yield of pharmaceutically important enzymes is another important domain of redesign.

For example, Salvat *et al.* aimed to stabilize and deimmunize β-lactamase (an enzyme used for antibody-directed prodrug activation in target tissue). This was done through designing a library by joint-optimization and screening for activity and MHC-II binding (Salvat et al., 2017). Another example is the redesign of a pyocyanin demethylase to improve its stability and recombinant production yield, which created enzyme variants that could inhibit biofilm formation by the pathogenic bacterium *P. aeruginosa*. Biofilm formation (which depends on pyocyanin) endows *P. aeruginosa* with increased drug tolerance, whereas the combined treatment with tobramycin and the designed enzyme disrupted biofilm formation and led to synergistic bactericidal effect (VanDrisse et al., 2021).

3.4 Switch design

Switchable proteins, that can undergo a configuration change based on an external trigger, are designed to exist within a bi-stable energy landscape. The value of switchability for protein-based therapeutics comes in tunable control over the effective dose, providing a failsafe against potential toxicities, or may potentially allow the design of self-regulating medication.

Chimeric antigen receptor-T cells (CAR-T-cells) therapies have recently demonstrated programmable and diverse clinical utility, but may in some cases result in off-target toxicity (Lim & June, 2017). In order to introduce an *off switch* in CAR-T-cells in case side effects arise, Giordano-Attianese *et al.* used computational design to create a small molecule triggered *STOP-CAR-T-cells*. This was achieved by splitting the intracellular segment of the CAR receptor before the signaling domain, and reconstituting the split-receptor by inserting a protein-protein complex. One partner of this inserted complex is designed, and can be out-competed by a small molecule, creating a STOP-CAR receptor that functions only in the absence of the complex-disrupting small molecule (Fig. 1D) (Giordano-Attianese et al., 2020). This layout was extended to create an *on switch* that dimerizes only in the presence of the small molecule. By fusing the former protein-protein complex into a single chain, the designed binder domain will remain occluded in the absence of the small molecule, yet in its presence, will be available to a third small molecule-insensitive domain (Shui et al., 2021).

Baker's lab has created a range of elegant coiled coil switches built on top of *de novo* designed heterodimers (DHD) and latching, orthogonal cage/key protein (LOCKR) systems that could serve different, complex switching functions in different contexts, better reviewed in (ElGamacy & Hernandez Alvarez, 2021). The LOCKR switching system could actuate two- and three-input logic gating on the cancer cell surface. By recognizing the conditional co-localization of two or three surface antigens, the LOCKR switch exposes a peptide motif that can guide CAR-T-cells to the target cancer cell type (Lajoie et al., 2020). Two LOCKR-like systems were similarly built to exhibit the switching behavior of phosphorylation-dependent binding or binding-dependent phosphorylation behavior, where the latter could conceivably be used as a negative feedback regulator against CAR-T-cell exhaustion (Woodall et al., 2021). Another coiled coil-based switchable inhibitor is that of the anti-apoptotic protein MCL-1. The MCL-1-binding motif was grafted on a single chain of a heterodimeric coiled coil, whereby, the MCL-1 is conditionally inhibited by the presence of the two components of the designed heterodimer (Fletcher et al., 2018).

4. Outstanding challenges in protein design

The accuracy and speed of the design calculations remain the biggest barriers in the way of further progress, and either of these factors is often improved at the cost of the other. Both factors greatly influence each other and directly impact the success rates of experimentally tested designs, as well as the computing cost required for design. The simultaneous improvement of both accuracy and speed necessitate optimization across a stack of disciplines; spanning the theoretical framework, the mathematical formulation of the energy calculations, the sampling algorithms, and even the hardware architecture used for design calculations. The accuracy and speed also dictate an upper limit of design complexity, where the capacity to model subtle dynamic properties, conformational allostery, or multi-stable proteins is very limited in a design context. The unyielding difficulty of accurately evaluating the energies of electrostatics, hydrogen bonding, and solvation is a pressing challenge, where the parameters of these terms are highly sensitive to the local environment and induced polarization. These terms however are extremely critical to the one-sided *de novo* binder design, as it entails the design of large protein surfaces that will be desolvated and yet must involve a decent number of polar-contacts across the interface to dictate the interaction specificity. This difficulty is hence reflected in the currently low success rates and overall effectiveness of *de novo* binder design. The current inability to model the electronic structures as well as the lack of reliable methods for accurate, on-the-fly parametrization of small molecule also hampers the efforts to design *de novo* enzyme or *de novo* small molecule binders, leaving these problems largely unsolved.

Additional clinical challenges of designing proteins with maximum functional information per molecular weight is a critical factor for accelerating protein drug development. Moreover, tackling of the immunogenicity risks through design is yet another important goal to be pursued.

5. Perspectives on accelerating therapeutic protein design

Despite the groundbreaking progress in the field of protein design, more advancements can still be readily made with the already existing fundamental knowledge and technological capacities. Specifically, efforts to mitigate existing bottlenecks along the entire pipeline of protein drug design can lead to the broader use of workflows that can deliver designer proteins from computer to bedside. Moreover, addressing factors such as increasing the throughput of the *in silico* design, improving design success rates, designing more efficient proteins, and mitigating the immunogenicity risks of designed proteins are crucial milestones toward the broader adoption of therapeutic protein design

5.1 Accelerated design calculations

Given the rotamer sampling load in a typical design problem, significant computing cost and time is spent on the computational phase of the protein design pipeline. The sampling size (i.e., number of rotamers evaluated) *per se* can greatly influence the outcomes of the design process, where a finer granularity in conformational sampling per sequence improves the overall sequence scoring accuracy, while sampling more sequences improves the chances of capturing lower energy minima. Broader sampling is thus limited by the available computing power and the narrow set of representative rotamers available in the rotamer library, which is particularly pronounced with complex structural motifs (Kim et al., 2009). Acceleration of the most time-consuming step, the scoring step, is thus the highest priority, particularly the calculation of the non-bonded energy terms, as internal energies of the sampled rotamers are mostly precomputed.

Several approaches can be combined to achieve such speedups. We have developed a scoring framework that tensorizes all of the non-bonded energy terms. This relies on framing the energy calculation problem in a hardware-optimal form, that avoids atom-atom interaction calculation loops, but rather calculates the interaction energy between two bodies (i.e., an inbound rotamer; the first body, and a chemical environment; the second body) through a single matrix operation.[2] This concept is easily generalizable to the calculation of electrostatics, LJ, and solvation energy terms, and entails that the interaction fields of a rotamer are precomputed. This *single instruction, multiple data* processing framework not only can provide more than two orders of magnitude of speed-up by reducing memory call-ups on standard central processing units (CPU), but can also benefit from efficiencies of modern stream processors, as well as enables parallelization of the sampling problem.

The *energy gap* driving a specific folding or an interaction state, partitioning it apart from any other undesired, non-specific state, is a notion related to evaluating the absolute free energy of a protein structure (Fleishman & Baker, 2012). The capacity to approximately deduce this energy gap can serve as an *in silico* validation method of a designed protein sequence; providing a more rigorous orthogonal scoring scheme. For instance, template-free (i.e., *ab initio*) structure prediction (Kinch et al., 2011) of a designed sequence can serve to rank designs, prioritizing those with predicted structures closest to the designed coordinates (Fig. 2A) (Bradley et al., 2005). Similarly, MD simulations can also be used to pseudo-systematically query the energy gap, albeit for smaller protein structures (Grigoryan, 2013). These methods however can be fairly slow for small- to medium-sized proteins and are not even practical for large-sized proteins. Hence, it is more useful to deploy faster MD simulations to probe the propensity of a designed sequence to diffuse away from the designed structure into alternate configurations. This conformational search rather focuses the sampling on the local vicinity of the designed structure to explore the presence of deeper or easily accessible minima, instead of performing exhaustive *ab initio* sampling starting from an extended peptide. In either case, the magnitude of structural deviations from the designed coordinates can serve as a scoring metric. Relatively short or accelerated MD simulations could be used effectively to query the stability of designed inter- or intra-molecular interfaces (Fig. 2B) (Chevalier et al., 2017; ElGamacy et al., 2018a, 2018b).

[2] A pre-release of the Damietta software is available at: https://bio.mpg.de/damietta.

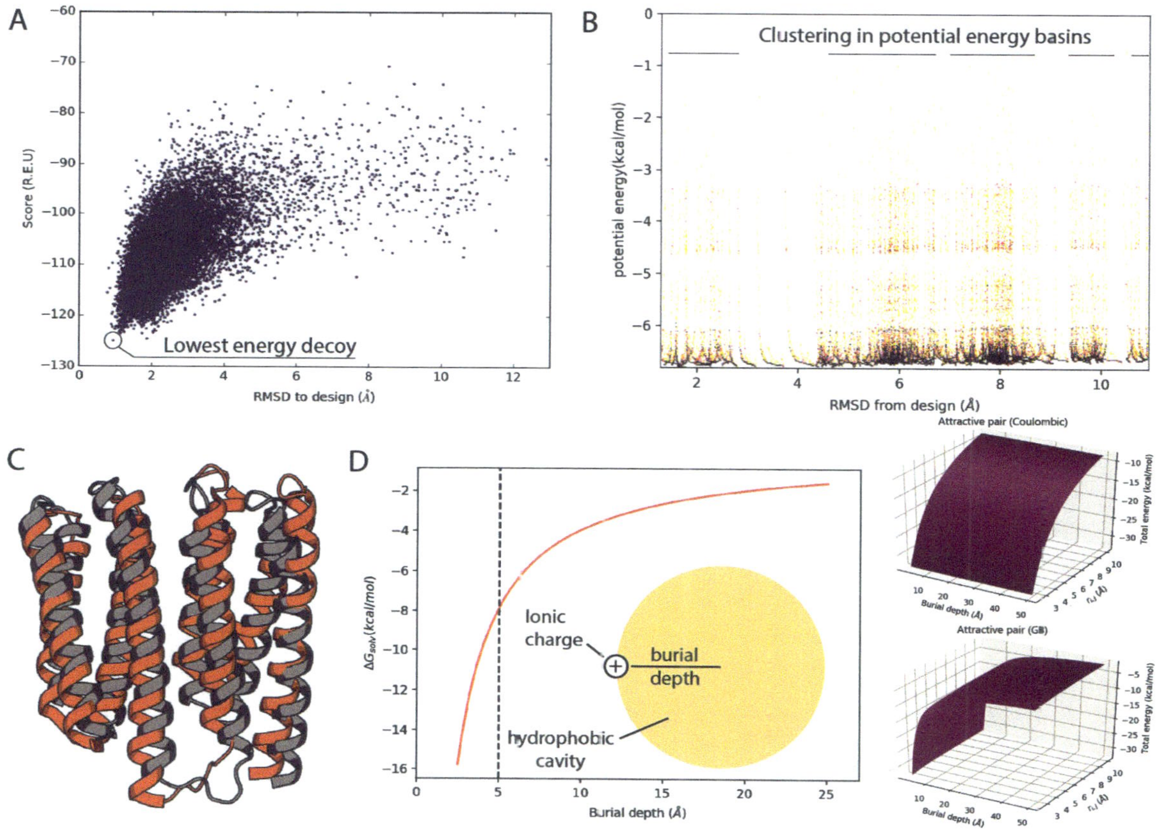

Fig. 2 See figure legend on next page.

5.2 Improving design success rates

The protein drug design success rate (i.e., number of successful designs/tested designs) can vary significantly according to the difficulty of the design problem and the accuracy of the design process. The difficulty relates to the dimensionality of the problem (e.g., novel fold design is more difficult than loop design) (Fleishman & Baker, 2012), design accuracy on the other hand is a function of the sampling breadth and scoring error, opening two fronts for such improvements. Many of the functional designed proteins so far made are the product of combining computational design of sequence libraries, and medium- to large-scale *in vitro* screening procedures, owing to the low success rates. While these design-guided libraries are superior to comparable random libraries, screening still constitutes an added experimental step that precedes the individual testing of a small number of top designs. While there remains an unquantifiable failure rate that stems from the impossibility of adequately estimating the energy gap (Fleishman & Baker, 2012), improving design accuracy can promise a direct *in silico*-to-*in vivo* pipeline, experimentally testing only a few design candidates.

Although more intensive rotamer sampling improves the solution accuracy (Kim et al., 2009), more locally focused sampling of small conformational changes can lead to large improvements in energy minimization.

Fig. 2 Factors to aid in accelerating protein design pipelines. (A) *Ab initio* folding simulations are template-free structure predictions, which are used to validate whether a designed sequence is (inversely-) predicted to fold into a structure close to the designed coordinates. This is visualized for thousands of such simulations that start from a linear polypeptide and end by predicted structure, from among this ensemble of predictions the lowest energy structure should match the lowest RMSD to the design model. (B) Instead of constructing a fold from a linear peptide to validate the designed structure, AMD simulations evaluate the stability (i.e., ease of destruction) of a designed structure. Tempering simulation methods can rapidly explore the accessibility and depths of energy basins in the vicinity of the designed structure; exploring its ease of mis- or unfolding. (C) AlphaFold2 predicted structure (*red*) of the de novo protein BRIC2 (ElGamacy et al., 2018a) closely matches its crystal structure (*gray*). This may offer an alternative means to for either *ab initio* folding validation, or even initial model building (i.e., blueprint generation) at a smaller computational footprint. (D) Electrostatics are extremely sensitive to the dielectric environment, where the hydrophobic core represents a very low dielectric environment (*left*). If two opposite charges (i.e., an attractive pair) approach each other through a distance $r_{i,j}$, their interaction energy should be strongly influenced by how deep they are buried within a protein core. While using a Columbic model completely dismisses the impact of desolvation on charges and their interactions (*upper right*), a Generalized Born model is dominated by the burial effect (*lower right*).

Gainza *et al.* have demonstrated that hierarchical rotamer sampling (i.e., more intensive rotamer sampling around initially good solutions) can result in more accurate predictions (Gainza et al., 2012). Introducing backbone movements in the design sampler was also shown to greatly improve the accuracy. Such *backrub* movements would introduce small torsional movements to a short segment of residues along the axis extending between C_α atoms of the flanking residues, where these movements were deployed in an MC sampler (Smith & Kortemme, 2008). Alternatively, interlacing of molecular dynamics steps with design steps can also be invaluable in sampling backbone movements and their correlated motions more effectively, and can capture local deformations (Childers & Daggett, 2017). Another advantage of MD is that it also optimizes the subtle changes bond lengths and angles, which are dismissed in the design stage, but have been shown to significantly impact the *energy gap* between native and non-native states (Conway et al., 2014).

Compromising the scoring accuracy for speed has been an entrenched trade off, which has marred the scoring accuracy in design applications. Electrostatics is one example of a highly important, yet expensive to accurately compute, energy term. For instance, fixed-point charges (the dominant representation in current design software, where atomic charges are modelled as permanent and atom-centered) dismiss the anisotropic charge distributions of lone-pairs, σ-holes, and π-systems. Induced polarization is another important though omitted phenomenon, where the charge distribution depends on the local environment or the amino acid conformation. These problems could be practically addressed without substantially compounding the computing load by deploying polarizable force fields (Cisneros et al., 2014) at the rotamer library level, or during the final scoring stages (including in MD simulations, if used). Electrostatics are also currently calculated as a Columbic term (i.e., $\frac{q_i q_j}{\epsilon r_{i,j}}$), ignoring the strong impact of the dielectric cavity (i.e., the protein core) on the desolvation of charges and their interactions. A gain in accuracy here can be easily achieved using a Generalized Born correction, which can be approximated further into a computationally inexpensive form (Fig. 2D).

Scoring error can also arise from the imperfect balance between the different energy terms; such as the weighting factors balancing the different energy terms, or the mixing of knowledge-based and physics-based terms in the same energy function. Hence, the use of a self-consistent energy function can eliminate the need for such learned weighting factors and avoids overfitting (Cerutti et al., 2014). Also, combining the scores from orthogonal energy functions can lead to better ranking of lowest energy decoys and reduce the scoring error (Rubenstein et al., 2018). Incorporating MD-based routines toward the final ranking stages of the *in silico* design workflow has been shown to greatly complement the static scoring of decoys (Barros et al., 2019; Chevalier et al., 2017). This MD-based screening is particularly useful for ranking binder designs, for which *ab initio* folding validation is not possible. In this regard, accelerated MD (AMD) should play a bigger role, given track record of several AMD method in drastically reducing the computing cost without a marked reduction in accuracy (Markwick & McCammon, 2011). The recent progress in structure prediction by AlphaFold2 and its remarkable performance in predicting the structure of novel folds to a decent accuracy (Jumper et al., 2021) can potentially serve as alternative means of in silico design validation (Fig. 2C). The neural network architecture of AlphaFold2 efficiently makes use of hardware acceleration (e.g., GPU), and can serve as a filtering step complementary to validation by AMD.

5.3 Protein efficiency

The heavy use of high-throughput screening techniques to search for small molecule drugs with the highest potency has resulted in clinical candidates with successively growing molecular weights over the years (Lipinski, 2000). Maximizing potency strongly relates to improving a drug's binding affinity to its target, and thus increases the number of functional groups that improve the binding enthalpy (e.g., hydrogen bonding) or desolvation energy (e.g., hydrophobic contacts). This has rendered the classical measures of potency (i.e., K_d or IC_{50}) in molar terms less useful in comparing two ligands of different sizes, and has led to the proposition of the ligand efficiency notion (Hopkins et al., 2004). *Ligand efficiency* could be expressed as the free energy of binding (ΔG_{bind}) normalized by the number of non-hydrogen atoms of the ligand (i.e., $\Delta G_{bind}/N_{atoms}$), where the ΔG_{bind} value can be also casually derived from IC_{50}, similar to K_d (i.e., $\Delta G_{bind} = RT \cdot \ln(K_d)$). Small molecule drugs were surveyed to maximally achieve a *ligand efficiency* of $-6.3\,\text{kJ/mol/}$ atom; such a value, however, is restricted to a few low-molecular weight ligands (with <10 non-hydrogen atoms). The same survey data also show a steep loss of ligand efficiency with higher molecular weights (Kuntz et al., 1999).

Conversely, the more efficacious, albeit narrow, class of small molecule drugs that can effectively inhibit protein-protein interactions (PPIs) tends to show a linear relationship between their binding free energies and their respective molecular weights. This means that PPI inhibitors maintain a uniform ligand efficiency regardless of their size (Wells & McClendon, 2007). This is a result of the fact that PPIs constitute generally larger and flatter interfaces that are more difficult to inhibit, which selects for efficient ligands that can use most of their atoms in forming strong contacts with the protein.

From the point-of-view of protein drugs, the closest notion to quantify ΔG_{bind} per functional group- (or residue-) specific basis is that of *hot spots* or *hot segments*. This notion builds on the fact that most of the binding free energy across a PPI is contributed by a small fraction of residues at the interface (*hot spots*), that in some cases form a spatially continuous sub-epitope (*hot segment*) (London et al., 2013). While hot spot scanning (experimentally and computationally) has traditionally been used to identify small druggable epitopes, the reverse view of the efficiency of a protein as a binder has to be critically considered. Hence, the proposition of the concept of "*protein efficiency*" as a quantifiable metric of the information density encoded in a therapeutic protein. Unlike in small molecules, in an active protein most of the atoms engage in intramolecular folding contacts, and most of the molecular surface is not part of the binding interface. This should motivate the routine use of *protein efficiency* not only as a metric to prioritize therapeutic protein candidates, but in early blueprint conception and computational design stages.

The *protein efficiency* can likewise be derived to normalize different functional metrics by the molecular mass in Daltons (i.e., $\Delta G_{func}/M_{wt}$), where a ΔG_{func} can be casually derived for binding (K_d), inhibitory (IC_{50}), activation (EC_{50}), or enzymatic (K_M/k_{cat}; i.e., inverse catalytic efficiency) function. This definition of efficiency prioritizes proteins with highest potency and smallest number of scaffolding (i.e., non-functional) residues. However, there is a huge potential for designed proteins that host several functional epitopes simultaneously; such as multi-specific antibodies (which host 2 or more binding sites) (Labrijn et al., 2019) or multi-active enzymes (hosting two or more active sites) (Palomo, 2021). These *Swiss knife* proteins can serve more complex functions in comparison with traditional monoclonal antibodies or recombinant enzymes, and can play a diverse set of roles such as co-localization, cell targeting, novel catalytic pipelines, etc. Using more of their accessible molecular surface, such proteins can achieve higher functional:scaffolding residue ratios, and further stretch the protein efficiency limits. In this case, the efficiency metric for a set of functions (i.e., $\{func_1, func_2, \ldots, func_n\}$) can be defined as the normalized summation of their respective free energies (i.e., $\frac{G_{func_1} + G_{func_2} + \ldots + G_{func_n}}{M_{wt}}$).

To date, the best examples of highly efficient proteins that could deliver novel functions come from the non-immunoglobulin binders. These mini-binders have been arranged into multidomain fusions to achieve multi-specific or multi-epitopic chimeras, that could co-localize target proteins, recruit effector cells to target cells, and achieve avidity (Hober et al., 2019). While such chimeras are trivial to modularly engineer via flexible linkers, in many cases these chimeras: (i) cannot be rigidly juxtaposed to defined configuration, (ii) their functional sites may cross-interact or cross-occlude each other, (iii) and the mismatched stability between the constituting domains would lead to partial degradation biproducts *in vivo* with different spill-over activities. Overcoding of more than one function onto a single protein domain can avert all such disadvantages and guarantee a stoichiometric and geometric balance between the functional sites. Protein design effort can thus further advance the *protein efficiency* concept along this line (Fig. 3).

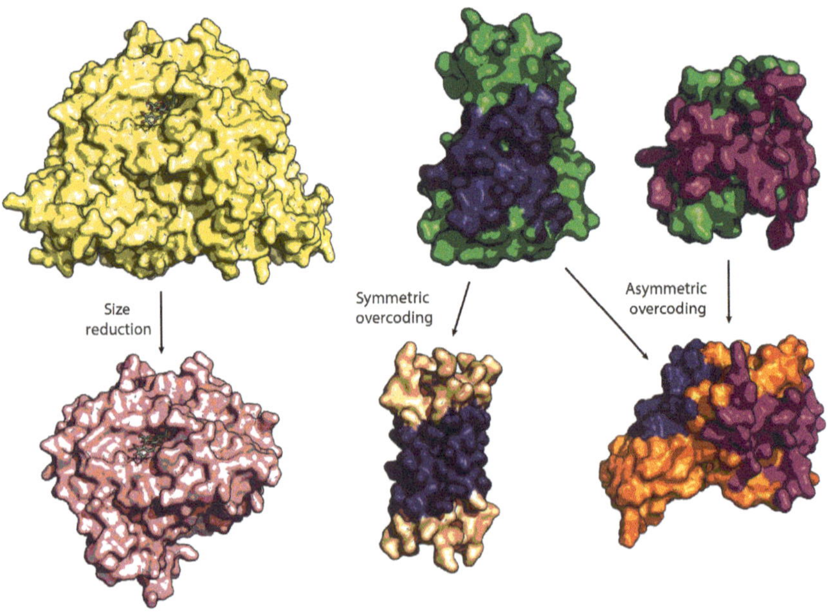

Fig. 3 Different ways of enhancing *protein efficiency*. *Protein efficiency* can be enhanced through miniaturization of an active protein, or encoding multiple active sites onto a single protein. Example of miniaturizing an enzyme to preserve the active site structure and dynamics (*left*). Another example is encoding symmetric copies of an active epitope onto a symmetric protein (*middle*). Asymmetric overcoding can also be achieved by imprinting two functional epitopes from different proteins onto a new scaffold (*right*).

5.4 Immunogenicity of *de novo* proteins

While the type and extent of an immune reaction to exogenous proteinaceous material is a complex biological response and can involve several possible pathways, T-cell-dependent responses constitute the main immune pathway directed against protein therapeutics. Such responses can result in high-affinity anti-drug antibodies (ADA) and a durable immune memory against a protein (Jawa et al., 2013). At the beginning of this process, five critical steps are necessary for antigen uptake, processing, and presentation. This starts by (i) the antigen presenting cells (APCs; either dendritic cells or specific B-cells) internalizing the therapeutic protein, then (ii) proteolytically digesting it into short peptide segments, which (iii) may bind to an MHC-II molecule, that (iv) is presented on the cell surface as a peptide-MHC-II complex, this is then followed by (v) recruitment, stimulation, and proliferation of helper T-cells (Rosenberg & Sauna, 2017). Although ADAs are generated from antigen-specific B-cells, the latter still rely on information relayed by helper T-cells. This, in addition to the fact that B-cell epitopes are conformational and extremely difficult to predict, has emphasized the importance of the early steps of antigen processing leading to the step of T-cell peptide recognition (Jawa et al., 2013).

A huge body of research explored computational and rapid *in vitro* techniques to assay and mitigate immunogenicity of therapeutic proteins, which mostly focuses on T-cell epitope analysis (Sauna et al., 2018). Alternatively, protein design can, from the start, integrate the objectives of either increasing or decreasing the immunogenicity of a protein into the design process itself. Such modulation of immunogenicity can be either achieved through the control of the protein's biophysical features and its sequence. Using a design scoring function that evaluates not just the energy-cost of a mutation, but also the immunogenicity score of the local peptide sequence (e.g., the predicted affinity of the mutated peptide to bind MHC-II), can mitigate the immunogenicity risk very early in the design process. Different protein design software deploying such joint-optimization could deplete T-cell epitopes this way (King et al., 2014; Salvat et al., 2017). The biophysical properties of a protein also greatly affect its susceptibility to immune surveillance in several ways. Highly stable proteins, for instance are less prone to proteolytic degradation and thus processing and presentation by APCs, affecting their availability for MHC-II loading. On the other hand, the structural rigidity of hyperstable proteins narrows the conformational search space for B-cell epitopes, albeit they are supported by less T-cell help (Scheiblhofer et al., 2017). Likewise, aggregation tendency and poor solubility often raise immunogenicity risks (Ratanji et al., 2014).

The counter objective of maximizing the immunogenicity of a protein drug is generally desired for vaccines development. The most straightforward application is the redesign of a marginally stable protein to make it amenable for bacterial expression or improve its production yield. This was done to the RH5 protein from *Plasmodium falciparum*; a malaria vaccine candidate (Campeotto et al., 2017), and the envelope glycoprotein from HIV-1 (Malladi et al., 2020). More advanced schemes of structural rescaffolding of immune-epitopes have been pioneered by Correia and co-workers, whereby continuous or discontinuous conformational epitopes are mounted on minimal protein scaffolds with more favorable biophysical features (Correia et al., 2014; Sesterhenn et al., 2020). Several geometric features have been established to enhance immunogenicity; these concern the antigen particle size (ideally 20–200 nm), internal repetitiveness, and antigen display patterns (Bachmann & Jennings, 2010). A large number of repetitive and virus-like assemblies have been computationally designed to mount various antigens, including antigens from HIV, RSV, influenza virus, or SARS-CoV-2 (Brouwer et al., 2019; Marcandalli et al., 2019; Ueda et al., 2020; Walls et al., 2020). These examples show that the objective control of protein sequence and structure through protein design can readily be used to either deimmunize therapeutic proteins or boost the immunogenicity of protein-based vaccines.

6. Conclusions

The design of proteins that can exert pharmacological function provides for a superior approach that minimizes the time and costs of the therapeutics development process. Moreover, it expands the range of the druggable targets given that bespoke proteins that can be built from first principles to possess novel structures and functions. Above all, therapeutic protein design improves our understanding of protein biophysics and molecular recognition phenomena, allowing the continual improvement of our descriptive and predictive models. Designed proteins can be deliberately made to overcome several challenges facing protein drugs in general. Several examples were highlighted of designed, protein-based inhibitors, activators, enzymes, and switches with therapeutic potential, outshining any comparable advancements that could be achieved through either small molecule therapeutics, or empirically-engineered proteins. Moving forward, advances in the computational models and computing technologies are poised to further upgrade the design capacity to undertake more difficult problems, and achieve better accuracy, leading to more effective discovery, and reduced pre-clinical and clinical development costs and timelines.

Acknowledgments

The author would like to thank Pak-Lee Chau, Kateryna Maksymenko, Birte Hernandez Alvarez, Andrei Lupas, Julia Skokowa, and Murray Coles for the invaluable scientific discussions.

References

Alford, R. F., et al. (2017). The Rosetta all-atom energy function for macromolecular modeling and design. *Journal of Chemical Theory and Computation*, *13*(6), 3031–3048.

Amstutz, P., et al. (2001). In vitro display technologies: Novel developments and applications. *Current Opinion in Biotechnology*, *12*(4), 400–405.

Andersen, D. C., & Krummen, L. (2002). Recombinant protein expression for therapeutic applications. *Current Opinion in Biotechnology*, *13*(2), 117–123.

Anguela, X. M., & High, K. A. (2019). Entering the modern era of gene therapy. *Annual Review of Medicine*, *70*(1), 273–288.

Ashworth, J., et al. (2006). Computational redesign of endonuclease DNA binding and cleavage specificity. *Nature*, *441*(7093), 656–659.

Asial, I., et al. (2013). Engineering protein thermostability using a generic activity-independent biophysical screen inside the cell. *Nature Communications*, *4*(1), 2901.

Bachmann, M. F., & Jennings, G. T. (2010). Vaccine delivery: A matter of size, geometry, kinetics and molecular patterns. *Nature Reviews Immunology*, *10*(11), 787–796.

Bansal, R., et al. (2021). Size-based degradation of therapeutic proteins—Mechanisms, modelling and control. *Biomolecular Concepts*, *12*(1), 68–84.

Barros, E. P., et al. (2019). Improving the efficiency of ligand-binding protein design with molecular dynamics simulations. *Journal of Chemical Theory and Computation*, *15*(10), 5703–5715.

Bloom, J. D., et al. (2006). Protein stability promotes evolvability. *Proceedings of the National Academy of Sciences*, *103*(15), 5869–5874.

Böttger, R., et al. (2017). Differential stability of therapeutic peptides with different proteolytic cleavage sites in blood, plasma and serum. *PLoS One*, *12*(6), 1–15.

Bradley, P., et al. (2005). Toward High-resolution de novo structure prediction for small proteins. *Science*, *309*(5742), 1868–1871.

Brekke, O. H., & Sandlie, I. (2003). Therapeutic antibodies for human diseases at the dawn of the twenty-first century. *Nature Reviews Drug Discovery*, *2*(1), 52–62.

Brewer, D. B. (1951). Renal clearances of dextrans of varying molecular weights. *Proceedings of the Royal Society of Medicine*, *44*(7), 561–563.

Brouwer, P. J. M., et al. (2019). Enhancing and shaping the immunogenicity of native-like HIV-1 envelope trimers with a two-component protein nanoparticle. *Nature Communications*, *10*(1), 4272.

Bryan, C. M., et al. (2021). Computational design of a synthetic PD-1 agonist. *Proceedings of the National Academy of Sciences*, *118*(29).

Campeotto, I., et al. (2017). One-step design of a stable variant of the malaria invasion protein RH5 for use as a vaccine immunogen. *Proceedings of the National Academy of Sciences*, *114*(5), 998–1002.

Canutescu, A. A., et al. (2003). A graph-theory algorithm for rapid protein side-chain prediction. *Protein Science*, *12*(9), 2001–2014

Cao, L., et al. (2020). De novo design of picomolar SARS-CoV-2 miniprotein inhibitors. *Science*, *370*(6515), 426–431.
Carone, F. A. (1978). Renal handling of proteins and peptides. *Annals of Clinical and Laboratory Science*, *8*(4), 287–294.
Carroll, J. (2006). Enzyme replacement therapies: Better lives versus the bottom line. *Biotechnology Healthcare*, *3*(6), 46–56.
Cerutti, D. S., et al. (2014). ff14ipq: A self-consistent force field for condensed-phase simulations of proteins. *Journal of Chemical Theory and Computation*, *10*(10), 4515–4534.
Chevalier, A., et al. (2017). Massively parallel de novo protein design for targeted therapeutics. *Nature*, *550*(7674), 74–79.
Childers, M. C., & Daggett, V. (2017). Insights from molecular dynamics simulations for computational protein design. *Molecular Systems Design & Engineering*, *2*, 9–33.
Cisneros, G. A., et al. (2014). Classical electrostatics for biomolecular simulations. *Chemical Reviews*, *114*(1), 779–814.
Conway, P., et al. (2014). Relaxation of backbone bond geometry improves protein energy landscape modeling. *Protein Science*, *23*(1), 47–55.
Correia, B. E., et al. (2014). Proof of principle for epitope-focused vaccine design. *Nature*, *507*(7491), 201–206.
Costanzo, J. A., et al. (2014). Conformational stability as a design target to control protein aggregation. *Protein Engineering, Design and Selection*, *27*(5), 157–167.
Dahiyat, B. I., & Mayo, S. L. (1997). De novo protein design: Fully automated sequence selection. *Science*, *278*(5335), 82–87.
Demain, A. L., & Vaishnav, P. (2009). Production of recombinant proteins by microbes and higher organisms. *Biotechnology Advances*, *27*(3), 297–306.
Desmet, J., et al. (1992). The dead-end elimination theorem and its use in protein side-chain positioning. *Nature*, *356*(6369), 539–542.
Dinner, A. R., & Karplus, M. (2001). The roles of stability and contact order in determining protein folding rates. *Nature Structural Biology*, *8*(1), 21–22.
Dobson, C. M. (2003). Protein folding and disease: A view from the first horizon symposium. *Nature Reviews Drug Discovery*, *2*(2), 154–160.
ElGamacy, M., & Hernandez Alvarez, B. (2021). Expanding the versatility of natural and de novo designed coiled coils and helical bundles. *Current Opinion in Structural Biology*, *68*, 224–234.
ElGamacy, M., et al. (2018a). An interface-driven design strategy yields a novel, corrugated protein architecture. *ACS Synthetic Biology*, 7(9), 2226–2235.
ElGamacy, M., et al. (2018b). Asymmetric protein design from conserved supersecondary structures. *Journal of Structural Biology*, *204*(3), 380–387.
Fernandez-Escamilla, A.-M., et al. (2004). Prediction of sequence-dependent and mutational effects on the aggregation of peptides and proteins. *Nature Biotechnology*, *22*(10), 1302–1306.
Fleishman, S. J., & Baker, D. (2012). Role of the biomolecular energy gap in protein design, structure, and evolution. *Cell*, *149*(2), 262–273.
Fleishman, S. J., et al. (2011). Computational design of proteins targeting the conserved stem region of influenza hemagglutinin. *Science*, *332*(6031), 816–821.
Fletcher, J. M., et al. (2018). De novo coiled-coil peptides as scaffolds for disrupting protein–protein interactions. *Chemical Science*, *9*, 7656–7665.
Frokjaer, S., & Otzen, D. E. (2005). Protein drug stability: A formulation challenge. *Nature Reviews Drug Discovery*, *4*(4), 298–306.
Froning, K., et al. (2020). Computational stabilization of T cell receptors allows pairing with antibodies to form bispecifics. *Nature Communications*, *11*(1), 2330.
Gainza, P., et al. (2012). Protein design using continuous rotamers. *PLoS Computational Biology*, *8*(1), 1–15.

Giordano-Attianese, G., et al. (2020). A computationally designed chimeric antigen receptor provides a small-molecule safety switch for T-cell therapy. *Nature Biotechnology*, *38*(4), 426–432.
Gordon, S. R., et al. (2012). Computational design of an α-gliadin peptidase. *Journal of the American Chemical Society*, *134*(50), 20513–20520.
Grigoryan, G. (2013). Absolute free energies of biomolecules from unperturbed ensembles. *Journal of Computational Chemistry*, *34*(31), 2726–2741.
Hawe, A., & Frieß, W. (2007). Formulation development for hydrophobic therapeutic proteins. *Pharmaceutical Development and Technology*, *12*(3), 223–237.
Hernandez Alvarez, B., et al. (2020). Design of novel granulopoietic proteins by topological rescaffolding. *PLoS Biology*, *18*(12), e3000919.
Hober, S., et al. (2019). Bispecific applications of non-immunoglobulin scaffold binders. *Methods*, *154*, 143–152.
Hopkins, A. L., et al. (2004). Ligand efficiency: A useful metric for lead selection. *Drug Discovery Today*, *9*(10), 430–431.
Huang, X., et al. (2020). Toward the accuracy and speed of protein side-chain packing: A systematic study on rotamer libraries. *Journal of Chemical Information and Modeling*, *60*(1), 410–420.
Ivankov, D. N., et al. (2003). Contact order revisited: Influence of protein size on the folding rate. *Protein Science*, *12*(9), 2057–2062.
Jawa, V., et al. (2013). T-cell dependent immunogenicity of protein therapeutics: Preclinical assessment and mitigation. *Clinical Immunology*, *149*(3), 534–555.
Jumper, J., et al. (2021). Highly accurate protein structure prediction with AlphaFold. *Nature*, *596*(7873), 583–589.
Kazlauskas, R. (2018). Engineering more stable proteins. *Chemical Society Reviews*, *47*, 9026–9045.
Kim, D. E., et al. (2009). Sampling bottlenecks in de novo protein structure prediction. *Journal of Molecular Biology*, *393*(1), 249–260.
Kinch, M. S. (2015). An overview of FDA-approved biologics medicines. *Drug Discovery Today*, *20*(4), 393–398.
Kinch, L., et al. (2011). CASP9 assessment of free modeling target predictions. *Proteins: Structure, Function, and Bioinformatics*, *79*(S10), 59–73.
King, C., et al. (2014). Removing T-cell epitopes with computational protein design. *Proceedings of the National Academy of Sciences*, *111*(23), 8577–8582.
Kuntz, I. D., et al. (1999). The maximal affinity of ligands. *Proceedings of the National Academy of Sciences*, *96*(18), 9997–10002.
Labrijn, A. F., et al. (2019). Bispecific antibodies: A mechanistic review of the pipeline. *Nature Reviews Drug Discovery*, *18*(8), 585–608.
Lajoie, M. J., et al. (2020). Designed protein logic to target cells with precise combinations of surface antigens. *Science*, *369*(6511), 1637–1643.
Leader, B., et al. (2008). Protein therapeutics: A summary and pharmacological classification. *Nature Reviews Drug Discovery*, 7(1), 21–39.
Leaver-Fay, A., et al. (2011). Chapter Nineteen—Rosetta3: An object-oriented software suite for the simulation and design of macromolecules. In M. L. Johnson, & L. Brand (Eds.), *Computer methods, Part C* (pp. 545–574). Academic Press.
Lee, C., & Subbiah, S. (1991). Prediction of protein side-chain conformation by packing optimization. *Journal of Molecular Biology*, *217*(2), 373–388.
Lim, W. A., & June, C. H. (2017). The principles of engineering immune cells to treat Cancer. *Cell*, *168*(4), 724–740.
Lipinski, C. A. (2000). Drug-like properties and the causes of poor solubility and poor permeability. *Journal of Pharmacological and Toxicological Methods*, *44*(1), 235–249.

Liu, Z., et al. (2003). Beyond the rotamer library: Genetic algorithm combined with the disturbing mutation process for upbuilding protein side-chains. *Proteins: Structure, Function, and Bioinformatics*, *50*(1), 49–62.
Liu, Y. D., et al. (2008). Human IgG2 antibody disulfide rearrangement in vivo. *Journal of Biological Chemistry*, *283*(43), 29266–29272.
London, N., et al. (2013). Druggable protein–protein interactions from hot spots to hot segments. *Current Opinion in Chemical Biology*, *17*(6), 952–959.
Lu, R.-M., et al. (2020). Development of therapeutic antibodies for the treatment of diseases. *Journal of Biomedical Science*, *27*(1), 1.
Malladi, S. K., et al. (2020). One-step sequence and structure-guided optimization of HIV-1 envelope gp140. *Current Research in Structural Biology*, *2*, 45–55.
Marcandalli, J., et al. (2019). Induction of potent neutralizing antibody responses by a designed protein nanoparticle vaccine for respiratory syncytial virus. *Cell*, *176*(6), 1420–1431.e17.
Markwick, P. R. L., & McCammon, J. A. (2011). Studying functional dynamics in bio-molecules using accelerated molecular dynamics. *Physical Chemistry Chemical Physics*, *13*, 20053–20065.
Martin, O. A., & Vila, J. A. (2020). The marginal stability of proteins: How the jiggling and wiggling of atoms is connected to neutral evolution. *Journal of Molecular Evolution*, *88*(5), 424–426.
McCammon, J. A., et al. (1977). Dynamics of folded proteins. *Nature*, *267*(5612), 585–590.
McConnell, A. D., et al. (2014). A general approach to antibody thermostabilization. *MAbs*, *6*(5), 1274–1282.
Metropolis, N., et al. (1953). Equation of state calculations by fast computing machines. *The Journal of Chemical Physics*, *21*(6), 1087–1092.
Miklos, A. E., et al. (2012). Structure-based design of supercharged, highly thermoresistant antibodies. *Chemistry & Biology*, *19*(4), 449–455.
Minchinton, A. I., & Tannock, I. F. (2006). Drug penetration in solid tumours. *Nature Reviews Cancer*, *6*(8), 583–592.
Mohan, K., et al. (2019). Topological control of cytokine receptor signaling induces differential effects in hematopoiesis. *Science*, *364*(6442), eaav7532.
Moussa, E. M., et al. (2016). Immunogenicity of therapeutic protein aggregates. *Journal of Pharmaceutical Sciences*, *105*(2), 417–430.
O'Brien, C. J., et al. (2016). Modulating non-native aggregation and electrostatic protein–protein interactions with computationally designed single-point mutations. *Protein Engineering, Design and Selection*, *29*(6), 231–243.
Pallasch, C. P., et al. (2014). Sensitizing protective tumor microenvironments to antibody-mediated therapy. *Cell*, *156*(3), 590–602.
Palomo, J. M. (2021). Artificial enzymes with multiple active sites. *Current Opinion in Green and Sustainable Chemistry*, *29*, 100452.
Plückthun, A. (2012). Ribosome display: A perspective. In J. A. Douthwaite, & R. H. Jackson (Eds.), *Ribosome display and related technologies: Methods and protocols* (pp. 3–28). New York, NY: Springer New York.
Polizzi, N. F., & DeGrado, W. F. (2020). A defined structural unit enables de novo design of small-molecule–binding proteins. *Science*, *369*(6508), 1227–1233.
Ponder, J. W., & Richards, F. M. (1987). Tertiary templates for proteins: Use of packing criteria in the enumeration of allowed sequences for different structural classes. *Journal of Molecular Biology*, *193*(4), 775–791.
Procko, E., et al. (2014). A computationally designed inhibitor of an Epstein-Barr viral Bcl-2 protein induces apoptosis in infected cells. *Cell*, *157*(7), 1644–1656.
Ratanji, K. D., et al. (2014). Immunogenicity of therapeutic proteins: Influence of aggregation. *Journal of Immunotoxicology*, *11*(2), 99–109.

Rego, N. B., et al. (2021). Identifying hydrophobic protein patches to inform protein interaction interfaces. *Proceedings of the National Academy of Sciences, 118*(6).
Renukuntla, J., et al. (2013). Approaches for enhancing oral bioavailability of peptides and proteins. *International Journal of Pharmaceutics, 447*(1–2), 75–93.
Rohl, C. A., et al. (2004). Protein structure prediction using Rosetta. In *Methods in enzymology* (pp. 66–93). Academic Press.
Romero, P. A., & Arnold, F. H. (2009). Exploring protein fitness landscapes by directed evolution. *Nature Reviews Molecular Cell Biology, 10*(12), 866–876.
Rosenberg, A. S., & Sauna, Z. E. (2017). Immunogenicity assessment during the development of protein therapeutics. *Journal of Pharmacy and Pharmacology, 70*(5), 584–594.
Rubenstein, A. B., et al. (2018). Systematic comparison of Amber and Rosetta energy functions for protein structure evaluation. *Journal of Chemical Theory and Computation, 14*(11), 6015–6025.
Salvat, R. S., et al. (2017). Computationally optimized deimmunization libraries yield highly mutated enzymes with low immunogenicity and enhanced activity. *Proceedings of the National Academy of Sciences, 114*(26), E5085–E5093.
Sauna, Z. E., et al. (2018). Evaluating and mitigating the immunogenicity of therapeutic proteins. *Trends in Biotechnology, 36*(10), 1068–1084.
Scheiblhofer, S., et al. (2017). Influence of protein fold stability on immunogenicity and its implications for vaccine design. *Expert Review of Vaccines, 16*(5), 479–489.
Schuster, J., et al. (2020). In vivo stability of therapeutic proteins. *Pharmaceutical Research, 37*(2), 23.
Sesterhenn, F., et al. (2020). De novo protein design enables the precise induction of RSV-neutralizing antibodies. *Science, 368*(6492), eaay5051.
Sethuraman, N., & Stadheim, T. A. (2006). Challenges in therapeutic glycoprotein production. *Current Opinion in Biotechnology, 17*(4), 341–346.
Sheffler, W., & Baker, D. (2009). RosettaHoles: Rapid assessment of protein core packing for structure prediction, refinement, design, and validation. *Protein Science, 18*(1), 229–239.
Shui, S., et al. (2021). A rational blueprint for the design of chemically-controlled protein switches. *Nature Communications, 12*(1), 5754.
Silva, D.-A., et al. (2019). De novo design of potent and selective mimics of IL-2 and IL-15. *Nature, 565*(7738), 186–191.
Singh, K., et al. (2019). Antibody delivery for intracellular targets: Emergent therapeutic potential. *Bioconjugate Chemistry, 30*(4), 1028–1041.
Smith, C. A., & Kortemme, T. (2008). Backrub-like backbone simulation recapitulates natural protein conformational variability and improves mutant side-chain prediction. *Journal of Molecular Biology, 380*(4), 742–756.
Solá, R. J., & Griebenow, K. (2010). Glycosylation of therapeutic proteins: An effective strategy to optimize efficacy. *BioDrugs: Clinical Immunotherapeutics, Biopharmaceuticals and Gene Therapy, 24*(1), 9–21.
Strauch, E.-M., et al. (2017). Computational design of trimeric influenza-neutralizing proteins targeting the hemagglutinin receptor binding site. *Nature Biotechnology, 35*(7), 667–671.
Taverna, D. M., & Goldstein, R. A. (2002). Why are proteins marginally stable? *Proteins: Structure, Function, and Bioinformatics, 46*(1), 105–109.
Tinberg, C. E., et al. (2013). Computational design of ligand-binding proteins with high affinity and selectivity. *Nature, 501*(7466), 212–216.
Ueda, G., et al. (2020). Tailored design of protein nanoparticle scaffolds for multivalent presentation of viral glycoprotein antigens. *eLife, 9*, e57659.
Valente, J. J., et al. (2005). Colloidal behavior of proteins: Effects of the second virial coefficient on solubility, crystallization and aggregation of proteins in aqueous solution. *Current Pharmaceutical Biotechnology, 6*(6), 427–436.

VanDrisse, C. M., et al. (2021). Computationally designed pyocyanin demethylase acts synergistically with tobramycin to kill recalcitrant Pseudomonas aeruginosa biofilms. *Proceedings of the National Academy of Sciences*, *118*(12), e2022012118.
Walls, A. C., et al. (2020). Elicitation of potent neutralizing antibody responses by designed protein nanoparticle vaccines for SARS-CoV-2. *Cell*, *183*(5), 1367–1382.e17.
Weinhold, F., & Klein, R. A. (2012). What is a hydrogen bond? Mutually consistent theoretical and experimental criteria for characterizing H-bonding interactions. *Molecular Physics*, *110*(9–10), 565–579.
Wells, J. A., & McClendon, C. L. (2007). Reaching for high-hanging fruit in drug discovery at protein\textendashprotein interfaces. *Nature*, *450*(7172), 1001–1009.
Wolf, D. H., & Menssen, R. (2018). Mechanisms of cell regulation—Proteolysis, the big surprise. *FEBS Letters*, *592*(15), 2515–2524.
Wolf, C., et al. (2015). Engineering of Kuma030: A gliadin peptidase that rapidly degrades immunogenic gliadin peptides in gastric conditions. *Journal of the American Chemical Society*, *137*(40), 13106–13113.
Woodall, N. B., et al. (2021). De novo design of tyrosine and serine kinase-driven protein switches. *Nature Structural & Molecular Biology*, *28*(9), 762–770.
Woolfson, D. N. (2021). A brief history of De novo protein design: Minimal, rational, and computational. *Journal of Molecular Biology*, *433*(20), 167160.
Wypych, J., et al. (2008). Human IgG2 antibodies display disulfide-mediated structural isoforms. *Journal of Biological Chemistry*, *283*(23), 16194–16205.
Xu, D., & Nussinov, R. (1998). Favorable domain size in proteins. *Folding and Design*, *3*(1), 11–17.
Yue, P., et al. (2005). Loss of protein structure stability as a major causative factor in monogenic disease. *Journal of Molecular Biology*, *353*(2), 459–473.
Zheng, F., et al. (2014). A highly efficient cocaine-detoxifying enzyme obtained by computational design. *Nature Communications*, *5*(1), 3457.
Zhou, J., et al. (2020). A general-purpose protein design framework based on mining sequence-structure relationships in known protein structures. *Proceedings of the National Academy of Sciences of the United States of America*, *117*(2), 1059–1068.

CHAPTER FIVE

Aromatic interactions directing peptide nano-assembly

Sajitha Sasidharan and Vibin Ramakrishnan*
Department of Biosciences and Bioengineering, Indian Institute of Technology Guwahati, Guwahati, Assam, India
*Corresponding author: e-mail address: vibin@iitg.ac.in

Contents

Advances in Protein Chemistry and Structural Biology, Volume 130
ISSN 1876-1623
https://doi.org/10.1016/bs.apcsb.2022.01.001

Abstract

Self-assembly is a process of spontaneous organization of molecules as a result of non-covalent interactions. Organized self-assembly at the nano level is emerging as a powerful tool in the bottom-up fabrication of functional nanostructures for targeted applications. Aromatic π–π stacking plays a significant role by facilitating the persistent supramolecular association of individual subunits to the self-assembled structures of high stability. Understanding, the supramolecular chemistry of the materials interacting through aromatic interactions, is of tremendous interest in not only constructing functional materials but also in revealing the mechanism of molecular assembly in living organisms. This chapter aims to focus on understanding the potential role of π–π interactions in directing and regulating the self-assembly of peptide nanostructures. The scope of the chapter starts with an outline of the history and mechanism of the aromatic π–π interactions. It progresses through the design strategy for the assembly of peptides containing aromatic rings, the conditions affecting the aromatic stacking interactions, their resulting nanoassemblies, properties, and applications. The properties and applications of the supramolecular materials formed through the aromatic stacking interactions are highlighted to provide an increased understanding of the role of weak interactions in the design and construction of novel functional materials.

1. Introduction

The molecular self-assembly process in biology was the main inspiration to study the organization of biomolecules in facilitating the desired physiological process. The association of the individual subunits through reversible, non-covalent interactions is dynamic, and hence the mechanism can be fine-tuned to generate smart and functional nanomaterials. The structures generated through the process not only represent the thermodynamic minima but also are intrinsically resistant to the incorporation of impurities. Any deliberate effort to design such nanostructures involves understanding of the fundamental forces that trigger and stabilize the intermolecular interactions. The stabilizing interactions are usually a large number of distributed individually weak interactions than a localized small number of strong covalent bonds. Molecular self-assembly, a spontaneous association observed in proteins, also leads to the pathogenic role underlying the formation of amyloid fibrils. Hence, understanding the self-assembly process and its associated non-covalent interactions is a central problem in structural biochemistry.

Significant efforts and careful examination of short functional fragments of amyloid-forming proteins led to the observation of a 'common factor', which it is the presence of aromatic amino acids in all the fragments. It raised the question of the possible role of π–π interactions in amyloid formation.

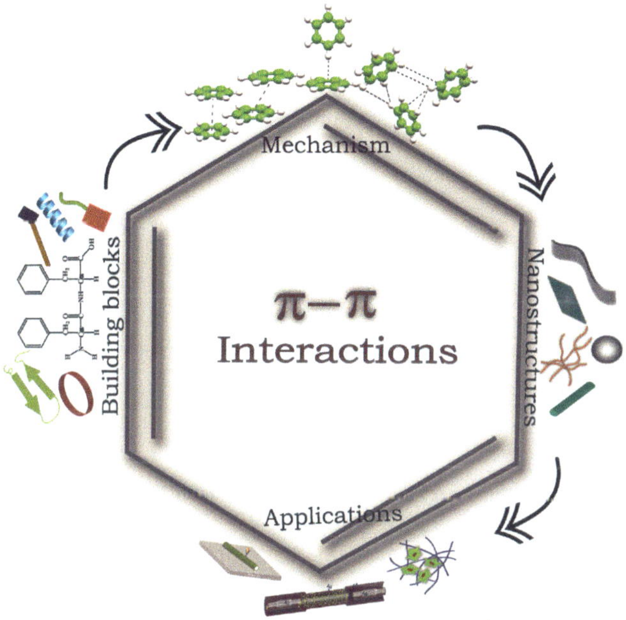

Fig. 1 Role of aromatic π-π interactions in directing the self-assembly of basic peptide building blocks to functional nanostructures.

Also, it was speculated that the π-stacking interactions might provide energetic distribution, order, and directionality to the self-assembly of amyloid fibrils (Gazit, 2002). The suggested π-stacking hypothesis inspired the design of new peptide sequences, which helped in understanding the self-assembly mechanism (Reches & Gazit, 2003) and the development of a series of novel functional peptides (Fig. 1) (Meital & Ehud, 2006; Reches & Gazit, 2004).

2. Aromatic interactions in biomolecules

Aromatic interactions, ubiquitous in nature, play an essential role in stabilizing the structure of biomolecules like double-stranded DNA (Kool, 2001) and various proteins (Gazit, 2002). It was considered among the less studied and less understood interactions until the importance of π-π stacking interactions in proteins was described by Burley and Petsko in 1985 (Burley, 1985). Extensive investigations about the nature and significance of aromatic interactions followed, employing theoretical (Hunter, Lawson, Perkins, & Urch, 2001) and experimental methods (Nolan, Shiels, Tuite, Cecere, & Baranger, 1999; Shiels, Tuite, Nolan, & Baranger, 2002).

Among the biomolecules, nucleic acids, owing to the controlled programmable interaction, were traditionally used for engineering nanostructures for applications in nanotechnology. The appeal for peptides and proteins intensified due to their characteristics of structural and functional versatility and specificity in molecular recognition. Also, in the biological world, many molecular mechanisms at nano, micro, and macro scales are governed by proteins and peptides. Though protein offers many advantages, certain drawbacks like (i) instability, (ii) inability to design and synthesize proteins at will and (iii) sensitiveness to environmental conditions like pH, solvent, and temperature; remain as challenges in designing bio-nano devices (Andersen, 2001). Unlike proteins, peptides are recognized as versatile building blocks for self-assembly pertaining to chemical diversity, ease of synthesis, biocompatibility, and design flexibility allowing the incorporation of functionality. Hence, self-assembled nanostructures by peptides remain as one of the most promising future materials for many reasons. Inspired by the mechanisms adopted by the biomolecules, many peptide based nanomaterials have been designed and fabricated (Gazit, 2002; Guler & Stupp, 2007; Ulijn & Smith, 2008).

3. History of aromatic stacking

Aromatic π-stacking interactions are well-known for their role in molecular recognition and self-assembly (Claessens & Stoddart, 1997; Gillard, Raymo, & Stoddart, 1997; McGaughey, Gagné, & Rappé, 1998). As the amyloid formation also involves the process of molecular recognition and self-assembly, it led to various investigations on the role of π-stacking in aggregation patterns of amyloid proteins. It was hypothesized that (i) the energetic contribution from the π-stacking can thermodynamically drive the self-assembly process and (ii) the directionality and orientation required for the ordered aggregation is provided by the specific pattern of π-stacking (Gazit, 2002). Also, analysis of the functional fragments of various aggregating sequences of amyloid peptides revealed an interesting common feature (Fig. 2). A notable frequent occurrence of aromatic residues in all the functional fragments raised the possibility of their role in the amyloid formation. With the known fact that the frequency of occurrence of aromatic amino acids in normal proteins is relatively very low, the observation suggested significant role of aromatic amino acids in the aggregating sequences.

A short functional octapeptide fragment (NFGSVQFV) of the 5.5 kDa peptide component of the aortic medial amyloid with phenylalanine in the sequence, formed fibrils as the parent peptide (Haggqvist et al., 1999).

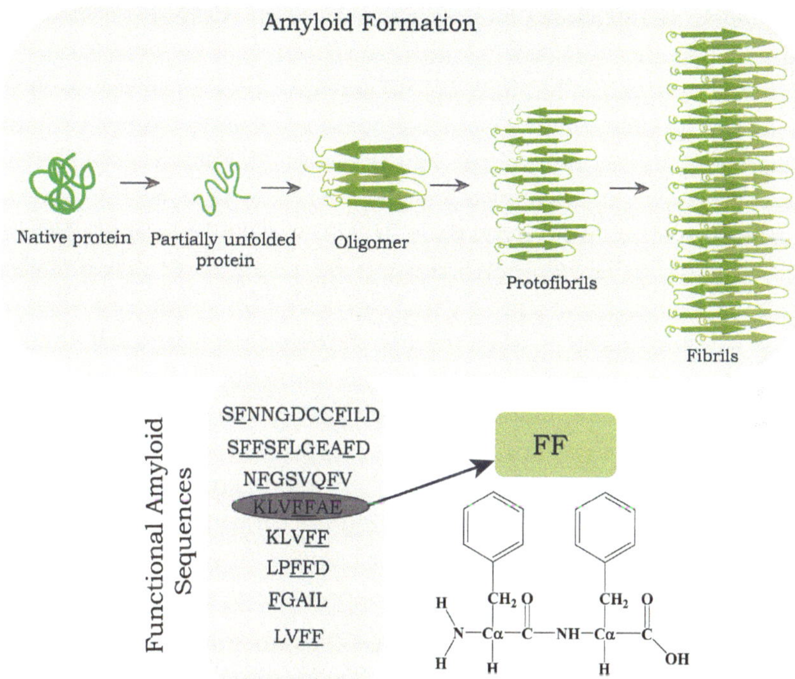

Fig. 2 Mechanism of amyloid formation through the stacking of β-sheets mediated by aromatic interactions. Common features of aromatic amino acids in different amyloid-forming sequences and how they led to the discovery of FF is shown.

Few other examples of the presence of phenylalanine residues are (i) near the carboxyl terminus of a 35 residue peptide fragment of α-synuclein (Han, Weinreb, & Lansbury, 1995), (ii) in calcitonin a short peptide which forms amyloid fibrils both *in vivo* and *in vitro* (Benvenga, Trimarchi, & Facchiano, 1994; Kedar, Ravid, & Sohar, 1976), and (iii) the amyloid fibrils formed by chromosome 13 of human BRI gene, which is associated with neuronal dysfunction and dementia (Vidal et al., 1999).

The first striking examples of the aromatic amino acids was observed in islet amyloid polypeptide (IAPP), or amylin (Kahn, Andrikopoulos, & Verchere, 1999). Shorter fragments of 6(NFGAIL) and 5(FGAIL) residues formed ordered amyloid fibrils, whereas, the sequence GAIL did not form any fibrils (Tenidis et al., 2000). Experiments like near-UV CD spectrum (Kayed et al., 1999) and alanine scan of the basic amyloidogenic unit of IAPP (Azriel & Gazit, 2001) demonstrated the physicochemical properties of the three short residues and the significant role of phenylalanine as a structural element.

Another example is the minimal fragment of the Alzheimer's β-amyloid (Aβ) polypeptides (QKLVFF) containing two phenylalanine residues. The short peptide specifically binds and inhibits the amyloid formation of the full-length Aβ polypeptide (Tjernberg et al., 1996). A few other inhibitors like LVFFA and its derivatives (Findeis et al., 1999) and LPFFD (Soto et al., 1998) also inhibited the Aβ peptide. Another study on the sequence KLVFFAE, derived from Aβ peptide forming well-ordered fibrils (Balbach et al., 2000), further confirmed the role of the FF motif as a molecular recognition element in the ordered amyloid fibril formation (Fig. 2). Diphenylalanine happens to be the first shortest dipeptide to self-assemble into rigid nanotubular structures (Reches & Gazit, 2003). Being recognized as the core recognition motif to drive self-assembly in the β-amyloid polypeptide, the dipeptide (FF) has been extensively studied. Owing to the different nanostructures formed, and physical properties exhibited by the FF dipeptide, excellent review articles on the properties and applications of diphenylalanine have been published (Silva, Araújo, Silva, Ando, & Alves, 2013; Yan, Zhu, & Li, 2010).

4. Mechanism of amyloid formation through aromatic π–π stacking

Understanding, the mechanism of aromatic stacking interactions, is of prime importance in peptide engineering, because it allows the rational design of peptides targeting specific applications. Each of the stacking interactions (Fig. 3), with precise planar orientation, regulates the well-ordered formation of the amyloid structures.

The suggested mechanism in stacking interactions is a stepwise assembly process. In the first stage, only two structural elements containing aromatic amino acid residues form a bimolecular structure. In the following stages, monomers with the same recognition elements are further added in a stepwise process. From the initial formation of the biomolecular structure to the stepwise addition of the monomers, the process is directed by the restricted geometries of the stacking interactions. The mechanism associated with π–π stacking plays a crucial role in the supramolecular association as it forms highly stable self-assembled structures with the desired precision.

4.1 The geometry of aromatic stacking

In protein folding, based on the interactions involved, aromatic rings can either serve as acceptors or donors in hydrogen bonding (Ciunik, Berski, Latajka, & Leszczyński, 1998). However, the most abundant and important interactions

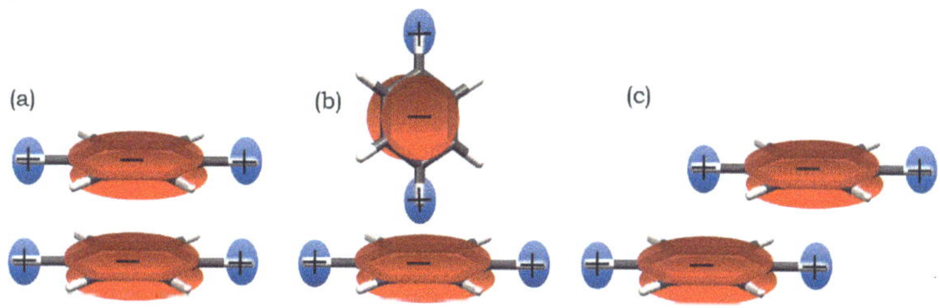

Fig. 3 Types of π–π stacking observed in aromatic systems. (A) Sandwich (B) T-shaped, and (C) parallel displaced.

involving aromatic moieties are the ones with other aromatic rings (McGaughey et al., 1998). Burley and Petsko reported that, aapproximately 60% of such interactions are aromatic pairing interactions. The preferential distance and dihedral angles for aromatic pairing interactions are reported to be in the distance range 4.5–7 A°, measured between the centroid of the phenyl rings and an angle of approximately ~90° (Burley, 1985; McGaughey, Gagné, & Rappé, 1998). Among the abundant aromatic contacts in peptide structures, at least one contact is reported to have an interacting distance less than 6 A°. Phenyl rings that make no contact with the other aromatic rings, also known as isolated phenyl rings with a centroid separation distance of 7 A° constitutes 5% in all the studied structures. The geometry of the parallel displaced interactions of the aromatic pairing groups due to the translation along the short crystallographic axis can be calculated using the vertical and horizontal offsets of the aromatic rings. The representation of the calculations in the aromatic rings was first reported by Gorbitz (Görbitz, 2012). The horizontal and vertical offsets of the aromatic rings in parallel displaced contacts are calculated using C sin θ, and C cos θ, described by H values in the range of 3.0–6.0 A° and V in the range 1.5–3.5 A° (Fig. 4).

The situation H = 0 A°, representing the direct stacking of amino acids on the top of each other, never occurred in the crystal structures, thus verifying the theoretical *ab initio* calculations with the benzene dimer (Van Der Avoird et al., 2010). Interactions with V = 0 A° was observed in very few cases that had distinct electrostatic characteristics. At conditions where the aromatic rings are not related by translation, the angle between the two aromatic rings may vary between 0° (parallel) −90° (perpendicular), and there will not be any correlation between the centroid distance and plane angle. In short, aromatic interactions with parallel displaced geometry, H-atoms will not be placed on the top of the aromatic ring, but displaces in such a way that leads to H••• C (π) interactions within a contact distance of 2.9 A°.

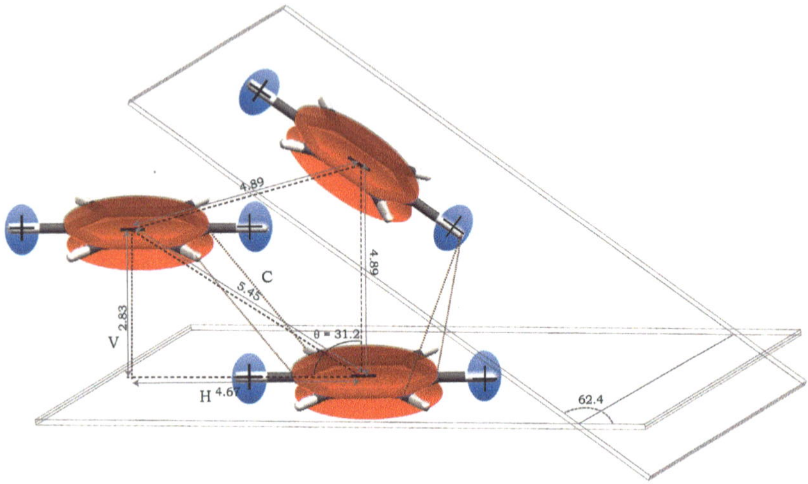

Fig. 4 An example of the different types of stacking interactions, as observed in the structure of cyclo (*N*-methyltaurine-Phe-D-Pro) (Van Der Avoird et al., 2010). Illustrated in the figure is the vertical (V) and horizontal (H) offset interactions (H), the centroid (C), and H•••C distances represented with *dotted lines* between the phenyl groups. The H and V values in the range of 3.0–6.0 A° and 1.5–3.5 A° are calculated using C sin θ, and C cos θ.

Structural reorganization in peptides by specific interactions, including aromatic stacking, can turn the ability of the peptide from forming aggregates to fibrillar structures. The interactions also incorporate the ability to recruit and orient new peptides acting as seeds of the fibrillogenic process. Hence, understanding the mechanism of such interactions through peptides will be central in developing strategies in the possible prevention and treatment of the diseases. Also, the peptides with inherent physical properties and as templates in fabricating materials can serve as ideal models for the design and synthesis of peptides targeting different applications.

5. Strategies adopted and types of building blocks for the construction of bioorganic nanoassemblies

Peptides can adopt different configurations based on their repeating units, and it has been the focus of scientific research for the construction of functional nanoassemblies. Upon stereochemical perturbation of the backbone, amino acids other than Gly can exist in both L and D configuration, and interacts through non-covalent interactions like ionic, hydrophobic, H-bonding, and π–π stacking to form supramolecular structures.

With the possibility of 2^n configurations (n = number of amino acids), the possible sequences are enormous. In general, the design strategies of peptides can be catergorised in two ways (i) natural and (ii) non-natural. Natural design is a frequently encountered design strategy by mimicking the characteristics of the biological assemblies in nature. Non-natural *de novo* designs of peptides are either linked through covalent or non-covalent interactions (Fig. 5). The simple peptide structures with different design strategies can help us in better understanding the complex mechanisms underlying the inter-molecular assembly of biological systems. The important role of secondary structures and the design parameters in directing the aromatic π–π interactions in the self-assembly of molecules is discussed in the following section.

5.1 Intermolecular associations: Concepts derived from nature

The limitations in the top-down approach have provided an alternative route of "bottom-up" strategies in the fabrication of nanomaterials. Initial attempts of man-made designs in the bottom-up process using peptides were not much successful. The multiple functions of the biomolecules incorporated created chaos by interfering with each other and generated unexpected functional outcome. In the vast amount of possible peptide sequences, nature has utilized only a fraction of the available chemical space. Also, the important design feature of nature is the hierarchical self-assembly of biomolecules into highly ordered functional supramolecular architectures. Hence, researchers started mimicking the natural designs of functional nanomachines, which are fabricated with high degree of organization and efficiency in accomplishing the designed tasks. The naturally occurring structures in the proteins, such as α-helices and β-sheets, were utilized to drive the self-assembly process (Panda & Chauhan, 2014).

5.1.1 Symmetry

Evolution in nature has created a range of nano-objects self-assembled by bottom-up approach. Among all the concepts adopted by nature, the concept of symmetry offers many functional advantages in the biomolecules. A few examples of symmetry observed in biomolecules are (i) Point group symmetry, further classified into the cyclic, dihedral and cubic symmetry, (ii) translation symmetry, and (iii) Quasi-symmetry (Goodsell, 2004).

Inspired by the concept of symmetry in directing self-organization in biological systems, symmetry along with aromatic π–π interactions, and end to end electrostatic interactions have been used as design variables in

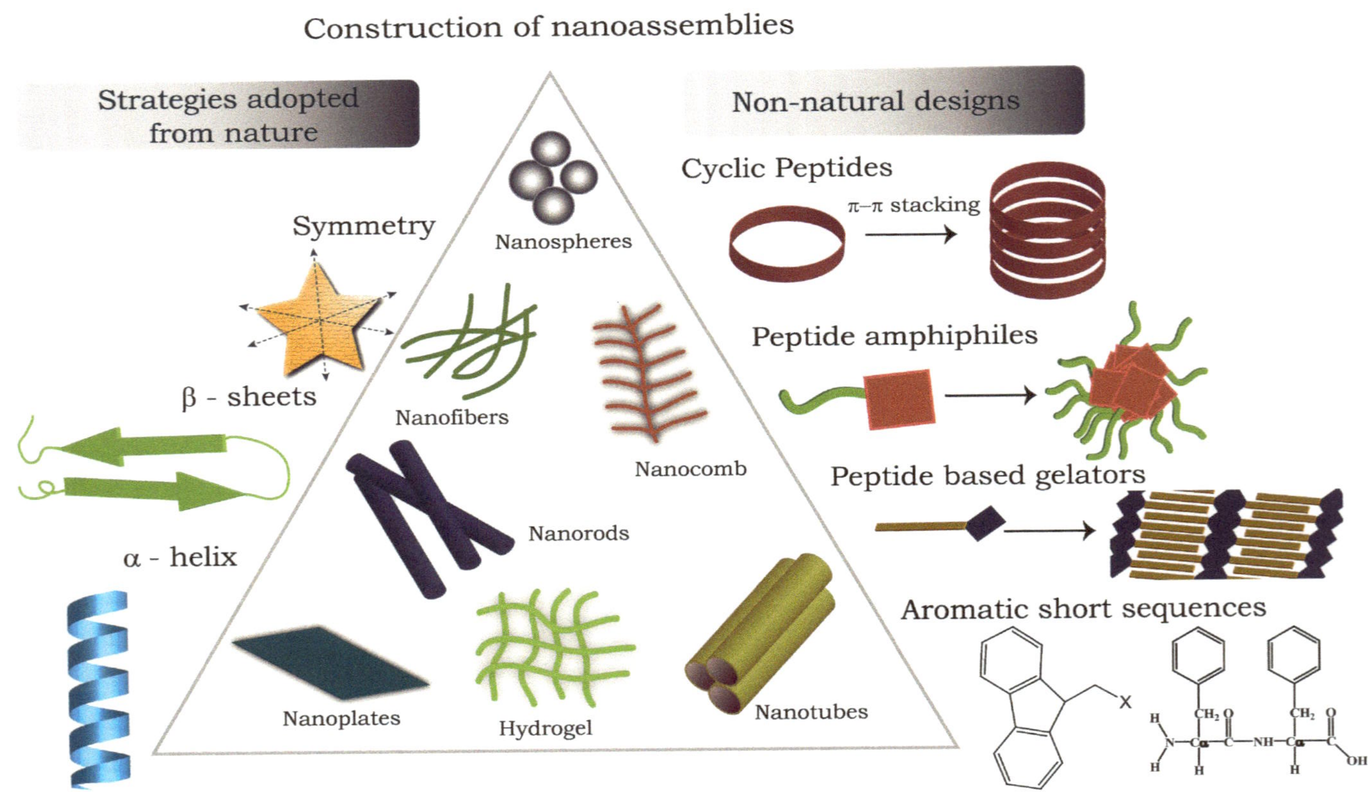

Fig. 5 Various strategies employed for the design of peptide-based nanoassemblies.

six peptide model systems by Ramakrishnan and coworkers. Peptide systems with the same electrostatics and symmetrical elements tend to form similar crystalline nanotube like structures. When the symmetrical element, along with the distance between the aromatic rings of the basic units are changed, it resulted in a nano comb-like architecture. The hypothesis that asymmetrical elements will prohibit the epitaxial growth of supramolecular assemblies was evident with the peptide model system having asymmetrical design forming no pronounced structural assembly. The peptide system designed based on the gramicidin helix with both intramolecular as well as intermolecular interactions, assembled to form nanofibers (Sasidharan, Shyni, Chaudhary, & Ramakrishnan, 2017). The observations from the experiments suggest that symmetry, along with charge distributions, can be used as a design variable for the design and modulation of nanoassemblies.

5.1.2 Secondary structures from proteins

5.1.2.1 β-Sheets

β-sheet structures contain parallel and antiparallel strands with an extended backbone, that permits hydrogen bonding between the backbone amides and carbonyls. The arrangement of C and N-termini of the beta strands (on the same side-parallel and alternate sides-antiparallel) has an important impact on the H-bonding, side chains, and orientations in the structure. From the literature studies, β-Sheets forming peptides both natural or synthesized ones are known to dominate α-helical peptides as they can form long-range interactions. β-sheets, well known for their ability to form ordered amyloid fibrils in amyloid diseases, was initially identified by Pauling, Corey, and others in the 1950s (Pauling & Corey, 1951). It was Zhang in 1900, who first demonstrated the use of β-sheets in the design of materials (Zhang, Holmes, Lockshin, & Rich, 1993). Self-assembled β-sheets formed a hierarchy of structures such as tapes, ribbons, fibrils, and fibers (Aggeli et al., 2001). Studies were reported on the systematic variation of peptide sequences containing a different number of hydrophobic residues, and their position helped in finding the key features like the critical concentration of aggregation in the peptides (Caplan, Schwartzfarb, Zhang, Kamm, & Lauffenburger, 2002). The hypothesis of aromatic stacking in the β-sheets of amyloid proteins (Chiti and Dobson, 2006), further inspired the design of a series of β-sheets forming peptides interacting through aromatic π-π stacking. However, the main challenge faced in designing a self-assembled system mediated through β-sheets is to control the assembly to form a uniform and reproducible architecture.

5.1.2.2 β-Hairpins

β-Hairpins are short sequences containing two short β sheets linked by a turn sequence. When the terminal amino acids in the β sheets do not interact with each other, β-hairpins ensures that the peptides stack in register with one other. A biological motif, peptide β-sheet, rationally designed with sequences interacting through aromatic interactions formed macroscopic hydrogels with nanoscale order (Aggeli et al., 1997). A C3-symmetric peptide conjugate, Trigonal-WTW, was synthesized based on the intramolecular tryptophan zipper (β-hairpin) forming peptide reported by Cochran et al. At a pH 11, the peptide self-assembled to form fibrous structures due to stacking interactions between the hydrophobic core (Matsuura, Hayashi, Murasato and Kimizuka, 2011). β-hairpins are also known to play an important role in amyloid self-assembly (Maity, Hashemi, & Lyubchenko, 2017).

5.1.2.3 α—helix

First described by Pauling, Corey, and Crick (Crick, 1953; Pauling & Corey, 1951, 1953), α-helical structures were used as components of coiled-coils, as an alternative to the usage of β-sheets for the formation of fibrous structures. With the aid of hydrogen bonding between backbone amides, and with a periodicity of 3.6 residues per turn, the α-helical peptides tend to stack on top of each other, to form extended columns. The extended structures are formed by the side chains of amino acids of the α-helical peptides, protruding outside from the helix to form interactions with other helices. Though it is challenging to produce α-helical structure in practice, owing to the longer length of amino acids required for making stable interactions, a few examples involving aromatic–aromatic interactions have also been reported (Aravinda et al., 2003).

The approach of designing peptides, though afforded a significant control over the critical features of the resulting nanoarchitectures, it lagged in terms of the development of functional nanoassemblies. The significant challenges faced in the natural peptide-based design has led to the development of the non-natural designs for the integration of more complex functionalities.

5.2 Non-natural peptide design

Non-natural peptides are designed based on the insights gleaned from natural proteins, or they are entirely new designs based on the peptide derivatives. For example, surfactant-like peptides are inspired by the lipid bilayers

(Zhao & Zhang, 2006), and aromatic peptides are derived from amyloidogenic sequences (Gazit, 2002; Meital & Ehud, 2006; Reches & Gazit, 2003, 2004). Several peptide-based building blocks with short sequences such as cyclic peptides, dendritic peptides, amphiphilic peptides, and aromatic peptides have been utilized for creating functional nanostructures.

5.2.1 Cyclic peptides

Cyclic peptides usually contain alternating L and D amino acids, and the building blocks stack upon each other to form a flat conformational structure. The final self-assembled structures are stabilized through H-bonding, which are arranged perpendicular to the ring. In peptide crystallization, cyclic peptides are introduced to address the challenges of conformational flexibility in peptides. Cyclic peptides have distinct properties like precise diameter control and functionalization through amino acid sequences. Among cyclic peptides, a novel family of aromatic-bridged cysteine cyclic peptides, with locked-in and well-defined secondary structures, are known to interact through aromatic interactions, can act as a model for studying the biological reactions. The peptides, also known as cystinophanes, had a general structure of cyclo(Ar-CONH-Cyst-NHCO−)$_n$ and cyclo (Ar-CONH-Xaa-Cyst-Xaa-NHCO−) (Ar=Ph or Pyr; Cyst=l-cystine dimethyl ester; n=2, 3, 4, and Xaa=amino acid X). The crystallographic analysis revealed the near-parallel face-to-face orientation of the two phenyl rings in the cyclic peptides (Ranganathan, Haridas, & Karle, 1998). Another group of peptides called cyclodepsipeptides, with alternating repeats of aromatic and Ser units stacks on the top of each other through π–π interactions between the aromatic groups like phenyl or pyridyl and H-bonding to form tubular structures (Ranganathan, Haridas, Gilardi, & Karle, 1998). Designs with cyclic peptides offers distinct properties like precise diameter control and functionalization through amino acid sequences.

5.2.2 Peptides with unnatural amino acids

Peptides, despite their significant potential, face challenges such as unstable conformation and low stability in the biological medium, and introducing unnatural amino acids in the sequence plays a crucial role in overcoming these problems.(Ramakrishnan, Ranbhor, Kumar, & Durani, 2006; Ranbhor, Ramakrishnan, Kumar, & Durani, 2006) Unnatural amino acids, when introduced in the peptide backbones, reduces the conformational flexibility of the peptides (Hazam, Akhil, Jerath, Saikia, & Ramakrishnan, 2019; Hazam, Goyal, & Ramakrishnan, 2019; Hazam, Phukan, Akhil,

Singh, & Ramakrishnan, 2021; Kumar, Ramakrishnan, Ranbhor, Patel, & Durani, 2009; Mehra, Jerath, Ramakrishnan, & Trivedi, 2015; Pandey, Morla, Kumar, & Ramakrishnan, 2020; Ranbhor et al., 2006). Also, it has a high tendency to adopt predetermined secondary structures and enhances metabolic stability (Hazam, Jerath, Kumar, Chaudhary, & Ramakrishnan, 2017; Ramakrishnan et al., 2006). Aib (aminoisobutyric acid or B), a non-proteinogenic α-amino acid, is well-known known for inducing helix formation in peptide sequences (Prakash, Ranbhor, & Ramakrishnan, 2020). The problem of conformational flexibility arises with the increase in the length of the peptide sequences, and unnatural amino acids like Aib are introduced in the sequences to address this issue. To understand the importance of aromatic π–π interactions in stabilizing helices and β-hairpins, three sets of peptides with phenylalanine at different positions were synthesized and crystallized. To restrict a helical conformation, Aib was incorporated at different positions in the peptide sequences. Structural determination of the crystals revealed that the peptide adopted helical conformations, and the aromatic rings have interacted through perpendicular, parallel-displaced, and inclined orientations (Aravinda et al., 2003). Ff ('f' stands for D-Phenylalanine) peptides forms nanotubular structures, and when Aib is introduced between diphenylalanine (FBf) peptide it restricts the conformation of the peptide in the α-helical region to form nano-comb like architectures (Sasidharan, Hazam, & Ramakrishnan, 2017). The approach of using non-proteinogenic amino acids will bring in enormous diversity in the basic building blocks of the assembled units.

5.2.3 Peptide amphiphiles

Inspired by the association and structural integrity of the lipid membranes, amphiphilic peptide molecules with charged hydrophilic head groups and hydrophobic tails were designed. Among the different types of peptides, aromatic peptide amphiphiles are considered to be a distinct class of building blocks. The self-assembly of the aromatic amphiphiles through parallel, antiparallel, and interlocked antiparallel stacking conformations are governed by the structural features of the four segments present in them. A typical aromatic amphiphile is comprised of (i) an N-terminal aromatic component (ii) a linker segment (iii) peptide sequence and (iv) a C-terminus.

The process of amphiphilic self-assembly can be controlled by modifying the aromatic component and by changing the external conditions like presence of ions, pH, temperature, and kinetics. The aromatic moieties impart both amphiphilicity and structure to the directed self-assembly in the aromatic stacking interactions. Further, the process of self-assembly is also influenced by the planarity and geometry associated with stacking interactions. Based on the structural trends and design rules, the resulting nanostructures can be fabricated, targeting a wide range of applications. Owing to the versatility in peptide design and its range of applications, excellent review articles dedicated to the design aspects and different types of stacking conformations in peptide amphiphiles are reported (Fleming & Ulijn, 2014).

5.2.4 Aromatic short peptide derivatives

Investigation on the reduced minimal aggregate of the amyloid protein revealed that, the stacking role of aromatic dipeptide diphenylalanine may play a key role in fibril formation (Gazit, 2002). Diphenylalanine, the core recognition motif of β amyloid sequence self-assembles through π–π stacking of the phenyl rings to form extremely rigid (Kol et al., 2005), discrete, hollow nanotubes with high persistence lengths and were cast into nanowires (Reches & Gazit, 2003). Experimental conditions strongly influence the aromatic stacking interactions and the self-assembly of FF nanotubes to form diverse nanostructures. For example, the nanotubes were aligned vertically to form "nano-forest" and were coated with magnetic nanoparticles to align horizontally in the presence of an external magnetic field (Reches & Gazit, 2006). Modulation of the biological stability of the peptide nanostructures were achieved by the incorporation of L and D-amino acids. Peptide nanowires designed with D chiral amino acids were found to be stable against proteolytic degradation (Reches & Gazit, 2003).

The versatility of the FF peptide has led to significant research progress towards understanding the design rules in short aromatic sequences, and their role in dictating nanoassemblies. A series of variants of FF peptides have been designed, synthesized, and many physicochemical characterizations were performed (Meital & Ehud, 2006; Reches & Gazit, 2004, 2010; Yan et al., 2010). FF-based peptides formed different nanostructures such as nanotubes, nanoforests, spherical vesicles, nanofibrils, nanorods, and ordered molecular chains under different conditions like vapor deposition, dilution and water-vapor-mediated self-assembly (Adler-Abramovich et al., 2009; Ryu & Park, 2008; Yan et al., 2008). A few analogues of diphenylalanine forming diverse nanostructures are listed in Table 1 (Reches & Gazit, 2003). Discussing, all the variants of the FF peptides, is out of the scope of this chapter, and many review articles discussing the variants of FF based peptides have already been published (Yan et al., 2010).

Table 1 Analogues of diphenylalanine and its structures.

Peptide	Solvent	Nanostructure
NH_2-Phg-Phg-COOH NH_2-Cys-Phe-Phe-COOH	HFIP-water	Nanospheres (Reches & Gazit, 2004)
Fmoc-Phe-Phe-COOH	DMSO-water (Dudukovic & Zukoski, 2015) HFIP-water (Mahler, Reches, Rechter, Cohen, & Gazit, 2006)	Hydrogel
Fmoc-Phe-Phe-COOH (Reches & Gazit, 2010) Cbz-Phe-Phe-COOH (Reches & Gazit, 2010) NH_2-(D-1-Nal-)$_2$-COOH (Meital & Ehud, 2006)	HFIP-water	Fibrils
(D-Phe-D-Phe)	Water	Nanotubes +Vesicles (Song et al., 2004)
(L-Phe-D-Phe)	HFIP-water	Nanotubes (Sasidharan, Hazam, & Ramakrishnan, 2017)
Ac-Phe-Phe-NH_2 NH_2-Phe-Phe-NH_2 Boc-Phe-Phe-OH Cyclo-Phe-Phe NH_2-(D-2-Nal-)$_2$-COOH (Meital & Ehud, 2006) NH_2-(p-fluoro-Phe)$_2$-COOH (Meital & Ehud, 2006)	HFIP-water	Nanotubes (Reches & Gazit, 2010)
NH_2-(p-nitro-Phe)$_2$-OH	HFIP-water	Fibrils + spheres (Meital & Ehud, 2006)
(NH_2-Phe-Aib-Phe- COOH)	HFIP-water	Nanocombs (Sasidharan, Hazam, & Ramakrishnan, 2017)
NH_2-(pentafluoro-Phe)$_2$-COOH NH_2-(p-iodo-Phe)$_2$-OH	DMSO-water	Nanotubes (Meital & Ehud, 2006)
NH_2-(4phenyl-Phe)$_2$-COOH	HFIP-water	Square plates (Meital & Ehud, 2006)
NH_2-D-Val-Phe-Phe-COOH (Marchesan, Easton, Kushkaki, Waddington, & Hartley, 2012) NH_2-D-Phe-Phe-Val-COOH (Marchesan, Easton, et al., 2012) NH_2-D-Leu-Phe-Phe-COOH (Marchesan et al., 2012)	pH switch ~ (12 to 7.4 in Sodium phosphate buffer)	Hydrogels

The molecular basis for the formation of nanostructures from small peptides containing aromatic moieties stems from the geometric restrictions between the aromatic groups. Introducing a variety of short aromatic groups, like carbobenzyloxy, naphthalene, anthracene, and fluorenylmethoxycarbonyl (Fmoc) with peptides, through chemical coupling, promotes aromatic π–π stacking to form nanoscale architectures, primarily directed by local electrostatics (Yang, Liang, Ma, Gao, & Xu, 2007). A few among the Fmoc–dipeptides reported are gelators driven by H-bonding and π–π interactions (Fleming & Ulijn, 2014). Diphenylalanine, when replaced and substituted with functional groups like naphthalene and halogen atoms, imparts properties like conductivity and rigidity (stiffness) to the nanostructures (Meital & Ehud, 2006). Other than FF, many short linear peptides, like KLVFF (Krysmann et al., 2008), DFNK and DFNKF were designed, based on aromatic interactions guiding the self-assembly forming nanofibrillar structures (Reches, Porat, & Gazit, 2002).

The aromatic short peptide derivatives were considered promising candidates from a scientific point of view, due to their versatility in forming different functional nanoassemblies and structural modulation with external stimulus. The advantage in the modulation of the peptide, manifested in their emergent nanostructures, and its properties, find applications in areas like bioimaging, biosensors, guest encapsulation, nanofabrication, and drug delivery (Yan et al., 2010).

6. Nanostructures formed through aromatic π–π interactions

Self-assembly has enabled the generation of supramolecular nanostructures with high order and complexity. Among the diverse materials, self-assembling peptides serve as potential building blocks for generating novel nanostructures with tailored features. Identifying sequences in which tailor-made modifications can be introduced to influence the intermolecular aromatic π–π stacking can give rise to structures with desirable properties. Based on the design, basic constituents, and experimental conditions, peptides can self-assemble into an array of nanostructures.

6.1 Nanotubes

Naturally existing tubular proteins, such as motor proteins, ion channels, and aquaporins, had inspired scientists to design artificial nanotubular structures. The hollowness of the nanotube that can be functionalized through chemical modifications has driven the synthesis and fabrication of many artificial nanotubes from a variety of biomolecules. Peptide-based nanotubes with the property of molecular recognition has gained significant importance and can be classified based on the building blocks.

6.1.1 Cyclic peptide nanotubes

In 1974, De Santis and coworkers theoretically predicted that D- and L-amino acids could self-assemble to form nanotubes by stacking cyclic peptide monomers (De Santis, Forni, & Rizzo, 1974). However, the first cyclic peptide nanotubes were engineered and developed by M. Ghadiri and coworkers in 1993 (Ghadiri, Granja, & Buehler, 1994; Ghadiri, Granja, Milligan, McRee, & Khazanovich, 1993). The cyclic peptides with alternating D- and L-amino acids were designed to form a planar ring that stacks one on top of the other, to form tubular structures. In another interesting work from the same group, a 1,4,5,8-naphthalene tetracarboxylic acid diimide (NDI) group was incorporated in the cyclic peptides to allow a charge transfer along with the tubular system (Horne, Ashkenasy, & Ghadiri, 2005). The diameter and function of the cyclic peptide nanotubes can be precisely controlled by tuning the length and sequence of the peptides. The surface properties of the nanotubes can be determined by the number and amino acid group in the side chain. The precisely tunable properties are the characteristic features of the cyclic peptide nanotubes distinguishes them from other self-assembled nanostructures. The features of these cyclic peptides was the driving force for designing and developing such sequences forming stable nanostructures (Chapman, Danial, Koh, Jolliffe, & Perrier, 2012).

6.1.2 Self-assembled peptide nanotubes

Nanostructure formation in dipeptides was first reported by Görbitz (2006). Phenylalanine dipeptide (FF), was reported as the structural motif in the β-amyloid peptide, responsible for Alzheimer's disease (Yan et al., 2010). In the self-assembly process, six FF units of the aromatic dipeptide come assemble to form a cyclic hexamer. The cyclic hexamers stack themselves through hexagonal packing to produce narrow channels leading to the formation of sheets, and the sheets coils themselves to form nanoscale tubes with external hydrophobic walls. The nanotubular structure self-assembles at larger scales to form bundles and are held together by the backbone hydrogen bonds and π–π interactions between the side chain aromatic rings (Silva, Araújo, Silva, Ando, & Alves, 2013). The molecular organization of the remarkably rigid peptide nanotubes was further validated by the crystal structure reported by Gorbitz, where the hierarchical array of stacked diphenylalanine molecules in a surrounding water cluster tightly holds the hydrogen-bonded peptide main chains in nanotubes (Görbitz, 2001, 2006; Silva et al., 2013). Further, the dynamics of the nanotube assembly, demonstrating that the self-assembly process is a nucleation-dependent process was performed by Gazit et al. using NMR spectroscopy. In the study, during nanotube formation, diphenylalanine molecules migrated out of their water solvation shells to form aggregates. It was reflected by a decrease in the integral of the diphenylalanine signal and a concomitant increase in the integral of the water signal (Carny, Shalev, & Gazit, 2006).

The molecular organization of FF is compared to that of graphene and other organic polymers (Sasidharan, Shyni, et al., 2017). Self-assembly pattern is similar to graphene as it forms a flat two-dimensional sheet stabilized by the interactions of aromatic ringside chains, which directs the order during the assembly by π–π stacking. A model of FF nanotube assembly also suggests that self-assembled nanotubes are similar to fullerene and carbon nanotubes. The robustness and the stability of the peptide nanotubes have been tested under extreme physical and chemical conditions like boiling, autoclave treatment, and exposure to various organic solvents (Adler-Abramovich et al., 2006). Two different methods demonstrated the alignment of the nanotubes in vertical and horizontal patterns. By solvent evaporation on siliconized glass, vertically aligned nanoforests were formed by axial unidirectional growth of a dense array of dipeptide nanotubes. And the horizontal alignment was achieved by coating the nanotubes with ferrofluid and exposing them to an external magnetic field (Reches & Gazit, 2006).

Modification of FF nanotubes with Boc, fluorine and amine groups also resulted in tubular nanostructures (Meital & Ehud, 2006). On replacing the phenyl with naphthalene group, a promising molecular conductor, di-D-2-Nal dipeptides formed tubular structures (Sedman et al., 2013). Inorder to study the role of electrostatic interactions in the formation of tubular structures, two different peptide sequences with N and C terminal modifications (Ac-Phe-Phe-NH_2 & NH_2-Phe-Phe-NH_2) were analyzed. Both the sequences self-assembled to form nanotube like structures further validating the hypothesis of the role of aromatic π–π interactions in amyloid formation(Reches & Gazit, 2010). A seven residue peptide CH_3CO-KLVFFAE-NH_2, the core segment of Aβ peptide, with a capped N- and C-termini, also self assembles into parallel β-sheet and forms micron-long highly homogeneous hollow tubes at pH2. By using the "salting out" technique, Lynn and coworkers directed the same peptide from homogeneous nanotubes to macroscale parallel arrays (Lu et al., 2007; Lu, Jacob, Thiyagarajan, Conticello, & Lynn, 2003). With such remarkable properties, further research is necessary to understand the dynamics and equilibria of the one-dimensional assemblies.

6.2 Nanofibers

Nanofibers, the fibrillar nanostructures with large aspect ratios, and specific surface area find applications in various fields such as biomedicine, sensors, electronic nanodevices, and catalysis. The formation of nanofibrils were initially reported in various analogues of the amyloid-forming sequences (Reches et al., 2002). The N-terminus of diphenylalanine, the shortest sequence derived from Aβ peptide on modification with aromatic groups such as Fmoc and CbZ, resulted in nanofibrils, with morphologies similar to that of amyloid fibrils (Meital & Ehud, 2006). Diphenylalanine, when replaced by naphthalene group, and repeating the experimental conditions similar to FF, the sequence (D-1-Nal)-(D-1-Nal), formed fibrils of approximately 10 nm in diameter (Meital & Ehud, 2006). In addition to the designed sequences, on varying the experimental conditions like solvents, pH, *etc.*, it resulted in fibrillar structures. For example, on introducing acetonitrile as a co-solvent, a structural transition of self-assembled diphenylalanine peptide from microtubes to uniform peptide nanofibers was observed (Huang, Qi, Su, Zhao, & He, 2011). In a similar study, a temperature-induced and substrate assisted transition from an organogel to uniform crystalline nanowires was also demonstrated (Huang, Wang, Qi, Su, & He, 2014). The aromatic contribution in self-assembly was studied by substituting them with aliphatic hydrophobic amino acids in the central domain of the multidomain peptides. However, the resulting nanofibrillar morphology remains unaffected, switching interaction from the antiparallel H-bonding to parallel H-bonding was observed. Also, the selection of amino acids affecting the non-covalent interactions played a crucial role in determining the length and quality of the nanofiber, which affected the rheological properties of the nanofibers (Bakota, Sensoy, Ozgur, Sayar, & Hartgerink, 2013). Hence the choice of amino acids influencing the aromatic stacking interactions is crucial in tailoring the properties of the nanofibrous structure.

6.3 Hydrogels

Gelation is a process to form molecular gels through weak interactions, using the designed peptide-based gelators. On inducing molecular chirality, the gelators are endowed with enhanced gelation capability, which in turn bestows it with chiral functions such as a chiral switch, sensing, and recognition. Peptide-based supramolecular hydrogels, an essential class of biomaterials are increasingly explored as it holds promise in a range of biological and non-biological applications. The process of gelation in the peptide systems can be controlled by alterations in the peptide sequence, concentration, and pH conditions. The role of aromatic amino acids in promoting self-assembly and improving gelation characteristics through π–π stacking and hydrophobic interactions have been extensively studied.

Diphenylalanine, one of the smallest peptide gelators, self-assembles into long organic fibrils in organic solvent systems, and entangles further, to form gels (Yan, Cui, He, Wang, & Li, 2008). On the addition of aromatic groups, carbobenzyloxy naphthalene, or fluorenyl-methoxycarbonyl (Fmoc) to the N-terminus of peptides, at preparative conditions forms stable hydrogels (Jayawarna, Smith, Gough, & Ulijn, 2007; Mahler et al., 2006; Saikia et al., 2021; Smith et al., 2008; Toledano, Williams, Jayawarna, & Ulijn, 2006; Yang, Gu, Zhang, Wang, & Xu, 2004). The results represent the role of different aromatic moieties in controlling the stacking effect. Diphenylalanine peptide protected by different aromatic stacking ligands, self-assemble into nanofibrous hydrogels of different dimensions *via* interlocked β-sheets or π-stacks. The effect of variation in the stacking group has reflected in the curvature, branching, and diameter of the resulting fibrous components (Jayawarna et al., 2007). A β-sheet peptide was designed based on a sequence related to the transmembrane domain of the IsK protein. The peptide self-assembled in methanol through aromatic interactions to form hydrogel with polymeric β-sheet tapes (Aggeli et al., 1997). The peptide Ac-C(FKFE)$_2$CG-NH$_2$, on cyclization through the disulfide bonds on the flanking cysteine residues, self-assembles to form fibrillar structures of β-sheet conformation. On increasing the concentration, the peptide formed rigid viscoelastic hydrogels (Bowerman & Nilsson, 2010; Jayawarna et al., 2006). Several supramolecular hydrogelators reported from small molecules are used for therapeutic applications (Zhao, Ma, & Xu, 2009). Implementing the principles of hydrogelator, a vast amount of work has been done on the "self-delivery systems for therapeutic purposes" (Fleming & Ulijn, 2014), and FDA has approved one hydrogelator for clinical use (Valéry et al., 2003).

The π electrons in the aromatic fluorenyl rings play a crucial role in the π stacking and act as a driving force to self-assemble a series of Fmoc dipeptides into hydrogels. Ulijn et al. reported a series of Fmoc dipeptide forming hydrogels and their application as biomimetic fibrous scaffolds for three-dimensional cell culture (Fleming & Ulijn, 2014). Non-covalent interactions responsible for the formation of hydrogelators, being dynamic, enables the three-dimensional networks to respond to external stimuli. Enzyme regulated hydrogelation of a series of Fmoc peptides endows the hydrogels for potential applications in drug delivery, detecting the activity of enzymes, typing bacteria, and screening of enzyme inhibitors (Yang, Liang, & Xu, 2008). Alteration in the pH, concentration, and charge distribution of the peptide can result in varying hydrogels with the stiffness that can be fine-tuned. The tunable morphology and material properties make hydrogels, as effective components with promising applications in fields like tissue engineering, 3D printing and injectable delivery vectors (Yan & Pochan, 2010).

6.4 Nanospheres

Nanovesicles, the spherical structures are potential candidates for encapsulation and delivery of drugs, biomolecules, and other chemical agents. For the sustained release of encapsulated materials, efforts have been made to develop smart vesicular systems responding to external stimuli like pH, temperature, and enzymatic conditions. Diphenylalanine known to form nanotubes, through aromatic π–π interactions, on dilution at pH 7, transformed into spherical vesicular structures (Yan, Cui, He, Wang, Li, et al., 2008). In another experiment to delineate the role of aromatic residues, a systematic replacement, and shuffling of FF sequences were attempted with cyclohexylalanine, and proline yielded nanovesicular structures (Mishra, Panda, Basu, & Chauhan, 2008). For example, on introducing a thiol group in diphenylalanine (CFF), a change in the nano assembly was observed from nanotubes to nanospheres. An aromatic peptide diphenylglycine, with molecular properties similar to that of diphenylalanine, but with a rigid molecular structure due to the lack of rotation around C – C bond and steric hindrance, self-assembled to form nanospheres (Reches & Gazit, 2004). When diphenylalanine modified with nitro and phenyl group-, di-para-nitro-Phe peptide, at a higher concentration of $10\,mg\,ml^{-1}$ form spherical structures (Meital & Ehud, 2006). Two sets of amphiphilic dipeptides α,β-dehydrophenylalanine (ΔPhe), H-Glu-ΔPhe-OH and H-Lys-ΔPhe-OH also self-assembled through aromatic interactions to form nanovesicles. The structures resistant to proteinase K were shown encapsulating small drug molecules such as riboflavin, vitamin B12, bioactive peptides, and small protein molecules (Mishra et al., 2008).

7. Factors controlling stimulus-response and the formation of nanoassemblies

Peptides with different functional groups are susceptible to changes involving non-covalent interactions, and the supramolecular self-assembly can be programmed to attain structural transitions in response to changes in environmental conditions. Considering the dynamic nature, they are envisioned for smart responsive materials and flexible systems in technological applications. Hence the ability to control and manipulate the self-assembly processes in peptides is an ongoing endeavor in biotechnology. Reconfiguration of the self-assembly process in peptides triggered by assembly/ disassembly command can be attained using triggers such as pH, ionic strength, temperature, electric, magnetic fields, and photon. The engineered stimulus-responsive peptide systems are of potential use in fabricating smart nanomaterials with enhanced complexities and fewer defects. Though an overview of the progress in the development of stimulus-responsive materials can be found in a number of recent reviews, through his section, we intend to discuss the changes in the stacking reactions in response to external factors.

7.1 Ionic strength/pH

The functional group of amino acids being charged entities, to induce a change in stacking interactions triggered through pH, is one of the methods to manipulate the peptide systems in self-assembly. The ionic strength of the solution is varied, which in turn masks the effect of charges in the amino acids. Few Fmoc-based hydrogels are pH-responsive and form different nanostructures at different pH conditions (Lange et al., 2015; Raeburn et al., 2012; Reddy, Shanmugam, Duraipandy, Kiran, & Mandal, 2015). The role chirality, along with pH as a trigger, was explored in peptides (D-Leu-Phe-Phe-OH), (D-Phe-Phe-Val-OH) and (D-Val-Phe-Phe-OH). On changing the pH from 12 to 7.4, the peptides instantly form hydrogels (Marchesan, Easton, et al., 2012; Marchesan, Waddington, et al., 2012).

7.2 Enzymes

Enzyme assisted self-assembly is an emerging concept which employs the catalytic activity of the enzymes in the bottom-up fabrication of nanomaterials.

In biological systems, the enzyme assisted spatially confined molecular mechanisms tightly regulate the reactions. To regulate and enhance the control in the nanofabrication process through self-assembly, researchers are trying to mimic and integrate the catalytic activity of enzymes in peptidic systems. Enzyme assisted process is achieving it either by catalyzing the synthesis or by removing the blocking group from a molecule to allow assembly. The degree of specificity of the enzyme-catalyzed reactions to the substrate ranges from highly specific to non-specific reactions. Moreover, the range of specificity towards the substrate can be utilized in reactions with different types of natural and unnatural molecules.

One such example of an enzyme assisted self-assembly is the use of proteases to selectively trigger the self-assembly of Fmoc-FF to form hydrogels by reversed hydrolysis. The reversible nature of the hydrogels prevents the formation of kinetic aggregates and results in a thermodynamically stable structure with fewer defects (Toledano et al., 2006). A kinase or phosphatase switch used to control the sol-gel transition was demonstrated in a hydrogelator Nap − FFGEY. On addition of kinase, in the presence of ATP, peptide becomes more hydrophilic (Nap − FFGEY − $P(O)(OH)_2$) and induces a gel-sol phase transition. Whereas, on treating the same solution with phosphatase restores the hydrogel structure. Such biomimetic enzyme assisted approaches may help in regulating and constructing new biomaterials (Yang, Liang, Wang, & Xu, 2006). Several Fmoc based enzyme assisted self-assembled complex nanostructures has been reported, which otherwise would have resulted in kinetically trapped misfolded aggregates (Das, Collins, & Ulijn, 2008). Enzyme hydrogelation hence offers a new platform for developing many novel smart biomaterials that can find applications in drug delivery, biomineralization, and biosensing (Ulijn, 2006; Yang et al., 2008).

7.3 Solvent system and concentration

The effect of solvent and solvent mixtures in controlling the self-assembly in supramolecular chemistry has a profound impact, because the nucleation process is often correlated with the solubility parameters. Solvent plays a crucial role in regulating the interactions as (i) participants in the H-bonded network and (ii) space fillers with no interactions between solvent and solute.

Diphenylalanine known to form diverse nanostructures has extensively been investigated for reversible transitions in different solvent systems. The reversible self-assembly of FF peptides in the polar solvents may be controlled by the electronic and aromatic interactions among the FF molecules, which is often associated with the dissociation equilibrium. The structural transition of FF from microtubes to uniform nanowires was observed on the introduction of acetonitrile as a co-solvent (Huang et al., 2011). In solvents such as toluene or chloroform, FF self-assembles to form amorphous nanofibers (Yan, Cui, He, Wang, & Li, 2008). An experiment on temperature-induced triggered self-assembly along with the change in the solvent led to the self-assembly of FF into different nanostructures such as microtubes, nanowires, and organogels (Huang et al., 2014). In another interesting experiment using peptide with a quaterthiophene (4T) core flanked by the sequence Asp-Phe-Ala-Gly, on evaporative self-assembly, in different solvent systems, the local molecular and supramolecular structures are affected. The interactions among the peptide units are affected by solvent polarity, and a change in the geometry of the assembled helices was observed in different solvent systems. The observed results also suggest that the dipole moment plays a crucial role in determining the self-assembly (Li et al., 2017).

The structural transition of the peptides can also be manipulated by the dilution and concentration of the peptide systems. Dilution of cationic FF nanotubes in water resulted in the formation of vesicles suggests that dilution and concentration as a critical factor in nano assembly formation. Also, in the process of transition, necklace shaped jointed nanovesicles were observed as intermediates. The cationic nanotubes have been demonstrated for oligonucleotide delivery of ssDNA, and the structural transition on dilution didn't affect the attachment of the ssDNA to the peptidic nanostructures (Yan et al., 2007).

7.4 Magnetic and electric field

The magnetic field can be used as a tool to induce orientation in anisotropically shaped macromolecular assemblies. The physical torque that originates in polypeptides or proteins is due to the diamagnetism of the oriented aromatic groups(Pauling, 1979) and planar peptide bonds(Pauling, 1979; Worcester, 1978). In proteins, the secondary structures involving α-helices will orient with the helices parallel to the magnetic field, whereas in β-pleated sheets, the orientation of aromatic side chains decides the orientation of the local structure. The magnetic field has also been used to grow highly oriented amyloid fibrils for X-ray diffraction (Inouye, Fraser, & Kirschner, 1993).

In nanotechnology, much efforts have been invested in controlling the spatial organization of objects at the nanoscale. In one such attempt, horizontal alignment of the peptide nanotubes was achieved by the non-covalent coating of the nanotubes using a ferrofluid. A magnetic field of 0.5 T was used to align the tubes in the direction of the magnetic field (Reches & Gazit, 2006). In another experiment of magneto-alignment, Richard et al. demonstrated the alignment of nanotubes without any coating at a magnetic field of 12 T. The alignment to the magnetic field was attributed to the effect of magnetic torque associated with the diamagnetic anisotropy of the aromatic phenyl rings involved in the π–π interactions (Hill et al., 2007).

The electric field as a tool to modulate the self-assembly, chemical, and structural properties of the nanostructures has been investigated (Pandey et al., 2017; Pandey, Morla, Nemade, Kumar, & Ramakrishnan, 2019; Pandey & Ramakrishnan, 2020; Saikia et al., 2019). The results suggest that the electric field could arrest the formation of structures in the presence of field (Pandey et al., 2017), and were in agreement with the earlier observations on amyloid fibrils (Baumketner, 2014). On exposure to the electric field, the peptides exhibited distinct morphological differences, and post-exposure, difference in the crystal habits were observed (Saikia et al., 2021).

8. Physical properties of Nanoassemblies

Self-assembling peptides have received significant attention due to their ability to assemble into well-ordered nanostructured biomaterials that display a range of physical properties (Fig. 6). Owing to the aromatic π–π interactions, change in the physical properties of the peptide nanostructures is widely reported by various groups.

8.1 Electrical properties

Peptide materials with the ability of charge transfer can act as an interface of electronics and biology and can be employed in bioelectronics. However, the display of poor electrical properties restricts their use in electronic devices. In the increased efforts to make peptide as a toolkit, charge transfer in few peptide systems has been investigated, and there are emerging reports on conductive peptides of such molecular systems.

Conductive protein fibers in bacterial systems inspired the design of many peptide sequences that can self-assemble into conductive nanostructures. The most commonly studied pilus system contains π-stacking amino acid

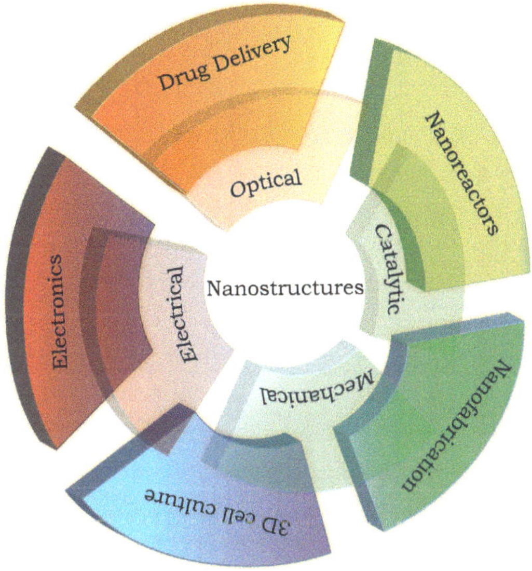

Fig. 6 Physical Properties and respective technology platforms developed from peptide nanoassemblies.

side chains responsible for the long-range conductivity (Ing & Spencer, 2018). Aromatic peptide systems like FF and its variants also exhibited piezoelectric and ferroelectric properties (Kim et al., 2020).

Electrical measurements on peptide monomer GFPRFAGFP forming films was found to significantly improve, once the peptide self-assembles to form fibers. The extended stacking of aromatic residues providing π–π conjugation along the fibers is considered to play important role in the conductivity of the self-assembled peptides (Creasey, Shingaya, & Nakayama, 2015). As observed, such soft semiconducting materials can be controlled through the local-ordering of the conductive subunits and $\pi-\pi$ stacking interactions. Even then, it is the electronic interactions through the self-assembled structures decide the ultimate device performance in such organic semiconductors. The electronic properties of the materials are defined by the specific chemical features underlying the long coherence length structures. The charge transfer properties in the peptides have enabled the development of any technologies ranging from field-effective transistors to piezoelectronics. Though significant progress has already been achieved, many attempts are under progress to understand the molecular interactions and the fundamental mechanisms that determine the device performance in molecular electronics.

8.2 Optical properties

The development of bio-organic molecules with optical properties has been under extensive research as intrinsic optical biomarkers that can contribute towards making novel diagnostic tools in medical research. The field of peptide self-assembly is undergoing a paradigm shift from simple to complex materials as the association of the building blocks can be explicitly controlled. Optical properties of the nanostructures dependent on molecular packing can advance the field to develop a new generation of optoelectronic materials for emerging nanophotonic applications.

Studies on the optical emission of fluorescent proteins revealed that on introducing Thr203Tyr substitution by mutation redshifts the Yellow Fluorescent Protein (YFP) relative to the Green Fluorescent Protein (GFP) due to the π–π stacking interactions between the tyrosine (Tyr) and the phenolate anion of the chromophore (Wachter, Elsliger, Kallio, Hanson, & Remington, 1998). Hence the packing of the molecules during self-assembly plays an important role in determining the optical properties of the self-assembled nanostructures. FF nanotubes and Boc-FF nanospheres exhibited optical properties similar to that of zero-dimensional quantum dots. FF nanotubes displayed second-order nonlinear optical response (Handelman et al., 2013) and, when cyclized and vertically aligned using vapor deposition, generated 2D quantum-wells. In addition to the above properties, Fmoc –FF self-assembled into gel-like architectures, displayed strong photoluminescence in the blue region along with UV absorbance of the exciton (Adler-Abramovich & Gazit, 2014). Park et al. reported a novel photoluminescent peptide-nanotube material by incorporating photosensitizers and lanthanide ions, such as terbium (Tb) and europium (Eu), into the FF nanotubes through a self-assembly process. FF nanotubes also act as an efficient host matrix for different luminescent complexes and developed colors based on the complexes (Ryu, Lim, & Park, 2009).

8.3 Mechanical and chemical stability

Mechanical characterization of individual insulin amyloid fibrils has shown that fibrils have a strength (0.6 ± 0.4 GPa) comparable to steel (0.6–1.8 GPa), and mechanical stiffness (3.3 ± 0.4 GPa) comparable to silk (1–10 GPa). In light of these results, the analysis of the mechanical properties can be viewed from two different perspectives. The intramolecular interactions responsible for the strength and stiffness of the fibrils explains (i) why it is challenging to degrade the amyloid fibrils and (ii) the possibility of using fibrils as a biomaterial for various applications (Smith, Knowles, Dobson, MacPhee, & Welland, 2006).

Self-assembly of the short aromatic dipeptide driven by the aromatic interactions resulted in one of the most rigid organic nanotubes, with Young's modulus of ~19 GPa (Kol et al., 2005). The physical strength of the diphenylalanine nanotubes makes them attractive building blocks for biocompatible devices. The dipeptide nanotubes are known for its stiffness, while the nanostructures made of dinapthylalanine peptides display greater flexibility. The two peptides, when mixed at different ratios, yielded nanostructures with different morphologies. The mechanical mapping of the nanostructures from different propositions revealed that as the percentage of dinapthylalanine peptides increases, a relative reduction in the stiffness was observed, suggesting the generation of materials with variable stiffness using their respective individual concentration as a variable (Sedman et al., 2013). Thermal and chemical conformational stability of the diphenylalanine nanotubes in both aqueous and dry conditions demonstrated their potential use in conventional microelectronic and microelectromechanical processes and fabrication into functional nanotechnological devices (Adler-Abramovich et al., 2006). The unique mechanical properties displayed by the peptide material may find broad applications related to environmental and medical sensing and also as templates for the fabrication of metallic and inorganic materials.

9. Applications

Peptides with the ability to mimic naturally existing structures self-assembling into diverse nanostructures are interesting building blocks, as they can be fine-tuned to achieve a wide variety of applications. Peptide-based nanostructures, along with its possibility of functionalization, has been utilized in several biological and non-biological applications.

9.1 Models to study the amyloid aggregation

A sharp increase in the occurrence of amyloid diseases and the increased reported instances of protein misfolding diseases is urging scientists to explore new ways to control the amyloid formation. Physical blockage of the self-assembly process by known aromatic drugs is one way to clinically control the amyloid formation. Considering the role of π-stacking in amyloid formation, it was suggested that any drug that can block the π-stacking interactions might be a potential candidate for controlling the amyloid diseases. A few molecules containing aromatic residues, namely Ro 65–8815/001 (Kuner et al., 2000), and 3-p-toluoyl-2-(4′-(3-diethylaminopropoxy)-phenyl]-benzofuran (Twyman & Allsop, 1999) have shown effectiveness in controlling the Aβ amyloid formation. Congo red, a specific amyloid binding dye-containing aromatic elements, also inhibit amyloid fibril formation in Aβ (Lorenzo & Yankner, 1994). Congo red and the aromatic drug Ro 65–8815/001 binds at the same site of Aβ. It was also shown that two anticancerous drugs with polyaromatic rings, 4′-iodo-4′-deoxydoxorubic, and the antibiotic tetracycline inhibits amyloid formation. The anti-cancer drug 4′-iodo-4′-deoxydoxorubicin, like Congo red, is a generic anti-amyloid agent and inhibits amyloid formation by five unrelated proteins (Merlini et al., 1995). But the lack of specificity in the drug molecules and their side effects demands other direct approaches like the usage of short aromatic molecular recognition elements. Various short peptide analogues containing the FF motif inhibits Aβ amyloid formation (Findeis et al., 1999; Soto et al., 1998; Tjernberg et al., 1996).

Aromatic recognition elements of amyloid-forming proteins can be conjugated to breakers of the amyloid self-assembly process. The breakers can be (i) charged amino acids to electrostatically block the Aβ self-assembly, (ii) bulky moieties that will sterically block molecular interactions, or (iii) β-breaker amino acids such as proline, glutamic acid, or aspartic acid, to affect the structure of the stacked assembly. Above mentioned approaches were successfully implemented to arrest the amyloid fibril formation by the Aβ peptide. The dipeptide FF was initially used as a model to study the amyloid aggregation, and then later on, based on its chemical and thermal stability and mechanical strength, they were explored for other applications.

9.2 Drug delivery

Research efforts are afoot in developing new treatment strategies to overcome the emergence of drug resistance and several disadvantages of the previously reported drug delivery systems. Peptide-based materials with properties such as biocompatibility, ability to transfer through the physiological barrier, high drug loading capacities, chemical diversity, specificity, and stimuli-responsive drug targeting at the disease site, show tremendous potential as drug delivery vehicles (Goyal et al., 2021, 2021; Jerath, Goyal, Trivedi, Santhoshkumar, & Ramakrishnan, 2019, 2020).

The importance of different types of peptide design in directing the formation of nanostructures such as nanoparticles, nanotubes, nanofibers, and hydrogels for drug delivery applications has been extensively investigated (Habibi, Kamaly, Memic, & Shafiee, 2016; Panda & Chauhan, 2014). The well-known aromatic dipeptide FF and its derivatives are explored as drug carriers due to their hydrophobic aromatic functional groups. Cationic FF nanotubes known for structural transition upon dilution has been demonstrated as an intracellular delivery vehicle (Yan et al., 2007). Also, conjugating drugs with FF is another strategy in drug delivery. Silva et al. conjugated Rhodamine to the FF microtubes, and the experimental analysis revealed that the drug release followed first-order kinetics, demonstrating the peptide arrays as potential drug carriers (Silva et al., 2013). In another experiment, Gazit et al. loaded doxorubicin and 5- fluoro-uracil into Fmoc-FF hydrogel nanoparticles and studied the drug release profiles (Ischakov, Adler-Abramovich, Buzhansky, Shekhter, & Gazit, 2013). Since discussing the application of all aromatic peptides in drug delivery is out of the scope of this chapter articles discussing the uses of peptide-based drug delivery can be referred for detailed information (Ni & Zhuo, 2019). On understanding the importance of aromatic interactions in directing the peptide nanostructures, the release profile can be tailored, and it can play a critical role in facilitating the clinical translation of the peptide systems.

9.3 Tissue engineering

Peptides can be decorated or tailored with the required functional group for promoting cell adhesion and growth. Peptide-based nanomaterials due to their ability to mimic extracellular cell-components and with properties like self-construction, degradability, and adaptability are much sought for tissue engineering applications. Designing peptides for three-dimensional cell culture and tissue repair has been a key objective in biomaterials science for decades. The dynamic nature of the nanomaterials formed through non-covalent interactions facilitate the adaptability of the nanostructures to the environment and also allows traceless dissociation of the material. With excellent hydrogelation properties and high aspect ratio, nanofibers are believed to support the growth of the cells mechanically. Also, on functionalization, peptidic and non-peptidic signals can be introduced at either ends of the molecule.

Among the widely explored functional groups for hydrogel formation and its application in tissue engineering, Fmoc, the protecting group of peptides, is well-known to promote the self-assembly through hydrophobic and π–π stacking interactions (Diaferia et al., 2019; Tao, Levin, Adler-Abramovich, & Gazit, 2016). In the process of assembly, Fmoc moiety is also known to introduce H-bonding and steric optimization from the linker. Fmoc-modified di and tripeptides forming highly ordered hydrogels through H-bonding and π–π interactions have been used for cell proliferation of chondrocytes in three dimensions (Jayawarna et al., 2007). Hydrogels formed through the co-assembly of two Fmoc containing peptides Fmoc-FF and Fmoc-RGD, acted as scaffolds for the growth of cells. Dermal fibroblasts, when combined with the hydrogels, formed dense fibrous networks through the secretion of extracellular matrix from cells (Zhou, Ulijn, & Gough, 2014). In another experiment, three diphenylalanine analogues with different aromatic ligands, Fmoc-Phe-Phe-OH, Nap (naphthalene)-Phe-Phe-OH and Cbz (benzyloxycarbonyl)-Phe-Phe-OH formed fibrous architectures with varying dimensions. The varying curvature of the fibrous dimensions is attributed to the difference in the protecting group. The fibrillar nature resembled the extracellular matrix, and the hydrogels supported chondrocyte cell culture, in both two and three dimensions (Jayawarna et al., 2007).

9.4 Electronics

Biologically conductive materials can emerge as next-generation electronic materials. Studies involving the conductive materials for application as organic electronic devices are focused with more emphasis on peptide-based materials, owing to the property of miniaturization, large surface area, functionalization, and degradability. The extensively studied FF were shown to have piezoelectric and ferroelectric properties (Bystrov et al., 2012). On solvent evaporation and vapor deposition techniques, peptides were aligned as vertical forests. Utilizing the three-dimensional arrangement of the FF peptides as vertical forests, peptide nanotubes were vertically assembled on carbon electrodes. On electrical measurements, the peptide displayed double layer capacitance in comparison to the carbon nanotubes of the same configuration (Adler-Abramovich et al., 2009).

The integration of peptides with probes is widely explored in biomolecular sensor applications for fabricating miniature biosensors. The peptide-based amperometric biosensor has been designed and fabricated by integrating conductive peptide nanotubes with electrodes to form composite electrodes. The biocomposite electrode exhibited excellent sensitivity in determining glucose and ethanol by monitoring the electrocatalytic oxidation of enzymatically liberated hydrogen peroxide and NADH (Yemini, Reches, Gazit, & Rishpon, 2005). A self-assembled peptide, upon functionalization with nanoparticles, can be used in detecting the concentration of enzyme protease. A tripeptide was designed with a Fmoc-group and it aggregated through π-stacking interactions in the absence of the enzyme. On addition of the enzyme, it cleaves the Fmoc group from the particle and reveals the amine group resulting in dispersion as a result of electrostatic repulsion. Due to the dispersion of nanoparticles, a visible color shift occurs from blue to red. The approach is straightforward and sensitive to detect the concentrations of the enzyme and can be tailored for other protease enzymes (Laromaine, Koh, Murugesan, Ulijn, & Stevens, 2007). Modulation of the inherent properties, along with high sensitivity to external stimuli, makes peptide as a novel platform to design and fabricate bioelectronics devices.

9.5 Nanoreactors

Nanoreactors are designed by the strategies employed by living cells. It primarily aims at confining a reaction environment, by hosting or encapsulating guest materials. They are generated by either positioning the required catalysts on the surface of the supramolecular architectures or by confining them inside closed structures. The amino acids and their spatial arrangement dictated by the tertiary structure plays an important role in the functionality of the biomolecular nanoreactors. Hence to design nanoreactors with high level of complexity, the choice of molecular candidate is an important parameter. The biomimetic approaches, experimented at low temperatures, yields crystals with fewer defects in comparison to the established high-temperature procedures in materials science. Pd^0 nanoparticles grown on the nanofibers using a multistep reduction technology was used for mild and efficient Suzuki–Miyaura coupling reactions in aqueous media at room temperature without additives (Khalily, Ustahuseyin, Garifullin, Genc, & Guler, 2012).

9.6 As templates for application-oriented materials

Peptides are interesting models to mimic the living components and to understand the process of biomineralization in nature. Also, bioinspired supramolecular chemistry allows for a better interface between inorganic and organic molecules due to surface functionalization. Nanostructures such as nanotubes or fibers can be used as templates for synthesizing metal nanowires. FF nanotubes known for its thermal and chemical stability were used as a mold for forming silver nanowires, and the peptide was degraded using proteinase K to form fine nanowires of 20 nm in diameter (Reches & Gazit, 2003).

Though most of the self-assembled nanostructures are non-conducting, functionalization through alterations at the sequence level, qualify them as templates for the fabrication of components in nano-circuitry. Histidine rich peptides are well known for their functional affinity towards metal ligands. In a reported study, assembled histidine nanotubes were coated with Cu nanocrystals for enhancing the electronic properties of the nanotubes. The size of the Cu crystals immobilized on the nanotubes is controlled by the conformation of the peptide nanotubes, which in turn is dictated by the pH conditions. Changes in the crystalline structure lead to significant changes in the electronic properties of the structure (Banerjee, Yu, & Matsui, 2003). Peptide nanotubes were used as scaffolds to fabricate a metal – insulator – metal, trilayered, "Teslaian" coaxial nanocables. The triaxial silver – peptide – gold nanowires were fabricated by linking FF peptides to linker peptides capable of binding gold nanoparticles and then reducing silver nitrate in the hollow pore of the peptide nanotubes (Carny et al., 2006).

10. Conclusions and future directions

Peptide nanotechnology with inherently biodegradable building blocks is speculated to be the next major green initiative. The field is transcending biology by taking advantage of the chemical versatility and functionalization of the peptides and hence introducing new features in the materials which are unexplored by evolution. Despite the attention gained by the peptide-based materials close to industrial applications, significant challenges are yet to overcome. An important challenge in the application of peptides in nanotechnology is the cost of production and change in functionalities in the non-biological environment. Nevertheless, with the advantages of easy reproducibility, and versatility in tailoring the chemical moieties as per the required application makes them stand apart in comparison to other nanomaterials.

The future direction of the field of peptide self-assembly should consider organizing the bioactive nanostructures at larger scales. Though the surplus amount of nanostructures has been reported, yet a significant amount of work has to be done in designing materials to produce groundbreaking work in terms of real-life applications. The design of complex nanostructures and a lot more innovations would still be possible. Aromatic interactions, one among the significant non-covalent interactions, are known to influence the design, arrangement, stability, and molecular recognition in the self-assembly process. While the design and studies of peptide-based assemblies through aromatic pi-pi-interactions are in progress, and considering the versatility of the field, we anticipate that this review will aid towards better understanding of aromatic interaction as one among the principal factor in the design rules governing the process of self-assembly. Also, it will help in implementing new design strategies in rationally tuning the fabrication of novel functional materials.

References

Adler-Abramovich, L., Aronov, D., Beker, P., Yevnin, M., Stempler, S., Buzhansky, L., et al. (2009). Self-assembled arrays of peptide nanotubes by vapour deposition. *Nature Nanotechnology*, *4*(12), 849.

Adler-Abramovich, L., & Gazit, E. (2014). The physical properties of supramolecular peptide assemblies: From building block association to technological applications. *Chemical Society Reviews*, *43*(20), 6881–6893.

Adler-Abramovich, L., Reches, M., Sedman, V. L., Allen, S., Tendler, S. J., & Gazit, E. (2006). Thermal and chemical stability of diphenylalanine peptide nanotubes: Implications for nanotechnological applications. *Langmuir*, *22*(3), 1313–1320.

Aggeli, A., Bell, M., Boden, N., Keen, J. N., Knowles, P. F., McLeish, T. C. B., et al. (1997). Responsive gels formed by the spontaneous self-assembly of peptides into polymeric β-sheet tapes. *Nature*, *386*, 259.

Aggeli, A., Nyrkova, I. A., Bell, M., Harding, R., Carrick, L., McLeish, T. C., et al. (2001). Hierarchical self-assembly of chiral rod-like molecules as a model for peptide β-sheet tapes, ribbons, fibrils, and fibers. *Proceedings of the National Academy of Sciences*, *98*(21), 11857–11862.

Andersen, N. H. (2001). In P. M. Kenneth (Ed.), *Vol. 168. Protein Structure, Stability, and Folding. Methods in Molecular Biology*. Totowa, New Jersey: (University of Iowa College of Medicine). Humana Press. ix + 252 pp. $89.50. ISBN 0-89603-682-0. Journal of the American Chemical Society, 2001, 123(51): 12933–12934.

Aravinda, S., Shamala, N., Das, C., Sriranjini, A., Karle, I. L., & Balaram, P. (2003). Aromatic– aromatic interactions in crystal structures of helical peptide scaffolds containing projecting phenylalanine residues. *Journal of the American Chemical Society*, *125*(18), 5308–5315.

Azriel, R., & Gazit, E. (2001). Analysis of the minimal amyloid-forming fragment of the islet amyloid polypeptide. *Journal of Biological Chemistry*, *276*(36), 34156–34161.

Bakota, E. L., Sensoy, O., Ozgur, B., Sayar, M., & Hartgerink, J. D. (2013). Self-assembling multidomain peptide fibers with aromatic cores. *Biomacromolecules*, *14*(5), 1370–1378.

Balbach, J. J., Ishii, Y., Antzutkin, O. N., Leapman, R. D., Rizzo, N. W., Dyda, F., et al. (2000). Amyloid fibril formation by Aβ16-22, a seven-residue fragment of the Alzheimer's β-amyloid peptide, and structural characterization by solid state NMR. *Biochemistry*, *39*(45), 13748–13759.

Banerjee, I. A., Yu, L., & Matsui, H. (2003). Cu nanocrystal growth on peptide nanotubes by biomineralization: Size control of Cu nanocrystals by tuning peptide conformation. *Proceedings of the National Academy of Sciences*, *100*(25), 14678–14682.

Baumketner, A. (2014). Electric field as a disaggregating agent for amyloid fibrils. *The Journal of Physical Chemistry B*, *118*(50), 14578–14589.

Benvenga, S., Trimarchi, F., & Facchiano, A. (1994). Homology of calcitonin with the amyloid-related proteins. *Journal of Endocrinological Investigation*, *17*(2), 119–122.

Bowerman, C. J., & Nilsson, B. L. (2010). A reductive trigger for peptide self-assembly and Hydrogelation. *Journal of the American Chemical Society*, *132*(28), 9526–9527.

Burley, S. K., & G. A. Petsko. (1985). Aromatic-aromatic interaction: a mechanism of protein structure stabilization. *Science*, *229*(4708), 23–28. https://doi.org/10.1126/science.3892686.

Bystrov, V., Paramonova, E., Bdikin, I., Kopyl, S., Heredia, A., Pullar, R., et al. (2012). BioFerroelectricity: Diphenylalanine peptide nanotubes computational modeling and ferroelectric properties at the nanoscale. *Ferroelectrics*, *440*(1), 3–24.

Caplan, M. R., Schwartzfarb, E. M., Zhang, S., Kamm, R. D., & Lauffenburger, D. A. (2002). Control of self-assembling oligopeptide matrix formation through systematic variation of amino acid sequence. *Biomaterials*, *23*(1), 219–227.

Carny, O., Shalev, D. E., & Gazit, E. (2006). Fabrication of coaxial metal Nanocables using a self-assembled peptide nanotube scaffold. *Nano Letters*, *6*(8), 1594–1597.

Chapman, R., Danial, M., Koh, M. L., Jolliffe, K. A., & Perrier, S. (2012). Design and properties of functional nanotubes from the self-assembly of cyclic peptide templates. *Chemical Society Reviews*, *41*(18), 6023–6041.

Chiti, F., & Dobson, C. M. (2006). Protein misfolding, functional amyloid, and human disease. *Annual Review of Biochemistry*, *75*, 333–366.

Ciunik, Z., Berski, S., Latajka, Z., & Leszczyński, J. (1998). New aspects of weak C– H ··· π bonds: Intermolecular interactions between alicyclic and aromatic rings in crystals of small compounds, peptides and proteins. *Journal of Molecular Structure*, *442*(1–3), 125–134.

Claessens, C. G., & Stoddart, J. F. (1997). π–π interactions in self-assembly. *Journal of Physical Organic Chemistry*, *10*(5), 254–272.

Creasey, R. C. G., Shingaya, Y., & Nakayama, T. (2015). Improved electrical conductance through self-assembly of bioinspired peptides into nanoscale fibers. *Materials Chemistry and Physics*, *158*, 52–59.

Crick, F. (1953). The packing of [alpha]-helices: Simple coiled-coils. *Acta Crystallographica*, *6*(8–9), 689–697.

Das, A. K., Collins, R., & Ulijn, R. V. (2008). Exploiting enzymatic (reversed) hydrolysis in directed self-assembly of peptide nanostructures. *Small*, *4*(2), 279–287.

De Santis, P., Forni, E., & Rizzo, R. (1974). Conformational analysis of DNA–basic polypeptide complexes: Possible models of nucleoprotamines and nucleohistones. *Biopolymers: Original Research on Biomolecules*, *13*(2), 313–326.

Diaferia, C., Ghosh, M., Sibillano, T., Gallo, E., Stornaiuolo, M., Giannini, C., et al. (2019). Fmoc-FF and hexapeptide-based multicomponent hydrogels as scaffold materials. *Soft Matter*, *15*(3), 487–496.

Dudukovic, N. A., & Zukoski, C. F. (2015). Gelation of Fmoc-diphenylalanine is a first order phase transition. *Soft Matter*, *11*(38), 7663–7673.

Findeis, M. A., Musso, G. M., Arico-Muendel, C. C., Benjamin, H. W., Hundal, A. M., Lee, J.-J., et al. (1999). Modified-peptide inhibitors of amyloid β-peptide polymerization. *Biochemistry*, *38*(21), 6791–6800.

Fleming, S., & Ulijn, R. V. (2014). Design of nanostructures based on aromatic peptide amphiphiles. *Chemical Society Reviews*, *43*(23), 8150–8177.

Gazit, E. (2002). A possible role for pi-stacking in the self-assembly of amyloid fibrils. *The FASEB Journal*, *16*(1), 77–83.

Ghadiri, M. R., Granja, J. R., & Buehler, L. K. (1994). Artificial transmembrane ion channels from self-assembling peptide nanotubes. *Nature*, *369*(6478), 301–304.

Ghadiri, M. R., Granja, J. R., Milligan, R. A., McRee, D. E., & Khazanovich, N. (1993). Self-assembling organic nanotubes based on a cyclic peptide architecture. *Nature*, *366*, 324.

Gillard, R. E., Raymo, F. M., & Stoddart, J. F. (1997). Controlling self-assembly. Chemistry–A. *European Journal*, *3*(12), 1933–1940.

Goodsell, D. S. (2004). *Bionanotechnology: Lessons from nature*. John Wiley & Sons.

Görbitz, C. H. (2001). Nanotube formation by hydrophobic dipeptides. Chemistry – A. *European Journal*, 7(23), 5153–5159.
Görbitz, C. H. (2006). The structure of nanotubes formed by diphenylalanine, the core recognition motif of Alzheimer's β-amyloid polypeptide. *Chemical Communications*, *22*, 2332–2334.
Görbitz, C. (2012). Theoretical basis of biological self-assembly. *Self-Assembled Peptide Nanostructures*, 1–3.
Goyal, R., Jerath, G., Akhil, R., Chandrasekharan, A., Puppala, E. R., Ponneganti, S., et al. (2021). Geometry encoded functional programming of tumor homing peptides for targeted drug delivery. *Journal of Controlled Release*, *333*, 16–27.
Goyal, R., Jerath, G., Chandrasekharan, A., Christian, Y., Kumar, T., & Ramakrishnan, V. (2021). Molecular hybridization combining tumor homing and penetrating peptide domains for cellular targeting. *Drug Delivery and Translational Research*, 1–8.
Guler, M. O., & Stupp, S. I. (2007). A self-assembled nanofiber catalyst for Ester hydrolysis. *Journal of the American Chemical Society*, *129*(40), 12082–12083.
Habibi, N., Kamaly, N., Memic, A., & Shafiee, H. (2016). Self-assembled peptide-based nanostructures: Smart nanomaterials toward targeted drug delivery. *Nano Today*, *11*(1), 41–60.
Haggqvist, B., Naslund, J., Sletten, K., Westermark, G. T., Mucchiano, G., Tjernberg, L. O., et al. (1999). Medin: An integral fragment of aortic smooth muscle cell-produced lactadherin forms the most common human amyloid. *Proceedings of the National Academy of Sciences of the United States of America*, *96*(15), 8669–8674.
Han, H., Weinreb, P. H., & Lansbury, P. T., Jr. (1995). The core Alzheimer's peptide NAC forms amyloid fibrils which seed and are seeded by beta-amyloid: Is NAC a common trigger or target in neurodegenerative disease? *Chemistry & Biology*, *2*(3), 163–169.
Handelman, A., Lavrov, S., Kudryavtsev, A., Khatchatouriants, A., Rosenberg, Y., Mishina, E., et al. (2013). Nonlinear optical bioinspired peptide nanostructures. *Advanced Optical Materials*, *1*(11), 875–884.
Hazam, P. K., Akhil, R., Jerath, G., Saikia, J., & Ramakrishnan, V. (2019). Topological effects on the designability and bactericidal potency of antimicrobial peptides. *Biophysical Chemistry*, *248*, 1–8.
Hazam, P. K., Goyal, R., & Ramakrishnan, V. (2019). Peptide based antimicrobials: Design strategies and therapeutic potential. *Progress in Biophysics and Molecular Biology*, *142*, 10–22.
Hazam, P. K., Jerath, G., Kumar, A., Chaudhary, N., & Ramakrishnan, V. (2017). Effect of tacticity-derived topological constraints in bactericidal peptides. *Biochimica et Biophysica Acta (BBA)-Biomembranes*, *1859*(8), 1388–1395.
Hazam, P. K., Phukan, C., Akhil, R., Singh, A., & Ramakrishnan, V. (2021). Antimicrobial effects of syndiotactic polypeptides. *Scientific Reports*, *11*(1), 1–10.
Hill, R. J. A., Sedman, V. L., Allen, S., Williams, P., Paoli, M., Adler-Abramovich, L., et al. (2007). Alignment of aromatic peptide tubes in strong magnetic fields. *Advanced Materials*, *19*(24), 4474–4479.
Horne, W. S., Ashkenasy, N., & Ghadiri, M. R. (2005). Modulating charge transfer through cyclic d, l α-peptide self-assembly. *Chemistry (Weinheim an der Bergstrasse, Germany)*, *11*(4), 1137.
Huang, R., Qi, W., Su, R., Zhao, J., & He, Z. (2011). Solvent and surface controlled self-assembly of diphenylalanine peptide: From microtubes to nanofibers. *Soft Matter*, 7(14), 6418–6421.
Huang, R., Wang, Y., Qi, W., Su, R., & He, Z. (2014). Temperature-induced reversible self-assembly of diphenylalanine peptide and the structural transition from organogel to crystalline nanowires. *Nanoscale Research Letters*, *9*(1), 653.

Hunter, C. A., Lawson, K. R., Perkins, J., & Urch, C. J. (2001). Aromatic interactions. *Journal of the Chemical Society, Perkin Transactions 2*, *5*, 651–669.
Ing, N. L., & Spencer, R. K. (2018). Electronic conductivity in biomimetic alpha-helical peptide nanofibers and gels. *ACS Nano*, *12*(3), 2652–2661.
Inouye, H., Fraser, P. E., & Kirschner, D. A. (1993). Structure of beta-crystallite assemblies formed by Alzheimer beta-amyloid protein analogues: Analysis by x-ray diffraction. *Biophysical Journal*, *64*(2), 502–519.
Ischakov, R., Adler-Abramovich, L., Buzhansky, L., Shekhter, T., & Gazit, E. (2013). Peptide-based hydrogel nanoparticles as effective drug delivery agents. *Bioorganic & Medicinal Chemistry*, *21*(12), 3517–3522.
Jayawarna, V., Ali, M., Jowitt, T. A., Miller, A. F., Saiani, A., Gough, J. E., et al. (2006). Nanostructured Hydrogels for Three-Dimensional Cell Culture Through Self-Assembly of Fluorenylmethoxycarbonyl–Dipeptides. *Advanced Materials*, *18*(5), 611–614.
Jayawarna, V., Smith, A., Gough, J. E., & Ulijn, R. V. (2007). Three-dimensional cell culture of chondrocytes on modified di-phenylalanine scaffolds. *Biochemical Society Transactions*, *35*(3), 535–537.
Jerath, G., Goyal, R., Trivedi, V., Santhoshkumar, T., & Ramakrishnan, V. (2019). Syndiotactic peptides for targeted delivery. *Acta Biomaterialia*, *87*, 130–139.
Jerath, G., Goyal, R., Trivedi, V., Santhoshkumar, T. R., & Ramakrishnan, V. (2020). Conformationally constrained peptides for drug delivery. *Journal of Peptide Science*, *26*(4–5), e3244.
Kahn, S. E., Andrikopoulos, S., & Verchere, C. B. (1999). Islet amyloid: A long-recognized but underappreciated pathological feature of type 2 diabetes. *Diabetes*, *48*(2), 241–253.
Kayed, R., Bernhagen, J., Greenfield, N., Sweimeh, K., Brunner, H., Voelter, W., et al. (1999). Conformational transitions of islet amyloid polypeptide (IAPP) in amyloid formation in vitro. *Journal of Molecular Biology*, *287*(4), 781–796.
Kedar, I., Ravid, M., & Sohar, E. (1976). In vitro synthesis of "amyloid" fibrils from insulin, calcitonin and parathormone. *Israel Journal of Medical Sciences*, *12*(10), 1137–1140.
Khalily, M. A., Ustahuseyin, O., Garifullin, R., Genc, R., & Guler, M. O. (2012). A supramolecular peptide nanofiber templated Pd nanocatalyst for efficient Suzuki coupling reactions under aqueous conditions. *Chemical Communications*, *48*(92), 11358–11360.
Kim, D., Han, S. A., Kim, J. H., Lee, J. H., Kim, S. W., & Lee, S. W. (2020). Biomolecular Piezoelectric Materials: From Amino Acids to Living Tissues. *Advanced Materials*, 1906989.
Kol, N., Adler-Abramovich, L., Barlam, D., Shneck, R. Z., Gazit, E., & Rousso, I. (2005). Self-assembled peptide nanotubes are uniquely rigid bioinspired supramolecular structures. *Nano Letters*, *5*(7), 1343–1346.
Kool, E. T. (2001). Hydrogen bonding, base stacking, and steric effects in dna replication. *Annual Review of Biophysics and Biomolecular Structure*, *30*, 1–22.
Krysmann, M. J., Castelletto, V., Kelarakis, A., Hamley, I. W., Hule, R. A., & Pochan, D. J. (2008). Self-assembly and hydrogelation of an amyloid peptide fragment. *Biochemistry*, *47*(16), 4597–4605.
Kumar, A., Ramakrishnan, V., Ranbhor, R., Patel, K., & Durani, S. (2009). Homochiral stereochemistry: The missing link of structure to energetics in protein folding. *The Journal of Physical Chemistry B*, *113*(51), 16435–16442.
Kuner, P., Bohrmann, B., Tjernberg, L. O., Näslund, J., Huber, G., Celenk, S., et al. (2000). Controlling polymerization of β-amyloid and prion-derived peptides with synthetic small molecule ligands. *Journal of Biological Chemistry*, *275*(3), 1673–1678.
Lange, S. C., Unsleber, J., Drücker, P., Galla, H.-J., Waller, M. P., & Ravoo, B. J. (2015). pH response and molecular recognition in a low molecular weight peptide hydrogel. *Organic & Biomolecular Chemistry*, *13*(2), 561–569.

Laromaine, A., Koh, L., Murugesan, M., Ulijn, R. V., & Stevens, M. M. (2007). Protease-triggered dispersion of nanoparticle assemblies. *Journal of the American Chemical Society, 129*(14), 4156–4157.
Li, B., Valverde, L. R., Zhang, F., Zhou, Y., Li, S., Diao, Y., et al. (2017). Macroscopic alignment and assembly of π-conjugated oligopeptides using colloidal microchannels. *ACS Applied Materials & Interfaces, 9*(47), 41586–41593.
Lorenzo, A., & Yankner, B. A. (1994). Beta-amyloid neurotoxicity requires fibril formation and is inhibited by Congo red. *Proceedings of the National Academy of Sciences, 91*(25), 12243–12247.
Lu, K., Guo, L., Mehta, A. K., Childers, W. S., Dublin, S. N., Skanthakumar, S., et al. (2007). Macroscale assembly of peptide nanotubes. *Chemical Communications, 26*, 2729–2731.
Lu, K., Jacob, J., Thiyagarajan, P., Conticello, V. P., & Lynn, D. G. (2003). Exploiting amyloid fibril lamination for nanotube self-assembly. *Journal of the American Chemical Society, 125*(21), 6391–6393.
Mahler, A., Reches, M., Rechter, M., Cohen, S., & Gazit, E. (2006). Rigid, self-assembled hydrogel composed of a modified aromatic dipeptide. *Advanced Materials, 18*(11), 1365–1370.
Maity, S., Hashemi, M., & Lyubchenko, Y. L. (2017). Nano-assembly of amyloid β peptide: Role of the hairpin fold. *Scientific Reports*, 7(1), 1–10.
Marchesan, S., Easton, C. D., Kushkaki, F., Waddington, L., & Hartley, P. G. (2012). Tripeptide self-assembled hydrogels: Unexpected twists of chirality. *Chemical Communications, 48*(16), 2195–2197.
Marchesan, S., Waddington, L., Easton, C. D., Winkler, D. A., Goodall, L., Forsythe, J., et al. (2012). Unzipping the role of chirality in nanoscale self-assembly of tripeptide hydrogels. *Nanoscale, 4*(21), 6752–6760.
Matsuura, K., Hayashi, H., Murasato, K., & Kimizuka, N. (2011). Trigonal tryptophane zipper as a novel building block for pH-responsive peptide nano-assemblies. *Chemical Communications, 47*(1), 265–267.
McGaughey, G. B., Gagné, M., & Rappé, A. K. (1998). π-Stacking interactions: Alive and well in proteins. *Journal of Biological Chemistry, 273*(25), 15458–15463.
Mehra, A., Jerath, G., Ramakrishnan, V., & Trivedi, V. (2015). Characterization of ICAM-1 biophore to design cytoadherence blocking peptides. *Journal of Molecular Graphics and Modelling, 57*, 27–35.
Meital, R., & Ehud, G. (2006). Designed aromatic homo-dipeptides: Formation of ordered nanostructures and potential nanotechnological applications. *Physical Biology, 3*(1), S10.
Merlini, G., Ascari, E., Amboldi, N., Bellotti, V., Arbustini, E., Perfetti, V., et al. (1995). Interaction of the anthracycline 4′-iodo-4′-deoxydoxorubicin with amyloid fibrils: Inhibition of amyloidogenesis. *Proceedings of the National Academy of Sciences of the United States of America, 92*(7), 2959–2963.
Mishra, A., Panda, J. J., Basu, A., & Chauhan, V. S. (2008). Nanovesicles based on self-assembly of Conformationally constrained aromatic residue containing amphiphilic dipeptides. *Langmuir, 24*(9), 4571–4576.
Ni, M., & Zhuo, S. (2019). Applications of self-assembling ultrashort peptides in bionanotechnology. *RSC Advances, 9*(2), 844–852.
Nolan, S. J., Shiels, J. C., Tuite, J. B., Cecere, K. L., & Baranger, A. M. (1999). Recognition of an essential adenine at a protein – RNA Interface: Comparison of the contributions of hydrogen bonds and a stacking interaction. *Journal of the American Chemical Society, 121*(38), 8951–8952.
Panda, J. J., & Chauhan, V. S. (2014). Short peptide based self-assembled nanostructures: Implications in drug delivery and tissue engineering. *Polymer Chemistry, 5*(15), 4418–4436.

Pandey, G., Morla, S., Kumar, S., & Ramakrishnan, V. (2020). Modulating Aβ fibrillogenesis with 'Trojan'peptides. *Neuropeptides*, *81*, 102030.
Pandey, G., Morla, S., Nemade, H. B., Kumar, S., & Ramakrishnan, V. (2019). Modulation of aggregation with an electric field; scientific roadmap for a potential non-invasive therapy against tauopathies. *RSC Advances*, *9*(9), 4744–4750.
Pandey, G., & Ramakrishnan, V. (2020). Invasive and non-invasive therapies for Alzheimer's disease and other amyloidosis. *Biophysical Reviews*, 1–12.
Pandey, G., Saikia, J., Sasidharan, S., Joshi, D. C., Thota, S., Nemade, H. B., et al. (2017). Modulation of peptide based nano-assemblies with electric and magnetic fields. *Scientific Reports*, 7(1), 1–9.
Pauling, L. (1979). Diamagnetic anisotropy of the peptide group. *Proceedings of the National Academy of Sciences*, *76*(5), 2293–2294.
Pauling, L., & Corey, R. B. (1951). The pleated sheet, a new layer configuration of polypeptide chains. *Proceedings of the National Academy of Sciences of the United States of America*, *37*(5), 251–256.
Pauling, L., & Corey, R. B. (1953). Compound helical configurations of polypeptide chains: Structure of proteins of the α-keratin type. *Nature*, *171*, 59.
Prakash, V., Ranbhor, R., & Ramakrishnan, V. (2020). De novo designed Heterochiral blue fluorescent protein. *ACS Omega*, *5*(41), 26382–26388.
Raeburn, J., Pont, G., Chen, L., Cesbron, Y., Lévy, R., & Adams, D. J. (2012). Fmoc-diphenylalanine hydrogels: Understanding the variability in reported mechanical properties. *Soft Matter*, *8*(4), 1168–1174.
Ramakrishnan, V., Ranbhor, R., Kumar, A., & Durani, S. (2006). The link between sequence and conformation in protein structures appears to be stereochemically established. *The Journal of Physical Chemistry B*, *110*(18), 9314–9323.
Ranbhor, R., Ramakrishnan, V., Kumar, A., & Durani, S. (2006). The interplay of sequence and stereochemistry in defining conformation in proteins and polypeptides. *Biopolymers: Original Research on Biomolecules*, *83*(5), 537–545.
Ranganathan, D., Haridas, V., Gilardi, R., & Karle, I. L. (1998). Self-assembling aromatic-bridged serine-based cyclodepsipeptides (serinophanes): A demonstration of tubular structures formed through aromatic $\pi-\pi$ interactions. *Journal of the American Chemical Society*, *120*(42), 10793–10800.
Ranganathan, D., Haridas, V., & Karle, I. L. (1998). Cystinophanes, a novel family of aromatic-bridged cystine cyclic peptides: Synthesis, crystal structure, molecular recognition, and conformational studies. *Journal of the American Chemical Society*, *120*(12), 2695–2702.
Reches, M., & Gazit, E. (2003). Casting metal nanowires within discrete self-assembled peptide nanotubes. *Science*, *300*(5619), 625–627.
Reches, M., & Gazit, E. (2004). Formation of closed-cage nanostructures by self-assembly of aromatic dipeptides. *Nano Letters*, *4*(4), 581–585.
Reches, M., & Gazit, E. (2006). Controlled patterning of aligned self-assembled peptide nanotubes. *Nature Nanotechnology*, *1*(3), 195–200.
Reches, M., & Gazit, E. (2010). Self-assembly of peptide nanotubes and amyloid-like structures by charged-termini-capped diphenylalanine peptide analogues. *Israel Journal of Chemistry*, *45*(3), 363–371.
Reches, M., Porat, Y., & Gazit, E. (2002). Amyloid fibril formation by pentapeptide and tetrapeptide fragments of human calcitonin. *Journal of Biological Chemistry*, *277*(38), 35475–35480.
Reddy, S. M., Shanmugam, G., Duraipandy, N., Kiran, M. S., & Mandal, A. B. (2015). An additional fluorenylmethoxycarbonyl (Fmoc) moiety in di-Fmoc-functionalized L-lysine induces pH-controlled ambidextrous gelation with significant advantages. *Soft Matter*, *11*(41), 8126–8140.
Ryu, J., Lim, S. Y., & Park, C. B. (2009). Photoluminescent peptide nanotubes. *Advanced Materials*, *21*(16), 1577–1581.

Ryu, J., & Park, C. B. (2008). Solid-phase growth of nanostructures from amorphous peptide thin film: Effect of water activity and temperature. *Chemistry of Materials*, *20*(13), 4284–4290.

Saikia, J., Dharmalingam, K., Anandalakshmi, R., Redkar, A. S., Bhat, V., & Ramakrishnan, V. (2021). Electric field modulated peptide based hydrogel nanocatalyst. *Soft Matter*, *17*(42), 9725–9735.

Saikia, J., Pandey, G., Sasidharan, S., Antony, F., Nemade, H. B., Kumar, S., et al. (2019). Electric field disruption of amyloid aggregation: Potential noninvasive therapy for Alzheimer's disease. *ACS Chemical Neuroscience*, *10*(5), 2250–2262.

Sasidharan, S., Hazam, P. K., & Ramakrishnan, V. (2017). Symmetry-directed self-organization in peptide nanoassemblies through aromatic π–π interactions. *The Journal of Physical Chemistry B*, *121*(2), 404–411.

Sasidharan, S., Shyni, P., Chaudhary, N., & Ramakrishnan, V. (2017). Single crystal organic nanoflowers. *Scientific Reports*, 7(1), 1–8.

Sedman, V. L., Chen, X., Allen, S., Roberts, C. J., Korolkov, V. V., & Tendler, S. J. B. (2013). Tuning the mechanical properties of self-assembled mixed-peptide tubes. *Journal of Microscopy*, *249*(3), 165–172.

Shiels, J. C., Tuite, J. B., Nolan, S. J., & Baranger, A. M. (2002). Investigation of a conserved stacking interaction in target site recognition by the U1A protein. *Nucleic Acids Research*, *30*(2), 550–558.

Silva, R. F., Araújo, D. R., Silva, E. R., Ando, R. A., & Alves, W. A. (2013). L-diphenylalanine microtubes as a potential drug-delivery system: Characterization, release kinetics, and cytotoxicity. *Langmuir*, *29*(32), 10205–10212.

Smith, J. F., Knowles, T. P. J., Dobson, C. M., MacPhee, C. E., & Welland, M. E. (2006). Characterization of the nanoscale properties of individual amyloid fibrils. *Proceedings of the National Academy of Sciences*, *103*(43), 15806–15811.

Smith, A. M., Williams, R. J., Tang, C., Coppo, P., Collins, R. F., Turner, M. L., et al. (2008). Fmoc-diphenylalanine self assembles to a hydrogel via a novel architecture based on π–π interlocked β-sheets. *Advanced Materials*, *20*(1), 37–41.

Song, Y., Challa, S. R., Medforth, C. J., Qiu, Y., Watt, R. K., Peña, D., et al. (2004). Synthesis of peptide-nanotube platinum-nanoparticle composites. *Chemical Communications*, *9*, 1044–1045.

Soto, C., Sigurdsson, E. M., Morelli, L., Kumar, R. A., Castano, E. M., & Frangione, B. (1998). β-Sheet breaker peptides inhibit fibrillogenesis in a rat brain model of amyloidosis: Implications for Alzheimer's therapy. *Nature Medicine*, *4*(7), 822–826.

Tao, K., Levin, A., Adler-Abramovich, L., & Gazit, E. (2016). Fmoc-modified amino acids and short peptides: Simple bio-inspired building blocks for the fabrication of functional materials. *Chemical Society Reviews*, *45*(14), 3935–3953.

Tenidis, K., Waldner, M., Bernhagen, J., Fischle, W., Bergmann, M., Weber, M., et al. (2000). Identification of a penta-and hexapeptide of islet amyloid polypeptide (IAPP) with amyloidogenic and cytotoxic properties. *Journal of Molecular Biology*, *295*(4), 1055–1071.

Tjernberg, L. O., Näslund, J., Lindqvist, F., Johansson, J., Karlström, A. R., Thyberg, J., et al. (1996). Arrest of-amyloid fibril formation by a pentapeptide ligand (*). *Journal of Biological Chemistry*, *271*(15), 8545–8548.

Toledano, S., Williams, R. J., Jayawarna, V., & Ulijn, R. V. (2006). Enzyme-triggered self-assembly of peptide hydrogels via reversed hydrolysis. *Journal of the American Chemical Society*, *128*(4), 1070–1071.

Twyman, L. J., & Allsop, D. (1999). A short synthesis of the β-amyloid (Aβ) aggregation inhibitor 3-p-toluoyl-2-[4′-(3-diethylaminopropoxy)-phenyl]-benzofuran. *Tetrahedron Letters*, *40*(52), 9383–9384.

Ulijn, R. V. (2006). Enzyme-responsive materials: A new class of smart biomaterials. *Journal of Materials Chemistry*, *16*(23), 2217–2225.

Ulijn, R. V., & Smith, A. M. (2008). Designing peptide based nanomaterials. *Chemical Society Reviews*, *37*(4), 664–675.
Valéry, C., Paternostre, M., Robert, B., Gulik-Krzywicki, T., Narayanan, T., Dedieu, J.-C., et al. (2003). Biomimetic organization: Octapeptide self-assembly into nanotubes of viral capsid-like dimension. *Proceedings of the National Academy of Sciences of the United States of America*, *100*(18), 10258–10262.
Van Der Avoird, A., Podeszwa, R., Szalewicz, K., Leforestier, C., van Harrevelt, R., Bunker, P., et al. (2010). Vibration–rotation-tunneling states of the benzene dimer: An ab initio study. *Physical Chemistry Chemical Physics*, *12*(29), 8219–8240.
Vidal, R., Frangione, B., Rostagno, A., Mead, S., Révész, T., Plant, G., et al. (1999). A stop-codon mutation in the BRI gene associated with familial British dementia. *Nature*, *399*, 776.
Wachter, R. M., Elsliger, M.-A., Kallio, K., Hanson, G. T., & Remington, S. J. (1998). Structural basis of spectral shifts in the yellow-emission variants of green fluorescent protein. *Structure*, *6*(10), 1267–1277.
Worcester, D. L. (1978). Structural origins of diamagnetic anisotropy in proteins. *Proceedings of the National Academy of Sciences*, *75*(11), 5475–5477.
Yan, X., Cui, Y., He, Q., Wang, K., & Li, J. (2008). Organogels based on self-assembly of diphenylalanine peptide and their application to immobilize quantum dots. *Chemistry of Materials*, *20*(4), 1522–1526.
Yan, X., Cui, Y., He, Q., Wang, K., Li, J., Mu, W., et al. (2008). Reversible transitions between peptide nanotubes and vesicle-like structures including theoretical modeling studies. *Chemistry*, *14*(19), 5974–5980.
Yan, X., He, Q., Wang, K., Duan, L., Cui, Y., & Li, J. (2007). Transition of cationic dipeptide nanotubes into vesicles and oligonucleotide delivery. *Angewandte Chemie International Edition*, *46*(14), 2431–2434.
Yan, C., & Pochan, D. J. (2010). Rheological properties of peptide-based hydrogels for biomedical and other applications. *Chemical Society Reviews*, *39*(9), 3528–3540.
Yan, X., Zhu, P., & Li, J. (2010). Self-assembly and application of diphenylalanine-based nanostructures. *Chemical Society Reviews*, *39*(6), 1877–1890.
Yang, Z., Gu, H., Zhang, Y., Wang, L., & Xu, B. (2004). Small molecule hydrogels based on a class of antiinflammatory agents. *Chemical Communications*, *2*, 208–209.
Yang, Z., Liang, G., Ma, M., Gao, Y., & Xu, B. (2007). Conjugates of naphthalene and dipeptides produce molecular hydrogelators with high efficiency of hydrogelation and superhelical nanofibers. *Journal of Materials Chemistry*, *17*(9), 850–854.
Yang, Z., Liang, G., Wang, L., & Xu, B. (2006). Using a kinase/phosphatase switch to regulate a supramolecular hydrogel and forming the supramolecular hydrogel in vivo. *Journal of the American Chemical Society*, *128*(9), 3038–3043.
Yang, Z., Liang, G., & Xu, B. (2008). Enzymatic Hydrogelation of small molecules. *Accounts of Chemical Research*, *41*(2), 315–326.
Yemini, M., Reches, M., Gazit, E., & Rishpon, J. (2005). Peptide nanotube-modified electrodes for enzyme – biosensor applications. *Analytical Chemistry*, 77(16), 5155–5159.
Zhang, S., Holmes, T., Lockshin, C., & Rich, A. (1993). Spontaneous assembly of a self-complementary oligopeptide to form a stable macroscopic membrane. *Proceedings of the National Academy of Sciences*, *90*(8), 3334–3338.
Zhao, F., Ma, M. L., & Xu, B. (2009). Molecular hydrogels of therapeutic agents. *Chemical Society Reviews*, *38*(4), 883–891.
Zhao, X., & Zhang, S. (2006). Molecular designer self-assembling peptides. *Chemical Society Reviews*, *35*(11), 1105–1110.
Zhou, M., Ulijn, R. V., & Gough, J. E. (2014). Extracellular matrix formation in self-assembled minimalistic bioactive hydrogels based on aromatic peptide amphiphiles. *Journal of Tissue Engineering*, *5*, 2041731414531593.

CHAPTER SIX

Understanding binding affinity and specificity of modular protein domains: A focus in ligand design for the polyproline-binding families

Jose C. Martinez[a,*], Francisco Castillo[b], Javier Ruiz-Sanz[a], Javier Murciano-Calles[a], Ana Camara-Artigas[c], and Irene Luque[a]

[a]Departamento de Química Física e Instituto de Biotecnología, Facultad de Ciencias, Universidad de Granada, Granada, Spain
[b]Fundación MEDINA, Centro de Excelencia en Investigación de Medicamentos Innovadores en Andalucía, Parque Tecnológico Ciencias de la Salud, Granada, Spain
[c]Departamento de Química Física, Universidad de Almería, Campus de Excelencia Internacional Agroalimentario ceiA3 y CIAMBITAL, Almeria, Spain
*Corresponding author: e-mail address: jcmh@ugr.es

Contents

Abstract

Within the modular protein domains there are five families that recognize proline-rich sequences: SH3, WW, EVH1, GYF and UEV domains. This chapter reviews the main strategies developed for the design of ligands for these families, including peptides, peptidomimetics and drugs. We also describe some studies aimed to understand the molecular reasons responsible for the intrinsic affinity and specificity of these domains.

Advances in Protein Chemistry and Structural Biology, Volume 130
ISSN 1876-1623
https://doi.org/10.1016/bs.apcsb.2021.12.002

Abbreviations

APP avian pancreatic polypeptide
DSC differential scanning calorimetry
EVH1 Ena/VASP homology 1
Fab antigen-binding fragment
HTS high-throughput screening
ITC isothermal titration calorimetry
PDZ postsynaptic density protein 95, disc-large, zonula occludens
PPII polyproline II
PTB phospho-tyrosine binding
SH2 Src-homology region 2
SH3 Src homology 3
UEV ubiquitin E2 variant
ΨPro pseudoproline

1. Modular domains for protein recognition

Modular protein-protein recognition domains provide a very efficient mechanism to control cellular activity, contributing to the localization of enzymes and substrates and interconnecting different signaling pathways. They are present in multidomain proteins in variable numbers and forming different combinations. The human proteome contains thousands of these domains, which have been classified into more than 70 different families. Each family presents a compact and characteristic folding, at least one binding site to recognize the ligands, and generally small size (between 30 and 150 amino acids). They recognize specific amino acid sequences on their natural targets (Freund, Schmalz, Sticht, & Kuhne, 2008). According to the nature of such recognition sequences, these domains can be classified into different groups: the GYF (named for the presence of this conserved motif in their sequence), the EVH1 (Ena/VASP homology 1), the UEV (ubiquitin E2 variant), the WW (named for the presence of two highly conserved tryptophan residues) and the SH3 (Src-homology region 3) domain families recognize proline-rich sequences, whereas the SH2 (Src-homology region 2) and PTB (phospho-tyrosine binding) families bind to peptide sequences displaying phosphorylated tyrosine residues; other families like the PDZ (postsynaptic density protein 95, disc-large, zonula occludens) interact with sequences corresponding to carboxyl termini (Cesareni, Gimona, Sudol, & Yaffe, 2005).

Protein-protein interactions generally involve highly extensive binding interfaces, up to 6000 Å^2. Instead, modular protein domains are characterized by recognizing small continuous epitopes, bringing comparatively reduced functional binding interfaces into play, about 1000 Å^2. In this regard, the design of inhibitors of protein-protein interactions for the binding sites of modular domains can be particularly affordable. On the other hand, modular domains are specialized in transient interactions, since they are involved in cellular signal transduction processes. Consequently, the interactions mediated by these modular domains are relatively weak, with dissociation constants that typically range within the micromolar order. In addition, these domains present a certain degree of promiscuity, since the same protein can be recognized by more than one domain with comparable affinities. Such characteristics can explain why the binding pockets of modular domains are relatively flat and adaptable (Freund et al., 2008). In this sense, the binding sites of the modular domains are antagonistic to the catalytic sites of enzymes, which are usually deep and highly structured. Therefore, although the interaction surfaces of the modular domains constitute an attractive intervention point for being relatively narrow, the design of high quality ligands by improving their affinity and specificity is more complex than for the catalytic sites of the enzymes.

2. Modular domains for the recognition of proline-rich sequences

Within the modular domains there are five families specialized in the recognition of proline-rich sequences: SH3, WW, EVH1, GYF and UEV domains (Fig. 1). Some authors also consider profilin, an actin-binding protein that recognizes proline-rich sequences, as the sixth family (Mahoney, Rozwarski, Fedorov, Fedorov, & Almo, 1999). The polyproline ligands are typically arranged as a polyproline II (PPII) helix. The binding mechanism is also similar, based on the interaction of aromatic residues from the domain binding surface which arrange in grooves, named as "xP," where the proline residues from the ligands dock (Fig. 2). The domains contain an additional binding groove which confers specificity within a family, and intervenes in the recognition of the positions that flank the central and usually well-conserved ligand motifs. Such central motifs are characteristic of each domain family (Fig. 1). Together, these elements define the intrinsic binding specificity. The proper functioning of the cellular signaling processes in which these domains participate also requires complex networks

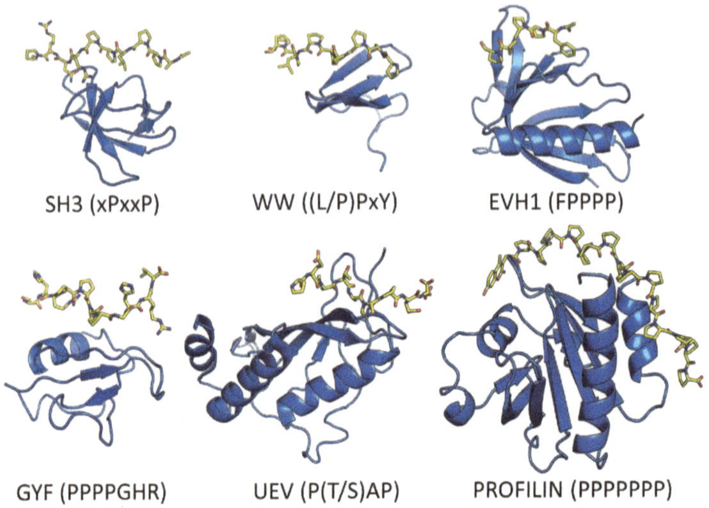

Fig. 1 Representative complexes of the six families of proline-rich recognition modules (SH3, WW, EVH1, GYF, UEV and Profilin). The domains are represented in ribbons, the ligand peptides in PPII conformation are shown in sticks, and relevant protein residues are represented as red sticks. The respective PDB codes are 4HVV (SH3), 2OEI (WW), 1EVH (EVH1), 1L2Z (GYF), 1M4Q (UEV) and 1CJF (Profilin). The consensus-binding motif of each family is shown. All the structures have been drawn using PyMol (The PyMOL Molecular Graphics System, Version 2.5 Schrödinger, LLC.).

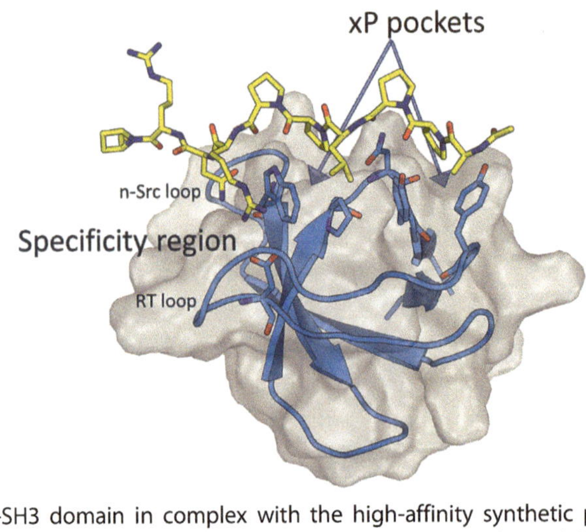

Fig. 2 c-Src-SH3 domain in complex with the high-affinity synthetic peptide APP12 (APPLPPRNRPRL) (PDB entry 4HVV). The xP pockets in the surface of the SH3 domain are represented with the residues in sticks. The APP12 peptide is represented in yellow sticks. The specificity region comprises residues in the n-Src and RT-loops, where Arg7 residue interacts with Asp99 and Trp118. The structure has been drawn using PyMol (The PyMOL Molecular Graphics System, Version 2.5 Schrödinger, LLC.).

of highly selective interactions, regulated by a contextual specificity, in which factors such as subcellular location or the cooperative effect of multiple interactions may play an important role (Stein & Aloy, 2008; Zarrinpar, Park, & Lim, 2003).

Families are subclassified by types based on the motif they recognize, but the molecular determinants governing intrinsic specificity within each type are unknown, which would require an individual and prioritized study (Opitz et al., 2015). In this sense, it has become particularly interesting the characterization of those proline-rich binding domains involved in the pathogenesis of viral diseases and hereditary syndromes of certain types of cancer, such as Liddle, Noone and Hopkins (Corbi-Verge & Kim, 2016).

Although the structural analysis has allowed rationalizing the main characteristics of these domain-peptide interactions, the understanding of the origins of the binding energy emerges controversial. Thus, isothermal titration calorimetry (ITC) experiments have revealed a thermodynamic signature for these interactions (very favorable enthalpic contributions opposed by an unfavorable binding entropy) inconsistent with the highly hydrophobic nature of the peptide ligands and their corresponding domain binding sites. Nevertheless, a deeper looking at the crystal structures may reveal some relevant details. Thus, for example, the complex between Abl-SH3 and the peptide p41 (APSYSPPPPP) (PDB code: 1BBZ (Pisabarro, Serrano, & Wilmanns, 1998)) shows the establishment of a complex hydrogen-bond network mediated by several water molecules buried at the binding interface (Fig. 3). This network would explain the observed thermodynamic behavior

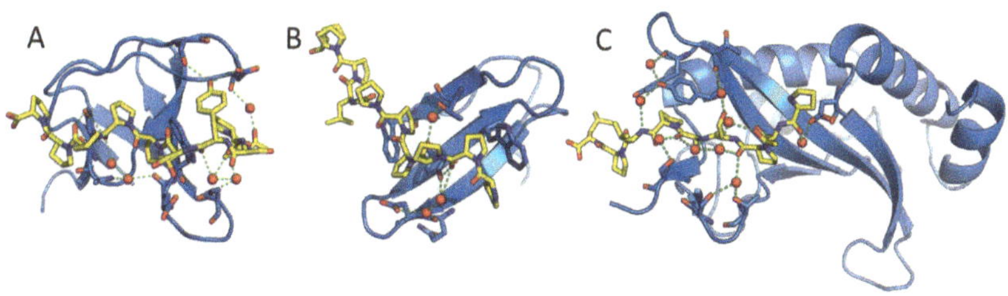

Fig. 3 Water-mediated interaction patterns in SH3, WW and UEV domains. The domains are shown in light red ribbons, ligands are depicted as marine blue sticks, and relevant protein residues are represented as light red sticks. Interfacial waters are shown as nonbonded red spheres. Hydrogen bonds are depicted as discontinuous green lines. (A) Abl-SH3 in complex with the rationally designed peptide p41 (APSYSPPPPP) (1BBZ). (B) Dystrophin–WW domain in complex with a β-dystroglycan peptide (1EG4). (C) Tsg101–UEV domain in complex with a proline-rich viral late domain sequence (3OBQ). Water molecules are those identified elsewhere (Martin-Garcia, Ruiz-Sanz, & Luque, 2012). All the structures have been drawn using PyMol (The PyMOL Molecular Graphics System, Version 2.5 Schrödinger, LLC.).

(Palencia, Cobos, Mateo, Martinez, & Luque, 2004). The origin of the binding energetics for proline-rich ligands to the Abl-SH3 domain was further investigated by a comparative calorimetric analysis of a set of p41-related ligands. The striking effects upon the enthalpic and entropic contributions provoked by conservative substitutions at solvent-exposed positions in the ligand confirmed the complexity of the interaction, as well as the energetic relevance that such network of water molecules possess in the interaction (Palencia et al., 2004). The design of mutations in the Abl-SH3 domain to alter the water-mediated network confirmed such a hypothesis through thermodynamic, structural and molecular dynamics experiments. In the light of these results, a new dual binding mechanism is proposed, where a network of water-mediated hydrogen bonds complements the canonical hydrophobic interactions. The mechanism provides a better description of proline-rich ligand recognition by Abl-SH3 and has important implications for rational design (Palencia, Camara-Artigas, Pisabarro, Martinez, & Luque, 2010).

The systematic analysis of the SH3 structural database reveals that this dual binding mode is universal to SH3 domains. Tightly bound buried-interfacial water molecules mediating the interaction between the peptide ligand and the domain were found in most of the SH3 complexes studied. Moreover, structural waters were also identified in a high percentage of the free SH3 domains. Thus, the analysis enabled the identification of conserved hydration sites in the polyproline-recognition region and the establishment of relationships between hydration profiles and the sequence of both, ligands and SH3 domains (Zafra-Ruano & Luque, 2012).

Water-mediated interactions have also been systematically observed in WW, UEV and EVH1 crystal structures (Fig. 3), where similar thermodynamic signatures have been observed for binding. This outcome strongly suggests that the current description of proline-rich sequence recognition by modular protein domains is incomplete and insufficient for a correct structural and energetic understanding of these systems. In the words of the authors, *a new binding paradigm is required that includes interfacial water molecules as relevant elements in polyproline-recognition domains* (Martin-Garcia, Ruiz-Sanz, & Luque, 2012).

3. High-throughput screening strategies for the design of binding partners for polyproline-binding domains

Since the first X-ray structure of an SH3 domain was published in 1992 (Musacchio, Noble, Pauptit, Wierenga, & Saraste, 1992), the first reported of a polyproline-binding domain, the main focus has been finding potential binding partners for these modular domains (Musacchio, Gibson, Lehto, & Saraste, 1992). From a practical point of view, high-throughput screening (HTS) methodologies have contributed the most to this point, since they allowed obtaining moderately or highly related peptide sequences for these domains. The analysis of common features among these sequences allowed knowing the binding abilities of polyproline-binding domains.

Focussing on SH3 domains, the most studied among the five families, phage display screening with fully randomized libraries revealed that the SH3 recognition sequences identified shared a conserved xPxxP motif. Step by step experimental protocols can be found elsewhere (Sparks, Adey, Quilliam, Thorn, & Kay, 1995; Tonikian, Zhang, Boone, & Sidhu, 2007). Biased libraries have been usually designed by including the consensus motif flanked by random sequences, and expanding recognition over a 12 amino acids window but exploring each library a sequence space of 6 amino acids (as an example: Gx_6G, x_6PPIP, $RSLRPLx_6$, $PPPYPPx_6$ (Rickles et al., 1994)). Screens of such bias against a representative number of SH3 domains effectively converged toward domain-specific sequences. Thus, each SH3 domain has a preferred sequence: Src-SH3 the sequence xxxRPLPPLPxP, Fyn xxxRPLPP(I/L)Pxx, Lyn RxxRPLPPLPxP, PI3K RxxRPLPPLPPP, while Abl-SH3 selects phages containing the sequence PPPYPPPP(I/V)Pxx (Rickles et al., 1994). Subsequent studies demonstrated that these high-affinity peptide ligands may serve as new tools to interfere with the cellular functions of SH3-mediated processes and constitute the basis for the design of SH3-specific inhibitors of disease pathways (Cheadle et al., 1994; Rickles et al., 1995; Sparks, Quilliam, Thorn, Der, & Kay, 1994).

Parallel studies with WW domains have also yielded a similar perspective. For example, a comprehensive screening of 42 WW domains against several cellulose-bound peptide libraries, including designed peptides and peptides based on naturally occurring sequences, and phosphorylated residues, was carried out. Thirty-two WW domains were classified into six groups according to detected ligand recognition preferences for binding the motifs PPx(Y/pY), (p/φ)P(p,g)PPpR, (p/φ)PPRgpPp, PPLPp, (p/ψ) PPPPP, and (pS/pT)P (Kasanov, Pirozzi, Uveges, & Kay, 2001; Linn et al., 1997; Otte et al., 2003).

The use of biased phage display libraries has been especially informative when applied to the c-Src-SH3 domain. Thus, Feng et al. (Feng, Kasahara, Rickles, & Schreiber, 1995) found two dodecapeptides that are oriented oppositely at the binding site (defined as class I and class II). Both peptide sequences are constituted by: (i) a core proline-rich sequence arranged as a PPII helix; and (ii) a flanking sequence that occupies a large pocket between the RT and n-Src loops of the SH3 domain (Fig. 1). However, detailed structural and mutational analyses illustrate how the two flanking sequences exploit the binding surface differently to increase the binding affinity and specificity (Bacarizo & Camara-Artigas, 2013; Bacarizo, Martinez-Rodriguez, & Camara-Artigas, 2015; Camara-Artigas, Ortiz-Salmeron, Andujar-Sanchez, Bacarizo, & Martin-Garcia, 2016; Martin-Garcia, Luque, Ruiz-Sanz, & Camara-Artigas, 2012). Interestingly, a similar study carried out with Lck-SH3 has shown that phage-display selected peptides represented mainly class I ligands, whereas the comparatively diverse Lck-SH3 binding sites of all analyzed natural binding partners emerged as class II proteins. An explanation for the observed variations between artificial and native ligands, which are not due to significant differences in affinity, suggest that phage display may not irrevocably mirror physiologically relevant domain-ligand interactions (Tran et al., 2005).

Definitely, these studies have provided relevant information at a molecular level on the interactions responsible for the binding affinity and specificity of SH3 and WW domains. In addition, the high-affinity ligands (sometimes improved over 40-fold compared to natural affinities) can be used as tools for the *in vivo* and *in vitro* blocking of natural protein-protein interactions. Nevertheless, as we have shown above, these analyses have produced ligands that in some cases resemble the natural ones but, in many others, such artificial binding motifs clearly differ from them (Cestra et al., 1999; Fazi et al., 2002; Mongiovi et al., 1999).

To overcome these discrepancies, Tong et al. (Tong et al., 2002) describe how to combine phage display and cDNA yeast-two-hybrid results to produce highly reliable interaction predictions, including both artificial and natural sequences, respectively. Such predictions were achieved computationally, by identifying potential natural binding partners within the yeast proteome for 20 selected yeast SH3 domains assayed experimentally and, thus, generating a possible natural network of interactions for these SH3 domains. Therefore, to map natural interaction networks, phage display improves when combined with other approaches (Sidhu, Bader, & Boone, 2003). An interesting alternative would be the cloning and further screening of a cDNA library into the surface of a bacteriophage, which profits the main advantages of both approaches (Kurakin & Bredesen, 2002).

The SPOT (Specificity Prediction Of Target) algorithm emerged as an attempt of integrating these "irrational" high-throughput approaches with "rational" structural and energetic information. SPOT was also developed to overcome the limitation of phage display studies which assume that residues contribute individually to the binding. Thus, the sequence alignment by means of tools as WebLogo (https://weblogo.berkeley.edu/) allows knowing the frequency of residues at individual positions; the overall height of the stack indicates the sequence conservation at that position, while the height of symbols within the stack indicates the relative frequency of each amino or nucleic acid at that position (Fig. 4). Nevertheless, it is well known that correlation among definite positions can exist, and the selection of one amino acid at a definite position can determine the selection at another one (Stein & Aloy, 2008). Thus, SPOT would predict peptide recognition specificity of any SH3 domain from its sequence. The known structures of SH3/peptide complexes and the domain sequence alignments are used to rationalize the experimental data obtained by panning peptide repertoires displayed in bacteriophages. The results of panning experiments are organized in contact matrices which describe the frequency of occurrence of any specific pair of residues in the SH3 and in the ligand peptide for each SH3/peptide contact position. The matrices are then used to evaluate the probability of interaction between different SH3 domains and a given peptide. Therefore, SPOT can be applied to any family of proteins for which both structural information about at least one protein/peptide complex and energetic data between a few domain/peptide sequences may exist.

```
---VLITYDELS-SWC
-EDFLIEYAEVSS---
----PPDYDAVVKGLL (x2)
----PPDYEAIYKFV-
MSDTLPSYQTLC----
KFTTLPSYLSIF----
NIVPLPSYASLF---- (x2)
WD--LPLYSEIILP-- (x3)
---SLPEYHDLVPYW-
----LPLYCELYFDTC (x6)
WFPMLPSYWDLF----
----LPLYEDLYYYTF
```

Fig. 4 Phage display results of the panning against a WW domain of a fully randomized X_{12} library displayed in M13-pVIII protein. The resulting sequences are aligned in the left and the corresponding Web-logo representation (https://weblogo.berkeley.edu/logo.cgi) in the right. In the Web-logo, the total height indicates the degree of conservation of that position in the alignment; within each position, the height of the symbol indicates the frequency of the corresponding amino acid. Color code corresponds to the nature of amino acids (black = apolar, green = polar, red = negative charge, blue = positive charge).

In our particular example, SPOT was designed to predict the preferred ligand of SH3 domains whose specificity was not known experimentally (Brannetti, Via, Cestra, Cesareni, & Helmer-Citterich, 2000). Today, the development of microarrays technology allows the synthesis and screening of thousands of peptide sequences arranged on cellulose membrane supports (Kramer & Schneider-Mergener, 1998).

SPOT analysis has also been developed for other polyproline-binding domains. Thus, the phage-display-derived recognition motif for CD2BP2-GYF is PPG(W/F/Y/M/L) and was confirmed by SPOT analysis too, which also confirmed that the GYF domain interacts with peptides from human proteins containing a consensus polyproline motif (Fig. 1). Epitope mapping by NMR spectroscopy performed for several peptides showed a conserved binding surface. The new interacting partners discovered were verified by yeast two-hybrid analysis (Kofler, Motzny, Beyermann, & Freund, 2005).

In the case of WW domains, and mainly due to their small size, SPOT platforms were developed to analyze the structure–function relationship of FBP28-WW domain through a complete substitution screening within the domain, with variants synthesized as a cellulose-bound peptide array. The functionality of the FBP28-WW domain variants was examined by probing a peptide array of potential ligands (Przezdziak et al., 2006). A similar approach has been used to select a high-affinity WW domain against the extracellular region of VEGF receptor isoform-2 from a combinatorial library using a cell-free molecular display platform. The isolate has nanomolar affinity to VEGFR-2 and inhibits the binding of human VEGF to its receptor (with nanomolar affinity as well). The structure is amenable to cyclization to improve its proteolytic stability and has the clear advantage over larger protein scaffolds of being synthesized chemically to high yields, offering the potential for therapeutic and nontherapeutic applications (Patel, Mathonet, Jaulent, & Ullman, 2013).

In any case, the strategy of designing artificial modular protein domains with definite affinities and specificities can also be achieved if the randomized residues in the domain are those responsible for these binding characteristics. Thus, SH3 binding properties can be profoundly altered by sequence modifications in the RT-loop and n-Src-loop regions (Bauer & Sticht, 2007; Hiipakka & Saksela, 2007). Going even further, Freund et al. have developed a novel SH3 domain scaffold engineered to bind the CD2BP2-GYF domain and the HIV capsid protein. Phage display screens resulted in the isolation of variants with changes in all randomized positions compared to wild-type SH3. The soluble scaffolds bind with

340 and 600 nM affinity to HIV capsid protein and CD2BP2-GYF, respectively. Similar approaches have been designed to target and inhibit molecular interactions in the context of disease, with both diagnostic and therapeutic potential. Apart from antibodies, several alternative scaffolds have been exploited over the years making use of the fact that individual domain families are best suited for certain target families (Piotukh & Freund, 2012). Antibodies for SH3 domains have also been obtained by phage display of an antigen-binding fragment (Fab) library. The authors have established a rapid and robust HTS methodology that can be applied to generate highly functional and renewable antibodies targeting protein domains on a proteome-wide scale. Affinity assays demonstrated that representative Fabs bind tightly and specifically to their targets. The Fabs were tested in common cell biology applications and confirmed recognition of the full-length antigen in immunoprecipitation, immunoblotting and immunofluorescence assays (Huang et al., 2015).

Particularly interesting would be the work carried out by Ranganathan et al., aimed at designing *de novo* artificial WW domains. Their method, termed statistical coupling analysis (SCA), does not use structural or physicochemical information but instead extracts information about essential patterns of amino acids from the evolutionary record. The statistical information used to compute the artificial WW domains is just the encoded in a multiple sequence alignment, where the computer extracts simple statistical energy functions capturing coevolution between amino acid residues. The artificial proteins showed thermodynamic stabilities and structural arrangements similar to natural WW domains (Socolich et al., 2005). In addition, the authors demonstrated that these artificial WW sequences function like their natural counterparts, showing class-specific recognition of proline-containing target peptides. Moreover, they identified the network of residues responsible for the functional specificity in WW domains. This work demonstrates that a relatively small quantity of sequence information is sufficient to specify the global energetics of amino acid interactions (Russ, Lowery, Mishra, Yaffe, & Ranganathan, 2005).

Definitely, the computational integration of several screening approaches, together with the technological development which allows them to expand the diversity of sequences to be explored, emerges as the most promising strategy to obtain integrative information on the specificity networks of modular protein domains (Teyra, Sidhu, & Kim, 2012). Coming back to SH3 domains, the combination of phage display, yeast two-hybrid and peptide array screening to independently identify SH3 domain binding partners, followed by the integration of all results using a Bayesian algorithm has allowed the authors to generate a high-confidence yeast SH3 domain interaction map. In addition, this analysis has permitted the prediction of new functionalities for concrete SH3 domains (Tonikian et al., 2009). An additional dimension that combined approaches can include in the protein/peptide binding studies would be the evolution analysis among species of these interaction networks. Specifically, an interesting study carried out in several yeast species points toward a remarkably conserved specificity profile for SH3 domains over a large evolutionary distance. The goal of these kinds of studies would be the unraveling of complex specificity networks of SH3 and other peptide recognition domains in higher eukaryotes, including mammals (Teyra et al., 2017; Verschueren et al., 2015).

On the other hand, the computational rationalization of HTS results has allowed the assembling of a database for peptide recognition modules (http://www.prm-db.org) that will enable many structural, functional, and biological studies (Teyra et al., 2020). These databases will also be enough information to map the specificity profiles in detail, including derived-position weight matrices and binding specificity logos based on multiple peptide ligands. These studies conclude that optimal peptide ligands resemble peptides observed in the existing structures of domain/ligand complexes, indicating that a large majority of the phage-derived peptides are likely to target natural peptide-binding sites and could thus act as inhibitors of natural protein-protein interactions.

3.1 High-throughput screening of cyclic peptides

The life cycle of enveloped viruses requires extensive assistance from the host cell proteins and pathways. Tsg101 is one of the cellular proteins involved in the budding process of virions. In this process, Tsg101 is recruited from the internal site of the infected cell to the budding site to aid in the release of the HIV-1 virus particles. Depletion of Tsg101 from virus-producing cells can lead to a budding defect. Therefore, Tsg101 is a potentially attractive target for therapeutic intervention in several viral infections. VP40 and Gag matrix proteins are the key viral proteins that drive the budding process by mediating specific virus/host interactions to facilitate the efficient release of virions from the infected cell, concretely, the interaction between some motifs displayed by these proteins (generically named as viral L domains), which contain the consensus sequences PTAP and PPxY (Fig. 1), with the human domains Tsg101-UEV and Nedd4-WW3, respectively (Harty, 2009).

Although the development of inhibitors of the interaction between viral L domains and their cellular target domains will drive the development of wide-spectrum antivirals, few strategies have been developed to discover functional compounds. Currently, these strategies are in their early stages, far from providing molecules with pharmacologically optimal properties and/or in the clinical phase. The works published during the last decade by Professors Tavassoli and Harty demonstrate that it is possible to block with small molecules these interactions. Specifically, Tavassoli et al. have published a strategy for the *in vivo* screening, in *Escherichia coli*, of libraries of cyclic peptides capable of inhibiting the interaction between the TSG101-UEV domain and the L-domain PTAP motif of the p6 region of the HIV-Gag protein. Cyclic peptides have a greater survival time than linear, which increases their therapeutic applicability. This methodology allowed the identification of cyclic penta-peptides that lack the canonical P(T/S)xP motif recognized by Tsg101-UEV. These inhibitors have an alternative specificity that allows them to block the release of HIV viral particles in a cellular model, without interfering in the cellular traffic mediated by the interaction between Tsg101-UEV and the PTAP motif of the cellular protein HGS (hepatocyte growth factor-regulated tyrosine kinase substrate). Thus, these findings suggest that small molecule therapeutics addressed to inhibit specific interactions between viral and host proteins may have general applicability in antiviral therapy (Lennard, Gardner, Doigneaux, Castillo, & Tavassoli, 2019; Tavassoli et al., 2008). More recently, cyclic peptide inhibitors of the hepatitis E virus have been found by following a similar strategy (Anang et al., 2018).

3.2 High-throughput screening of drug libraries

On the other hand, Harty et al. have carried out the computational screening of virtual libraries of commercially available molecules to identify inhibitors of the interaction between the viral L domains PTAP and PPxY with their respective cellular targets, Tsg101-UEV and Nedd4-WW3. They have also actively worked on developing assays in human cells that allowed them to validate the biological activity of the primary compounds identified in their *in silico* screening campaigns. These efforts have given promising results, showing that some of these compounds are capable of blocking the cellular release of several encapsulated viruses (Han et al., 2014).

More recently, a simple, robust, and reliable HTS strategy based on an enzyme-linked immunosorbent assay (ELISA) has allowed identifying compounds that inhibit HIV-1 budding, also by targeting the Tsg101-UEV/Gag interaction. Several hits were identified through screening of a 9600-compound library. Subsequent assays revealed some lead compounds, HSM-9 and HSM-10 (Siarot et al., 2018), or FC-10696 (Han et al., 2021), which have antiviral activity.

4. Rational design of ligands for polyproline-binding domains

Parallel to the high-throughput studies, rational design exercises have been carried out with polyproline-binding domains. Although these domains are quite promiscuous, a pioneering work by Pisabarro and Serrano demonstrated that it is possible to design mutations into a peptide ligand to improve their selectivity by using existing biocomputing tools and simple physicochemical reasoning. Thus, they designed mutations in a previously found peptide ligand, able to bind Abl-SH3 and Fyn-SH3 with similar affinity, so that the affinity for Abl-SH3 increased 20-fold, while that for Fyn-SH3 decreased 10-fold. This work demonstrated that both the RT and n-Src loops are responsible for regulating the specificity for proline-rich ligands. Thus, the first N-terminal positions in the peptide ligand would be important for determining the specificity for these SH3 domains, while the remaining at the C-terminus seem to be more important for the affinity and organize the classical PPII helix characterizing such ligands (Fig. 2) (Pisabarro & Serrano, 1996). A later analysis showed that for the above examples phage display approaches are in full agreement with rational design, but combinatorial peptide library screens may help rational design by improving affinity and specificity of definite regions, reporting new amino acids that were not selected by phage display and rational design (Santamaria, Wu, Boulegue, Pal, & Lu, 2003).

The ability of certain SH3 domains to bind specifically both class I and class II polyproline ligands (opposite orientations) can be designed as well. A detailed mutational and structural analysis of Fyn-SH3 domain revealed that the conserved Trp in the binding pocket of this family can adopt two different orientations that, in turn, determine the type of ligand (class I or class II) able to bind to the domain. The only exceptions would be ligands that deviate from standard polyproline angles. The motion of the conserved Trp depends on the presence of certain residues located in a key position, near the binding pocket. SH3 domains placing aromatic residues in such key position are promiscuous; by contrast, those presenting beta-branched or long aliphatic residues block the conserved Trp in one of the two possible orientations, preventing binding in a class I orientation. This was experimentally demonstrated by a single mutation in such a key position in Fyn-SH3 (Y132I) that effectively abolished class I ligand binding, while preserving binding to class II ones. This work demonstrates that conformational changes having noticeable effects on protein-protein interactions are governed by simple rules, highlighting the importance of structural and energetic details to predict protein-protein interactions (Fernandez-Ballester, Blanes-Mira, & Serrano, 2004).

4.1 Strategies to replace the highly-conserved proline residues without altering the polyproline II helical structure of the peptide ligands

Since the conformational arrangement of the ligands has revealed as a well-conserved feature of their successful interactions with polyproline-binding domains, a great effort has been invested into the natural or artificial replacement of Pro residues within these sequences which, actually, can encompass a significant part of the sequence. As was already demonstrated by Pisabarro and Serrano (Pisabarro & Serrano, 1996; Pisabarro et al., 1998), it is possible to replace some of the Pro residues, postulated to be essential for the interaction with SH3 domains, and still retain a significant affinity in the peptide ligand. Of course, such replacement has to be rationally designed. This fact indicates that the sequence repertoire that could interact with a specific SH3 domain could be larger than previously thought. Phage display may not be helpful in this way, since the highly-conserved xP pockets in the domains have been evolutionarily designed to accommodate Pro residues.

4.1.1 Design of miniprotein scaffolds to assemble the polyproline II binding epitope

An important limitation to all the above design strategies is that the PPII conformation needed to bind to a polyproline-binding domain is maintained in peptides by the presence of the already mentioned highly-conserved Pro residues. Thus, although other amino acids might confer higher specificity and/or affinity to the ligand, Pro residues cannot be mutated because of entropic destabilization (Palencia et al., 2004; Pisabarro & Serrano, 1996). An alternative strategy to simple mutagenesis will be the design of a stable polypeptide scaffold within which the PPII conformation is preserved by residues of the scaffold itself. Thus, a suitable universal scaffold aimed at any polyproline-binding domain should have a region adopting a PPII conformation. The family of the avian pancreatic polypeptide (APP) is suitable

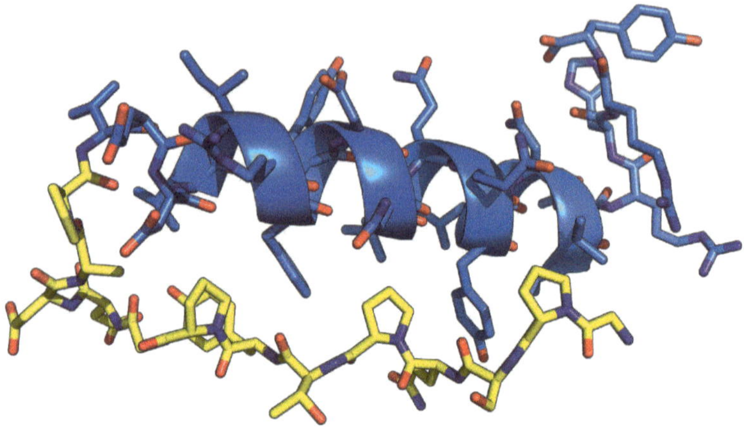

Fig. 5 Crystal structure of the Avian Pancreatic Polypeptide (APP) showing the hairpin arranged between an α-helix (blue ribbon and sticks) and a PPII helix (yellow sticks). The PDB code is 1PPT (Blundell, Pitts, Tickle, Wood, & Wu, 1981). The structure has been drawn using PyMol (The PyMOL Molecular Graphics System, Version 2.5 Schrödinger, LLC.).

for these purposes. APP is a small protein of 36 residues with a fold consisting of an α-helix packed against a PPII helix, without Cys residues (Fig. 5) (Blundell et al., 1981). On the solvent-exposed face of the PPII helix there are no Pro residues, which confirm that the PPII conformation is determined by packing against the α-helix. In fact, the interaction between the PPII helix and the α-helix forms a small hydrophobic core, established by an interdigitation of three Pro side-groups of the PPII helix between the mainly nonpolar side-chains of one face of the amphipathic α-helix. In this situation, the exposed residues in contact with the polyproline-binding domain can be replaced by other amino acids or analogs without affecting the PPII conformation.

This miniprotein has been extensively used by Schepartz et al. for the development of peptide ligands for several proteins, optimizing concrete residues on the structure by phage display (Chin, Grotzfeld, Fabian, & Schepartz, 2001). Thus, by optimizing the solvent-exposed residues of the α-helix, they have obtained APP-derivatives with DNA binding properties, as ligands that overcome the protective effects associated with upregulation of antiapoptotic Bcl-2 proteins, or phosphorylated peptide ligands that recognize the surface of CBP KIX domain; always achieving high affinities (nanomolar to low micromolar). In addition, since the natural APP miniprotein assembles as a dimer, Schepartz et al. identified, characterized, and replaced two structural elements responsible for APP dimerization by proline switches that single-handedly repack APP signature fold. The result was a monomeric and well-folded miniature protein that may serve as a starting point for the *in vitro* and *in vivo* applications of these molecules (Hodges & Schepartz, 2007).

By using the monomeric version of APP and by following a similar optimization strategy (named by the authors as *protein grafting* (Chin et al., 2001)), but exploring the PPII helix, they have developed pGolemi, which binds EVH1 domains. pGolemi is monomeric at 10^{-4} M concentration and well-folded, despite the fact that primary sequence differences exceed 50% from wild-type APP. This designed miniprotein binds with high affinity to the EVH1 domain of Mena (Mena1–112) but not to those of VASP (VASP1–115) or Evl (Evl1–115), and also causes an unusual defect in actin-driven *Listeria monocytogenes* motility. Scanning mutagenesis was used to rationalize affinity and specificity, miniature protein secondary structure, and *L. monocytogenes* motility. The NMR analysis confirmed that pGolemi contains the expected APP-like fold and binds Mena1–112 in a manner highly analogous to the proline-rich repeat region of *L. monocytogenes* ActA protein. Thus, this peptide shares the common pancreatic peptide fold with its scaffold, but shows key differences at the N-terminus. The interplay of spatial fixation and flexibility appears to be the reason for its high affinity toward Mena-EVH1. Together with earlier investigations, these structural data shed light also on the specificity determinants of pGolemi and the importance of additional binding epitopes around the residues Thr74 and Phe32 on EVH1 domains regulating paralog specificity (Holtzman, Woronowicz, Golemi-Kotra, & Schepartz, 2007; Link, Hunke, Mueller, Eichler, & Bayer, 2009).

The application of this strategy may help the design of ligands for other polyproline-recognition domains, and even for the *in vivo* application of these miniproteins. In fact, in our research group we rationally designed a structure in which some residues of the APP PPII helix were replaced by a sequence motif, named RP1 (APSYPPPPP), which interacts with the Abl-SH3 domain. This design, APP-RP1, was well folded and, as shown by circular dichroism, its structural content was similar to that of natural APP. The stability of both miniproteins was compared by unfolding experiments; surprisingly, the designed APP-RP1 is almost 20 °C more thermostable than the wild-type and has a higher Gibbs energy change at 25 °C too. This increase in stability has an entropic origin. Isothermal titration calorimetry and fluorescence spectroscopy showed that the binding thermodynamics of the APP-RP1 molecule to Abl-SH3 is comparable to that of the shorter RP1 peptide. Furthermore, the mutation by Tyr of the two Pro residues from the consensus xPxxP motif in APP-RP1, demonstrates the effectiveness of the scaffold in enhancing the variability in the design of high-affinity and high-specificity ligands for any SH3 domain (Cobos et al., 2004).

4.1.2 Design of peptidomimetics

An alternative way to achieve the replacement of Pro residues will be the development of peptidomimetic compounds. This approach is also convenient for developing therapeutic molecules, since natural peptides may suffer natural proteolysis and their cellular penetration is also limited when used *in vivo*.

The cycle of protein-structure-based combinatorial chemistry followed by structure determination of the highest affinity ligands constitutes a powerful tool to discover ligands containing nonpeptide binding elements to polyproline-binding domains. In the case of c-Src-SH3 the encoded library used had the form Cap-M1-M2-M3-PLPPLP, in which the Cap and Mi's were composed of a diverse set of organic monomers. The PLPPLP portion provided a structural bias directing the nonpeptide fragment Cap-M1-M2-M3 to the SH3 specificity pocket. Fifteen ligands were selected from more than 1 million of different compounds. The NMR solution structures of the c-Src-SH3 domain complexed with two of such selected ligands showed that the nonpeptide moieties of the ligands interact with the specificity pocket of c-Src-SH3, as predicted, but differently to known peptides complexed with this and other SH3 domains. Structural information about the ligands was used to design various homologs, whose affinities for the SH3 domain were measured (Feng, Kapoor, Shirai, Combs, & Schreiber, 1996). The Leu-Pro binding pocket of Src-SH3 was also explored by structure-based split-pool synthesis to discover nonpeptide binding elements. Binding characteristics of the protein xP pocket were then explored by comparing a series of ligands that contain subtle variants of the parent nonpeptide binding structure. Further insights into this receptor/ligand interaction were provided by multidimensional NMR structure determination of one of the nonpeptide ligands (Morken, Kapoor, Feng, Shirai, & Schreiber, 1998).

A significant advance in the design of peptoids for SH3 and WW domains was achieved by Lim et al. (Nguyen, Turck, Cohen, Zuckermann, & Lim, 1998) by exploring amide N-substituted Pro residues. These domains broadly accept amide N-substituted residues, being Pro the only endogenous N-substituted amino acid. This discriminatory mechanism may explain how these domains achieve specific but low-affinity recognition, a property that is necessary for transient signaling interactions. The authors have exploited the mechanism by screening a series of ligands in which some key prolines were replaced by nonnatural N-substituted residues. The results yielded a ligand that selectively bound the Grb2-SH3 domain with 100-times greater affinity. Moreover, the replacement of Pro residues by the N-substituted ones might be a new source to improve ligand specificity (Nguyen et al., 1998). The same authors have tested the effects of combining multiple peptoid substitutions with specific flanking sequences on ligand affinity and specificity, showing that ligands can be selectively tuned to target a single SH3 domain. In addition, by making multiple substitutions, high-affinity ligands can be generated so that they lack entirely the canonical xPxxP motif. The resulting ligands can potently disrupt natural SH3-mediated interactions. Thus, these hybrid scaffolds yield SH3 ligands with markedly improved domain selectivity, overcoming one of the main challenges in designing inhibitors against these domains. Definitely, these compounds represent important leads in the search for orthogonal inhibitors of SH3 domains (Nguyen et al., 2000).

Even more, it has been published elsewhere (Vidal et al., 2004) the design of the highest affinity peptidomimetic ligands reported so far for the Grb2-SH3 domain. These compounds were designed by combining *N*-alkyl amino acid incorporation in a proline-rich sequence with subsequent dimerization of the peptoid sequence based on structural data and molecular modeling. Because the affinity for Grb2 of the optimized compounds was too high to be measured using the fluorescent modifications that they induce on the Grb2 emission spectrum, a competition assay was developed. In this test, Grb2 was pulled down from a cellular extract by an initial VPPPVPPRRR peptide bound to Sepharose beads. In the presence of competitors, the test quantified the amount of Grb2 displaced from the beads. This methodology enabled the determination of a K_d value in the 10^{-10} M range for the highest affinity Grb2 peptoid analog dimer (Vidal et al., 2004).

The N-substitution of amino acids has also been demonstrated as useful for improving some natural ligand interactions. As an example, the phospholipase Cγ1 (PLCγ1) has a regulatory SH3 domain that can specifically recognize and interact with a defined xPxxP-containing decapeptide segment (185-QP**P**VP**P**QRPM-194) coming from the adaptor protein

SLP76. The isolated peptide binds the PLCγ1-SH3 domain with a moderate affinity due to the lack of protein context support. The peptide affinity was noticeably improved by replacing the two key Pro residues, Pro187 and Pro190, of the xPxxP motif with nonnatural N-substituted amino acids. Two N-substituted peptides, N-Leu187/N-Gln190 and N-Thr187/N-Gln190, were designed that improved affinity by, respectively, 8.5- and 3.4-fold ($K_d = 0.67 \pm 0.18$ and $1.7 \pm 0.3\,\mu M$, respectively) relative to native peptide ($K_d = 5.7 \pm 1.2\,\mu M$) (Tang, Zhao, Wang, Ye, & Yang, 2019).

The incorporation of substituents of variable size and polarity at the C2 position of pseudoprolines (ΨPro) offers another powerful tool for the design of nonpeptide ligands. ΨPro consist of Ser-, Thr-, or Cys-derived proline-like structures with enhanced inherent properties of natural L-Pro (Tuchscherer et al., 2001). ΨPro building blocks exert a dual functionality in: (i) inducing and stabilizing the relevant PPII conformation; and (ii) increasing and optimizing the van der Waals contacts and setting hydrogen bonds with the receptor. Most notably, the generation of a library of different substituents at C2 allows the factors contributing to affinity and specificity in protein-protein interactions to be explored, and to further elucidate ligand recognition by polyproline-binding domains at a molecular level (Tuchscherer et al., 2001).

WW domains have also been used as targets for peptidomimetics design. Apart from the Yap-WW domain studied by Lim et al. (Nguyen et al., 1998), the Pin1-WW domain has been extensively used, since is involved in cancer developments and is considered as a pharmaceutical target. Thus, finding a high-affinity inhibitor of Pin1 has become an attractive topic. The WW domain of human Pin1 can recognize phosphoserine/phosphothreonine-proline (pS/pT)P motifs. A series of four-substituted proline derivatives were incorporated into the phosphopeptides and investigated their affinities for the WW domain of Pin1. Isothermal titration calorimetry and fluorescence anisotropy analyses showed that the replacement of Pro with (2S,4R)-4-fluoroproline increased the binding affinity of the peptides, whereas circular dichroism measurements suggested that a more PPII-like structure of phosphopeptides would be the reason. Chemical shift perturbation experiments also indicated that (2S,4R)-4-fluoroproline interacts with Trp34 of the WW domain in the binding site, revealing a strong C—H⋯π interaction (Huang & Horng, 2015).

To explore the druggability of these modular protein domains, a modular toolkit has been developed to generate small molecules able to compete with defined domain/peptide interactions (Opitz et al., 2015). The modular strategy collects a comprehensive toolkit of chemical fragments (ProMs) designed to replace pairs of conserved Pro residues in recognition motifs. The authors developed a small, selective, peptidomimetic inhibitor of Ena/VASP-EVH1 domain interactions as proof of principle. Highly invasive MDA MB 231 breast-cancer cells treated with this ligand showed displacement of VASP from focal adhesions, as well as from the front of lamellipodia, and strongly reduced cell invasion. Based on this previously developed nonpeptidic micromolar inhibitor, the authors determined 22 crystal structures of Ena-EVH1 in complex with different inhibitors and rationally extended the library of conformationally defined Pro-derived modules to succeed in developing a nanomolar inhibitor of lower size ($K_d = 120\,nM$; $MW = 734\,Da$). In contrast to the previous inhibitor, the optimized compounds reduced extravascularization of invasive breast cancer cells in a zebrafish model. In words of the authors, *this study represents an example of successful, structure-guided development of low molecular weight inhibitors specifically and selectively addressing a proline-rich sequence-recognizing domain that is characterized by a shallow epitope lacking defined binding pockets*. Thus, the evolved high-affinity inhibitor may now serve as a tool in validating the basic therapeutic concept, i.e., the suppression of cancer metastasis by inhibiting a crucial protein-protein interaction involved in actin filament processing and cell migration (Barone et al., 2020). The general applicability of this strategy was also illustrated by the design of an ErbB4-derived ligand containing two ProM fragments, targeting the YAP1-WW domain with a fivefold higher affinity (Opitz et al., 2015).

4.2 Design of chimeric proteins

Chimeric proteins have been traditionally designed to target and inhibit molecular interactions in the context of disease, with both diagnostic and therapeutic potential. Apart from antibodies, several alternative scaffolds have been exploited over the years using the fact that individual domain families are best suited for certain target families.

With the aim of improving the NMR NOE pattern for protein structure determination of low/medium affinity complexes of interacting molecules, covalent linkage can be a way to shift the equilibrium between the interacting partners to the bound state. This approach has been developed with the CD2BP2-GYF domain and the target peptide SHRPPPPGHRV from CD2. The peptide was covalently linked at the N-terminus of the domain through a 21 residues linker. In conjunction with general recognition rules for proline-rich sequence recognition the NOEs obtained allowed the accurate modeling of the protein/peptide complex (Freund et al., 2003).

Nature has also been exploited a similar approach among the SH3 domains of the Tec family of tyrosine kinases (Roberts et al., 2016). Since the binding site is far away from the N- and C-terminal tails of these domains, an alternative to shorten the connecting linker has been the design of a chimeric protein (SPCp41) by connecting a circular permutant of the α-spectrin SH3 (Spc-SH3) domain to the proline-rich decapeptide p41 (APSYSPPPPP) with a simple three-residues link. The aim was to obtain a single-chain protein with a tertiary fold that would imitate the binding between SH3 domains and proline-rich peptides. The high-resolution NMR structure of the chimera reproduces perfectly the interactions typically found in SH3/peptide complexes and is remarkably similar to that of the complex between the Spc-SH3 domain and the ligand. NMR relaxation data confirmed the tight binding between the ligand and the SH3 part of the chimera (Candel, Conejero-Lara, Martinez, van Nuland, & Bruix, 2007). Accordingly, DSC experiments indicated that the interactions at the binding interface develop high cooperativity with the rest of the structure since the protein unfolds by a two-state process. The chimera is more stable than the circular permutant by 6–8 $kJ\,mol^{-1}$ of Gibbs energy at 25 °C; in addition, the difference in unfolding enthalpy is approximately 32 $kJ\,mol^{-1}$. These values coincide with those found for the binding of proline-rich peptides to SH3 domains (Martin-Sierra et al., 2003). In full agreement, the folding/unfolding kinetics can be correctly interpreted by a two-state process, where the folding transition state produces essentially the same picture shown by the circular permutant S19-P20s (the "nucleus" of the design), and the ligand will dock at the latter stages of folding (Candel, Cobos, Conejero-Lara, & Martinez, 2009). Thus, all conclusions corroborate the effectiveness of the chimera SPCp41 to study energetic, dynamic and structural aspects of SH3/ligand interactions. In fact, the contribution of each Pro residue of the ligand sequence (APSYSPPPPP) to the SH3/peptide interaction was evaluated by producing six single Pro-Ala mutants of the chimeric protein. Structural analyses of the mutant chimeras by circular dichroism, fluorescence and NMR, together with NMR-relaxation measurements, indicate conformational flexibility at the binding interface, which is strongly affected by the different Pro-Ala mutations. A three-state unfolding model was developed to evaluate the unfolding thermodynamics by DSC. The model assumes equilibrium between the "unbound" and "bound" states at the SH3/peptide binding interface and, therefore, allows distinguishing the thermodynamic magnitudes of the interaction at the binding interface from the ones of the unfolding. The resulting thermodynamic magnitudes classify the different Pro residues according to their relevance in the interaction as P2 ∼ P7 ∼ P10 > P9 ∼ P6 > P8, which agrees well with Lim's model for the interaction between SH3 domains and proline-rich peptides. These results demonstrate that this chimeric design may serve as a suitable tool to analyze the energetics of weak biomolecular interactions such as those involving SH3 domains and other polyproline-binding domains, using a combination of unfolding experiments and site-directed mutagenesis (Candel, van Nuland, Martin-Sierra, Martinez, & Conejero-Lara, 2008).

The ability to design proteins with desired properties by using protein structural information will allow creating high-value therapeutic and diagnostic products. For example, using the protein structures of lambda lysozyme and the human Crk-SH3 domain, a synthetic protein switch that controls lysozyme activity by sterically hindering its active cleft through the binding of SH3 to its CB1 peptide-binding partner was designed. The fusion protein, including lysozyme and CB1, was tested *in vitro*. In the absence of SH3, the lysozyme-CB1 fusion protein functioned normally. In the presence of SH3, the lysozyme activity was inhibited, and with the addition of excess CB1 peptides to compete for SH3 binding the lysozyme activity was restored. Lastly, this structure-based strategy can be used to engineer synthetic regulation by peptide/domain binding interfaces into a variety of proteins (Pham & Truong, 2012).

References

Anang, S., Kaushik, N., Hingane, S., Kumari, A., Gupta, J., Asthana, S., et al. (2018). Potent inhibition of hepatitis E virus release by a cyclic peptide inhibitor of the interaction between viral open reading frame 3 protein and host tumor susceptibility gene 101. *Journal of Virology, 92*(20).

Bacarizo, J., & Camara-Artigas, A. (2013). Atomic resolution structures of the c-Src SH3 domain in complex with two high-affinity peptides from classes I and II. *Acta Crystallographica. Section D, Biological Crystallography, 69*(Pt. 5), 756–766.

Bacarizo, J., Martinez-Rodriguez, S., & Camara-Artigas, A. (2015). Structure of the c-Src-SH3 domain in complex with a proline-rich motif of NS5A protein from the hepatitis C virus. *Journal of Structural Biology, 189*(1), 67–72.

Barone, M., Muller, M., Chiha, S., Ren, J., Albat, D., Soicke, A., et al. (2020). Designed nanomolar small-molecule inhibitors of Ena/VASP EVH1 interaction impair invasion and extravasation of breast cancer cells. *Proceedings of the National Academy of Sciences of the United States of America, 117*(47), 29684–29690.

Bauer, F., & Sticht, H. (2007). A proline to glycine mutation in the Lck SH3-domain affects conformational sampling and increases ligand binding affinity. *FEBS Letters, 581*(8), 1555–1560.

Blundell, T. L., Pitts, J. E., Tickle, I. J., Wood, S. P., & Wu, C. W. (1981). X-ray analysis (1. 4-A resolution) of avian pancreatic polypeptide: Small globular protein hormone. *Proceedings of the National Academy of Sciences of the United States of America*, *78*(7), 4175–4179.

Brannetti, B., Via, A., Cestra, G., Cesareni, G., & Helmer-Citterich, M. (2000). SH3-SPOT: An algorithm to predict preferred ligands to different members of the SH3 gene family. *Journal of Molecular Biology*, *298*(2), 313–328.

Camara-Artigas, A., Ortiz-Salmeron, E., Andujar-Sanchez, M., Bacarizo, J., & Martin-Garcia, J. M. (2016). The role of water molecules in the binding of class I and II peptides to the SH3 domain of the Fyn tyrosine kinase. *Acta Crystallographica. Section F, Structural Biology Communications*, *72*(Pt. 9), 707–712.

Candel, A. M., Cobos, E. S., Conejero-Lara, F., & Martinez, J. C. (2009). Evaluation of folding co-operativity of a chimeric protein based on the molecular recognition between polyproline ligands and SH3 domains. *Protein Engineering, Design & Selection*, *22*(10), 597–606.

Candel, A. M., Conejero-Lara, F., Martinez, J. C., van Nuland, N. A., & Bruix, M. (2007). The high-resolution NMR structure of a single-chain chimeric protein mimicking a SH3-peptide complex. *FEBS Letters*, *581*(4), 687–692.

Candel, A. M., van Nuland, N. A., Martin-Sierra, F. M., Martinez, J. C., & Conejero-Lara, F. (2008). Analysis of the thermodynamics of binding of an SH3 domain to proline-rich peptides using a chimeric fusion protein. *Journal of Molecular Biology*, *377*(1), 117–135.

Cesareni, G., Gimona, M., Sudol, M., & Yaffe, M. (2005). *Modular protein domains*. Weinheim: Wiley-VCH.

Cestra, G., Castagnoli, L., Dente, L., Minenkova, O., Petrelli, A., Migone, N., et al. (1999). The SH3 domains of endophilin and amphiphysin bind to the proline-rich region of synaptojanin 1 at distinct sites that display an unconventional binding specificity. *The Journal of Biological Chemistry*, *274*(45), 32001–32007.

Cheadle, C., Ivashchenko, Y., South, V., Searfoss, G. H., French, S., Howk, R., et al. (1994). Identification of a Src SH3 domain binding motif by screening a random phage display library. *The Journal of Biological Chemistry*, *269*(39), 24034–24039.

Chin, J. W., Grotzfeld, R. M., Fabian, M. A., & Schepartz, A. (2001). Methodology for optimizing functional miniature proteins based on avian pancreatic polypeptide using phage display. *Bioorganic & Medicinal Chemistry Letters*, *11*(12), 1501–1505.

Cobos, E. S., Pisabarro, M. T., Vega, M. C., Lacroix, E., Serrano, L., Ruiz-Sanz, J., et al. (2004). A miniprotein scaffold used to assemble the polyproline II binding epitope recognized by SH3 domains. *Journal of Molecular Biology*, *342*(1), 355–365.

Corbi-Verge, C., & Kim, P. M. (2016). Motif mediated protein-protein interactions as drug targets. *Cell Communication and Signaling*, *14*, 8.

Fazi, B., Cope, M. J., Douangamath, A., Ferracuti, S., Schirwitz, K., Zucconi, A., et al. (2002). Unusual binding properties of the SH3 domain of the yeast actin-binding protein Abp1: Structural and functional analysis. *The Journal of Biological Chemistry*, *277*(7), 5290–5298.

Feng, S., Kapoor, T. M., Shirai, F., Combs, A. P., & Schreiber, S. L. (1996). Molecular basis for the binding of SH3 ligands with non-peptide elements identified by combinatorial synthesis. *Chemistry & Biology*, *3*(8), 661–670.

Feng, S., Kasahara, C., Rickles, R. J., & Schreiber, S. L. (1995). Specific interactions outside the proline-rich core of two classes of Src homology 3 ligands. *Proceedings of the National Academy of Sciences of the United States of America*, *92*(26), 12408–12415.

Fernandez-Ballester, G., Blanes-Mira, C., & Serrano, L. (2004). The tryptophan switch: Changing ligand-binding specificity from type I to type II in SH3 domains. *Journal of Molecular Biology*, *335*(2), 619–629.

Freund, C., Kuhne, R., Park, S., Thiemke, K., Reinherz, E. L., & Wagner, G. (2003). Structural investigations of a GYF domain covalently linked to a proline-rich peptide. *Journal of Biomolecular NMR*, *27*(2), 143–149.

Freund, C., Schmalz, H. G., Sticht, J., & Kuhne, R. (2008). Proline-rich sequence recognition domains (PRD): Ligands, function and inhibition. *Handbook of Experimental Pharmacology*, (186), 407–429.

Han, Z., Lu, J., Liu, Y., Davis, B., Lee, M. S., Olson, M. A., et al. (2014). Small-molecule probes targeting the viral PPxY-host Nedd4 interface block egress of a broad range of RNA viruses. *Journal of Virology*, *88*(13), 7294–7306.

Han, Z., Ye, H., Liang, J., Shepley-McTaggart, A., Wrobel, J. E., Reitz, A. B., et al. (2021). Compound FC-10696 inhibits egress of Marburg virus. *Antimicrobial Agents and Chemotherapy*, *65*(7), e0008621.

Harty, R. N. (2009). No exit: Targeting the budding process to inhibit filovirus replication. *Antiviral Research*, *81*(3), 189–197.

Hiipakka, M., & Saksela, K. (2007). Versatile retargeting of SH3 domain binding by modification of non-conserved loop residues. *FEBS Letters*, *581*(9), 1735–1741.

Hodges, A. M., & Schepartz, A. (2007). Engineering a monomeric miniature protein. *Journal of the American Chemical Society*, *129*(36), 11024–11025.

Holtzman, J. H., Woronowicz, K., Golemi-Kotra, D., & Schepartz, A. (2007). Miniature protein ligands for EVH1 domains: Interplay between affinity, specificity, and cell motility. *Biochemistry*, *46*(47), 13541–13553.

Huang, H., Economopoulos, N. O., Liu, B. A., Uetrecht, A., Gu, J., Jarvik, N., et al. (2015). Selection of recombinant anti-SH3 domain antibodies by high-throughput phage display. *Protein Science*, *24*(11), 1890–1900.

Huang, K. Y., & Horng, J. C. (2015). Modulating the affinities of phosphopeptides for the human Pin1 WW domain using 4-substituted proline derivatives. *Biochemistry*, *54*(40), 6186–6194.

Kasanov, J., Pirozzi, G., Uveges, A. J., & Kay, B. K. (2001). Characterizing Class I WW domains defines key specificity determinants and generates mutant domains with novel specificities. *Chemistry & Biology*, *8*(3), 231–241.

Kofler, M., Motzny, K., Beyermann, M., & Freund, C. (2005). Novel interaction partners of the CD2BP2-GYF domain. *The Journal of Biological Chemistry*, *280*(39), 33397–33402.

Kramer, A., & Schneider-Mergener, J. (1998). Synthesis and screening of peptide libraries on continuous cellulose membrane supports. *Methods in Molecular Biology*, *87*, 25–39.

Kurakin, A., & Bredesen, D. (2002). Target-assisted iterative screening reveals novel interactors for PSD95, Nedd4, Src, Abl and Crk proteins. *Journal of Biomolecular Structure & Dynamics*, *19*(6), 1015–1029.

Lennard, K. R., Gardner, R. M., Doigneaux, C., Castillo, F., & Tavassoli, A. (2019). Development of a cyclic peptide inhibitor of the p6/UEV protein-protein interaction. *ACS Chemical Biology*, *14*(9), 1874–1878.

Link, N. M., Hunke, C., Mueller, J. W., Eichler, J., & Bayer, P. (2009). The solution structure of pGolemi, a high affinity Mena EVH1 binding miniature protein, suggests explanations for paralog-specific binding to Ena/VASP homology (EVH) 1 domains. *Biological Chemistry*, *390*(5–6), 417–426.

Linn, H., Ermekova, K. S., Rentschler, S., Sparks, A. B., Kay, B. K., & Sudol, M. (1997). Using molecular repertoires to identify high-affinity peptide ligands of the WW domain of human and mouse YAP. *Biological Chemistry*, *378*(6), 531–537.

Mahoney, N. M., Rozwarski, D. A., Fedorov, E., Fedorov, A. A., & Almo, S. C. (1999). Profilin binds proline-rich ligands in two distinct amide backbone orientations. *Nature Structural Biology*, *6*(7), 666–671.

Martin-Garcia, J. M., Luque, I., Ruiz-Sanz, J., & Camara-Artigas, A. (2012). The promiscuous binding of the Fyn SH3 domain to a peptide from the NS5A protein. *Acta Crystallographica. Section D, Biological Crystallography*, *68*(Pt. 8), 1030–1040.

Martin-Garcia, J. M., Ruiz-Sanz, J., & Luque, I. (2012). Interfacial water molecules in SH3 interactions: A revised paradigm for polyproline recognition. *The Biochemical Journal*, *442*(2), 443–451.

Martin-Sierra, F. M., Candel, A. M., Casares, S., Filimonov, V. V., Martinez, J. C., & Conejero-Lara, F. (2003). A binding event converted into a folding event. *FEBS Letters*, *553*(3), 328–332.

Mongiovi, A. M., Romano, P. R., Panni, S., Mendoza, M., Wong, W. T., Musacchio, A., et al. (1999). A novel peptide-SH3 interaction. *The EMBO Journal*, *18*(19), 5300–5309.

Morken, J. P., Kapoor, T. M., Feng, G. S., Shirai, F., & Schreiber, S. L. (1998). Exploring the leucine-proline binding pocket of the Src SH3 domain using structure-based, split-pool synthesis and affinity-based selection. *Journal of the American Chemical Society*, *120*(1), 30–36.

Musacchio, A., Gibson, T., Lehto, V. P., & Saraste, M. (1992). SH3—An abundant protein domain in search of a function. *FEBS Letters*, *307*(1), 55–61.

Musacchio, A., Noble, M., Pauptit, R., Wierenga, R., & Saraste, M. (1992). Crystal structure of a Src-homology 3 (SH3) domain. *Nature*, *359*(6398), 851–855.

Nguyen, J. T., Porter, M., Amoui, M., Miller, W. T., Zuckermann, R. N., & Lim, W. A. (2000). Improving SH3 domain ligand selectivity using a non-natural scaffold. *Chemistry & Biology*, *7*(7), 463–473.

Nguyen, J. T., Turck, C. W., Cohen, F. E., Zuckermann, R. N., & Lim, W. A. (1998). Exploiting the basis of proline recognition by SH3 and WW domains: Design of N-substituted inhibitors. *Science*, *282*(5396), 2088–2092.

Opitz, R., Muller, M., Reuter, C., Barone, M., Soicke, A., Roske, Y., et al. (2015). A modular toolkit to inhibit proline-rich motif-mediated protein-protein interactions. *Proceedings of the National Academy of Sciences of the United States of America*, *112*(16), 5011–5016.

Otte, L., Wiedemann, U., Schlegel, B., Pires, J. R., Beyermann, M., Schmieder, P., et al. (2003). WW domain sequence activity relationships identified using ligand recognition propensities of 42 WW domains. *Protein Science*, *12*(3), 491–500.

Palencia, A., Camara-Artigas, A., Pisabarro, M. T., Martinez, J. C., & Luque, I. (2010). Role of interfacial water molecules in proline-rich ligand recognition by the Src homology 3 domain of Abl. *The Journal of Biological Chemistry*, *285*(4), 2823–2833.

Palencia, A., Cobos, E. S., Mateo, P. L., Martinez, J. C., & Luque, I. (2004). Thermodynamic dissection of the binding energetics of proline-rich peptides to the Abl-SH3 domain: Implications for rational ligand design. *Journal of Molecular Biology*, *336*(2), 527–537.

Patel, S., Mathonet, P., Jaulent, A. M., & Ullman, C. G. (2013). Selection of a high-affinity WW domain against the extracellular region of VEGF receptor isoform-2 from a combinatorial library using CIS display. *Protein Engineering, Design & Selection*, *26*(4), 307–315.

Pham, E., & Truong, K. (2012). Engineered regulation of lysozyme by the SH3-CB1 binding interaction. *Protein Engineering, Design & Selection*, *25*(6), 307–311.

Piotukh, K., & Freund, C. (2012). A novel hSH3 domain scaffold engineered to bind folded domains in CD2BP2 and HIV capsid protein. *Protein Engineering, Design & Selection*, *25*(10), 649–656.

Pisabarro, M. T., & Serrano, L. (1996). Rational design of specific high-affinity peptide ligands for the Abl-SH3 domain. *Biochemistry*, *35*(33), 10634–10640.

Pisabarro, M. T., Serrano, L., & Wilmanns, M. (1998). Crystal structure of the abl-SH3 domain complexed with a designed high-affinity peptide ligand: Implications for SH3-ligand interactions. *Journal of Molecular Biology*, *281*(3), 513–521.

Przezdziak, J., Tremmel, S., Kretzschmar, I., Beyermann, M., Bienert, M., & Volkmer-Engert, R. (2006). Probing the ligand-binding specificity and analyzing the folding state of SPOT-synthesized FBP28 WW domain variants. *Chembiochem*, 7(5), 780–788.

Rickles, R. J., Botfield, M. C., Weng, Z., Taylor, J. A., Green, O. M., Brugge, J. S., et al. (1994). Identification of Src, Fyn, Lyn, PI3K and Abl SH3 domain ligands using phage display libraries. *The EMBO Journal*, *13*(23), 5598–5604.

Rickles, R. J., Botfield, M. C., Zhou, X. M., Henry, P. A., Brugge, J. S., & Zoller, M. J. (1995). Phage display selection of ligand residues important for Src homology 3 domain binding specificity. *Proceedings of the National Academy of Sciences of the United States of America*, *92*(24), 10909–10913.

Roberts, J. M., Tarafdar, S., Joseph, R. E., Andreotti, A. H., Smithgall, T. E., Engen, J. R., et al. (2016). Dynamics of the Tec-family tyrosine kinase SH3 domains. *Protein Science*, *25*(4), 852–864.

Russ, W. P., Lowery, D. M., Mishra, P., Yaffe, M. B., & Ranganathan, R. (2005). Natural-like function in artificial WW domains. *Nature*, *437*(7058), 579–583.

Santamaria, F., Wu, Z., Boulegue, C., Pal, G., & Lu, W. (2003). Reexamination of the recognition preference of the specificity pocket of the Abl SH3 domain. *Journal of Molecular Recognition*, *16*(3), 131–138.

Siarot, L., Chutiwitoonchai, N., Sato, H., Chang, H., Fujino, M., Murakami, T., et al. (2018). Identification of human immunodeficiency virus type-1 Gag-TSG101 interaction inhibitors by high-throughput screening. *Biochemical and Biophysical Research Communications*, *503*(4), 2970–2976.

Sidhu, S. S., Bader, G. D., & Boone, C. (2003). Functional genomics of intracellular peptide recognition domains with combinatorial biology methods. *Current Opinion in Chemical Biology*, 7(1), 97–102.

Socolich, M., Lockless, S. W., Russ, W. P., Lee, H., Gardner, K. H., & Ranganathan, R. (2005). Evolutionary information for specifying a protein fold. *Nature*, *437*(7058), 512–518.

Sparks, A. B., Adey, N. B., Quilliam, L. A., Thorn, J. M., & Kay, B. K. (1995). Screening phage-displayed random peptide libraries for SH3 ligands. *Methods in Enzymology*, *255*, 498–509.

Sparks, A. B., Quilliam, L. A., Thorn, J. M., Der, C. J., & Kay, B. K. (1994). Identification and characterization of Src SH3 ligands from phage-displayed random peptide libraries. *The Journal of Biological Chemistry*, *269*(39), 23853–23856.

Stein, A., & Aloy, P. (2008). Contextual specificity in peptide-mediated protein interactions. *PLoS One*, *3*(7), e2524.

Tang, W., Zhao, Z., Wang, C., Ye, T., & Yang, B. (2019). Molecular design and optimization of hepatic cancer SLP76-derived PLCgamma1 SH3-binding peptide with the systematic N-substitution of peptide PXXP motif. *Journal of Molecular Recognition*, *32*(12), e2806.

Tavassoli, A., Lu, Q., Gam, J., Pan, H., Benkovic, S. J., & Cohen, S. N. (2008). Inhibition of HIV budding by a genetically selected cyclic peptide targeting the Gag-TSG101 interaction. *ACS Chemical Biology*, *3*(12), 757–764.

Teyra, J., Huang, H., Jain, S., Guan, X., Dong, A., Liu, Y., et al. (2017). Comprehensive analysis of the human SH3 domain family reveals a wide variety of non-canonical specificities. *Structure*, *25*(10). 1598–1610.e3.

Teyra, J., Kelil, A., Jain, S., Helmy, M., Jajodia, R., Hooda, Y., et al. (2020). Large-scale survey and database of high affinity ligands for peptide recognition modules. *Molecular Systems Biology*, *16*(12), e9310.

Teyra, J., Sidhu, S. S., & Kim, P. M. (2012). Elucidation of the binding preferences of peptide recognition modules: SH3 and PDZ domains. *FEBS Letters*, *586*(17), 2631–2637.

Tong, A. H., Drees, B., Nardelli, G., Bader, G. D., Brannetti, B., Castagnoli, L., et al. (2002). A combined experimental and computational strategy to define protein interaction networks for peptide recognition modules. *Science*, *295*(5553), 321–324.

Tonikian, R., Xin, X., Toret, C. P., Gfeller, D., Landgraf, C., Panni, S., et al. (2009). Bayesian modeling of the yeast SH3 domain interactome predicts spatiotemporal dynamics of endocytosis proteins. *PLoS Biology*, 7(10), e1000218.

Tonikian, R., Zhang, Y., Boone, C., & Sidhu, S. S. (2007). Identifying specificity profiles for peptide recognition modules from phage-displayed peptide libraries. *Nature Protocols*, *2*(6), 1368–1386.

Tran, T., Hoffmann, S., Wiesehan, K., Jonas, E., Luge, C., Aladag, A., et al. (2005). Insights into human Lck SH3 domain binding specificity: Different binding modes of artificial and native ligands. *Biochemistry*, *44*(45), 15042–15052.

Tuchscherer, G., Grell, D., Tatsu, Y., Durieux, P., Fernandez-Carneado, J., Hengst, B., et al. (2001). Targeting molecular recognition: Exploring the dual role of functional pseudoprolines in the design of SH3 ligands. *Angewandte Chemie (International Ed. in English)*, *40*(15), 2844–2848.

Verschueren, E., Spiess, M., Gkourtsa, A., Avula, T., Landgraf, C., Mancilla, V. T., et al. (2015). Evolution of the SH3 domain specificity landscape in yeasts. *PLoS One*, *10*(6), e0129229.

Vidal, M., Liu, W. Q., Lenoir, C., Salzmann, J., Gresh, N., & Garbay, C. (2004). Design of peptoid analogue dimers and measure of their affinity for Grb2 SH3 domains. *Biochemistry*, *43*(23), 7336–7344.

Zafra-Ruano, A., & Luque, I. (2012). Interfacial water molecules in SH3 interactions: Getting the full picture on polyproline recognition by protein-protein interaction domains. *FEBS Letters*, *586*(17), 2619–2630.

Zarrinpar, A., Park, S. H., & Lim, W. A. (2003). Optimization of specificity in a cellular protein interaction network by negative selection. *Nature*, *426*(6967), 676–680.

CHAPTER SEVEN

The structural basis of histone modifying enzyme specificity and promiscuity: Implications for metabolic regulation and drug design

Tim J. Kamerzell[a,b,c,d,e,*], Brittney Mikell[c], Lei Chen[d], Harold Elias[d], Buddhadeb Dawn[d], Calum MacRae[c], and C. Russell Middaugh[a]

[a]Department of Pharmaceutical Chemistry, The University of Kansas, Lawrence, KS, United States
[b]Division of Internal Medicine, HCA MidWest Health, Overland Park, KS, United States
[c]Cardiovascular Division, Brigham and Women's Hospital and Harvard Medical School, Boston, MA, United States
[d]Division of Cardiovascular Diseases, Cardiovascular Research Institute, University of Kansas Medical Center, Kansas City, KS, United States
[e]Applied AI Technologies, LLC, Overland Park, KS, United States
*Corresponding author: e-mail address: tkamerzell@ku.edu

Contents

Advances in Protein Chemistry and Structural Biology, Volume 130
ISSN 1876-1623
https://doi.org/10.1016/bs.apcsb.2022.02.004

Abstract

Histone modifying enzymes regulate chromatin architecture through covalent modifications and ultimately control multiple aspects of cellular function. Disruption of histone modification leads to changes in gene expression profiles and may lead to disease. Both small molecule inhibitors and intermediary metabolites have been shown to modulate histone modifying enzyme activity although our ability to identify successful drug candidates or novel metabolic regulators of these enzymes has been limited. Using a combination of large scale *in silico* screens and *in vivo* phenotypic analysis, we identified several small molecules and intermediary metabolites with distinctive HME activity. Our approach using unsupervised learning identifies the chemical fingerprints of both small molecules and metabolites that facilitate recognition by the enzymes active sites which can be used as a blueprint to design novel inhibitors. Furthermore, this work supports the idea that histone modifying enzymes sense intermediary metabolites integrating genes, environment and cellular physiology.

Diverse post-translational modifications of histone core proteins alter the structure of chromatin and regulate DNA accessibility. The patterns of these covalent histone modifications define a cellular code that ultimately regulates gene expression, embryological development and a wide range of cellular functions (Bhaumik, Smith, & Shilatifard, 2007; Kouzarides, 2007; Strahl & Allis, 2000). The most widely recognized histone modifications associated with transcriptional activation and repression include the catalytic transfer of acetyl and methyl groups to and from lysine and arginine residues. The enzymes that catalyze these modifications, histone modifying enzymes (HMEs), are implicated in diverse disease processes. Not surprisingly, the past several decades witnessed significant interest in developing therapeutics targeting the active sites of individual histone modifying enzymes (Arrowsmith, Bountra, Fish, Lee, & Schapira, 2012; Cole, 2008; Copeland, Solomon, & Richon, 2009; Haberland, Montgomery, & Olson, 2009; Helin & Dhanak, 2013; Hojfeldt, Agger, & Helin, 2013; Kelly, De Carvalho, & Jones, 2010; McKinsey & Kass, 2007). For example, several histone deacetylase inhibitors have been discovered and applied in the therapy of various types of cancer and cardiovascular disease (Falkenberg & Johnstone, 2014; McKinsey & Kass, 2007; West & Johnstone, 2014). Inhibitors of histone acetyltransferases, demethylases and methyltransferases have also been described and many are currently being evaluated in clinical trials (Arrowsmith et al., 2012; Helin & Dhanak, 2013). The lysine specific demethylase (LSD1) inhibitor tranylcypromine is a promising therapeutic for leukemia, depression and Huntington's disease (Helin & Dhanak, 2013; Schenk et al., 2012) and several co-factor competitors of the methyltransferase DOT1L are under evaluation for treatment of leukemia and lymphoma (Schapira, 2016a).

Emerging evidence also suggests a central role of endogenous intermediary metabolites in chromatin regulation via covalent histone modification (Gut & Verdin, 2013; Kaelin Jr. & McKnight, 2013; Lu & Thompson, 2012; Zhao et al., 2010). Many known small molecule inhibitors are structural derivatives of metabolites and most histone modifying enzymes require small molecule metabolic co-factors for catalysis. Several of these cofactors include acetyl-CoA, alpha-ketoglutarate, S-adenosylmethionine (SAM) and flavin adenine dinucleotide (FAD), among others. Studies suggest that the activities of chromatin associated enzymes are a function of the concentration and possibly the intracellular compartmentalization of specific intermediary metabolites, and that this information is integrated at a genomic or epigenomic level to direct cellular activity (Ladurner, 2006). For example, β-hydroxybutyrate (βOHB) and short chain fatty acids have been shown to directly inhibit HDACs and fasting or administration of exogenous βOHB correlated with changes in global transcriptional networks that are protective against oxidative stress (Shimazu et al., 2013; Waldecker, Kautenburger, Daumann, Busch, & Schrenk, 2008). Histone methyltransferases and demethylases are also required for physiologic cellular energy homeostasis. A combination of amino acid and one carbon metabolism facilitates the generation of the substrates FAD+ and SAM which are employed by histone methyltransferases (HMTs) and demethylases (HDMs) and the endogenous metabolite methylthioadenosine is an inhibitor of arginine methyltransferase (Kaelin Jr. & McKnight, 2013; Kryukov et al., 2016; Mavrakis, McDonald 3rd, Schlabach, Billy, & Hoffman, 2016; Serefidou, Venkatasubramani, & Imhof, 2019). Finally, both succinate and 2-hydroxyglutarate are metabolites that also have been shown to inhibit histone demethylases and are thought to alter tumor cell properties through epigenetic mechanisms and help maintain cells in an undifferentiated state (Tarhonskaya et al., 2017).

Herein, we highlight the structural and functional consequences of histone modifying enzymes. Because of their ubiquitous role in various disease states, we also describe several drug discovery and design efforts including novel small molecules not previously described. Using a combination of large scale *in silico* screens and *in vivo* phenotypic analysis, we identified several small molecules and intermediary metabolites with distinctive HME activity. Our approach using unsupervised learning identifies the chemical fingerprints of both small molecules and metabolites that facilitate recognition by the enzymes active sites which can be used as a blueprint to design novel inhibitors. Furthermore, this work supports the idea that histone modifying enzymes sense intermediary metabolites integrating genes, environment and cellular physiology.

1. The structure of histone modifying enzymes

1.1 Histone acetyltransferase (HAT) structure and catalysis

There are at least five families of histone acetyltransferases based on sequence differences of the HAT domain (Lee & Workman, 2007). The HAT domain adopts a generalized globular α/β-fold containing a central region that is associated with Acyl CoA cofactor binding (Fig. 1). The core domain is formed by antiparallel β-strands which lie over an amphipathic α-helix. Variable regions flank the cofactor binding site and are thought to be partially responsible for the specificity of histone substrates. The N- and C-terminal segments assist in binding histone substrates. Histone acetyltransferases are enzymes that catalyze the direct transfer of an acyl group from a donor substrate (acyl-CoA variants) onto the amino group of an acceptor substrate (lysine side chain of histones). The substrate amino group is deprotonated which enables nucleophilic attack on the carbonyl carbon of the Acyl CoA (Marmorstein & Zhou, 2014). Crystal structure and mutational analysis of GCN5, PCAF and ESA1 histone acetyltransferases highlight a structurally conserved glutamic acid reside that has been shown to function as a general catalytic base to deprotonate the amino group of reactive lysine residues (Clements et al., 1999; Rojas et al., 1999; Tanner et al., 1999; Yan, Barlev, Haley, Berger, & Marmorstein, 2000). The deprotonated lysine attacks the carbonyl carbon of the acetyl moiety forming a tetrahedral intermediate which eventually collapses to form CoA and acetylated lysine.

Histone acetyltransferase 1 (HAT1) is the quintessential member of the family of histone acetyltransferases and can be described in terms of three domains including the N-terminal domain, central domain and C-terminal domain. The crystal structure of yeast HAT1 was the first HAT structure to be solved (Dutnall, Tafrov, Sternglanz, & Ramakrishnan, 1998). The structure consists of an elongated α-β region with a conserved core containing a three strand β-sheet and long helix that flanks the strand (Fig. 1). The central domain of HAT1 displays significant structural similarity to the GCN5 and MYST families of histone acetyltransferases which contain a twisted β-sheet formed by five anti-parallel β-strands (β_9-β_{13}) centered between α-helices. The Acetyl CoA cofactor binding site involves a cleft between the central and C-terminal domains making multiple direct hydrophobic and hydrogen

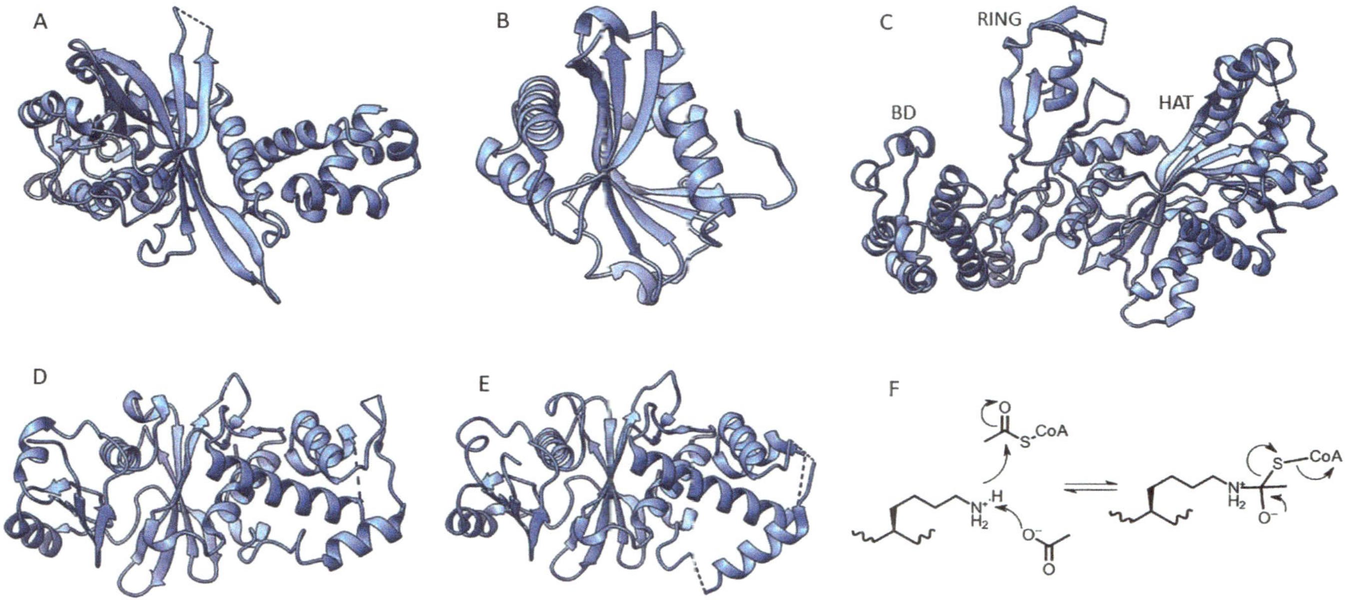

Fig. 1 Representative structures of histone acetyltransferases and associated PDB codes. (A) HAT1 (PDB 2P0W), (B) GCN5 (PDB 1Z4R), (C) P300 (PDB 5LKU), (D) MYST1 (PDB 2PQ8) and (E) MYST3 (PDB 2RC4). (F) Histone acetyltransferase catalytic mechanism of action. Generally, a Glu residue acts as a general base responsible for activating the attacking lysine nucleophile. The direct attack of lysine on the carbonyl carbon of acetyl-CoA leads to the formation of a tetrahedral intermediate. The tetrahedral intermediate collapses to form CoA-SH and acetylated product. Abbreviations: BD, Bromodomain; RING, Really Interesting New Gene; HAT, Histone Acetyltransferase; PDB, Protein Data Bank.

bonding interactions with C-terminal domain residues along the α_7 helix and the β_{14}-α_7 loop as well as the β-strand β_{13}, α_6 helix and η_4-α_6 loop of the central domain (Wu et al., 2012). The acetyl moiety is anchored by the interactions with Ile, Pro and Tyr. The crystal structure of the human HAT1 complex with Acetyl CoA and Histone H4 identified several critical residues for Lysine acetylation within a canyon between the central and C-terminal domains. The highly conserved Asp62 and Glu64 residues form a negatively charged surface that complement the positively charged Arg17 and Arg19 of the H4 peptide while residues Trp199, Tyr225, Tyr 55 and Ile186 of HAT1 contribute to hydrophobic interactions with the substrate. The human HAT1 structure reveals that three residues, Glu187, Glu276 and Asp277 are in close proximity to the nitrogen of H4K12 and can act as a general base for catalysis (Wu et al., 2012).

The general control non-repressible 5 (GCN5)-related *N*-acetyltransferase (GNAT) superfamily of enzymes is comprised of approximately 17 distinct families with a characteristic conserved fold incorporating 6–7 β-strands and 7 α-helices (Lee & Workman, 2007; Schuetz et al., 2007). The N-terminal motifs C and D are relatively conserved. Motif C consists of one β-strand (β_1) and an α-helix (α_1) while motif D includes two β-strands (β_2-β_3). The C and D motifs are connected by a loop with variable numbers and lengths of α-helices which allows different acceptor substrate binding sites and substrate specificities. The β_4 strand of motif A is parallel to the short β_5 strand of motif B and together create a V-shaped opening in the center of the catalytic domain of the protein to accommodate acyl-CoA. The HAT domain of human GCN5 folds around a central core that contains a mixed β-sheet of antiparallel β-strands (Fig. 1). The coenzyme is bound in a cleft on the surface via hydrogen bonds between the oxygen atoms of the pyrophosphate of Acetyl CoA and the backbone amide nitrogen of Val587, Gly589, Gly591 and Thr592. Additional important interactions with the HAT domain include Val581, Gln586 and Tyr617. The crystal structure of GCN5 reveals the presence of a conserved Glu570 in the active site that acts as a general base for catalysis (Schuetz et al., 2007). The reaction is consistent with a sequential mechanism in which acetyl-CoA, followed by the lysine substrate, binds to the active site to form a ternary complex. An active site general base assists in deprotonating the ε-amino group on the substrate lysine to increase its nucleophilicity. The uncharged lysine then attacks the acetyl-moiety forming a tetrahedral intermediate and collapse of the intermediate result in acetylated lysine and Coenzyme-A products.

An oxyanion hole is formed at the center of strand β_4 near the end of β_5 that stabilizes the tetrahedral reaction intermediates (Albaugh & Denu, 2021; Jiang et al., 2012).

The p300 histone methyltransferase is comprised of four different domains including the HAT domain, Bromodomain, RING and PHD domain. The HAT domain displays conservation with GNATs within the central core region associated with cofactor binding but with significant differences in the regions flanking the central core (Kaczmarska et al., 2017). The catalytic domain has a loop and a β-bulge in the β_4 strand with a V-shaped cleft between the β_4 and β_5 strands where the CoA binds. Residues 580–587 in the β_4-loop-α_3 region interact extensively via direct and water mediated hydrogen bonds with the pyrophosphate group. Furthermore, the aliphatic side chains of Gln, Cys, Ala and Val near the top of the β_4 strand make van der Waals contacts along the pantetheine arm. Of note, an unusually long substrate binding loop connecting β_5 and α_4 encapsulates the cofactor making van der Waals contacts with the β-mercaptoethylamine segment. The histone substrate binding pocket is more shallow and highly acidic compared to other HATs formed by the side chains of Glu, Asp and backbone carbonyls of Ile, Val and Tyr. In addition, there is no glutamate residue that is analogous to the key residues that function as general base residues for catalysis. Mutational studies showed that Tyr1467 and Trp1436 are important for catalysis and it has been proposed that the Tyrosine functions as a general acid (Liu et al., 2008).

The MYST (MOZ, Ybf2, Sas2 and TIP60) family of histone acetyltransferases display a structurally and functionally homologous central region but divergent N- and C-terminal regions. The structures of MYST1 (MOF) and MYST3 (MOZ) are shown in Fig. 1 (Holbert et al., 2007; MacPherson et al., 2020). The N-terminal segment is β-rich and the C-terminal segment is α-helix rich. The N-terminal segment contains a TFIIIA type zinc finger fold with three cysteine's and one histidine residue that wraps around and binds the zinc ion. Similar to other HATs, the greatest homology maps to the cofactor binding and catalytic domains. The central core region binds cofactor and contains the catalytic Glu residue. Esa1 has been used as to thoroughly investigate the catalytic mechanism of the MYST family of HATs. Similar to Gcn5 HATs, a tetrahedral intermediate is formed and the conserved Glu of Esa1 functions as a general base to remove the proton from lysine and promote nucleophilic attack on the acetyl carbonyl of acetyl-CoA (Berndsen, Albaugh, Tan, & Denu, 2007).

1.2 Histone deacetylase (HDAC) structure and catalysis

Histone modifying enzymes that deacetylate lysine residues are comprised of four separate classes. Class I, II and IV HDACs are zinc-dependent metallo-enzymes with a conserved globular α/β domain that consists of a multistranded parallel β-sheet flanked by several α-helices (Fig. 2) (Ficner, 2009). The Class I HDACs include HDAC1–3 and HDAC8 while Class II HDACs include HDAC4–7 and 9–10. The active site contains a narrow tunnel formed by several hydrophobic residues leading to a cavity with the catalytic Zn^{2+}-ion located at the bottom of the pocket. The Zn^{2+}-ion is coordinated by the side chains of conserved Asp and His residues and water molecules. The hydrophobic residues of the channel and conformational changes by the tyrosine residue are thought to provide enzyme specificity for the acyl lysine side chain. Class III histone deacetylases or sirtuins are structurally unrelated NAD-dependent deacetylase enzymes (Grozinger & Schreiber, 2002). The Class IV HDACs are comprised of a single member, HDAC11.

Class I HDAC catalytic activity is dependent on two separate pairs of His-Asp residues (Lauffer et al., 2013; Somoza et al., 2004; Watson, Fairall, Santos, & Schwabe, 2012). Class I HDACs possess multiple conserved residues essential for catalytic activity including two histidine residues and a tyrosine adjacent to the Zn^{2+} ion. Catalysis is thought to proceed through nucleophilic attack of a Zn^{2+} bound water molecule on the carbonyl of the acetyl lysine substrate (Fig. 2) (Lombardi et al., 2011). The carbonyl is also activated by zinc ion coordination and hydrogen bonding. One of the histidine residues functions as a general base and both are influenced by hydrogen bonds with aspartic acid residues. The enzymes active site and catalytic machinery are surrounded by long loops that connect the various secondary structures. Class I HDACs 1–3 possess a C-terminal domain that extends from the catalytic domain and is thought to recruit additional enzymes to protein complexes (Ayer, 1999). The crystal structures of Class I HDAC 2, 3 and 8 are shown in Fig. 2 (Lauffer et al., 2013; Somoza et al., 2004). HDAC8 is comprised of a single α/β domain that includes an eight stranded parallel β-sheet and 13 α-helices (Somoza et al., 2004). The overall structure is similar to other class I HDACs, however, HDAC8 is unique due to the absence of a C-terminal domain of approximately 100 residues used to recruit complexes that modulates enzyme activity and localization.

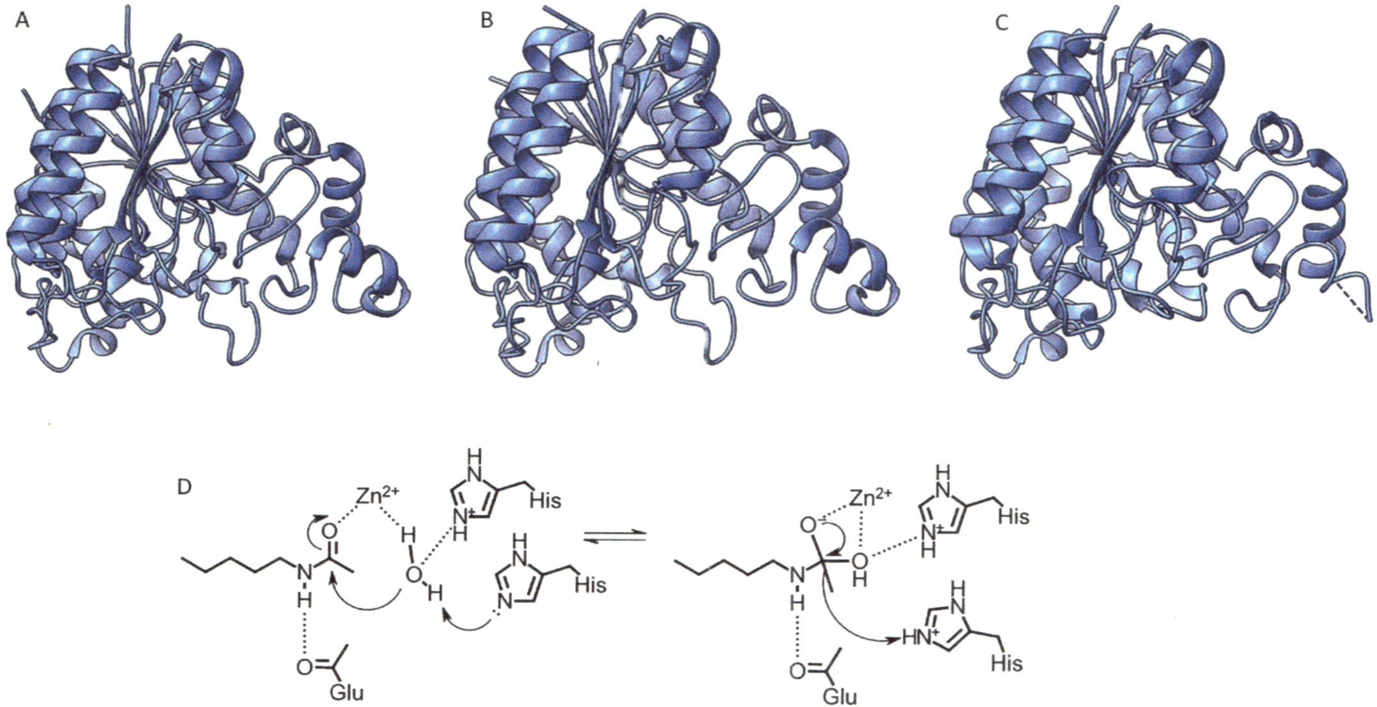

Fig. 2 Representative structures of Class I HDACs and associated PDB codes. (A) HDAC2 (PDB 4LXZ), (B) HDAC3 (PDB 4A69) and (C) HDAC8 (PDB 1T69). (D) Proposed mechanism of histone deacetylase catalysis (Lombardi, Cole, Dowling, & Christianson, 2011). The general base Histidine and transition metal ion Zn^{2+} promote the nucleophilic attack of a water molecule at the carbonyl group of the acetylated lysine substrate. The oxyanion of the tetrahedral intermediate and its flanking transition states are stabilized by the metal ion and hydrogen bond interactions with Tyr (not shown) and two His residues. The tandem His residues also forms hydrogen bonds with aspartic acid residues (not shown).

Class II HDACs are subdivided into class IIa (HDAC4, 5, 7, and 9) and IIb (HDAC6 and 10). Class IIa enzymes are characterized by the presence of a conserved C-terminal catalytic domain and an additional N-terminal regulatory domain. In contrast, class IIb enzymes contain two catalytic domains. The catalytic domain of HDAC4 and 7 are similar to that of HDAC8 including a central parallel β-sheet of 8–10 β-strands flanked by several α-helices and central Zn^{2+} ion (Bottomley et al., 2008; Schuetz et al., 2008). The central catalytic zinc ion coordinates to the side chains of Asp, His and Asp residues. However, HDAC4 and 7 are different from class I enzymes since several distinct segments coordinate additional zinc ions (Bottomley et al., 2008; Schuetz et al., 2008). The first segment involves a loop that connects two α-helices and coordinates zinc via histidine and cysteine residues. The second region involves a helix-turn-helix followed by a β-hairpin and terminal Cys that coordinates another Zn^{2+}.The crystal structure of class IIb HDAC6 is unique in that it contains two catalytic domains (CD1 and CD2) and a ubiquitin binding domain (Hai & Christianson, 2016). The HDAC6 catalytic zinc ion is also situated at the base of a deep tunnel and is liganded by two Asp residues and a histidine with a serine residue crucial for positioning the acetyl lysine residue in the enzyme-substrate complex.

1.3 Histone methyltransferase (HMT) structure and catalysis

Histone methyltransferases are comprised of two separate classes that include the lysine specific methyltransferases (PKMTs) and arginine specific methyltransferases (PRMTs). The PKMTs are further subdivided into SET domain and non-SET domain containing methyltransferase while PRMTs are subdivided into three categories based on arginine methylation patterns. Protein arginine methyltransferases catalyze the transfer of a methyl group from S-adenosyl-L-methionine (SAM) to arginine residues on histone tails. Type I PRMTs transfer one methyl group (mono-methylation) or two methyl groups to a single nitrogen (asymmetric methylation) of the guanidine moiety of arginine. Type II PRMTs either transfer one or two methyl groups to two different nitrogen atoms (symmetric demethylation). Type III PRMTs produce mono-methylation patterns.

Histone lysine methyltransferases transfer methyl groups from S-adenosyl-L-methionine to a lysine amino group resulting in the cofactor S-adenosyl-L-homocysteine and lysine residues with various methylation patterns (Cheng, Collins, & Zhang, 2005). Lysine residues can be mono-, di-, or trimethylated with product specificity regulated by steric hindrance to nucleophilic attack (Hu & Zhang, 2006). An example of the mono-methyltransferase catalytic reaction mechanism is shown in Fig. 3. Quantum mechanical/molecular mechanical calculations and kinetic isotope effects suggest a prototypical linear S_N2 nucleophilic substitution reaction with a dissociative or loose transition state (Hu & Zhang, 2006; Linscott et al., 2016; Poulin et al., 2016; Wang, Hu, & Zhang, 2007). Two separate tyrosine residues are important for catalytic activity through hydrogen bonding and steric hindrance (Hu & Zhang, 2006; Xiao et al., 2003). Furthermore, a water molecule is shown to act as a single hydrogen bond acceptor to help orient the amino group and increase nucleophilicity.

Several structural studies and sequence alignment of lysine specific methyltransferase SET proteins have revealed four conserved sequence motifs that form the active site (Cheng et al., 2005; Wu et al., 2010). The SET domain forms a novel β-fold with a series of β-strands packed closely with pre-SET, post-SET and variable I-SET insert domains. The SET domain consists of multiple β-sheets that surround a knot structure and α-helix. The cofactor involved in methyl group transfers, S-adenosyl-methionine, binds in a pocket formed between the SET and post-SET domains. Several structures of PKMTs in complex with histone lysine peptides show the target lysine inserted into a channel with the target nitrogen in close proximity to the methyl donor SAM at the opposite end of the channel (Qian & Zhou, 2006; Wu et al., 2010; Xiao et al., 2003, 2005). The post-SET domain contains cysteine residues that coordinate the binding of a zinc ion. The zinc-coordinated post-SET domain forms one side of the SAM binding pocket and is critical for the catalytic competency of the enzyme. The post-SET domains of G9a and GLP consist of an α-helix that contributes to peptide binding whereas other PKMTs have a loop (Wu et al., 2010). The loops connecting the SET and post-SET domains in different enzymes have distinct conformations. The loop in NSD1 and SET2 is ordered and occupies a position that binds substrate peptides in all known PKMTs (Huang et al., 2020; Zheng et al., 2012). Several PKMTs also contain an I-SET domain inserted between the Pre- and Post-SET domains. The I-SET domain is thought to act as a platform for the substrate histone tail, making several critical contacts with side chains upstream of the methyl acceptor (Wu et al., 2010).

Protein arginine methyltransferases share four conserved sequence motifs (I, post-I, II and III) and a loop that contains the SAM binding pocket (Cheng et al., 2005). The core catalytic domain of PRMTs is well conserved and is comprised of a methyltransferase domain, β-barrel domain and dimerization arm (Fig. 4) (Cheng et al., 2005). The N-terminal domain includes a

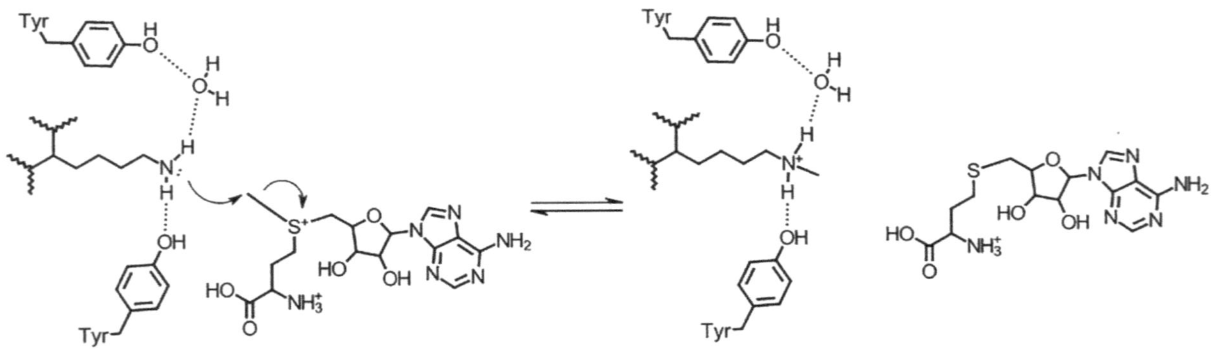

Fig. 3 Representative single methyl group transfer reaction catalyzed by the SET7/9 monomethyltransferase from AdoMet to a histone lysine residue.

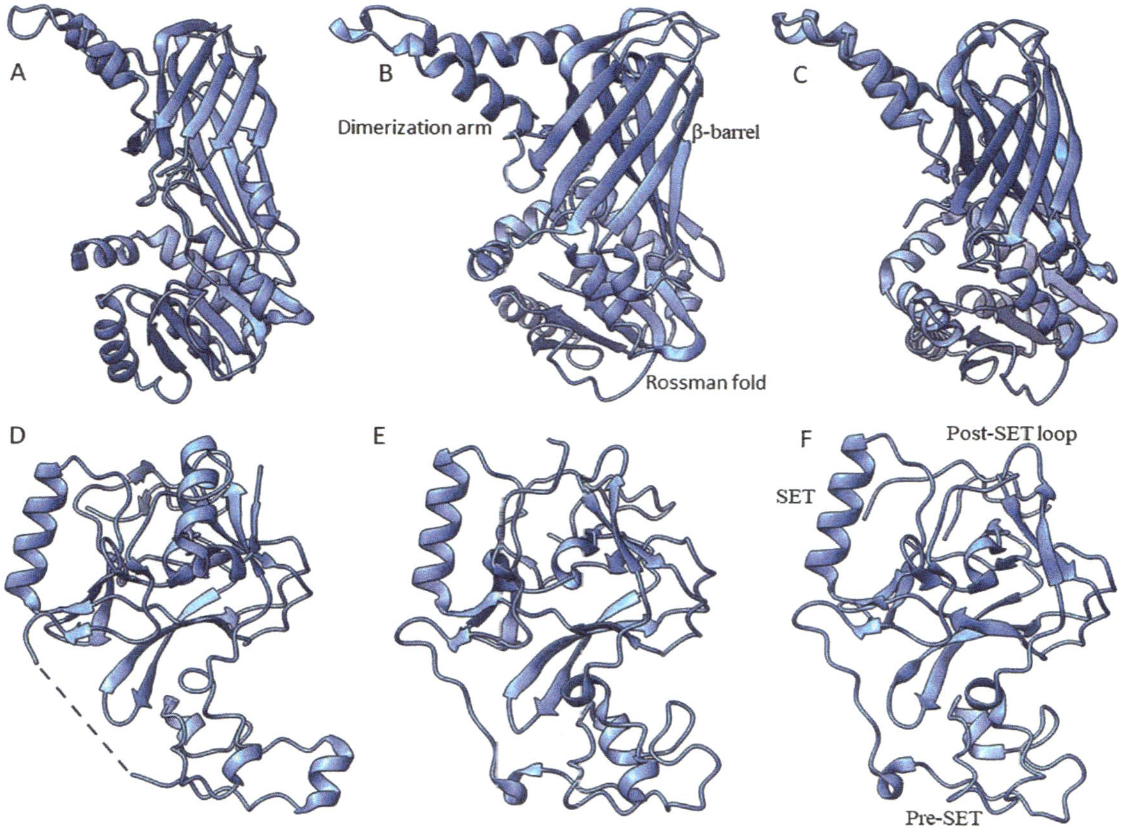

Fig. 4 Representative structures of the arginine specific methyltransferases (PRMTs) including (A) PRMT3 (PDB 3SMQ), (B) PRMT4 (PDB 6IZQ), and (C) PRMT6 (PDB 6P7I). Representative structures of the lysine specific methyltransferases (PKMTs) including (D) SETD2 (PDB 4H12), (E) NSD1 (PDB 6KQP) and (F) NSD3 (PDB 5UPD).

Rossmann fold and two helices followed by a β-barrel and four-helix extension. The Rossmann fold is characterized as a repeat of the βαβαβ structure which has twofold symmetry and binds to a dinucleotide such as nicotinamide adenine dinucleotide (NAD) or flavin adenine dinucleotide (FAD) to catalyze multiple types of reactions (Shin & Kihara, 1958). The target arginine is located in a deep acidic pocket between the methyltransferase and β-barrel domains. The active site contains two glutamate residues that help neutralize the positive charge and methionine thought to influence the methylation pattern of arginine through steric hindrance.

1.4 Histone demethylase (HDM) structure and catalysis

Histone lysine demethylase (HDM) enzymes belong to either the flavin adenine dinucleotide (FAD)-dependent LSD family or the Fe^{2+} and α-ketoglutarate (αKG) dependent Jumonji C-terminal domain family. Lysine specific demethylase 1 (LSD1) was the founding member of a class of enzymes that directly reverse histone H3K4 or H3K9 modifications by an oxidation demethylation reaction in which flavin is a cofactor. The LSD family of enzymes require association with other proteins such as the restin corepressor complex (Lee, Wynder, Cooch, & Shiekhattar, 2005; Shi et al., 2005). In contrast to the LSD family, the largest class of histone demethylase enzymes contain a Jumonji C (JmjC) domain and catalyze lysine demethylation of histones through an oxidative reaction that requires iron and α-ketoglutarate as cofactors.

The structures of two LSD family of enzymes highlights both similarities and differences between LSD1 and its only mammalian ortholog LSD2 (Fig. 5) (Fang et al., 2013; Mimasu, Sengoku, Fukuzawa, Umehara, & Yokoyama, 2008). LSD1 interacts with several other proteins including the CoRest complex which is required for demethylase activity (Lee et al., 2005; Shi et al., 2005). The LSD family of proteins are comprised of an N-terminal unstructured region followed by closely packed SWIRM and oxidase domains (Fang et al., 2013; Mimasu et al., 2008). The SWIRM domain is found in multiple histone modifying enzymes and consists of a long central α-helix separated by two separate helix-turn-helix motifs. The oxidase domain contains the active site which includes a large cavity and two lobes that bind FAD and substrate. The amine oxidase-like domain is shared by a number of metabolic enzymes (Fitzpatrick, 2010). Several residues of the histone tail fit within the cavity to facilitate positioning and methylation of the substrate lysine. The isoalloxazine ring of the FAD cofactor is positioned

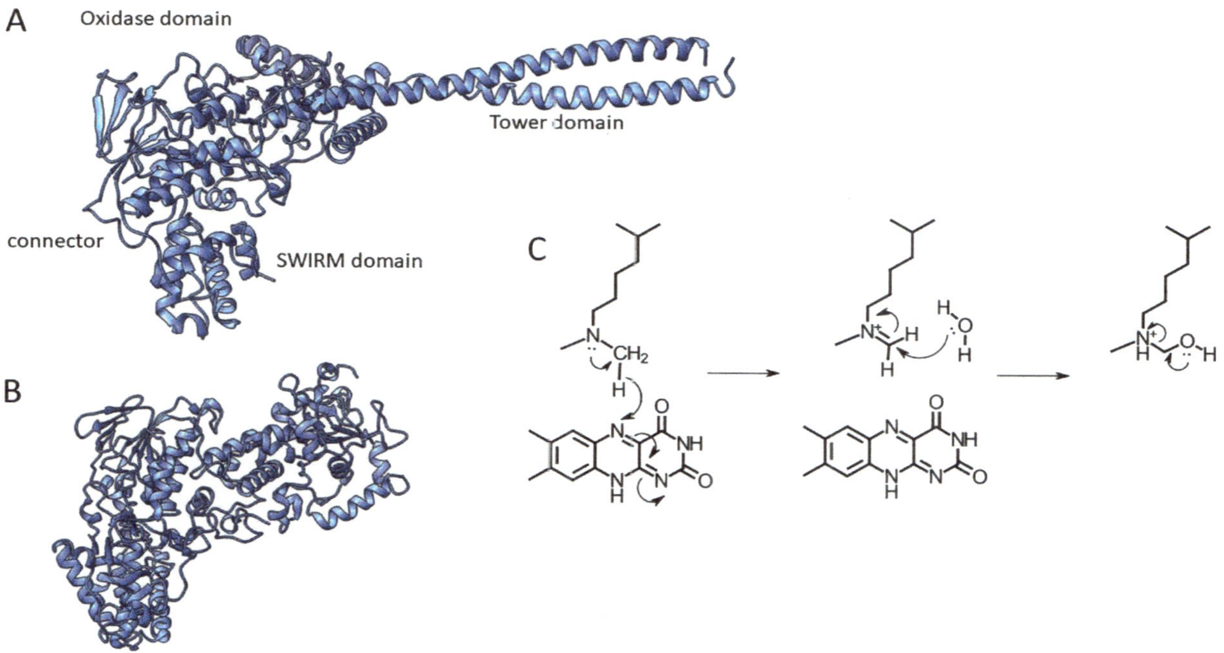

Fig. 5 Structures of flavin adenine dinucleotide (FAD)-dependent LSD family enzymes (A) LSD1 (PDB 2H94) and (B) LSD2 (PDB 4GU1) with domains highlighted. (C) Proposed hydride transfer mechanism of flavin dependent LSD family of histone methyltransferases.

within a hydrophobic pocket. An extensive hydrophobic interface is formed between the SWIRM and oxidase domains by an extensive network of van der Waals contacts and connector region thought to engage the histone tail substrate. An insertion in LSD1 but not LSD2 forms the tower domain comprised of two long α-helices that bind tightly to each other by hydrophobic interactions (Mimasu et al., 2008). The structure of LSD2 is compact and comprised of two zinc finger domains, a SWIRM domain and oxidase domain (Fang et al., 2013). The catalytic domains of LSD1 and LSD2 are well conserved and nearly indistinguishable. However, several differences in structure between LSD1 and LSD2 exist. The tower domain in LSD1 that functions as the binding site for CoREST is absent in LSD2 and the two zinc finger domains present in LSD2 are absent in LSD1. In LSD2, the two zinc fingers wrap around the SWIRM domain and form a structure that contacts the oxidase domain.

The LSD family of enzymes catalyze the demethylation of mono- or dimethylated histone H3 lysine 4 (H3K4) and H3 lysine 9 (H3K9). Similar to other flavin dependent amine oxidases, the catalytic mechanism is thought to involve a direct hydride transfer (Kong et al., 2011; Yu, Higashi, Cembran, Gao, & Truhlar, 2013). The proposed mechanism includes oxidation of the methyl amine coupled to the reduction of FAD to form a hydrolytically labile imine and FADH2. Imine formation likely proceeds via a hydride shift from the aminomethyl group to FAD.

The second family of histone lysine demethylases is characterized by the presence of the Jumonji C (JmjC) domain (Fig. 6). Jumonji C (JmjC) enzymes are a family of nonheme Fe (II) and 2-oxoglutarate dependent dioxygenases that share a conserved fold (Klose, Kallin, & Zhang, 2006). Fe (II) and 2-oxoglutarate react with dioxygen to form an oxoferryl intermediate that hydroxylates methyl groups of the methylated lysine substrate (Hausinger, 2004). The resulting product is unstable and releases the methyl group from nitrogen in the form of formaldehyde. The JmjC domain consists of a double stranded beta-helix fold with the active site located at one end where the catalytic Fe is bound by three conserved residues called the facial triad (HxxD/E..H) and 2-oxoglutarate bound to two conserved residues (F/T/Y…K) (Klose, Kallin, & Zhang, 2006).

Based on homology of the JmjC domain, HDMs are divided into seven subfamilies (Klose, Kallin, & Zhang, 2006). The JmjC domain only family includes human MINA53/NO66, JMJD4 and 5, PLA2G4B, FIH, HSPBAP1 and PTDSR and as the name implies contain no other known protein domains (Klose, Kallin, & Zhang, 2006). All other JmjC demethylase enzymes consist

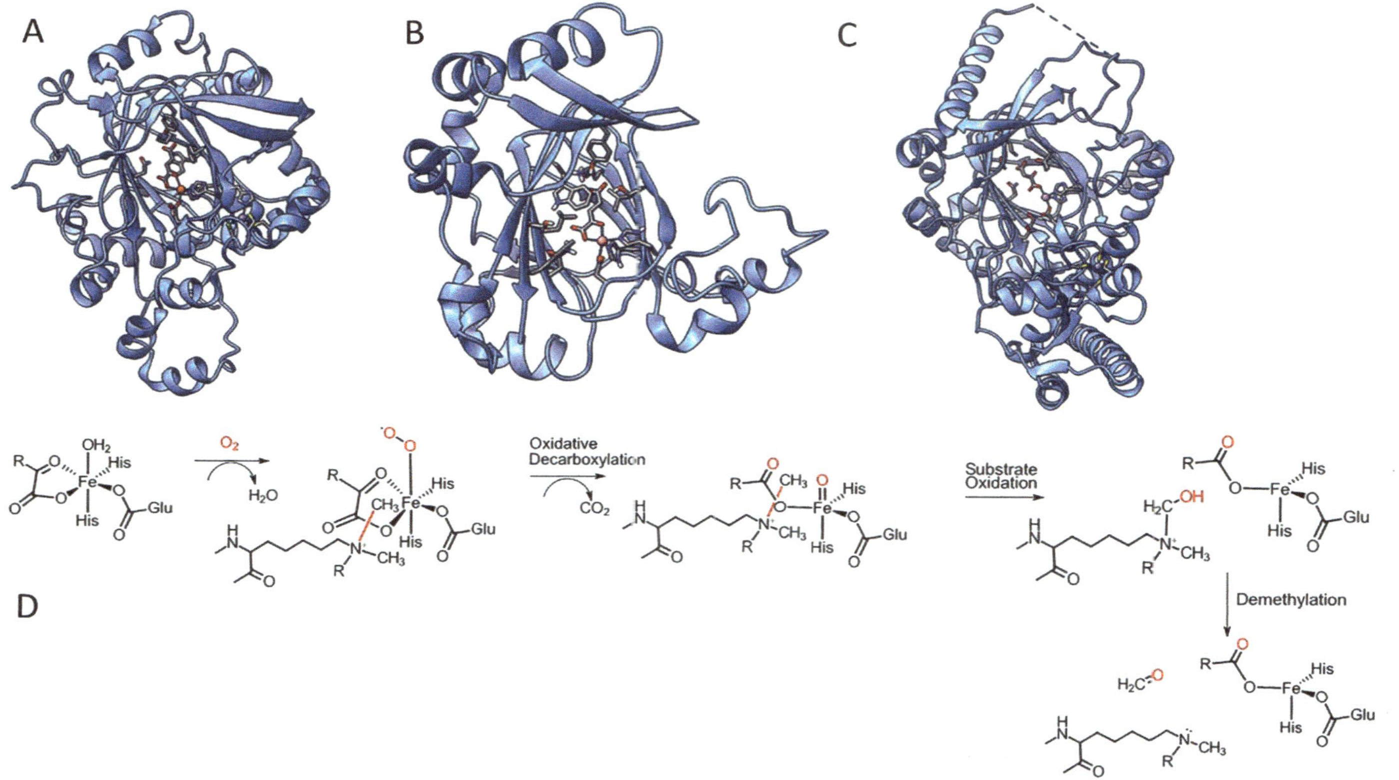

Fig. 6 Representative structures of the α-ketoglutarate (αKG) dependent Jumonji C-terminal domain family (A) JMJD2A (PDB 2GP5), (B) JMJD5 (PDB 4GJZ) and (C) UTY (PDB 5FXX). (D) Proposed mechanism of JmjC oxygenase catalysis. Oxidative decarboxylation of 2-oxoglutarate (2OG) in the active site forms a reactive iron-oxo intermediate which hydroxylates the substrate. The product is unstable and collapses to form the demethylated residue and formaldehyde.

of multiple domains with varying arrangements of zinc finger, JmjN, pHD and FBOX domains among others (Klose, Kallin, & Zhang, 2006). The JARID group consists of two subgroups, JARID1 and JARID2 because of the high level of homology despite differences in the spatial arrangement of the conserved domains.

The JMJD2 family of histone demethylase enzymes target H3K9 and H3K36 and require both the JmjN and JmjC domains for activity (Fodor et al., 2006; Klose et al., 2006; Whetstine et al., 2006). Crystal structures of JMJD2A, JMJD2C and JMJD2D have been solved and highlight the core catalytic domains of active histone demethylases (Fig. 6) (Chen et al., 2006). Proteins from this family contain JmjC, JmjN, PHD and Tudor domains. The catalytic core of JMJD2 consists of the N-terminal JmjN domain, a long β-hairpin motif followed by a mix of α-helices and β-strands, the JmjC domain and finally the C-terminal domain which contains several short helices and one loop (Chen et al., 2006). The N-terminal JmjN domain is comprised of three helices surrounding two β-strands and located opposite the catalytic core. The JmjC domain is a conserved structure comprised of 8 β-strands that form two four stranded sides in a jellyroll like structure. The core contains an iron atom that is chelated by the conserved residues His188, Glu 190 and His276. Additional chelating sites are provided by the carboxylate and ketone groups of α-ketoglutarate. The C-terminal domain follows the JmjC domain and is a mixture of α-helices and coils with two cysteine residues at the tip of the loop that coordinate a zinc ion. A zinc finger motif brings the C-terminal domain in tight association with the JmjC domain forming a rigid conformation around the catalytic center.

Members of the JHDM1 and JHDM2 families of histone demethylases are also comprised of multiple domains. In addition to the JmjC domain, JHDM1 family members consist of a CXXC zinc finger domain, PHD domain, leucine rich repeats and F-box domain (Klose, Kallin, & Zhang, 2006; Tsukada et al., 2006). The JmjC domain is sandwiched between several alpha helices and contains the catalytic core. Similar to other jumonji domain proteins, the Fe atom makes contacts with Asp, αKG and two His residues and αKG sits in a pocket surrounded by Tyr, Thr, Asn and Lys. JHDM2 family members possess a highly conserved JmjC domain and a modified zing finger domain (Hu et al., 2001). Another unique structural component includes a 20–32 amino acid sequence insertion within the JmjC domain between β-sheets 4 and 5 which is thought to form a helical secondary structure.

The UTX and UTY family of histone H3K27me3/2 demethylase enzymes are characterized by a JmjC domain and several tetratricopeptide

(TPR) repeats. The closely related JMJD3 which lacks the TRP domain is included because of significant sequence homology. The structures of these related members of the Jumonji family of demethylases, provide insight into their sequence specific demethylation mechanism (Sengoku & Yokoyama, 2011; Walport et al., 2014). The overall structure consists of a helical domain, liker region, JmjC domain and zinc binding domain. The JmjC domain structure is similar to other histone demethylases and consists of a jellyroll like fold flanked by α-helices (Sengoku & Yokoyama, 2011; Walport et al., 2014). The structures show that the catalytic domain cooperates with the zinc binding domain in the recognition of H3. The jumonji domain of UTX accommodates H3K27 by specific side chain interactions including histone Arg26 (H3R26) with Asp, Glu and Asn but also recognizes H3K27me3, H3A29 and H3P30. The analog of α-ketoglutarate (n-oxalylglycine) forms hydrogen bonds with Lys, Thre, Ser and Asn in the central cavity similar to JMJD2A. The zinc binding domain is unique and inserted in the helical domain.

The crystal structures of the Jumonji enzymes JMJD5, JMJD6 and JMJD7 reveal a catalytic domain whose fold is most similar to known hydroxylases of the JmjC subfamily compared to the JmjC histone lysine specific demethylases (Fig. 6) (Del Rizzo, Krishnan, & Trievel, 2012; Islam et al., 2019). For example, the JMJD5 and JMJD6 enzymes do not contain JmjN, PHD or tudor domains seen in other Jumonji histone demethylases. Nevertheless, the catalytic core of JMJD5 and JMJD6 comprises a JmjC domain similar to JMJD2 family demethylases and preceded by an N-terminal extension composed of mixed helical and β-sheet (Del Rizzo et al., 2012; Islam et al., 2019). The active site is located within the β-barrel fold and contains the Fe(II) site coordinated with His, Asp and His. The oxoglutarate compound participates in hydrogen bonding to Tyr, Ser/Thr, Asn and Lys and forms van der Waals interactions with Trp, Leu and Val. A long acidic channel is formed by several helices from the JmjC domain and N-terminal extension next to the active site entrance and may be a possible site for recognizing a histone tail.

2. Small molecule inhibitors of histone modifying enzymes

Post-translational modifications of histones tails have been identified in myriad disease states ranging from cancer, inflammation, neuropsychiatric disorders, cardiovascular disease and metabolic derangements. Not surprisingly, the small molecule inhibitors of the enzymes that catalyze these modifications have been the subject of intense investigation. The identification and characterization of small molecule inhibitors of histone modifying enzymes has led to promising new therapeutics and chemical probes. In this section, we highlight several examples of known small molecule HME inhibitors including representative structural scaffolds and physicochemical properties.

2.1 Histone acetyltransferase inhibitors

The first inhibitor of a histone acetyltransferase enzyme was identified in 1982 (Cullis, Wolfenden, Cousens, & Alberts, 1982). Cullis et al. developed a library of bisubstrate compounds by joining coenzyme A with acetyl lysine via an acetic acid linkage (Cullis et al., 1982). Since the initial discovery, several bisubstrate inhibitors have been designed with the general structure of a polyamide coenzyme A tail coupled to various simple peptide or organic structures via aliphatic chains of varying length (Table 1) (Huang, Huang, Zheng, & Sun, 2019). The bisubstrate compounds are designed to engage each individual substrate (the lysine and the CoA) simultaneously.

The second class of HAT inhibitors includes natural compounds with promiscuous activity. Anacardic acids are phenolic lipids of varying alkyl chain lengths derived from plants that have been shown to inhibit many different enzymes including acetyltransferases (Grazzini et al., 1991; Kubo, Kinst-Hori, & Yokokawa, 1994; Wang et al., 1998). Anacardic acid and several derivatives have been shown to inhibit the histone acetyltransferases p300, PCAF and Tip60 (Balasubramanyam, Swaminathan, Ranganathan, & Kundu, 2003; Ghizzoni et al., 2012; Ghizzoni, Boltjes, Graaf, Haisma, & Dekker, 2010). The salicylate group of anacardic acid is thought to mimic the pyrophosphate group of CoA with the carboxylic group acting as a hydrogen bond acceptor from the backbone Lys, Gly, Gly and Trp residues. The salicylate hydroxyl group acts as a hydrogen bond donor to the Thr side chain in the pyrophosphate binding pocket and the long aliphatic chain extends into the hydrophobic binding pocket. Quantitative structure activity relationships show that aliphatic side chains with 10–15 carbon atoms in the salicylate 6-position provide improved inhibitory potency in contrast to aliphatic chains with 5 carbon atoms (Ghizzoni et al., 2012). Other studies have shown that benzamide derivatives of anacardic acid have increased potency against acetyltransferases (Chandregowda, Kush, & Reddy, 2009).

Curcumin is another natural product polyphenol with nonspecific inhibitor activity against several enzyme classes. Curcumin is a potent inhibitor

Table 1 Representative histone acetyltransferase inhibitors.

HAT inhibitors	Representative structure	Other scaffolds
Curcumin derivatives	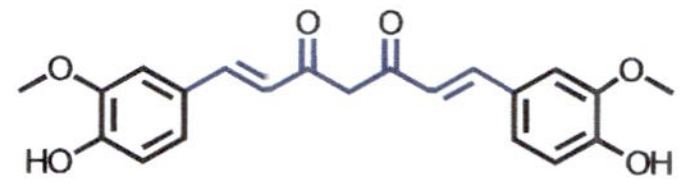	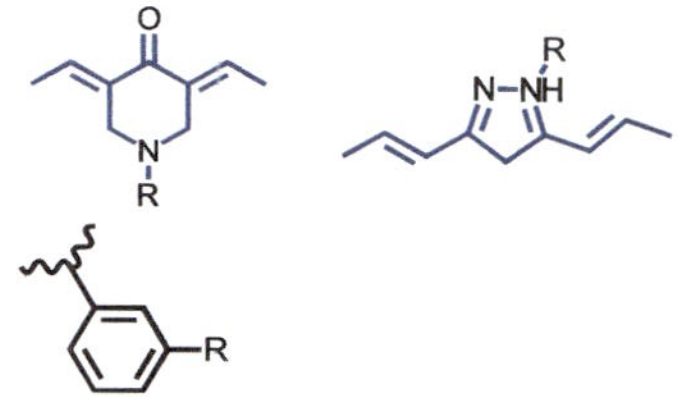
Anacardic acid derivatives	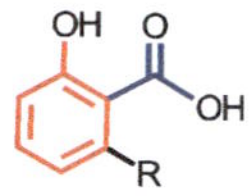	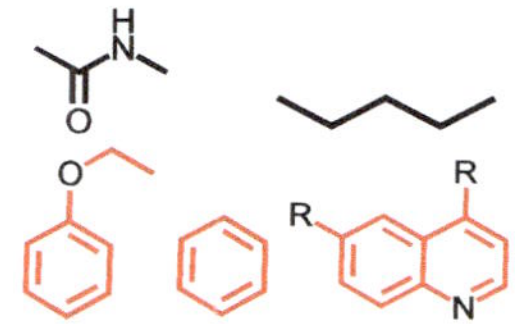
Polyamine-CoA derivatives	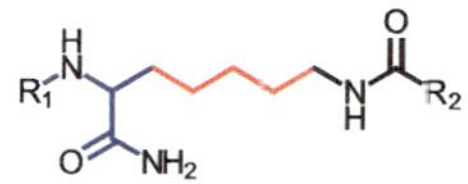	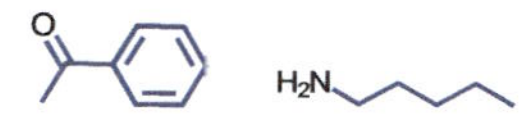 Polyamide SCoA
Catechin derivatives	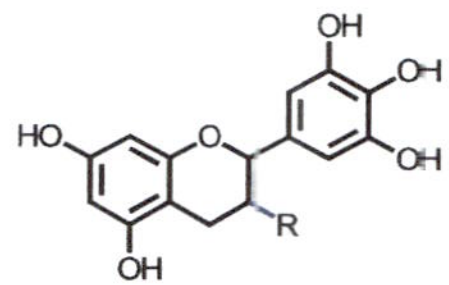	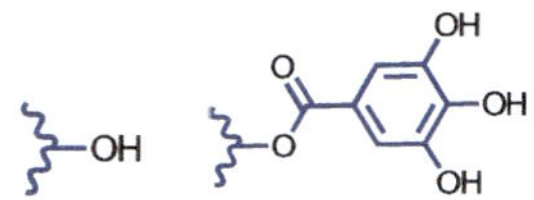

Continued

Table 1 Representative histone acetyltransferase inhibitors.—cont'd

HAT inhibitors	Representative structure	Other scaffolds
Alkaloid derivatives	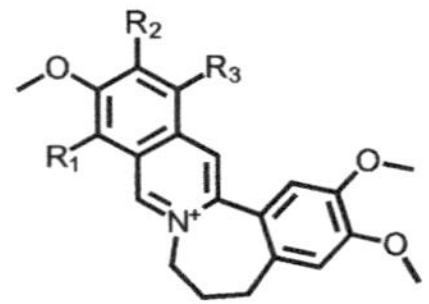	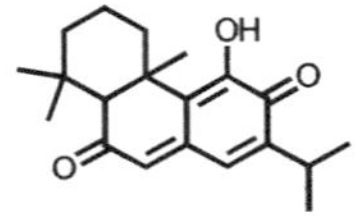
Thiazole derivatives	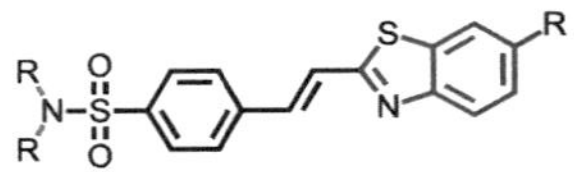 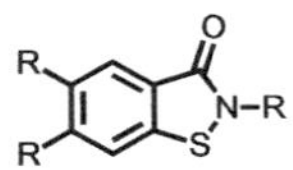 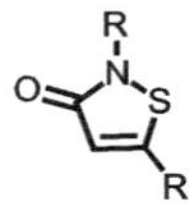	

of several acetyltransferases, lipoxygenase, xanthine dehydrogenase, protein kinase C and EGF receptor tyrosine kinase (Heery & Fischer, 2007; Lin, 2007). The scaffold is comprised of a central unsaturated structure that connects two terminal aromatic moieties (Table 1). Curcumin has been shown to inhibit p300 via electrophilic attack of the ketone by a cysteine thiol leading to a covalent product with an IC50 value of 25 μM (Balasubramanyam et al., 2004; Marcu et al., 2006). Several curcumin derivatives have since been developed with modifications at the central region including cycloketo heterocycles and terminal benzene ring substitutions. In particular, bromine and iodine substitution at aromatic terminals significantly increased p300 inhibitory activity (Madia et al., 2017).

The natural plant-based alkaloids and their derivatives have also been reported to be potent inhibitors of histone acetyltransferases. In one study, a virtual screening protocol was used to identify novel p300 inhibitors from a natural product library (Li et al., 2016). The most promising candidates were selected and tested for in vitro activity. Compounds with palmatine, spinosine and taxodione scaffolds were identified in the screen and all shown to be potent inhibitors of p300. Structure activity relationships revealed that the shift of a methoxy group from R 1 to R 2 increased the inhibitory activity of p300 and compounds with benzyl groups substituted on the quaternary amine significantly decreased potency (Yang et al., 2018).

Finally, additional natural products have been shown to inhibit acetyltransferases including garcinol, butyrolactone and epigallocatechin-3-gallate (EGCG) (Dancy & Cole, 2015; Huang et al., 2019). Small molecules comprised of the catechin scaffold (e.g., EGCG) inhibit p300, CBP, PCAF and Tip60 acetyltransferases with variable activity. Similarly, butyrolactones and their derivatives were of varying specificities against PCAF and GCN5 with IC50 values near 100 μM while garcinol is a more potent inhibitor with demonstrated IC50 values of 7 and 5 μM against p300 and PCAF, respectively.

2.2 Histone deacetylase inhibitors

The most extensively studied histone modifying enzyme inhibitors target the five phylogenetic classes of HDACs (I, IIa, IIb, IV, V). Suberoylanilide hydroxamic acid (SAHA), also known as vorinostat, was the first HDAC inhibitor to receive FDA approval for the treatment of T-cell lymphoma (Marks & Breslow, 2007). Since the early 2000s, several additional inhibitors of HDACs have been evaluated in various stages of development as drug

Table 2 Table of known HDAC inhibitors and representative structures.

HDAC Inhibitors	Representative structure	Ring scaffold
TSA 4-PBHA SAHA Tubacin		
4,251,365 4,250,302 4,239,232 4,237,636	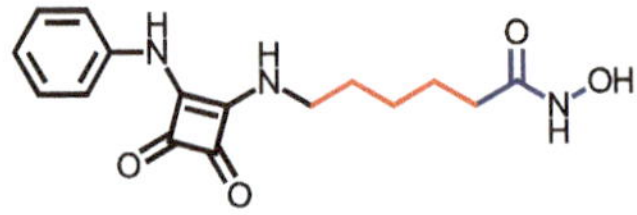	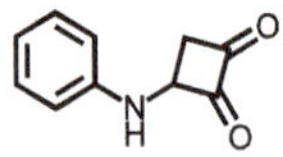
Quisinostat Panobinostat Dacinostat Pracinostat	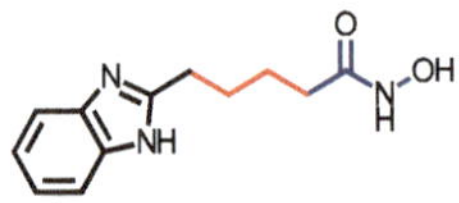	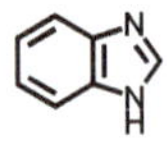
Valproic acid β-OH butyrate 2-Oxoglutarate Phenylacetate Propionate	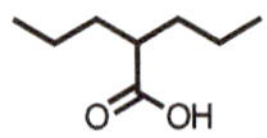	
Entinostat Tacedinaline Mocetinostat Tucidinostat Domatinostat	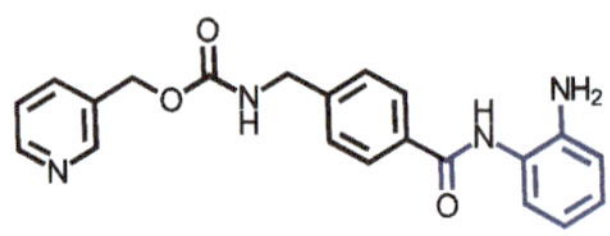	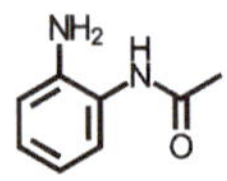

The hydroxamic acid and aminoanilide moieties that coordinate the zinc cation in the active site are highlighted in blue and aliphatic linker shown in red. The aliphatic linker may be variable in length and composition contributing to specificity.

candidates and clinical trials (Arrowsmith et al., 2012). HDAC inhibitors were initially designed to mimic the tetrahedral oxyanion enzyme intermediate (Fig. 2). Many HDAC inhibitors, including SAHA, are comprised of three scaffolding components, (1) a hydroxamic acid that coordinates zinc, (2) aliphatic linker that occupies the lysine channel and (3) surface specific ring structure or "cap" (Table 2). The aminoanilides are a class of HDAC inhibitors that coordinate zinc cations through the carbonyl oxygen and aniline nitrogen. Aminoanilides are weak binders compared to hydroxamic acids and display slow on and off kinetics. Other common HDAC inhibitors include various short chain fatty acids, hydroxamic acids, oxime amides, ketones and peptides all of which chelate the catalytic Zn^{2+} ion via

4, 5, 6 or 7 membered ring structures. The zinc ion facilitates hydrolysis of the acetyl lysine bond and the cap is generally a hydrophobic moiety that interacts with amino acids at the opening of the deep pocket surrounding the catalytic site.

Most HDAC inhibitors target multiple histone deacetylase enzymes often with only modest if any class selectivity (Falkenberg & Johnstone, 2014). For example, the small molecules trichostatin A (TSA), trapoxin B (TPX), SB939 and SAHA have all been shown to inhibit multiple different classes of HDACs which has made it difficult to completely understand their mechanism of action and biological effects (Arrowsmith et al., 2012; Bradner et al., 2010; Falkenberg & Johnstone, 2014; Ho, Chan, & Ganesan, 2020; Novotny-Diermayr et al., 2010; Yoshida, Kijima, Akita, & Beppu, 1990). Nevertheless, some success developing isoform selective HDAC inhibitors has been reported (Falkenberg & Johnstone, 2014; Haggarty, Koeller, Wong, Grozinger, & Schreiber, 2003; Yu, Chang, Hsin, & Chern, 2013). For example, a series of quinazolin-4-one compounds containing a hydroxamic acid moiety were designed with linkers of varying lengths for HDAC6 selectivity (Yu, Chang, et al., 2013). Substitution patterns at the 2/3 positions of the novel quinazolin-4-one hydroxamic derivatives enabled optimized selectivity thought secondary to flexibility. In general, selectivity and potency of HDAC inhibitors are thought to be regulated through interactions at the surface outlet next to the catalytic site and flexibility of the aliphatic linker. The length, bulkiness and flexibility of the aliphatic linker is variable and contributes to differences in fit within the tunnel. The linker may be comprised of a long aliphatic chain, polyene or ring structure.

To help understand class specific differences and improve predictions of small HDAC inhibitor activity, we developed classification models using multiple machine learning methods. We calculated the chemical fingerprint and more than 700 molecular descriptors of 200 different small molecule HDAC inhibitors identified from the literature and ChEMBL chemical database (Davies et al., 2015; Mendez et al., 2019). The molecular descriptors for each small molecule were included in a density matrix with individual compounds represented as rows and descriptors as columns. Principal components analysis (PCA) and neighborhood component analysis (NCA) were used to reduce dimensionality and select the most important features. Classification and regression models were constructed using the calculated chemical descriptors as predictors and Ki values (regression) or HDAC class (classification) as response variables.

Two separate classification models highlight the promiscuity of HDAC inhibitors. Small molecules were assigned a number based on the class (I, II, III, IV) of HDAC enzyme that the small molecule inhibits. For example, inhibitors of HDAC1,2,3 and 8 were labeled as class I compounds and inhibitors of HDAC4, 5,6 7, 9 and 10 labeled as class II compounds. Several compounds were shown to inhibit multiple classes of HDACs. Compounds that were nonselective between HDAC classes were labeled class 5. We compared classification models using all classes of small molecule inhibitors (Class I-5 models) and models where the promiscuous compounds were excluded (Class I–III models). Table 3 summarizes and compares all classification models used in this study. All classification models were relatively well performing with accuracies ranging from approximately 70–90% and receiver operating characteristic (ROC) curves with areas under the curve ranging from 0.74 to 0.98.

The nearest neighbor classifier, support vector machine and ensemble classifier models display the greatest accuracy although most models are comparable. Naïve Bayes classifiers and discriminant analysis models are the least accurate. As expected, classification models that include compounds with multi-class inhibitor activity (class V) do not perform as well as models that exclude promiscuous inhibitors. Nevertheless, all HDAC inhibitor classification models show impressive accuracy considering the HDAC enzyme active site homology and small molecule self-similarity.

The features (molecular descriptors) contributing to class specific differences are shown in Table 4. Molecular descriptors that most strongly represent HDAC class I inhibitors include hydrophobicity or LogP, number of 3 and 4 membered rings and the electropological state maximum atom type (=CH2). The LogP values of HDAC I inhibitors reveal a multimodal probability distribution with a large peak near LogP 2.3 and smaller shoulder around 5.3. Interestingly, promiscuous inhibitors are also hydrophobic with two distinct peaks in the probability density function near LogP 1.9 and 4. HDAC I specific inhibitors are comprised of several compounds with 3 and 4 membered rings that are not observed in the other classes of inhibitors. The 3 membered ring consists of an epoxide moiety and the 4 membered ring is a cyclobutenedione group shown as a scaffold in Table 2. HDAC class II inhibitors are best characterized by the ETA shape index, maximal electrotopological negative variation, electrotopological state number of quaternary carbons, and the molecular distance edge between all secondary and quaternary carbons and all tertiary carbons. The electrotopological negative variation measures the field effect on atoms due to the perturbation of

Table 3 Classification models of HDAC inhibitors using more than 700 molecular descriptors.

Class I–V classification models	Acc	AUC	Class I–III classification models	Acc	AUC
Neural Network Activation: Tanh Lambda: 5.086e-08	75%	I = 0.91 II = 0.88 III = 0.85 V = 0.85	*Neural Network* Activation: Tanh Lambda: 2.459e-06	87%	I = 0.97 II = 0.96 III = 0.96
Decision Tree Maximum # splits: 6 Split criterion: Twoing rule	76%	I = 0.87 II = 0.92 III = 0.91 V = 0.79	*Decision Tree* Maximum # splits: 3 Split criterion: Twoing rule	89%	I = 0.95 II = 0.96 III = 0.95
Discriminant Analysis Diagonal Linear	72%	I = 0.87 II = 0.88 III = 0.75 V = 0.78	*Discriminant Analysis* Diagonal Linear	79%	I = 0.84 II = 0.92 III = 0.71
Naïve Bayes Classifiers Gaussian Kernel	68%	I = 0.86 II = 0.81 III = 0.82 V = 0.74	*Naïve Bayes Classifiers* Gaussian Kernel	80%	I = 0.86 II = 0.88 III = 0.85
Support Vector Machine Kernel function: Linear Box constraint level: 2.05 Multiclass method: one vs all Kernel scale: 1	78%	I = 0.93 II = 0.94 III = 0.87 V = 0.86	*Support Vector Machine* Kernel function: Quadratic Box constraint level: 0.1887 Multiclass method: one vs all Kernel scale: 1	89%	I = 0.95 II = 0.93 III = 0.94
Nearest Neighbor Classifiers Number of neighbors: 1 Distance metric: Cosine Distance weight: Inverse	79%	I = 0.90 II = 0.84 III = 0.84 V = 0.80	*Nearest Neighbor Classifiers* Number of neighbors: 3 Distance metric: City block Distance weight: Inverse	90%	I = 0.95 II = 0.95 III = 0.90
Ensemble Classifiers Learner type: Decision tree Ensemble method: bag Number of learners: 10 Maximum number of splits: 97	79%	I = 0.94 II = 0.91 III = 0.97 V = 0.87	*Ensemble Classifiers* Learner type: Decision tree Ensemble method: bag Number of learners: 10 Maximum number of splits: 39	89%	I = 0.98 II = 0.98 III = 0.97

All models used fivefold cross validation and Bayesian optimization for hyperparameter tuning. *Abbreviations*: ACC, accuracy; AUC, area under the curve; I, class I HDAC inhibitors; II, class II HDAC inhibitors; III, Class III HDAC inhibitors; V, promiscuous HDAC inhibitors.

Table 4 Table of HDAC inhibitor molecular descriptors (features) contributing to differences between class as determined by neighborhood component analysis (NCA) and principal component analysis (PCA).

Inhibitor class	Class specific features (descriptors)
HDAC I	SCH-5, VCH-4, VCH-5, CrippenLogP, minHBint3, minHBint9, minHsOH, maxdCH2, hmin, n3Ring, nT4Ring
HDAC II	SC-4, SC-5, nssssC, SHBint7, minHBint4, minHBint5, minHBint6, minHBint7, MAXDN, ETA_Shape_X, MDEC-24, MDEC-34, MDEC-44
HDAC III	nAcid, nB, C3SP3, ECCEN, nsssCH, ETA_Eta, ETA_Eta_L, nHBDon, MLFER_A, nRotB, WPATH, nHBint2, nHBint3, nHBint5
Non-specific (V)	ATSc4, nHBint10, SHBint10, maxHBint10, ETA_dEpsilon_D, nAtomLAC, nAtomLC, nAtomP, nT7ring, n7ring, minssCH2

For a list of descriptor abbreviations see Supplementary file.

all other atoms. Several HDAC II specific inhibitors consist of multiple fluorine substituents on the alkyl chain connector which is the primary factor contributing to classification of class II inhibitors. The features contributing to class III HDAC inhibitors include the overall solute hydrogen bond acidity, number of hydrogen bond donors, number of acidic groups and number of rotatable bonds. Class III inhibitors target the Sirtuin enzymes and are therefore expected to be the most dissimilar to the other classes of inhibitors. The molecular descriptors most strongly related to promiscuous HDAC inhibitors include hydrophobicity, the number of atoms in the length of aliphatic chain, autocorrelation weighted by charge, count and sum of *E*-state descriptors of strength for potential hydrogen bonds of path length 10 and a measure of the contribution of hydrogen bond donor atoms. The promiscuous inhibitors were hydrophobic with varying degrees of flexible aliphatic chains. A group of cyclic peptides were included in this class as well.

2.3 Histone methyltransferase inhibitors

Histone methyltransferase inhibitors have been designed to target both the methyl donating cofactor and methyl accepting substrate binding sites. The compound BIX-01294 was the first selective inhibitor of a lysine methyltransferase but showed relatively limited binding affinity and significant toxicity (Table 5). Several histone methyltransferase inhibitors have been

Table 5 Representative scaffolds of histone methyltransferase inhibitors.

HMT Inhibitors	Representative structure	Substitution examples
Quinazoline derivatives BIX-01294 E72 UNC 321 UNC 638 UNC 224	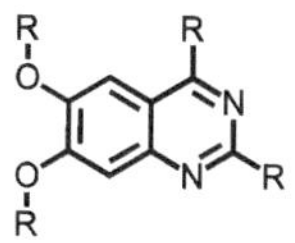	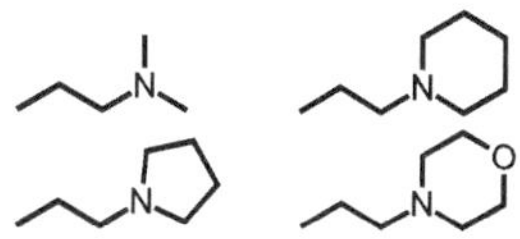
Indole Benzenecarbothiamide AS-5 AS-6 AS-85 AS-99	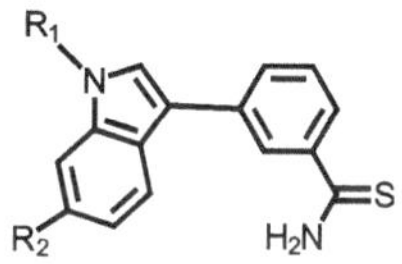	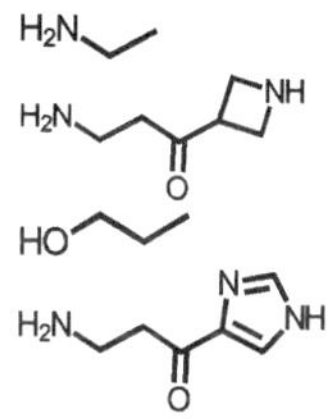
Benzothiazole-2-amine	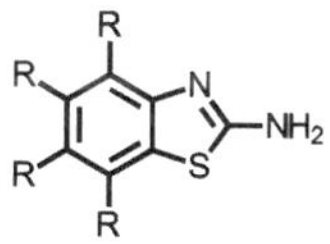	-Cl, -Br, -OH 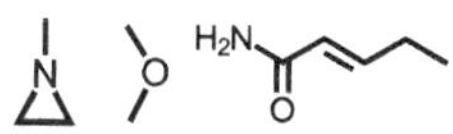
Adenine derivatives Sinefungin SAM/SAH derivatives NSC 46613 NSC 113989 EPZ 4777 EPZ 5676	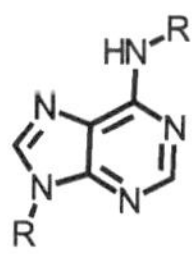	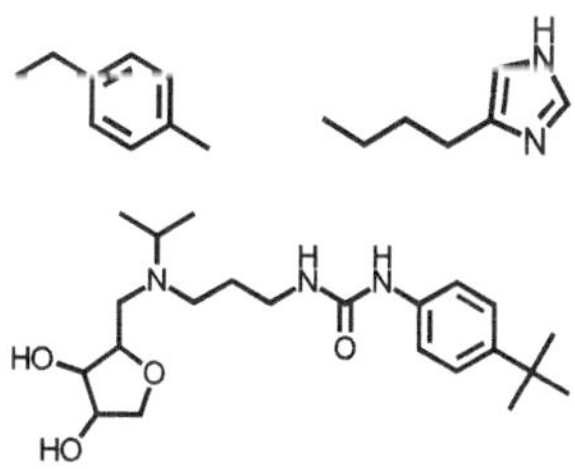
Diterpene derivatives Tanshindiones	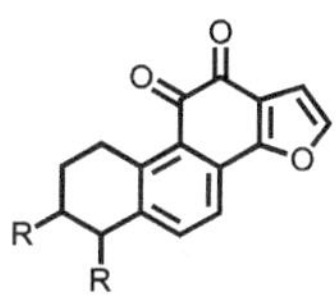	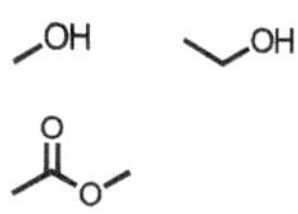

Scaffolds are shown with -R groups indicating patterns with various substituents. Representative substituents are highlighted in the final column.

described since the discovery of the first small molecule HMT inhibitor in 2007 (Kubicek et al., 2007; Schapira, 2016b). Increased affinity was observed with an alkoxyamine moiety linked to the quinazoline core (Chang et al., 2010; Liu et al., 2010; Vedadi et al., 2011). These compounds inhibit

G9A and GLP1 with the quinazoline core occupying the peptide groove and alkoxyamine substituent binding within the lysine channel. Most other compounds bind to the SAM binding site and several were designed as SAM analogs or adenine derivatives including methylthio-adenosine, adenosine dialdehyde, sinefungin, EPZ004777 and chaetocin among others (Arrowsmith et al., 2012). A series of compounds were also developed that bind to the autoinhibitory loop region in the SET domain of ASH1L (Rogawski et al., 2021). Initially, a thioamide compound was identified to bind the autoinhibitory loop with relatively weak binding affinity (Kd > 1 mM) using an NMR based screen of 1600 fragment like compounds (Rogawski et al., 2021). However, the replacement of the pyrrole with an indole, resulted in significantly improved binding affinity. The addition of a hydroxyethyl group to the indole nitrogen and aminoethyl at position six of the benzene ring led to a 200-fold increase in activity compared to the original compound fragment. Similarly, Huang et al. used a fragment like in house library screen to identify several compounds that bind to the NSD1 SET domain (Huang et al., 2020). A promising benzothiazole-2-amine was identified and several analogs subsequently synthesized. The most promising fragment candidate was a Benzothiazole-2-amine with 4-hydroxyl and 6-Bromo substituents. Several analogs were shown to bind the autoinhibitory loop based on changes in NMR chemical shifts and a subsequent covalently linked analog enabled the crystal structure to be solved. Lastly, a natural product library was screened against EZH2 activity in an effort to identify compounds with anti-tumor activity. Interestingly, several tanshinone derivatives isolated from Chinese sage demonstrated dose-dependent inhibition of EZH2 (Woo et al., 2014). Several derivates showed weak inhibitory activity, however, the 1,2-diol moiety in the A ring showed superior activity. Computational studies suggest that tanshindiols and tanshinones dock in the SAM binding pocket.

2.4 Histone demethylase inhibitors

The well-defined active site of histone demethylases provides optimal structural features for the design and development of small molecule inhibitors. Nevertheless, relatively few inhibitors of demethylases have been developed compared to other histone modifying enzymes. The monoamine oxidase (MAO) inhibitor tranylcypromine was one of the first known methyltransferase inhibitors initially shown to inhibit LSD1 via covalent adduct with the FAD cofactor in the amine oxidase like domain (AOL) (Table 6; Lee, Wynder, Schmidt, McCafferty, & Shiekhattar, 2006). Derivatives of

Table 6 Representative scaffolds of histone demethylase inhibitors.

HDM Inhibitors	Representative structure	Substitution examples
Tranylcypromine derivatives ORY-1001 S1201	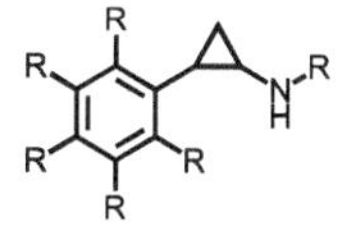	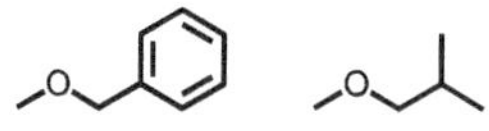 -Cl, -F
Polyamines		
N-oxalyl glycine derivatives	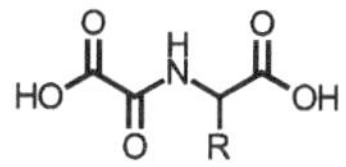	-H 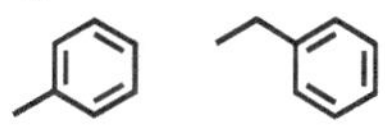
Hydroxyquinolines	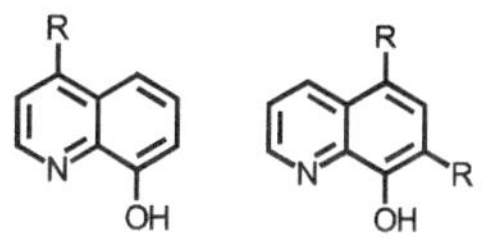	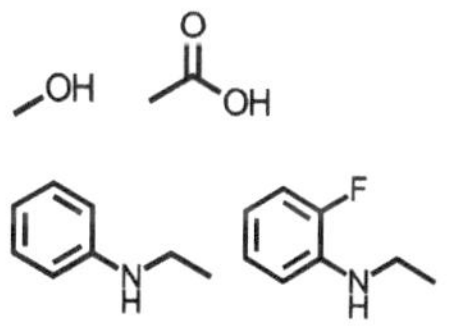

Scaffolds are shown with -R groups indicating patterns with various substituents. Representative substituents are highlighted in the final column.

tranylcypromine were subsequently developed with increased specificity and potency through substitutions on both the phenyl ring and nitrogen (Mimasu et al., 2010; Ueda et al., 2009). Another class of LSD inhibitor is the bisguanide polyamine analogs which have been shown to inhibit other amine oxidases (Huang et al., 2007;). The polyamine derivatives were shown to inhibit LSD1 in human colon carcinoma cells and results in increased active chromatin marks and decreased repressive marks in the promoter regions of several genes (Huang et al., 2007). Diverse inhibitors of JMJC demethylases also have been reported and most are metal chelators that bind competitively with the 2-oxoglutarate cofactor. Rose et al. highlight a variety of inhibitor scaffolds and show that N-oxalyl glycine derivatives and hydroxamic acid HDAC inhibitors also inhibit JMJD2 histone demethylases likely via 2-oxoglutarate dependent competition (Lohse et al., 2011; Rose et al., 2008). In a separate study, a large scale high throughput screen for inhibitors of the 2OG dependent histone demethylases identified multiple flavonoids and catechols that inhibit JMJD2E (Sakurai et al., 2010). The dihydroxybenze compounds methyldopa, carbidopa, dopamine, oxidopamine and isoproterenol inhibit JMJD2E with activities ranging from 4 to 10 μM.

Similarly, high throughput screening of more than 200,000 small molecules identified 8-hydroxyquinolines as JMJD2 inhibitors *in vitro* and cell-based assays (King et al., 2010). The crystal structure of 5-carboxy-8-hydroxyquinoline in complex with JMJD2A show that the inhibitor is positioned in a similar location to oxalylglycine and that the C-5 carboxylate interacts with the side chains Lys206 and Tyr132. In addition, the quinoline nitrogen and 8-hydroxy group is positioned to coordinate the active site metal ion (King et al., 2010).

3. Discovery of small molecule and metabolite inhibitors of histone modifying enzymes

3.1 Machine learning identifies novel small molecules and metabolites that inhibit histone modifying enzymes

We used large scale *in silico* screens and chemical informatics to identify new small molecule and metabolite inhibitors of histone modifying enzymes including molecular scaffolds with both nonspecific (promiscuous) and specific activity. A library of 50,000 drug-like small molecules from the Zinc database was screened against the active sites of nearly 70 different HMEs using a well validated docking program AutoDock Vina (Sterling & Irwin, 2015; Trott & Olson, 2010). Several machine learning methods were subsequently applied to identify small molecules and metabolites with similar chemical fingerprints. The activity profile of select small molecules and metabolites were then validated in zebrafish phenotypic screens and *in vitro* activity assays. Using this approach, we discovered several small molecules and metabolites that inhibit histone modifying enzymes with prediction of activity and phenotypic response. We hypothesized that metabolites with physicochemical properties similar to small molecules from the *in silico* screen and known small molecule inhibitors may also bind and inhibit the same enzyme. The chemical similarity measures among active small molecules and metabolites would then allow us to identify novel native metabolic regulators of HMEs that in turn may "sense" the metabolic state and facilitate changes in downstream transcriptional networks.

We used various methods for dimensionality reduction and multidimensional visualization to identify metabolites with physicochemical fingerprints similar to the small molecules from *in silico* screening and known HME inhibitors from the literature. Over 700 different molecular descriptors were calculated using the program PaDEL for the screening library and 2000 metabolites in the human metabolome database (HMDB) (Wishart et al., 2022; Yap, 2011). Many of the descriptors (e.g., electrotopological state index) have been used extensively for QSAR studies and chemoinformatics applications. A density matrix of chemical descriptors for each small molecule and metabolite was constructed with the columns represented as numeric variables from descriptors (e.g., ALogP, number of acidic groups, maximum electrotopological state, etc.) and rows as individual small molecules or metabolites. Dimensionality reduction and machine learning algorithms were subsequently used to explore the relationships between all variables. Principal component analysis identified the features that contribute most to the variation in the dataset. Multidimensional scaling, t-distributed Stochastic Neighbor Embedding and self-organizing feature maps were also compared to identify robust patterns and cluster of like compounds in the datasets.

Fig. 7A–D highlights the chemical space and self-similarity of small molecules and metabolites evaluated. Compounds with similar physicochemical properties are clustered in space as shown in the t-SNE scores plots and self-organizing maps. The scores plot shows that many previously known inhibitors from the literature overlap in chemical space with novel metabolites and small molecules identified here using *in silico* docking (Fig. 7A). The proximity in space of compounds suggests a high degree of chemical similarity and potential for novel histone modifying enzyme activity. For example, green circles are small molecules that bind with highest affinity from the virtual screen and overlap with several metabolites (blue circles) and known inhibitors from the literature (black circles). We used this information to help select new compounds with optimal properties for further testing *in vitro*. The SOMs cluster compounds using all 700 descriptors or cluster compounds according to a single descriptor (e.g., molecular polarizability). From the nearly 700 individual descriptors, Fig. 7B–D highlight three representative SOMs that organize clusters of compounds according to a single chemical property: (a) molecular polarizability, (b) unsaturation content relative to molecular size and (c) the count of states for strong hydrogen bond acceptors. The SOM clusters small molecules with similar values which are shown as differences in spatial location and color. For example, the red cluster in the bottom right corner of Fig. 6B is primarily composed of cholesterol and sterol lipid backbones or derivatives. Similar to t-SNE analysis, self-organizing maps show that several metabolites cluster with known HME inhibitors. Interestingly, our analysis shows that metabolites enriched in human heart failure share chemical fingerprints with known

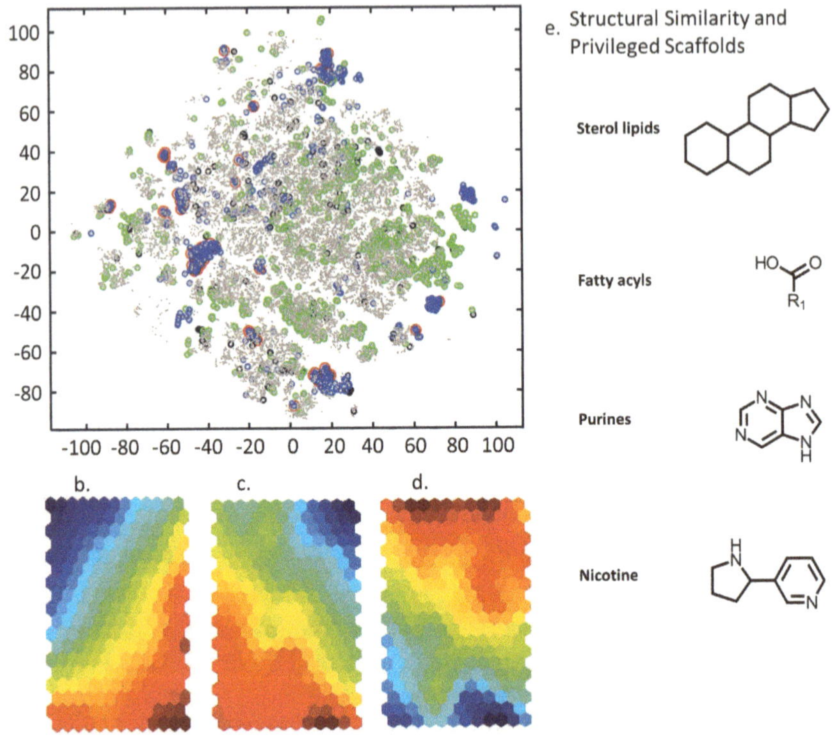

Fig. 7 (A) t-SNE scatter plot of small molecules and metabolites. Gray and green circles represent individual small molecules from the library used for docking studies and the top 50 inhibitors with greatest affinity for HMEs from docking. Black circles represent known HME inhibitors identified from the literature. Metabolites enriched in heart failure and all other metabolites from the human metabolome database shown in red and blue, respectively. (B–D) Representative SOMS that organize clusters of compounds according to chemical descriptors (B) molecular polarizability, (C) unsaturation content relative to molecular size and (D) the count of estates for strong hydrogen bond acceptors. (E) Structure of scaffolds from metabolites enriched in heart failure with self-similarity to known HME inhibitors.

inhibitors of histone modifying enzymes including fatty acids, succinate, ketone bodies, cholesterol sulfate, androsterone, nicotinamide, uric acid and xanthine. Several of the most common structural scaffolds are shown in Fig. 7E suggesting potential scaffolds to screen for HME inhibition.

Principal component analysis was also used to reduce dimensionality and identify the features contributing to the greatest variation in the dataset. Fig. 8A shows the PCA plots which contain the coordinates of the original data in the new coordinate system defined by the principal components.

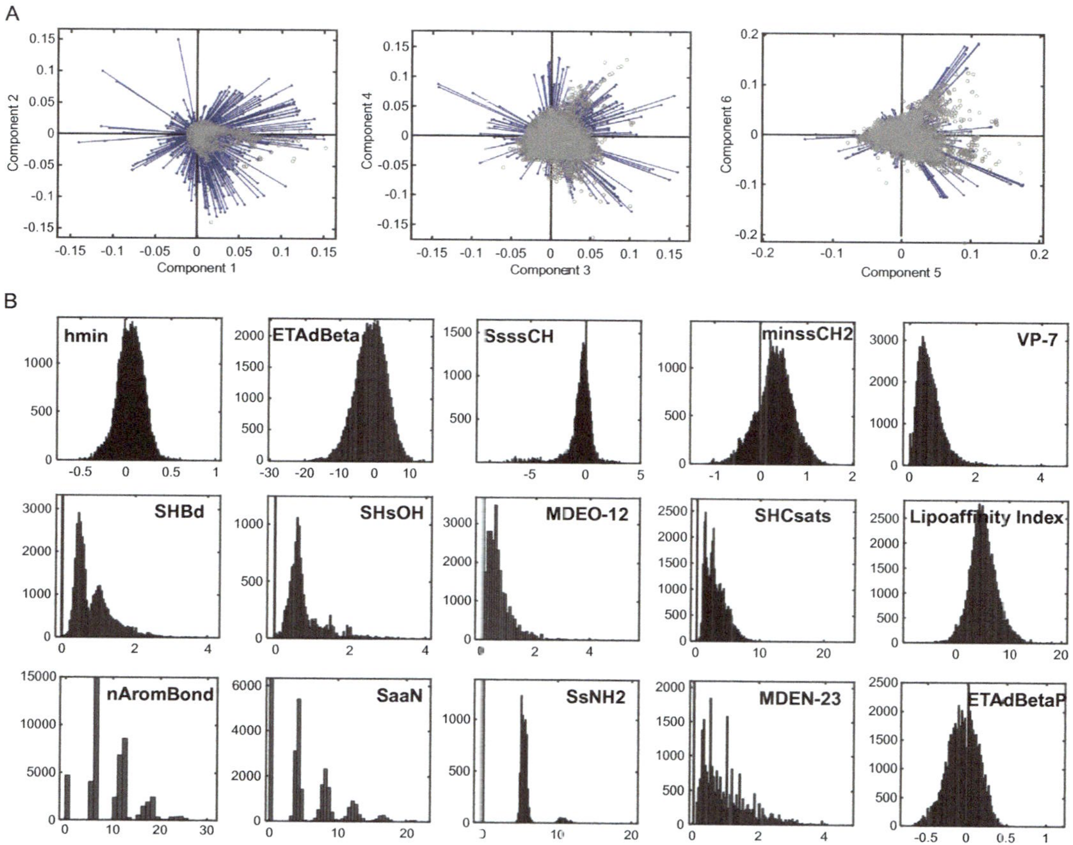

Fig. 8 (A) Principal component biplot with individual small molecules or metabolites represented by gray circles and descriptors as blue vectors. The PCA scores are the individual coordinates. The direction and length of the vector indicates how each variable contributes to the two components in the plot. (B) Frequency histograms of individual descriptors with the largest coefficients contributing to the principal components.

The individual small molecules (observations) are represented as gray circles and the descriptors (features) shown as vectors. The biplots show the contribution of individual descriptors to the separation or differences in the physicochemical properties of small molecules and metabolites. The relative importance of each physicochemical property is based on the magnitude and sign of each variable's contribution to the various principal components. The most important physicochemical properties contributing to the separation of individual clusters of small molecules and metabolites are shown in Fig. 7B as frequency histograms. For example, the distribution of ETAdBeta values represents each molecule's relative unsaturation content with a broad and mostly normal distribution with mean—1.26 and variance 22. In contrast, other descriptors are multimodal including the sum of *E*-States for strong hydrogen bond donors (SHBd).

3.2 Target activity of chemical probes and metabolites in zebrafish phenotypic screens

We selected several novel small molecules and metabolites from machine learning analysis to further evaluate target activity using zebrafish phenotype screens. Phenotypic screens provide a fast and reliable method for validating new chemical probes and active metabolites. Given the requirement for biological endpoints which integrate chromatin remodeling, metabolism and physiology, we explored developmental phenotypes in relevant tissues in the zebrafish. We focused on cardiac phenotype scoring due to the large number of HMEs contributing to cardiovascular development and disease and because our initial screen suggested significant cardiovascular effects and hear failure metabolites with similar chemical fingerprints to small molecule inhibitors. Perhaps most importantly, we chose cardiovascular phenotypes because they are quantitative. Specific phenotypes that were quantitatively scored as outlined in the methods section include embryonic heart looping, atrial contractility (AC), atrial morphogenesis (AM), ventricular contractility (VC) and ventricular morphogenesis (VM). Individual phenotypes were scored on a scale of 1 to 4 based on severity at 72 h post small molecule exposure as previously described (Musso et al., 2014).

Scored phenotypes were clustered using hierarchical cluster analysis to identify compounds with similar activity. Fig. 9 shows a hierarchical dendrogram plot and corresponding heat map of small molecule and metabolite cardiovascular phenotypic activity. The dendrogram defines the hierarchical binary cluster tree of zebrafish phenotypes in the presence of metabolite or chemical probe using similarity measures. The heat map shows increasingly

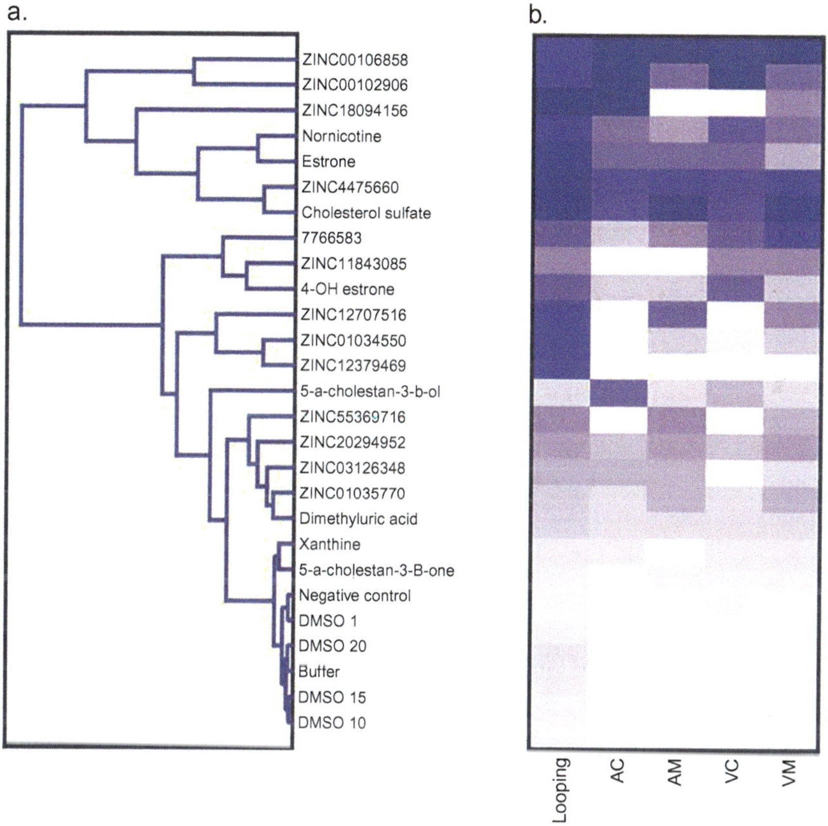

Fig. 9 Zebrafish phenotypic hierarchical cluster analysis. The following zebrafish phenotypes were scored on a scale of 1 to 4 based on severity at 72 h post drug exposure with 4 being more sever: Embryonic heart looping, atrial contractility (AC), atrial morphogenesis (AM), ventricular contractility (VC) and ventricular morphogenesis (VM). Scored phenotypes were then clustered using hierarchical cluster analysis to identify compounds with similar phenotypic activity. (A) Hierarchical dendrogram plot and corresponding (B) heat map of small molecule and metabolite cardiovascular phenotypic activity. Increasingly severe cardiac phenotypes are shown as increasingly darker shades of blue.

severe cardiac phenotypes (high scores) as increasingly darker shades of blue. Compounds with the most severe phenotypes are clustered near the top of the map. Multiple compounds including Zinc4475660, cholesterol sulfate and estrone and their metabolites influence all aspects of scored cardiovascular phenotype while others affect only looping (Zinc12379469) or chamber morphogenesis (Zinc12707516), respectively.

Based on the phenotypic screens, we were able to prioritize compound "hits" with the most significant biology and activity for additional studies. Small molecules with cardiovascular activity were subsequently tested *in vitro*. Interestingly, metabolites enriched or depleted in heart failure including cholesterol and estrogen derivatives, nornicotine, dimethyluric acid and xanthine contribute to cardiovascular phenotypes in a similar pattern but with varying severity. Nornicotine, xanthine and uric are examples of pyridine and purine alkaloids that were noted to cluster with HME inhibitors from our machine learning analysis.

3.3 Biochemical assays confirm the activity of small molecules and metabolites

We used multiple independent biochemical assays to directly and indirectly validate the activity of small molecules and metabolites against select histone modifying enzymes. Direct histone modifying enzyme inhibition was measured *in vitro* in the presence of endogenous cofactors using commercially available fluorescent and chemiluminescent assays. The percent activity of each histone modifying enzyme was measured as a function of small molecule or metabolite concentration. Known inhibitors of each enzyme were used as positive controls. The small molecules and metabolites tested were selected based on enzymes thought to be involved with cardiovascular development and the relative affinity for the enzymes they were predicted to inhibit. Fig. 10A shows a representative dose response curve highlighting the inhibitory activity of one active small molecule identified in our screen (Zinc14879494) against LSD1 compared to the known inhibitor Tranylcypromine. Figs. 10B–F show the percent inhibition of representative HMEs by three different metabolites (cholesterol sulfate, 5-alpha-cholestan-3-beta-ol and estrone) and two small molecules (Zinc14879494 and Zinc4475660). Dose response curves were measured for each enzyme and small molecule tested. Fig. 10D highlights the promiscuous activity of cholesterol sulfate and direct inhibition of multiple HMEs including MLL1, SETD2, NSD1, SIRT1, SETD7 and PARP1. Another steroid hormone and cholesterol metabolite, 5-α-cholestan-3β-ol, also inhibits NSD1 but not SETD2 as predicted (Fig. 10E) while estrone is shown to inhibit NSD1 and to a lesser extent MLL1 (Fig. 10F). Overall, we tested the inhibitory activity of 75 different small molecule-metabolite and enzyme combinations and of those combinations tested, 65% were accurately predicted.

Changes in histone methylation and acetylation in the presence and absence of small molecule or metabolite were also evaluated in cell culture

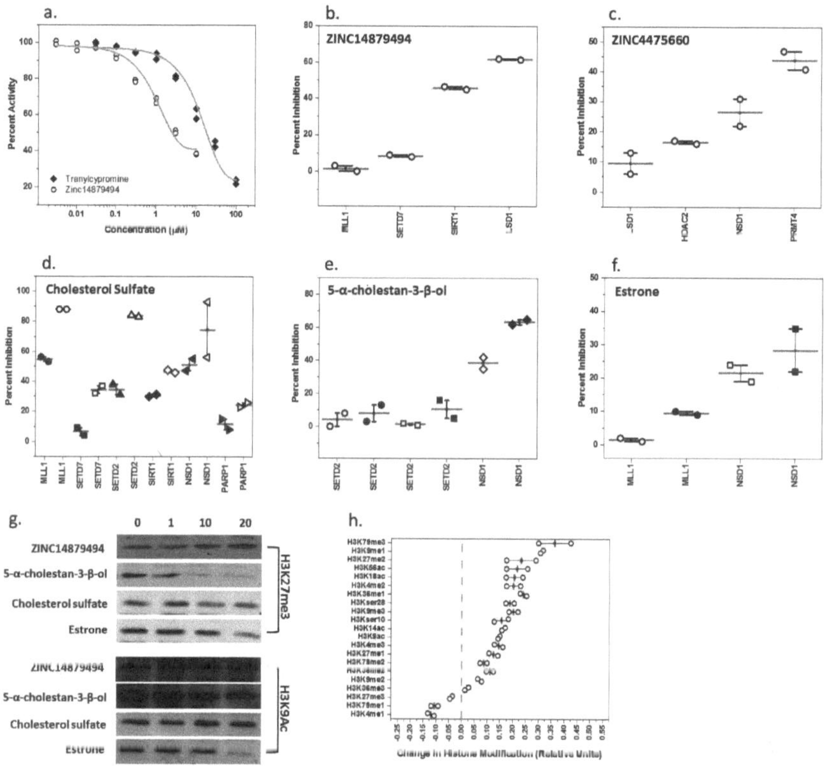

Fig. 10 (A) Representative dose response curve showing direct inhibition of LSD1 by Zinc14879494 ($IC_{50} = 3.5\,\mu M$) and known inhibitor Tranylcypromine ($IC_{50} = 20.5\,\mu M$). (B and C) Percent inhibition of various HMEs by $10\,\mu M$ Zinc14879494 and Zinc4475660. (D–F) Percent inhibition of various HMEs by cholesterol sulfate, 5-α-cholestan-3-β-ol, and estrone. Closed and open symbols represent metabolite concentrations of 30 and $100\,\mu M$, respectively. All experiments were completed in duplicate (biological replicates) and both values for percent activity at the respective concentration are shown on the graphs. The solid red line represents the median value. (G) Western immunoblot analysis of H9C2 cellular H3K27 trimethylation and H3K9 acetylation in the presence of varying concentrations (0–$20\,\mu M$) of Zinc14879494, 5-α-cholestan-3-β-ol, cholesterol sulfate and estrone. (H) The relative change in 21 different histone H3 modifications in the presence of $20\,\mu M$ cholesterol sulfate. The graph shows both data points determined from biological replicates in H9C2 cells. The dashed gray line indicates no change.

using H9C2 myoblasts. Fig. 10G shows an immunoblot of H3K27 tri-methylation and H3K9 acetylation in the presence of varying concentrations of metabolites and Zinc14879494. This data supports modulation of HME activity including decreased H3K27 tri-methylation in the presence of cholesterol sulfate, 5-alpha-cholestan-3-beta-ol and estrone and increased

H3K27 tri-methylation in the presence of Zinc14879494. Both cholesterol sulfate and Zinc14879494 increase H3K9 acetylation while estrone decreases acetylation. A commercially available colorimetric multiplex assay was also used to measure diverse histone H3 modifications in H9C2 cells in the presence and absence of cholesterol sulfate. Fig. 10H shows a plot of the relative change in 21 different histone H3 modifications in the presence of cholesterol sulfate as compared to control. These results confirm the findings from the immunoblot analysis and suggest promiscuous HME activity consistent with our previous findings and predictions.

4. Discussion

Histone modifying enzymes are an essential component of the molecular machines that through covalent modification regulate chromatin architecture and ultimately write or erase the chemical code responsible for changes in gene expression. As one might expect, disruption of this process leads to disease. Aberrant patterns of both histone acetylation and methylation have been observed in cancer, cardiovascular disease, metabolic disorders, inflammation and autoimmune disease, and neurodevelopmental disorders (Falkenberg & Johnstone, 2014; Haberland et al., 2009; Portela & Esteller, 2010). Therefore, our ability to understand the metabolic control of HMEs and to develop chemical probes and small molecule therapeutics that target these critical enzymes is of great interest. Although significant progress has been made in epigenetic research, relatively few small molecules have been identified that modulate the specific activities of nearly 100 different histone modifying enzymes (HMEs) with available crystal structures. Our approach identifies the physicochemical basis of small molecules and metabolites with activity against all classes of histone modifying enzymes that can be used as a chemical blueprint to guide further drug discovery and design.

In this study, we identified *in silico* a series of small molecules with predicted high affinity binding to 70 different histone modifying enzymes, several of which were subsequently validated using both *in vivo* and *in vitro* assays. We also defined the chemical fingerprint of small molecules that cluster based on their physicochemical properties and relate this profile to HME targets. This chemical profile can be used to design novel inhibitors with desired properties of promiscuity or specificity. For example, LSD1 cooperates with HDACs as part of the coREST complex allowing for a transition from an active to a repressed state and simultaneous inhibition of HDAC and LSD1 in glioblastoma cell lines, acting synergistically on cell proliferation (Lee et al., 2006; Singh et al., 2011; You, Tong, Grozinger, & Schreiber, 2001). Using our approach, we are able to identify small molecules with activity against both HDAC2 and LSD1. Similarly, both histone deacetylase and methyltransferase activity have been shown to influence vascular remodeling and heart development (Guarani et al., 2011; Han, Hang, Yang, & Chang, 2011; McKinsey, Zhang, Lu, & Olson, 2000; Zhang et al., 2002). We have also shown that compounds targeting these histone modifying enzymes have strong cardiovascular phenotypes. Numerous examples exist of aberrant epigenetic alterations in disease states. Our approach which identifies compounds with targeted activity toward multiple histone modifications should provide novel classes of therapies with diverse activity.

The critical role of intermediary metabolism in chromatin regulation has become increasingly clear over the past decade (Gut & Verdin, 2013). For example, fluctuations in acetyl CoA during fasting or feeding are essential modulators of histone acetylation and have been shown to affect several HATs and HDACs orchestrating a coordinated network that regulates energy storage and proliferation (Choudhary, Weinert, Nishida, Verdin, & Mann, 2014; Wellen et al., 2009; Zhao et al., 2010). The ketone body β-hydroxybutyrate, a major source of energy during starvation, has been shown to increase global histone acetylation and directly inhibit HDACs (Shimazu et al., 2013). However, the identification of novel metabolic regulators of covalent histone modifications has had few systematic tools and the number of known examples of metabolites that directly inhibit HMEs is sparse. Using chemical informatics and unsupervised learning, we identified several key metabolites with novel histone modifying enzyme activity and subsequently confirmed their *in vivo* and *in vitro* activity. Our predictions also confirmed previously known metabolite activities such as that of β-hydroxybutryate inhibition of HDACs.

Integration of metabolism and gene regulation is of particular importance in tissues with high energetic demands, such as myocardium. The heart consumes more energy than any other organ in the human body and is the most metabolically active (i.e., resting metabolic rates, K_i, for heart and kidneys $440 >$ brain $240 >$ liver $200 >$ skeletal muscle 13). Cardiac metabolism relies on the cellular uptake of free fatty acids and glucose, the breakdown of these components by beta-oxidation and glycolysis, and the entry of the resulting intermediary metabolites into the TCA cycle. While the primary myocardial energy substrate is fatty acids, it can also use glucose, ketone bodies, lactate and amino acids depending on availability. As a result, the heart's ability to rapidly switch between metabolic fuels is crucial to maintain energy homeostasis.

We sought to identify many metabolites that directly inhibit histone modifying enzymes as a way for the cell to "sense" the metabolic state and regulate cellular function and the energetic demands of the heart seemed a logical target. Indeed, changes in histone acetylation, methylation and ADP-ribosylation have been implicated in myriad cardiovascular diseases including congenital anomalies, arrhythmias, cardiac hypertrophy, atherosclerosis, angiogenesis, and heart failure (Baccarelli, Rienstra, & Benjamin, 2010; Chang & Bruneau, 2012; Han et al., 2011; McKinsey & Kass, 2007; Ordovas & Smith, 2010; Zhang et al., 2002). Despite the fact that changes in methylation and acetylation are observed in cardiovascular disease, we often do not know the driving force behind histone modifying enzyme regulation. Our approach identifies several novel classes of intermediary metabolites with HME activity leading to cardiovascular phenotypes, providing additional support for the emerging idea that chromatin-associated enzymes "sense" intracellular small molecules and transduce this information leading to changes in genome regulation of cellular functions.

Using unsupervised learning, we were able to predict that various sterol lipids and cholesterol metabolites directly bind histone modifying enzyme active sites leading to their inhibition. Three sterol lipids with a cholestane skeleton identified by cluster analysis were selected to further validate our predictions using *in vivo* and *in vitro* histone modifying enzyme assays and cardiovascular phenotypic analysis. We demonstrated that cholesterol sulfate directly inhibits the activity of multiple HMEs, alters both methylation and acetylation profiles in myoblast cell culture models and dramatically modifies cardiovascular developmental phenotypes in zebrafish. Interestingly, cholesterol sulfate is present in a variety of cells, plasma, lipoproteins and atherosclerotic lesions of the human aorta (Drayer & Lieberman, 1967; Merten, Dong, Lopez, & Thiagarajan, 2001). It is also thought to support the pro-thrombotic potential of atherosclerotic plaques and has been found at high levels in patients with hypercholesterolemia (Epstein Jr., Krauss, & Shackleton, 1981; Tamasawa, Tamasawa, & Takebe, 1993). Two additional cholesterol derivatives, 5-α-cholestan-3-ol and 5-α-cholestan-3-one, which are important in fuel and energy storage and membrane structure, have similar histone modifying effects but to a lesser extent than cholesterol sulfate.

Another major disorder of cardiac energy metabolism is heart failure and isolated metabolic derangements alone may lead to heart failure or arrhythmias (Hunter, Kelly, McGarrah, Kraus, & Shah, 2016; Lindman et al., 2014; Paulus & Tschope, 2013; van Heerebeek et al., 2008). Several studies have identified metabolites enriched in patients with heart failure (Dunn et al., 2007; Hunter, Kelly, McGarrah, Kraus, et al., 2016; Hunter, Kelly, McGarrah, Khouri, et al., 2016;), including derivatives of pyrimidine and purine, nucleotides, and various short chain fatty acids. For example, uric acid is a well-known biomarker in patients with severely decompensated heart failure (Okazaki et al., 2016). Furthermore, xanthosine and other nucleotide metabolites have been shown to correlate with increased cardiometabolic risk in human subjects from the Framingham Heart Study (Cheng et al., 2012). In this line of exploration, we identified two clusters of compounds comprised of various nucleotide metabolites and small molecules (clusters 18 and 22) that are predicted to target both HDACs and HATs. Caffeine, a derivative of the purine base xanthine, is included in cluster 18 and has previously been shown to induce hyperacetylation of histones at MEF2 on the Glut4 promoter (Mukwevho et al., 2008). Using phenotypic profiling in zebrafish, we also show that exposure to both xanthine and the purine compound dimethyluric acid result in cardiac contractile phenotypes.

In summary, our approach identifies small molecule chemical probes and intermediary metabolites with both promiscuous and specific activity. Using chemical informatics and unsupervised learning, we define a series of chemical fingerprints that can be used to identify native ligands and design novel inhibitors. We also show that various novel metabolites modulate histone modifying enzyme activity and provide relevant context with regard to different types of cardiometabolic diseases. Our results provide support to the idea that histone modifying enzymes sense the metabolic state of a cell which ultimately leads to changes in gene expression.

References

Albaugh, B. N., & Denu, J. M. (2021). Catalysis by protein acetyltransferase Gcn5. *Biochimica et Biophysica Acta, Gene Regulatory Mechanisms, 1864*(2), 194627. https://doi.org/10.1016/j.bbagrm.2020.194627 (Epub 2020/08/26. PubMed PMID: 32841743; PubMed Central PMCID: PMCPMC7854473).

Arrowsmith, C. H., Bountra, C., Fish, P. V., Lee, K., & Schapira, M. (2012). Epigenetic protein families: A new frontier for drug discovery. *Nature Reviews. Drug Discovery, 11*(5), 384–400. https://doi.org/10.1038/nrd3674 (PubMed PMID: 22498752).

Ayer, D. E. (1999). Histone deacetylases: Transcriptional repression with SINers and NuRDs. *Trends in Cell Biology, 9*(5), 193–198. Epub 1999/05/14 https://doi.org/10.1016/s0962-8924(99)01536-6. 10322454.

Baccarelli, A., Rienstra, M., & Benjamin, E. J. (2010). Cardiovascular epigenetics: Basic concepts and results from animal and human studies. *Circulation. Cardiovascular Genetics*, *3*(6), 567–573. https://doi.org/10.1161/CIRCGENETICS.110.958744 (PubMed PMID: 21156932; PubMed Central PMCID: PMC3030456).

Balasubramanyam, K., Swaminathan, V., Ranganathan, A., & Kundu, T. K. (2003). Small molecule modulators of histone acetyltransferase p300. *The Journal of Biological Chemistry*, *278*(21), 19134–19140. Epub 2003/03/08 https://doi.org/10.1074/jbc.M301580200. 12624111.

Balasubramanyam, K., Varier, R. A., Altaf, M., Swaminathan, V., Siddappa, N. B., Ranga, U., et al. (2004). Curcumin, a novel p300/CREB-binding protein-specific inhibitor of acetyltransferase, represses the acetylation of histone/nonhistone proteins and histone acetyltransferase-dependent chromatin transcription. *The Journal of Biological Chemistry*, *279*(49), 51163–51171. Epub 2004/09/24 https://doi.org/10.1074/jbc.M409024200. 15383533.

Berndsen, C. E., Albaugh, B. N., Tan, S., & Denu, J. M. (2007). Catalytic mechanism of a MYST family histone acetyltransferase. *Biochemistry*, *46*(3), 623–629. https://doi.org/10.1021/bi602513x (Epub 2007/01/17. PubMed PMID: 17223684; PubMed Central PMCID: PMCPMC2752042).

Bhaumik, S. R., Smith, E., & Shilatifard, A. (2007). Covalent modifications of histones during development and disease pathogenesis. *Nature Structural & Molecular Biology*, *14*(11), 1008–1016. https://doi.org/10.1038/nsmb1337 (PubMed PMID: 17984963).

Bottomley, M. J., Lo Surdo, P., Di Giovine, P., Cirillo, A., Scarpelli, R., Ferrigno, F., et al. (2008). Structural and functional analysis of the human HDAC4 catalytic domain reveals a regulatory structural zinc-binding domain. *The Journal of Biological Chemistry*, *283*(39), 26694–26704 (Epub 2008/07/11. PubMed PMID: 18614528; PubMed Central PMCID: PMCPMC3258910).

Bradner, J. E., West, N., Grachan, M. L., Greenberg, E. F., Haggarty, S. J., Warnow, T., et al. (2010). Chemical phylogenetics of histone deacetylases. *Nature Chemical Biology*, *6*(3), 238–243. https://doi.org/10.1038/nchembio.313 (Epub 2010/02/09. PubMed PMID: 20139990; PubMed Central PMCID: PMCPMC2822059).

Chandregowda, V., Kush, A., & Reddy, G. C. (2009). Synthesis of benzamide derivatives of anacardic acid and their cytotoxic activity. *European Journal of Medicinal Chemistry*, *44*(6), 2711–2719. Epub 2009/02/28 https://doi.org/10.1016/j.ejmech.2009.01.033. 19246131.

Chang, C. P., & Bruneau, B. G. (2012). Epigenetics and cardiovascular development. *Annual Review of Physiology*, *74*, 41–68. https://doi.org/10.1146/annurev-physiol-020911-153242 (PubMed PMID: 22035349).

Chang, Y., Ganesh, T., Horton, J. R., Spannhoff, A., Liu, J., Sun, A., et al. (2010). Adding a lysine mimic in the design of potent inhibitors of histone lysine methyltransferases. *Journal of Molecular Biology*, *400*(1), 1–7. https://doi.org/10.1016/j.jmb.2010.04.048 (Epub 2010/05/04. PubMed PMID: 20434463; PubMed Central PMCID: PMCPMC2895764).

Chen, Z., Zang, J., Whetstine, J., Hong, X., Davrazou, F., Kutateladze, T. G., et al. (2006). Structural insights into histone demethylation by JMJD2 family members. *Cell*, *125*(4), 691–702. Epub 2006/05/09 https://doi.org/10.1016/j.cell.2006.04.024. 16677698.

Cheng, X., Collins, R. E., & Zhang, X. (2005). Structural and sequence motifs of protein (histone) methylation enzymes. *Annual Review of Biophysics and Biomolecular Structure*, *34*, 267–294. https://doi.org/10.1146/annurev.biophys.34.040204.144452 (Epub 2005/05/05. PubMed PMID: 15869391; PubMed Central PMCID: PMCPMC2733851).

Cheng, S., Rhee, E. P., Larson, M. G., Lewis, G. D., McCabe, E. L., Shen, D., et al. (2012). Metabolite profiling identifies pathways associated with metabolic risk in humans. *Circulation*, *125*(18), 2222–2231. https://doi.org/10.1161/CIRCULATIONAHA.111.067827 (PubMed PMID: 22496159; PubMed Central PMCID: PMC3376658).

Choudhary, C., Weinert, B. T., Nishida, Y., Verdin, E., & Mann, M. (2014). The growing landscape of lysine acetylation links metabolism and cell signalling. *Nature Reviews. Molecular Cell Biology*, *15*(8), 536–550. https://doi.org/10.1038/nrm3841 (PubMed PMID: 25053359).

Clements, A., Rojas, J. R., Trievel, R. C., Wang, L., Berger, S. L., & Marmorstein, R. (1999). Crystal structure of the histone acetyltransferase domain of the human PCAF transcriptional regulator bound to coenzyme a. *The EMBO Journal*, *18*(13), 3521–3532. https://doi.org/10.1093/emboj/18.13.3521 (Epub 1999/07/07. PubMed PMID: 10393169; PubMed Central PMCID: PMCPMC1171431).

Cole, P. A. (2008). Chemical probes for histone-modifying enzymes. *Nature Chemical Biology*, *4*(10), 590–597. https://doi.org/10.1038/nchembio.111 (PubMed PMID: 18800048; PubMed Central PMCID: PMC2908280).

Copeland, R. A., Solomon, M. E., & Richon, V. M. (2009). Protein methyltransferases as a target class for drug discovery. *Nature Reviews. Drug Discovery*, *8*(9), 724–732. https://doi.org/10.1038/nrd2974 (PubMed PMID: 19721445).

Cullis, P. M., Wolfenden, R., Cousens, L. S., & Alberts, B. M. (1982). Inhibition of histone acetylation by N-[2-(S-coenzyme a)acetyl] spermidine amide, a multisubstrate analog. *The Journal of Biological Chemistry*, *257*(20), 12165–12169 (Epub 1982/10/25. PubMed PMID: 7118937).

Dancy, B. M., & Cole, P. A. (2015). Protein lysine acetylation by p300/CBP. *Chemical Reviews*, *115*(6), 2419–2452. https://doi.org/10.1021/cr500452k (Epub 2015/01/17. PubMed PMID: 25594381; PubMed Central PMCID: PMCPMC4378506).

Davies, M., Nowotka, M., Papadatos, G., Dedman, N., Gaulton, A., Atkinson, F., et al. (2015). ChEMBL web services: Streamlining access to drug discovery data and utilities. *Nucleic Acids Research*, *43*(W1), W612–W620. https://doi.org/10.1093/nar/gkv352 (Epub 2015/04/18. PubMed PMID: 25883136; PubMed Central PMCID: PMCPMC4489243).

Del Rizzo, P. A., Krishnan, S., & Trievel, R. C. (2012). Crystal structure and functional analysis of JMJD5 indicate an alternate specificity and function. *Molecular and Cellular Biology*, *32*(19), 4044–4052. https://doi.org/10.1128/MCB.00513-12 (Epub 2012/08/02. PubMed PMID: 22851697; PubMed Central PMCID: PMCPMC3457536).

Drayer, N. M., & Lieberman, S. (1967). Isolation of cholesterol sulfate from human aortas and adrenal tumors. *The Journal of Clinical Endocrinology and Metabolism*, *27*(1), 136–139. https://doi.org/10.1210/jcem-27-1-136 (PubMed PMID: 6019396).

Dunn, W. B., Broadhurst, D. I., Deepak, S. M., Buch, M. H., McDowell, G., Spasic, I., et al. (2007). Serum metabolomics reveals many novel metabolic markers of heart failure, including pseudouridine and 2-oxoglutarate. *Metabolomics*, *3*(4), 413–426.

Dutnall, R. N., Tafrov, S. T., Sternglanz, R., & Ramakrishnan, V. (1998). Structure of the histone acetyltransferase Hat1: A paradigm for the GCN5-related N-acetyltransferase superfamily. *Cell*, *94*(4), 427–438. https://doi.org/10.1016/s0092-8674(00)81584-6. PubMed PMID: 9727486.

Epstein, E. H., Jr., Krauss, R. M., & Shackleton, C. H. (1981). X-linked ichthyosis: Increased blood cholesterol sulfate and electrophoretic mobility of low-density lipoprotein. *Science*, *214*(4521), 659–660 (PubMed PMID: 6945674).

Falkenberg, K. J., & Johnstone, R. W. (2014). Histone deacetylases and their inhibitors in cancer, neurological diseases and immune disorders. *Nature Reviews. Drug Discovery*, *13*(9), 673–691. Epub 2014/08/19 https://doi.org/10.1038/nrd4360. 25131830.

Fang, R., Chen, F., Dong, Z., Hu, D., Barbera, A. J., Clark, E. A., et al. (2013). LSD2/KDM1B and its cofactor NPAC/GLYR1 endow a structural and molecular model for regulation of H3K4 demethylation. *Molecular Cell*, *49*(3), 558–570. https://doi.org/10.1016/j.molcel.2012.11.019 (Epub 2012/12/25. PubMed PMID: 23260659; PubMed Central PMCID: PMCPMC3625064).

Ficner, R. (2009). Novel structural insights into class I and II histone deacetylases. *Current Topics in Medicinal Chemistry*, *9*(3), 235–240. Epub 2009/04/10 https://doi.org/10.2174/156802609788085304. 19355988.

Fitzpatrick, P. F. (2010). Oxidation of amines by flavoproteins. *Archives of Biochemistry and Biophysics*, *493*(1), 13–25. https://doi.org/10.1016/j.abb.2009.07.019 (Epub 2009/08/05. PubMed PMID: 19651103; PubMed Central PMCID: PMCPMC2812625).

Fodor, B. D., Kubicek, S., Yonezawa, M., O'Sullivan, R. J., Sengupta, R., Perez-Burgos, L., et al. (2006). Jmjd2b antagonizes H3K9 trimethylation at pericentric heterochromatin in mammalian cells. *Genes & Development*, *20*(12), 1557–1562. https://doi.org/10.1101/gad.388206 (Epub 2006/06/02. PubMed PMID: 16738407; PubMed Central PMCID: PMCPMC1482475).

Ghizzoni, M., Boltjes, A., Graaf, C., Haisma, H. J., & Dekker, F. J. (2010). Improved inhibition of the histone acetyltransferase PCAF by an anacardic acid derivative. *Bioorganic & Medicinal Chemistry*, *18*(16), 5826–5834. Epub 2010/07/27 https://doi.org/10.1016/j.bmc.2010.06.089. 20655754.

Ghizzoni, M., Wu, J., Gao, T., Haisma, H. J., Dekker, F. J., & George, Z. Y. (2012). 6-alkylsalicylates are selective Tip60 inhibitors and target the acetyl-CoA binding site. *European Journal of Medicinal Chemistry*, *47*(1), 337–344. https://doi.org/10.1016/j.ejmech.2011.11.001 (Epub 2011/11/22. PubMed PMID: 22100137; PubMed Central PMCID: PMCPMC3399519).

Grazzini, R., Hesk, D., Heininger, E., Hildenbrandt, G., Reddy, C. C., Cox-Foster, D., et al. (1991). Inhibition of lipoxygenase and prostaglandin endoperoxide synthase by anacardic acids. *Biochemical and Biophysical Research Communications*, *176*(2), 775–780. https://doi.org/10.1016/s0006-291x(05)80252-9. PubMed PMID: 1902673.

Grozinger, C. M., & Schreiber, S. L. (2002). Deacetylase enzymes: Biological functions and the use of small-molecule inhibitors. *Chemistry & Biology*, *9*(1), 3–16. Epub 2002/02/14 https://doi.org/10.1016/s1074-5521(02)00092-3. 11841934.

Guarani, V., Deflorian, G., Franco, C. A., Kruger, M., Phng, L. K., Bentley, K., et al. (2011). Acetylation-dependent regulation of endothelial notch signalling by the SIRT1 deacetylase. *Nature*, *473*(7346), 234–238. https://doi.org/10.1038/nature09917 (PubMed PMID: 21499261; PubMed Central PMCID: PMC4598045).

Gut, P., & Verdin, E. (2013). The nexus of chromatin regulation and intermediary metabolism. *Nature*, *502*(7472), 489–498. Epub 2013/10/25 https://doi.org/10.1038/nature12752. 24153302.

Haberland, M., Montgomery, R. L., & Olson, E. N. (2009). The many roles of histone deacetylases in development and physiology: Implications for disease and therapy. *Nature Reviews. Genetics*, *10*(1), 32–42. https://doi.org/10.1038/nrg2485 (PubMed PMID: 19065135; PubMed Central PMCID: PMC3215088).

Haggarty, S. J., Koeller, K. M., Wong, J. C., Grozinger, C. M., & Schreiber, S. L. (2003). Domain-selective small-molecule inhibitor of histone deacetylase 6 (HDAC6)-mediated tubulin deacetylation. *Proceedings of the National Academy of Sciences of the United States of America*, *100*(8), 4389–4394. https://doi.org/10.1073/pnas.0430973100 (Epub 2003/04/05. PubMed PMID: 12677000; PubMed Central PMCID: PMCPMC153564).

Hai, Y., & Christianson, D. W. (2016). Histone deacetylase 6 structure and molecular basis of catalysis and inhibition. *Nature Chemical Biology*, *12*(9), 741–747. https://doi.org/10.1038/nchembio.2134 (Epub 2016/07/28. PubMed PMID: 27454933; PubMed Central PMCID: PMCPMC4990478).

Han, P., Hang, C. T., Yang, J., & Chang, C. P. (2011). Chromatin remodeling in cardiovascular development and physiology. *Circulation Research*, *108*(3), 378–396. https://doi.org/10.1161/CIRCRESAHA.110.224287 (PubMed PMID: 21293009; PubMed Central PMCID: PMC3079363).

Hausinger, R. P. (2004). FeII/alpha-ketoglutarate-dependent hydroxylases and related enzymes. *Critical Reviews in Biochemistry and Molecular Biology*, *39*(1), 21–68. Epub 2004/05/04 https://doi.org/10.1080/10409230490440541. 15121720.

Heery, D. M., & Fischer, P. M. (2007). Pharmacological targeting of lysine acetyltransferases in human disease: A progress report. *Drug Discovery Today*, *12*(1–2), 88–99. Epub 2007/01/03 https://doi.org/10.1016/j.drudis.2006.11.012. 17198977.

Helin, K., & Dhanak, D. (2013). Chromatin proteins and modifications as drug targets. *Nature*, *502*(7472), 480–488. https://doi.org/10.1038/nature12751 (PubMed PMID: 24153301).

Ho, T. C. S., Chan, A. H. Y., & Ganesan, A. (2020). Thirty years of HDAC inhibitors: 2020 insight and hindsight. *Journal of Medicinal Chemistry*, *63*(21), 12460–12484. Epub 2020/07/02 https://doi.org/10.1021/acs.jmedchem.0c00830. 32608981.

Hojfeldt, J. W., Agger, K., & Helin, K. (2013). Histone lysine demethylases as targets for anticancer therapy. *Nature Reviews. Drug Discovery*, *12*(12), 917–930. https://doi.org/10.1038/nrd4154 (PubMed PMID: 24232376).

Holbert, M. A., Sikorski, T., Carten, J., Snowflack, D., Hodawadekar, S., & Marmorstein, R. (2007). The human monocytic leukemia zinc finger histone acetyltransferase domain contains DNA-binding activity implicated in chromatin targeting. *The Journal of Biological Chemistry*, *282*(50), 36603–36613. Epub 2007/10/11 https://doi.org/10.1074/jbc.M705812200. 17925393.

Hu, Z., Gomes, I., Horrigan, S. K., Kravarusic, J., Mar, B., Arbieva, Z., et al. (2001). A novel nuclear protein, 5qNCA (LOC51780) is a candidate for the myeloid leukemia tumor suppressor gene on chromosome 5 band q31. *Oncogene*, *20*(47), 6946–6954. Epub 2001/11/01 https://doi.org/10.1038/sj.onc.1204850. 11687974.

Hu, P., & Zhang, Y. (2006). Catalytic mechanism and product specificity of the histone lysine methyltransferase SET7/9: An ab initio QM/MM-FE study with multiple initial structures. *Journal of the American Chemical Society*, *128*(4), 1272–1278. Epub 2006/01/26 https://doi.org/10.1021/ja056153+. 16433545.

Huang, Y., Greene, E., Murray Stewart, T., Goodwin, A. C., Baylin, S. B., Woster, P. M., et al. (2007). Inhibition of lysine-specific demethylase 1 by polyamine analogues results in reexpression of aberrantly silenced genes. *Proceedings of the National Academy of Sciences of the United States of America*, *104*(19), 8023–8028. https://doi.org/10.1073/pnas.0700720104 (Epub 2007/04/28. PubMed PMID: 17463086; PubMed Central PMCID: PMCPMC1857229).

Huang, H., Howard, C. A., Zari, S., Cho, H. J., Shukla, S., Li, H., et al. (2020). Covalent inhibition of NSD1 histone methyltransferase. *Nature Chemical Biology*, *16*(12), 1403–1410. https://doi.org/10.1038/s41589-020-0626-6 (Epub 2020/09/02. PubMed PMID: 32868895; PubMed Central PMCID: PMCPMC7669657).

Huang, M., Huang, J., Zheng, Y., & Sun, Q. (2019). Histone acetyltransferase inhibitors: An overview in synthesis, structure-activity relationship and molecular mechanism. *European Journal of Medicinal Chemistry*, *178*, 259–286. https://doi.org/10.1016/j.ejmech.2019.05.078 (PubMed PMID: 31195169).

Hunter, W. G., Kelly, J. P., McGarrah, R. W., 3rd, Khouri, M. G., Craig, D., Haynes, C., et al. (2016). Metabolomic profiling identifies novel circulating biomarkers of mitochondrial dysfunction differentially elevated in heart failure with preserved versus reduced ejection fraction: Evidence for shared metabolic impairments in clinical heart failure. *Journal of the American Heart Association*, *5*(8). https://doi.org/10.1161/JAHA.115.003190 (PubMed PMID: 27473038; PubMed Central PMCID: PMC5015273).

Hunter, W. G., Kelly, J. P., McGarrah, R. W., 3rd, Kraus, W. E., & Shah, S. H. (2016). Metabolic dysfunction in heart failure: Diagnostic, prognostic, and pathophysiologic insights from metabolomic profiling. *Current Heart Failure Reports*, *13*(3), 119–131. https://doi.org/10.1007/s11897-016-0289-5 (PubMed PMID: 27216948).

Islam, M. S., McDonough, M. A., Chowdhury, R., Gault, J., Khan, A., Pires, E., et al. (2019). Biochemical and structural investigations clarify the substrate selectivity of the 2-oxoglutarate oxygenase JMJD6. *The Journal of Biological Chemistry*, *294*(30), 11637–11652. https://doi.org/10.1074/jbc.RA119.008693 (Epub 2019/05/31. PubMed PMID: 31147442; PubMed Central PMCID: PMCPMC6663879).
Jiang, J., Lu, J., Lu, D., Liang, Z., Li, L., Ouyang, S., et al. (2012). Investigation of the acetylation mechanism by GCN5 histone acetyltransferase. *PLoS One*, 7(5). https://doi.org/10.1371/journal.pone.0036660, e36660 (Epub 2012/05/11. PubMed PMID: 22574209; PubMed Central PMCID: PMCPMC3344931).
Kaczmarska, Z., Ortega, E., Goudarzi, A., Huang, H., Kim, S., Marquez, J. A., et al. (2017). Structure of p300 in complex with acyl-CoA variants. *Nature Chemical Biology*, *13*(1), 21–29. https://doi.org/10.1038/nchembio.2217 (Epub 2016/11/08. PubMed PMID: 27820805; PubMed Central PMCID: PMCPMC5757799).
Kaelin, W. G., Jr., & McKnight, S. L. (2013). Influence of metabolism on epigenetics and disease. *Cell*, *153*(1), 56–69. https://doi.org/10.1016/j.cell.2013.03.004 (Epub 2013/04/02. PubMed PMID: 23540690; PubMed Central PMCID: PMCPMC3775362).
Kelly, T. K., De Carvalho, D. D., & Jones, P. A. (2010). Epigenetic modifications as therapeutic targets. *Nature Biotechnology*, *28*(10), 1069–1078. https://doi.org/10.1038/nbt.1678 (PubMed PMID: 20944599; PubMed Central PMCID: PMC3022972).
King, O. N., Li, X. S., Sakurai, M., Kawamura, A., Rose, N. R., Ng, S. S., et al. (2010). Quantitative high-throughput screening identifies 8-hydroxyquinolines as cell-active histone demethylase inhibitors. *PLoS One*, *5*(11). https://doi.org/10.1371/journal.pone.0015535, e15535 (Epub 2010/12/03. PubMed PMID: 21124847; PubMed Central PMCID: PMCPMC2990756).
Klose, R. J., Kallin, E. M., & Zhang, Y. (2006). JmjC-domain-containing proteins and histone demethylation. *Nature Reviews. Genetics*, 7(9), 715–727. Epub 2006/09/21 https://doi.org/10.1038/nrg1945. 16983801.
Klose, R. J., Yamane, K., Bae, Y., Zhang, D., Erdjument-Bromage, H., Tempst, P., et al. (2006). The transcriptional repressor JHDM3A demethylates trimethyl histone H3 lysine 9 and lysine 36. *Nature*, *442*(7100), 312–316. Epub 2006/05/30 https://doi.org/10.1038/nature04853. 16732292.
Kong, X., Ouyang, S., Liang, Z., Lu, J., Chen, L., Shen, B., et al. (2011). Catalytic mechanism investigation of lysine-specific demethylase 1 (LSD1): A computational study. *PLoS One*, *6*(9). https://doi.org/10.1371/journal.pone.0025444, e25444 (Epub 2011/10/11. PubMed PMID: 21984927; PubMed Central PMCID: PMCPMC3184146).
Kouzarides, T. (2007). Chromatin modifications and their function. *Cell*, *128*(4), 693–705. https://doi.org/10.1016/j.cell.2007.02.005 (PubMed PMID: 17320507).
Kryukov, G. V., Wilson, F. H., Ruth, J. R., Paulk, J., Tsherniak, A., Marlow, S. E., et al. (2016). MTAP deletion confers enhanced dependency on the PRMT5 arginine methyltransferase in cancer cells. *Science*, *351*(6278), 1214–1218. https://doi.org/10.1126/science.aad5214 (Epub 2016/02/26. PubMed PMID: 26912360; PubMed Central PMCID: PMCPMC4997612).
Kubicek, S., O'Sullivan, R. J., August, E. M., Hickey, E. R., Zhang, Q., Teodoro, M. L., et al. (2007). Reversal of H3K9me2 by a small-molecule inhibitor for the G9a histone methyltransferase. *Molecular Cell*, *25*(3), 473–481. Epub 2007/02/10 https://doi.org/10.1016/j.molcel.2007.01.017. 17289593.
Kubo, I., Kinst-Hori, I., & Yokokawa, Y. (1994). Tyrosinase inhibitors from Anacardium occidentale fruits. *Journal of Natural Products*, *57*(4), 545–551. https://doi.org/10.1021/np50106a021. PubMed PMID: 8021657.
Ladurner, A. G. (2006). Rheostat control of gene expression by metabolites. *Molecular Cell*, *24*(1), 1–11. Epub 2006/10/05 https://doi.org/10.1016/j.molcel.2006.09.002. 17018288.

Lauffer, B. E., Mintzer, R., Fong, R., Mukund, S., Tam, C., Zilberleyb, I., et al. (2013). Histone deacetylase (HDAC) inhibitor kinetic rate constants correlate with cellular histone acetylation but not transcription and cell viability. *The Journal of Biological Chemistry, 288*(37), 26926–26943. https://doi.org/10.1074/jbc.M113.490706 (Epub 2013/07/31. PubMed PMID: 23897821; PubMed Central PMCID: PMCPMC3772242).

Lee, K. K., & Workman, J. L. (2007). Histone acetyltransferase complexes: One size doesn't fit all. *Nature Reviews. Molecular Cell Biology, 8*(4), 284–295. Epub 2007/03/24 https://doi.org/10.1038/nrm2145. 17380162.

Lee, M. G., Wynder, C., Bochar, D. A., Hakimi, M. A., Cooch, N., & Shiekhattar, R. (2006). Functional interplay between histone demethylase and deacetylase enzymes. *Molecular and Cellular Biology, 26*(17), 6395–6402. https://doi.org/10.1128/MCB.00723-06 (PubMed PMID: 16914725; PubMed Central PMCID: PMC1592851).

Lee, M. G., Wynder, C., Cooch, N., & Shiekhattar, R. (2005). An essential role for CoREST in nucleosomal histone 3 lysine 4 demethylation. *Nature, 437*(7057), 432–435. Epub 2005/08/05 https://doi.org/10.1038/nature04021. 16079794.

Lee, M. G., Wynder, C., Schmidt, D. M., McCafferty, D. G., & Shiekhattar, R. (2006). Histone H3 lysine 4 demethylation is a target of nonselective antidepressive medications. *Chemistry & Biology, 13*(6), 563–567. Epub 2006/06/24 https://doi.org/10.1016/j.chembiol.2006.05.004. 16793513.

Li, G. B., Huang, L., Li, H., Ji, S., Li, L., & Yang, S. (2016). Identification of new p300 histone acetyltransferase inhibitors from natural products by a customized virutal screening method. *RSC Advances, 6*, 61137–61140.

Lin, J. K. (2007). Molecular targets of curcumin. *Advances in Experimental Medicine and Biology, 595*, 227–243. Epub 2007/06/16 https://doi.org/10.1007/978-0-387-46401-5_10. 17569214.

Lindman, B. R., Davila-Roman, V. G., Mann, D. L., McNulty, S., Semigran, M. J., Lewis, G. D., et al. (2014). Cardiovascular phenotype in HFpEF patients with or without diabetes: A RELAX trial ancillary study. *Journal of the American College of Cardiology, 64*(6), 541–549. https://doi.org/10.1016/j.jacc.2014.05.030 (PubMed PMID: 25104521; PubMed Central PMCID: PMC4133145).

Linscott, J. A., Kapilashrami, K., Wang, Z., Senevirathne, C., Bothwell, I. R., Blum, G., et al. (2016). Kinetic isotope effects reveal early transition state of protein lysine methyltransferase SET8. *Proceedings of the National Academy of Sciences of the United States of America, 113*(52), E8369–E8378. https://doi.org/10.1073/pnas.1609032114 (Epub 2016/12/13. PubMed PMID: 27940912; PubMed Central PMCID: PMCPMC5206543).

Liu, F., Chen, X., Allali-Hassani, A., Quinn, A. M., Wigle, T. J., Wasney, G. A., et al. (2010). Protein lysine methyltransferase G9a inhibitors: Design, synthesis, and structure activity relationships of 2,4-diamino-7-aminoalkoxy-quinazolines. *Journal of Medicinal Chemistry, 53*(15), 5844–5857. https://doi.org/10.1021/jm100478y (Epub 2010/07/10. PubMed PMID: 20614940; PubMed Central PMCID: PMCPMC2920043).

Liu, X., Wang, L., Zhao, K., Thompson, P. R., Hwang, Y., Marmorstein, R., et al. (2008). The structural basis of protein acetylation by the p300/CBP transcriptional coactivator. *Nature, 451*(7180), 846–850. Epub 2008/02/15 https://doi.org/10.1038/nature06546. 18273021.

Lohse, B., Kristensen, J. L., Kristensen, L. H., Agger, K., Helin, K., Gajhede, M., et al. (2011). Inhibitors of histone demethylases. *Bioorganic & Medicinal Chemistry, 19*(12), 3625–3636. https://doi.org/10.1016/j.bmc.2011.01.046. 21596573.

Lombardi, P. M., Cole, K. E., Dowling, D. P., & Christianson, D. W. (2011). Structure, mechanism, and inhibition of histone deacetylases and related metalloenzymes. *Current Opinion in Structural Biology, 21*(6), 735–743. https://doi.org/10.1016/j.sbi.2011.08.004 (Epub 2011/08/30. PubMed PMID: 21872466; PubMed Central PMCID: PMCPMC3232309).

Lu, C., & Thompson, C. B. (2012). Metabolic regulation of epigenetics. *Cell Metabolism*, *16*(1), 9–17. https://doi.org/10.1016/j.cmet.2012.06.001 (Epub 2012/07/10. PubMed PMID: 22768835; PubMed Central PMCID: PMCPMC3392647).

MacPherson, L., Anokye, J., Yeung, M. M., Lam, E. Y. N., Chan, Y. C., Weng, C. F., et al. (2020). HBO1 is required for the maintenance of leukaemia stem cells. *Nature*, *577*(7789), 266–270. Epub 2019/12/13 https://doi.org/10.1038/s41586-019-1835-6. 31827282.

Madia, V. N., Benedetti, R., Barreca, M. L., Ngo, L., Pescatori, L., Messore, A., et al. (2017). Structure-activity relationships on cinnamoyl derivatives as inhibitors of p300 histone acetyltransferase. *ChemMedChem*, *12*(16), 1359–1368. Epub 2017/03/28 https://doi.org/10.1002/cmdc.201700040. 28346821.

Marcu, M. G., Jung, Y. J., Lee, S., Chung, E. J., Lee, M. J., Trepel, J., et al. (2006). Curcumin is an inhibitor of p300 histone acetylatransferase. *Medicinal Chemistry*, *2*(2), 169–174. Epub 2006/06/22 https://doi.org/10.2174/157340606776056133. 16787365.

Marks, P. A., & Breslow, R. (2007). Dimethyl sulfoxide to vorinostat: Development of this histone deacetylase inhibitor as an anticancer drug. *Nature Biotechnology*, *25*(1), 84–90. Epub 2007/01/11 https://doi.org/10.1038/nbt1272. 17211407.

Marmorstein, R., & Zhou, M. M. (2014). Writers and readers of histone acetylation: Structure, mechanism, and inhibition. *Cold Spring Harbor Perspectives in Biology*, *6*(7). https://doi.org/10.1101/cshperspect.a018762 (a018762. Epub 2014/07/06. PubMed PMID: 24984779; PubMed Central PMCID: PMCPMC4067988).

Mavrakis, K. J., McDonald, E. R., 3rd, Schlabach, M. R., Billy, E., & Hoffman, G. R. (2016). deWeck a, et al. disordered methionine metabolism in MTAP/CDKN2A-deleted cancers leads to dependence on PRMT5. *Science*, *351*(6278), 1208–1213. Epub 2016/02/26 https://doi.org/10.1126/science.aad5944. 26912361.

McKinsey, T. A., & Kass, D. A. (2007). Small-molecule therapies for cardiac hypertrophy: Moving beneath the cell surface. *Nature Reviews. Drug Discovery*, *6*(8), 617–635. https://doi.org/10.1038/nrd2193 (PubMed PMID: 17643091).

McKinsey, T. A., Zhang, C. L., Lu, J., & Olson, E. N. (2000). Signal-dependent nuclear export of a histone deacetylase regulates muscle differentiation. *Nature*, *408*(6808), 106–111. https://doi.org/10.1038/35040593 (PubMed PMID: 11081517; PubMed Central PMCID: PMC4459600).

Mendez, D., Gaulton, A., Bento, A. P., Chambers, J., De Veij, M., Felix, E., et al. (2019). ChEMBL: Towards direct deposition of bioassay data. *Nucleic Acids Research*, *47*(D1), D930–D940. https://doi.org/10.1093/nar/gky1075 (Epub 2018/11/07. PubMed PMID: 30398643; PubMed Central PMCID: PMCPMC6323927).

Merten, M., Dong, J. F., Lopez, J. A., & Thiagarajan, P. (2001). Cholesterol sulfate: A new adhesive molecule for platelets. *Circulation*, *103*(16), 2032–2034 (PubMed PMID: 11319189).

Mimasu, S., Sengoku, T., Fukuzawa, S., Umehara, T., & Yokoyama, S. (2008). Crystal structure of histone demethylase LSD1 and tranylcypromine at 2.25 a. *Biochemical and Biophysical Research Communications*, *366*(1), 15–22. Epub 2007/11/28 https://doi.org/10.1016/j.bbrc.2007.11.066. 18039463.

Mimasu, S., Umezawa, N., Sato, S., Higuchi, T., Umehara, T., & Yokoyama, S. (2010). Structurally designed trans-2-phenylcyclopropylamine derivatives potently inhibit histone demethylase LSD1/KDM1. *Biochemistry*, *49*(30), 6494–6503. Epub 2010/06/24 https://doi.org/10.1021/bi100299r. 20568732.

Mukwevho, E., Kohn, T. A., Lang, D., Nyatia, E., Smith, J., & Ojuka, E. O. (2008). Caffeine induces hyperacetylation of histones at the MEF2 site on the Glut4 promoter and increases MEF2A binding to the site via a CaMK-dependent mechanism. *American Journal of Physiology. Endocrinology and Metabolism*, *294*(3), E582–E588. https://doi.org/10.1152/ajpendo.00312.2007 (PubMed PMID: 18198354).

Musso, G., Tasan, M., Mosimann, C., Beaver, J. E., Plovie, E., Carr, L. A., et al. (2014). Novel cardiovascular gene functions revealed via systematic phenotype prediction in zebrafish. *Development*, *141*(1), 224–235. https://doi.org/10.1242/dev.099796 (PubMed PMID: 24346703; PubMed Central PMCID: PMC3865760).

Novotny-Diermayr, V., Sangthongpitag, K., Hu, C. Y., Wu, X., Sausgruber, N., Yeo, P., et al. (2010). SB939, a novel potent and orally active histone deacetylase inhibitor with high tumor exposure and efficacy in mouse models of colorectal cancer. *Molecular Cancer Therapeutics*, *9*(3), 642–652. Epub 2010/03/04 https://doi.org/10.1158/1535-7163.MCT-09-0689. 20197387.

Okazaki, H., Shirakabe, A., Kobayashi, N., Hata, N., Shinada, T., Matsushita, M., et al. (2016). The prognostic impact of uric acid in patients with severely decompensated acute heart failure. *Journal of Cardiology*. https://doi.org/10.1016/j.jjcc.2016.04.013 (PubMed PMID: 27261247).

Ordovas, J. M., & Smith, C. E. (2010). Epigenetics and cardiovascular disease. *Nature Reviews. Cardiology*, 7(9), 510–519. https://doi.org/10.1038/nrcardio.2010.104 (PubMed PMID: 20603647; PubMed Central PMCID: PMC3075976).

Paulus, W. J., & Tschope, C. (2013). A novel paradigm for heart failure with preserved ejection fraction: Comorbidities drive myocardial dysfunction and remodeling through coronary microvascular endothelial inflammation. *Journal of the American College of Cardiology*, *62*(4), 263–271. https://doi.org/10.1016/j.jacc.2013.02.092 (PubMed PMID: 23684677).

Portela, A., & Esteller, M. (2010). Epigenetic modifications and human disease. *Nature Biotechnology*, *28*(10), 1057–1068. https://doi.org/10.1038/nbt.1685 (PubMed PMID: 20944598).

Poulin, M. B., Schneck, J. L., Matico, R. E., McDevitt, P. J., Huddleston, M. J., Hou, W., et al. (2016). Transition state for the NSD2-catalyzed methylation of histone H3 lysine 36. *Proceedings of the National Academy of Sciences of the United States of America*, *113*(5), 1197–1201. https://doi.org/10.1073/pnas.1521036113 (Epub 2016/01/21. PubMed PMID: 26787850; PubMed Central PMCID: PMCPMC4747696).

Qian, C., & Zhou, M. M. (2006). SET domain protein lysine methyltransferases: Structure, specificity and catalysis. *Cellular and Molecular Life Sciences*, *63*(23), 2755–2763. Epub 2006/10/03 https://doi.org/10.1007/s00018-006-6274-5. 17013555.

Rogawski, D. S., Deng, J., Li, H., Miao, H., Borkin, D., Purohit, T., et al. (2021). Discovery of first-in-class inhibitors of ASH1L histone methyltransferase with anti-leukemic activity. *Nature Communications*, *12*(1), 2792. https://doi.org/10.1038/s41467-021-23152-6 (Epub 2021/05/16. PubMed PMID: 33990599; PubMed Central PMCID: PMCPMC8121805).

Rojas, J. R., Trievel, R. C., Zhou, J., Mo, Y., Li, X., Berger, S. L., et al. (1999). Structure of tetrahymena GCN5 bound to coenzyme a and a histone H3 peptide. *Nature*, *401*(6748), 93–98. Epub 1999/09/15 https://doi.org/10.1038/43487. 10485713.

Rose, N. R., Ng, S. S., Mecinovic, J., Lienard, B. M., Bello, S. H., Sun, Z., et al. (2008). Inhibitor scaffolds for 2-oxoglutarate-dependent histone lysine demethylases. *Journal of Medicinal Chemistry*, *51*(22), 7053–7056. Epub 2008/10/24 https://doi.org/10.1021/jm800936s. 18942826.

Sakurai, M., Rose, N. R., Schultz, L., Quinn, A. M., Jadhav, A., Ng, S. S., et al. (2010). A miniaturized screen for inhibitors of Jumonji histone demethylases. *Molecular BioSystems*, *6*(2), 357–364. https://doi.org/10.1039/b912993f (Epub 2010/01/23. PubMed PMID: 20094655; PubMed Central PMCID: PMCPMC3401096).

Schapira, M. (2016a). Structural chemistry of human RNA methyltransferases. *ACS Chemical Biology*, *11*(3), 575–582. https://doi.org/10.1021/acschembio.5b00781 (PubMed PMID: 26566070).

Schapira, M. (2016b). Chemical inhibition of protein methyltransferases. *Cell Chemical Biology*, *23*(9), 1067–1076. Epub 2016/08/30 https://doi.org/10.1016/j.chembiol.2016.07.014. 27569753.

Schenk, T., Chen, W. C., Gollner, S., Howell, L., Jin, L., Hebestreit, K., et al. (2012). Inhibition of the LSD1 (KDM1A) demethylase reactivates the all-trans-retinoic acid differentiation pathway in acute myeloid leukemia. *Nature Medicine*, *18*(4), 605–611. https://doi.org/10.1038/nm.2661 (PubMed PMID: 22406747; PubMed Central PMCID: PMC3539284).

Schuetz, A., Bernstein, G., Dong, A., Antoshenko, T., Wu, H., Loppnau, P., et al. (2007). Crystal structure of a binary complex between human GCN5 histone acetyltransferase domain and acetyl coenzyme a. *Proteins*, *68*(1), 403–407. Epub 2007/04/06 https://doi.org/10.1002/prot.21407. 17410582.

Schuetz, A., Min, J., Allali-Hassani, A., Schapira, M., Shuen, M., Loppnau, P., et al. (2008). Human HDAC7 harbors a class IIa histone deacetylase-specific zinc binding motif and cryptic deacetylase activity. *The Journal of Biological Chemistry*, *283*(17), 11355–11363. https://doi.org/10.1074/jbc.M707362200 (Epub 2008/02/21. PubMed PMID: 18285338; PubMed Central PMCID: PMCPMC2431080).

Sengoku, T., & Yokoyama, S. (2011). Structural basis for histone H3 Lys 27 demethylation by UTX/KDM6A. *Genes & Development*, *25*(21), 2266–2277. https://doi.org/10.1101/gad.172296.111 (Epub 2011/10/18. PubMed PMID: 22002947; PubMed Central PMCID: PMCPMC3219231).

Serefidou, M., Venkatasubramani, A. V., & Imhof, A. (2019). The impact of one carbon metabolism on histone methylation. *Frontiers in Genetics*, *10*, 764. https://doi.org/10.3389/fgene.2019.00764 (Epub 2019/09/27. PubMed PMID: 31555321; PubMed Central PMCID: PMCPMC6722216).

Shi, Y. J., Matson, C., Lan, F., Iwase, S., Baba, T., & Shi, Y. (2005). Regulation of LSD1 histone demethylase activity by its associated factors. *Molecular Cell*, *19*(6), 857–864. Epub 2005/09/06 https://doi.org/10.1016/j.molcel.2005.08.027. 16140033.

Shimazu, T., Hirschey, M. D., Newman, J., He, W., Shirakawa, K., Le Moan, N., et al. (2013). Suppression of oxidative stress by beta-hydroxybutyrate, an endogenous histone deacetylase inhibitor. *Science*, *339*(6116), 211–214. https://doi.org/10.1126/science.1227166 (Epub 2012/12/12. PubMed PMID: 23223453; PubMed Central PMCID: PMCPMC3735349).

Shin, W. H., & Kihara, D. (1958). 55 years of the Rossmann fold. *Methods in Molecular Biology*, *2019*, 1–13. Epub 2019/04/05 https://doi.org/10.1007/978-1-4939-9161-7_1. 30945211.

Singh, M. M., Manton, C. A., Bhat, K. P., Tsai, W. W., Aldape, K., Barton, M. C., et al. (2011). Inhibition of LSD1 sensitizes glioblastoma cells to histone deacetylase inhibitors. *Neuro-Oncology*, *13*(8), 894–903. https://doi.org/10.1093/neuonc/nor049 (PubMed PMID: 21653597; PubMed Central PMCID: PMC3145466).

Somoza, J. R., Skene, R. J., Katz, B. A., Mol, C., Ho, J. D., Jennings, A. J., et al. (2004). Structural snapshots of human HDAC8 provide insights into the class I histone deacetylases. *Structure*, *12*(7), 1325–1334. Epub 2004/07/10 https://doi.org/10.1016/j.str.2004.04.012. 15242608.

Sterling, T., & Irwin, J. J. (2015). ZINC 15- -ligand discovery for everyone. *Journal of Chemical Information and Modeling*, *55*(11), 2324–2337. https://doi.org/10.1021/acs.jcim.5b00559 (Epub 2015/10/20. PubMed PMID: 26479676; PubMed Central PMCID: PMCPMC4658288).

Strahl, B. D., & Allis, C. D. (2000). The language of covalent histone modifications. *Nature*, *403*(6765), 41–45. https://doi.org/10.1038/47412 (PubMed PMID: 10638745).

Tamasawa, N., Tamasawa, A., & Takebe, K. (1993). Higher levels of plasma cholesterol sulfate in patients with liver cirrhosis and hypercholesterolemia. *Lipids*, *28*(9), 833–836 (PubMed PMID: 8231659).

Tanner, K. G., Trievel, R. C., Kuo, M. H., Howard, R. M., Berger, S. L., Allis, C. D., et al. (1999). Catalytic mechanism and function of invariant glutamic acid 173 from the histone acetyltransferase GCN5 transcriptional coactivator. *The Journal of Biological Chemistry*, *274*(26), 18157–18160. Epub 1999/06/22 https://doi.org/10.1074/jbc.274.26.18157. 10373413.

Tarhonskaya, H., Nowak, R. P., Johansson, C., Szykowska, A., Tumber, A., Hancock, R. L., et al. (2017). Studies on the interaction of the histone demethylase KDM5B with tricarboxylic acid cycle intermediates. *Journal of Molecular Biology*, *429*(19), 2895–2906. https://doi.org/10.1016/j.jmb.2017.08.007 (Epub 2017/08/23. PubMed PMID: 28827149; PubMed Central PMCID: PMCPMC5636616).

Trott, O., & Olson, A. J. (2010). AutoDock Vina: Improving the speed and accuracy of docking with a new scoring function, efficient optimization, and multithreading. *Journal of Computational Chemistry*, *31*(2), 455–461. https://doi.org/10.1002/jcc.21334 (Epub 2009/06/06. PubMed PMID: 19499576; PubMed Central PMCID: PMCPMC3041641).

Tsukada, Y., Fang, J., Erdjument-Bromage, H., Warren, M. E., Borchers, C. H., Tempst, P., et al. (2006). Histone demethylation by a family of JmjC domain-containing proteins. *Nature*, *439*(7078), 811–816. Epub 2005/12/20 https://doi.org/10.1038/nature04433. 16362057.

Ueda, R., Suzuki, T., Mino, K., Tsumoto, H., Nakagawa, H., Hasegawa, M., et al. (2009). Identification of cell-active lysine specific demethylase 1-selective inhibitors. *Journal of the American Chemical Society*, *131*(48), 17536–17537. Epub 2009/12/03 https://doi.org/10.1021/ja907055q. 19950987.

van Heerebeek, L., Hamdani, N., Handoko, M. L., Falcao-Pires, I., Musters, R. J., Kupreishvili, K., et al. (2008). Diastolic stiffness of the failing diabetic heart: Importance of fibrosis, advanced glycation end products, and myocyte resting tension. *Circulation*, *117*(1), 43–51. https://doi.org/10.1161/CIRCULATIONAHA.107.728550 (PubMed PMID: 18071071).

Vedadi, M., Barsyte-Lovejoy, D., Liu, F., Rival-Gervier, S., Allali-Hassani, A., Labrie, V., et al. (2011). A chemical probe selectively inhibits G9a and GLP methyltransferase activity in cells. *Nature Chemical Biology*, 7(8), 566–574. https://doi.org/10.1038/nchembio.599 (Epub 2011/07/12. PubMed PMID: 21743462; PubMed Central PMCID: PMCPMC3184254).

Waldecker, M., Kautenburger, T., Daumann, H., Busch, C., & Schrenk, D. (2008). Inhibition of histone-deacetylase activity by short-chain fatty acids and some polyphenol metabolites formed in the colon. *The Journal of Nutritional Biochemistry*, *19*(9), 587–593. Epub 2007/12/07 https://doi.org/10.1016/j.jnutbio.2007.08.002. 18061431.

Walport, L. J., Hopkinson, R. J., Vollmar, M., Madden, S. K., Gileadi, C., Oppermann, U., et al. (2014). Human UTY(KDM6C) is a male-specific N-methyl lysyl demethylase. *The Journal of Biological Chemistry*, *289*(26), 18302–18313. https://doi.org/10.1074/jbc.M114.555052 (Epub 2014/05/07. PubMed PMID: 24798337; PubMed Central PMCID: PMCPMC4140284).

Wang, D., Girard, T. J., Kasten, T. P., LaChance, R. M., Miller-Wideman, M. A., & Durley, R. C. (1998). Inhibitory activity of unsaturated fatty acids and anacardic acids toward soluble tissue factor-factor VIIa complex. *Journal of Natural Products*, *61*(11), 1352–1355. https://doi.org/10.1021/np980117p. PubMed PMID: 9834151.

Wang, S., Hu, P., & Zhang, Y. (2007). Ab initio quantum mechanical/molecular mechanical molecular dynamics simulation of enzyme catalysis: The case of histone lysine methyltransferase SET7/9. *The Journal of Physical Chemistry. B*, *111*(14), 3758–3764. https://doi.org/10.1021/jp067147i (Epub 2007/03/29. PubMed PMID: 17388541; PubMed Central PMCID: PMCPMC2527688).

Watson, P. J., Fairall, L., Santos, G. M., & Schwabe, J. W. (2012). Structure of HDAC3 bound to co-repressor and inositol tetraphosphate. *Nature*, *481*(7381), 335–340.

https://doi.org/10.1038/nature10728 (Epub 2012/01/11. PubMed PMID: 22230954; PubMed Central PMCID: PMCPMC3272448).

Wellen, K. E., Hatzivassiliou, G., Sachdeva, U. M., Bui, T. V., Cross, J. R., & Thompson, C. B. (2009). ATP-citrate lyase links cellular metabolism to histone acetylation. *Science, 324*(5930), 1076–1080. https://doi.org/10.1126/science.1164097 (PubMed PMID: 19461003; PubMed Central PMCID: PMC2746744).

West, A. C., & Johnstone, R. W. (2014). New and emerging HDAC inhibitors for cancer treatment. *The Journal of Clinical Investigation, 124*(1), 30–39. https://doi.org/10.1172/JCI69738 (PubMed PMID: 24382387; PubMed Central PMCID: PMC3871231).

Whetstine, J. R., Nottke, A., Lan, F., Huarte, M., Smolikov, S., Chen, Z., et al. (2006). Reversal of histone lysine trimethylation by the JMJD2 family of histone demethylases. *Cell, 125*(3), 467–481. Epub 2006/04/11 https://doi.org/10.1016/j.cell.2006.03.028. 16603238.

Wishart, D. S., Guo, A., Oler, E., Wang, F., Anjum, A., Peters, H., et al. (2022). HMDB 5.0: The human metabolome database for 2022. *Nucleic Acids Research, 50*(D1), D622–D631. https://doi.org/10.1093/nar/gkab1062 (Epub 2022/01/07. PubMed PMID: 34986597; PubMed Central PMCID: PMCPMC8728138).

Woo, J., Kim, H. Y., Byun, B. J., Chae, C. H., Lee, J. Y., Ryu, S. Y., et al. (2014). Biological evaluation of tanshindiols as EZH2 histone methyltransferase inhibitors. *Bioorganic & Medicinal Chemistry Letters, 24*(11), 2486–2492. Epub 2014/04/29 https://doi.org/10.1016/j.bmcl.2014.04.010. 24767850.

Wu, H., Min, J., Lunin, V. V., Antoshenko, T., Dombrovski, L., Zeng, H., et al. (2010). Structural biology of human H3K9 methyltransferases. *PLoS One, 5*(1). https://doi.org/10.1371/journal.pone.0008570, e8570 (Epub 2010/01/20. PubMed PMID: 20084102; PubMed Central PMCID: PMCPMC2797608).

Wu, H., Moshkina, N., Min, J., Zeng, H., Joshua, J., Zhou, M. M., et al. (2012). Structural basis for substrate specificity and catalysis of human histone acetyltransferase 1. *Proceedings of the National Academy of Sciences of the United States of America, 109*(23), 8925–8930. https://doi.org/10.1073/pnas.1114117109 (Epub 2012/05/23. PubMed PMID: 22615379; PubMed Central PMCID: PMCPMC3384170).

Xiao, B., Jing, C., Kelly, G., Walker, P. A., Muskett, F. W., Frenkiel, T. A., et al. (2005). Specificity and mechanism of the histone methyltransferase pr-Set7. *Genes & Development, 19*(12), 1444–1454. https://doi.org/10.1101/gad.1315905 (Epub 2005/06/04. PubMed PMID: 15933069; PubMed Central PMCID: PMCPMC1151661).

Xiao, B., Jing, C., Wilson, J. R., Walker, P. A., Vasisht, N., Kelly, G., et al. (2003). Structure and catalytic mechanism of the human histone methyltransferase SET7/9. *Nature, 421*(6923), 652–656. Epub 2003/01/24 https://doi.org/10.1038/nature01378. 12540855.

Yan, Y., Barlev, N. A., Haley, R. H., Berger, S. L., & Marmorstein, R. (2000). Crystal structure of yeast Esa1 suggests a unified mechanism for catalysis and substrate binding by histone acetyltransferases. *Molecular Cell, 6*(5), 1195–1205. Epub 2000/12/07 https://doi.org/10.1016/s1097-2765(00)00116-7. 11106757.

Yang, Z., Shang, Y., Chen, X., Li, W., Li, G., & Wu, Y. (2018). Total synthesis and evaluation of B-homo Palmatine and berberine derivatives as p300 histone acetyltransferase inhibitors. *European Journal of Organic Chemistry*, 1041–1052.

Yap, C. W. (2011). PaDEL-descriptor: An open source software to calculate molecular descriptors and fingerprints. *Journal of Computational Chemistry, 32*(7), 1466–1474. Epub 2011/03/23 https://doi.org/10.1002/jcc.21707. 21425294.

Yoshida, M., Kijima, M., Akita, M., & Beppu, T. (1990). Potent and specific inhibition of mammalian histone deacetylase both in vivo and in vitro by trichostatin a. *The Journal of Biological Chemistry, 265*(28), 17174–17179 (Epub 1990/10/05. PubMed PMID: 2211619).

You, A., Tong, J. K., Grozinger, C. M., & Schreiber, S. L. (2001). CoREST is an integral component of the CoREST- human histone deacetylase complex. *Proceedings of the National Academy of Sciences of the United States of America, 98*(4), 1454–1458. https://doi.org/10.1073/pnas.98.4.1454 (PubMed PMID: 11171972; PubMed Central PMCID: PMC29278).

Yu, C. W., Chang, P. T., Hsin, L. W., & Chern, J. W. (2013). Quinazolin-4-one derivatives as selective histone deacetylase-6 inhibitors for the treatment of Alzheimer's disease. *Journal of Medicinal Chemistry, 56*(17), 6775–6791. Epub 2013/08/03 https://doi.org/10.1021/jm400564j. 23905680.

Yu, T., Higashi, M., Cembran, A., Gao, J., & Truhlar, D. G. (2013). Concerted hydrogen atom and electron transfer mechanism for catalysis by lysine-specific demethylase. *The Journal of Physical Chemistry. B, 117*(28), 8422–8429. https://doi.org/10.1021/jp404292t (Epub 2013/06/04. PubMed PMID: 23725223; PubMed Central PMCID: PMCPMC4477199).

Zhang, C. L., McKinsey, T. A., Chang, S., Antos, C. L., Hill, J. A., & Olson, E. N. (2002). Class II histone deacetylases act as signal-responsive repressors of cardiac hypertrophy. *Cell, 110*(4), 479–488. 12202037. PMC4459650.

Zhao, S., Xu, W., Jiang, W., Yu, W., Lin, Y., Zhang, T., et al. (2010). Regulation of cellular metabolism by protein lysine acetylation. *Science, 327*(5968), 1000–1004. https://doi.org/10.1126/science.1179689 (Epub 2010/02/20. PubMed PMID: 20167786; PubMed Central PMCID: PMCPMC3232675).

Zheng, W., Ibanez, G., Wu, H., Blum, G., Zeng, H., Dong, A., et al. (2012). Sinefungin derivatives as inhibitors and structure probes of protein lysine methyltransferase SETD2. *Journal of the American Chemical Society, 134*(43), 18004–18014. https://doi.org/10.1021/ja307060p (Epub 2012/10/10. PubMed PMID: 23043551; PubMed Central PMCID: PMCPMC3504124).

CHAPTER EIGHT

Designing synthetic transcription factors: A structural perspective

Aparna Boral[†], Madhurima Khamaru[†], and Devrani Mitra*
Department of Life Sciences, Presidency University, Kolkata, West Bengal, India
*Corresponding author: e-mail address: devrani.dbs@presiuniv.ac.in

Contents

[†] Equal contribution.

Advances in Protein Chemistry and Structural Biology, Volume 130
ISSN 1876-1623
https://doi.org/10.1016/bs.apcsb.2021.12.003

Abstract

In this chapter, we discuss different design strategies of synthetic proteins, especially synthetic transcription factors. Design and engineering of synthetic transcription factors is particularly relevant for the need-based manipulation of gene expression. With recent advances in structural biology techniques and with the emergence of other precision biochemical/physical tools, accurate knowledge on structure-function relations is increasingly becoming available. Besides discussing the underlying principles of design, we go through individual cases, especially those involving four major groups of transcription factors—basic leucine zippers, zinc fingers, helix-turn-helix and homeodomains. We further discuss how synthetic biology can come together with structural biology to alter the genetic blueprint of life.

Abbreviations

bZIP basic leucine zipper
HTH helix-turn-helix
STF synthetic transcription factor
ZnF zinc finger

1. Introduction

Regulation of gene expression is essential for the sustenance of all life forms. Gene activation or silencing is effectively controlled by a dense network of regulatory proteins—transcription factors (TFs) are prominent among them. TF binds to a specific DNA sequence, usually the operator in the promoter/enhancer region, and regulates the fate of transcription of a desired gene. TFs act in a coordinated fashion and are responsible for regulating cell growth, cell division, metabolic activities etc. A TF must possess at least one DNA Binding Domain (DBD) that binds to the specific substrate DNA sequence upstream or downstream of the gene sequence it regulates. Further, the presence of activation domain (AD) and signal sensing domain (SSD) is also a common sight in several TFs besides DBD. TFs employ different architectural motifs for the recognition of various DNA sequences. The specific DNA sequence, to which the DBD of a TF binds, is commonly known as Response elements. The TFs are generally classified on the basis of these DBDs. The classification is as follows: basic

Helix-Loop-Helix (bHLH); C-terminal effector domain of the bipartite response regulators; basic leucine zipper (bZIP); AP2/ERF/GCC Box; Helix-Turn-Helix (HTH) superclass including homeodomains; paired box class; winged helix; Heat Shock Factors (HSFs); tryptophan clusters; and transcriptional enhancer factor domains. Some important transcriptional regulations mediated by TFs include but not limited to basal transcription regulation by TFIIB, TFIIA etc.; development and cellular differentiation in multicellular organism (a prominent example would be the role of Hox TF family towards right body pattern formation in a diverse range of animals from fruit flies to humans); response to intercellular signals (e.g. a short cascade of signaling in estrogen signaling pathway involves the estrogen receptor transcription factor); gene expression in epigenetic control (e.g. regulation by HSFs, which enhance the expression of genes needed for survival when exposed to higher temperatures); cell cycle control (e.g. the Myc oncogene responsible for cell growth and apoptosis); pathogenesis (for example, in *Xanthomonas sp.*, Transcription Activators Like effectors (TAL effectors) are secreted that affect pathogenesis).

While we see numerous applications of TFs extending to almost all facets of transcriptional regulations supporting life, design and construction of synthetic TFs (STFs) contribute profoundly in developing tools for synthetic biology as well as translational research. Given the virtue of "modularity," STFs can be constructed by assembling interchangeable sensor/effector domains from the existing naturally occurring TFs. STFs with new regulatory activities and novel DNA sequence specificity are immensely beneficial for gene therapy with the need-based alteration of gene expression pattern. TFs rely on a general mechanism of DNA binding through the interaction of the protruding amino acid side chains present at the recognition helix within the major groove of DNA. It is therefore anticipated that the engineering of these DNA binding domains can bring customized regulation of genes by binding to various novel sequences of our choice (Moore & Ullman, 2003). STFs are also useful in synthetic biology because the construction and designing of complex gene regulatory networks requires programmable TFs. While naturally occurring TFs are amply used in biological circuit design, STFs are helpful in overcoming the limitations associated with the use of natural ones. The STFs are expected to provide enormous advantages over their natural counterparts with respect to precise trans-gene expression and its controlled regulation. On one hand, STFs can help in deciphering the macromolecular interactions behind the gene expression phenomenon, while on the other, they can be used as therapeutic agents to bring back the normal gene expression pattern in a diseased condition. However, the complexity of eukaryotic genome makes the development of gene specific TFs challenging. Insights into the principles of protein design and TFs' DNA recognition/interaction specificity would aid in the development of site specific TFs for controlled transcriptional activity. In this chapter, not only we attempt to discuss protein design and engineering principles but also discuss the examples and applications of several STFs, especially those which are derived from major TF classes. Before we specifically discuss STFs, in the following section, we discuss more general principles of synthetic protein designs.

2. General principles of de novo protein designing

De novo protein designing aims to generate new amino acid sequences against a specific 3D structure, following the existing principles of protein biophysics (Huang, Boyken, & Baker, 2016; Pan & Kortemme, 2021). While predicting or creating new protein structures is one aspect of protein design and engineering, crafting functionality into a non-functional but stable 3D protein scaffold is another major aspect. It is possible to design synthetic protein sequences with a correct fold provided there is a precise methodology for the calculation of the thermodynamics parameters of a protein chain and efficient methods for sampling the space for probable protein sequences and structures (Huang et al., 2016).

2.1 Backbone generation

Backbone conformation decides the entire shape of the protein structure so that it plays a very important role in protein functions. A well folded protein needs a well packed hydrophobic core with correct hydrogen bonding interactions. This means that only a small fraction of optimum background structure space is designable (Helling et al., 2001). *De novo* protein designing can be done by modular assembly. The first step of designing is to correctly determine a blueprint, which identifies the lengths and corresponding orientations (Bowers, Strauss, & Baker, 2000; Simons, Kooperberg, Huang, & Baker, 1997). In this blueprint method, three to nine residue fragments are assembled into desired structures with different topologies (Bowers et al., 2000; Pan & Kortemme, 2021; Simons et al., 1997). Another strategy that uses fragments of proteins from nature is "Structure Extension with Native Substructure Graphs" (SEWING) (Jacobs et al., 2016). This method first draws out small substructures from the proteins available in Protein Data

Bank (PDB). This is followed by the combination of substructures, which share high similarity in local region. Lastly, the gaps of the discontinuous elements are filled by designing loops. A recent method AbDesign (Lipsh-Sokolik, Listov, & Fleishman, 2021) considers the "modularity" aspect of a given protein structural family, and then recombines several modular fragments into a new backbone. Online gaming programs like Foldit (Cooper et al., 2010) and TopoBuilder (Sesterhenn et al., 2020; Yang et al., 2021), part of Rossetta software package, are also used for protein backbone generation de novo. In Foldit, the players start from a peptide chain and a stable structure is eventually generated. In TopoBuilder, the players take a bottom-up approach—where the functional site is first designed followed by the construction of entire protein scaffold. Even with similar folds, naturally existing proteins might have different functions. Inserting functional motifs in available protein structure backbone, which is stable yet non-functional, is extremely challenging—TopoBuilder addresses this by building stable 3D structural assembly of functional motifs with customized protein topologies. The information on protein's structural hierarchy is obtained by integrating SCOPe and ASTRAL data (Fox, Brenner, & Chandonia, 2014). Machine learning as well as deep learning also contribute a lot in designing backbone conformations. A recent method (Anishchenko, Chidyausiku, Ovchinnikov, Pellock, & Baker, 2020) uses "neural network hallucination," where the information stored within the different parameters of trained networks is used to design "ideal" protein structures with *de novo* sequences. The resultant hallucinated proteins have been successfully expressed and purified in stable forms *in vitro*. Awaiting further X-ray/NMR validations, the proteins generated by "hallucination" bear all expected secondary structural elements as confirmed from circular dichroism (CD) spectroscopy.

2.2 Backbone stabilization

The next step after generating protein backbone is selecting the correct amino acid pattern and conformations that stabilizes the backbone conformation to specify optimized 3D active site in the *de novo* protein design. LayerDesign is a well known strategy to accurately constrain the correct amino acid types in the 3D structure of the protein (Dahiyat & Mayo, 1997). The structure is divided into three positions: hydrophobic core, polar surface and mixed boundary region (Marcos et al., 2017, 2018; Pan et al., 2020; Rocklin et al., 2017). This method speeds up sampling and reduces errors due to insufficient sampling or scoring errors. Monte carlo sampling using property filters like core packing (Pan et al., 2020) and hydrogen bond satisfaction (Sheffler & Baker, 2009), further increases the accuracy of this method. Local sequence structure compatibility screening also stabilizes the design (Rocklin et al., 2017). Backbone flexibility poses a challenge in de novo protein designing. To solve this problem, methods take into consideration both side-chain rotamers and local backbone conformations (Georgiev & Donald, 2007; Ollikainen, de Jong, & Kortemme, 2015). Methods like CoupledMoves, BackrubEnsemble and FastDesign use flexible backbone design approach and they are definitely better than the fixed backbone approach. Methods like CoupledMoves sequentially optimize sequence and backbone structures making them superior than the ones who do it sequentially (Loshbaugh & Kortemme, 2020). Hydrogen bonding interactions are important determinants of protein-ligand including protein-protein interactions. It is very important for all hydrogen donor and acceptor groups to form hydrogen bonds either with themselves or with the solvent molecules to avoid the huge energy forfeit caused by unbounded hydrogen bonds (McDonald & Thornton, 1994). The HBNet method searches for possible hydrogen bond in the protein structure (Boyken et al., 2016). HBNet forms a graph with the nodal rotamers having hydrogen bond donor or acceptors. Then the two nodes are joined by an edge forming a possible hydrogen bond. Traversing this graph results in the generation of a hydrogen bonding network. In a new approach, a Monte-carlo variation of HBNet follows a stochastic algorithm to traverse the HBNet graph. This not only increases the sampling speed but also finds an easier solution to larger problems (Maguire, Boyken, Baker, & Kuhlman, 2018). This method was successfully implemented in designing helical bundle oligomers with precision (Pan & Kortemme, 2021).

2.3 Scoring Function optimization

After optimizing the backbone and sequences for the *de novo* protein, the next step is to calculate the scoring function. The scoring functions value help in differentiating between the desired *de novo* protein structures from the erroneous ones by usually identifying the low energy sequence-structure samples. A membrane scoring function franklin2019 (Alford, Fleming, Fleming, & Gray, 2020) was recently developed. This membrane scoring model takes into consideration the heterogeneous environment of the membrane protein and helps in designing *de novo* membrane proteins with complex topological features, accurate juxta membrane domain-lipid interactions and protein pores (Alford et al., 2020). This function also effectively distinguishes between native structures and incorrect models. Recent development in designing scoring function for protein and nonprotein interactions is also seen. The simple physical model has been constructed for the prediction and design of DNA-protein interactions (Havranek, Duarte, & Baker, 2004). This model successfully predicts DNA binding specificity for fixed protein sequences as well as fixed DNA sequences. A rapid energy computation is also done to distinguish between all the interactive samples. The lowest energy samples chosen are rich in side-chain based hydrogen bonds and they are well packed (Havranek et al., 2004). The Fragment Assembly of RNA (FARNA) model (Das & Baker, 2007), led by knowledge based energy function, considers both backbone conformational and side chain interaction preferences from experimentally determined RNA structures. This model does not take under consideration information like phylogeny, experimental data, structures of direct homologs and secondary structure predictions—making this a *de novo* approach (Das & Baker, 2007). Protein scoring functions usually search for low energy *de novo* protein structures for accurate protein designing. However, these approaches do not consider other possible structures, which a protein sequence might acquire with even lesser energies. One of the effective ways to solve this problem is direct calculation of the fitness of the given structure in a particular sequence space. The given protein structures when broken into 3D local fragments, are known as tertiary structural motifs (TERMs) (Zheng, Zhang, & Grigoryan, 2015). It is seen that the sequence preferences of each of the TERMs can be used in calculation of sequence fitness in a particular local structure. The TERM-derived scores and protein structure model from Critical Assessment of Structure Prediction (CASP) are found to be strongly linked (Kryshtafovych, Schwede, Topf, Fidelis, & Moult, 2019).

3. Synthetic structural biology

Cryo-electron microscopy (Cryo-EM), X-ray crystallography, NMR spectroscopy are perhaps the most important experimental techniques to study macromolecular interactions at the 3D level. Mass spectrometry is especially useful in omics approaches (McCafferty, Verbeke, Marcotte, & Taylor, 2020). Despite difficulties in obtaining protein crystals, X-ray diffraction accounts for most of the macromolecular structures deposited in PDB till date. However, recent advances in Cryo-EM technique like direct electron detectors, ultra-stable electron microscopes, automated data collection and real time processing of data has brought in "resolution revolution" in Cryo-EM. Despite its limitation in solving larger molecules/molecular assembly, Cryo-EM advances have remarkably enriched structural biology (McCafferty et al., 2020). Mass spectrometry in combination with electron microscopy is suitable for the characterization of protein structures, bypassing the need of even purifying the target molecule. Mass spectrometry coupled with the advances of structure determination methods, is effectively utilized to study protein-protein/ligand interactions. Further, the structures of protein complexes from cell lysate and even isolated cells (McCafferty et al., 2020) can also be studied. Structural modeling of active protein assemblies requires a prior knowledge about protein-protein/ligand interaction of the subunits as well as the stoichiometry of those interactions. The stoichiometry if not known can be determined by mass spectrometry (McCafferty et al., 2020). Besides classical structural biology techniques, data obtained from other techniques like chemical cross-linking, mass spectrometry, small angle X-ray scattering etc. can be effectively combined for "integrative modeling." In "multiscale structural biology" (Arai, 2018), uncovering the structure-function-dynamics of biological macromolecules/assembly often need synthetic protein designs. Perhaps the best possible way to understand a biological system (e.g. protein or molecular assembly) is to "create" the system following underlying biophysical principles. "Synthetic Structural Biology" emerges right at this juncture and could be profoundly useful in producing synthetic proteins with new folds and functions. As we proceed through the text, we discuss more about the effective couplings of structural and synthetic biology towards protein designs, especially of STFs.

4. General grammar for the structural assembly of STFs

As discussed above, one of the major characteristics of designing synthetic proteins including STF, especially those with multiple domains, is modularity. If we consider such modular domains as "biological parts," then proper order or orientation of these parts with suitable spacer sequences must be generated to achieve desired functionality. How the domain order could be arranged in STFs, is generally defined based on the location of their DBDs. Backed with experimental knowledgebase, Purcell, Peccoud and Lu (2014) proposes general grammar behind STF designs, which might be used directly for automated design of STFs or refined according to the need of the user. Under this framework for STF designs, the protein interaction domain (PID) remains present at the 3′ end while rest of the domains are located at the 5′ of the DBD. Cooperativity in STFs can be enabled by combining leucine zipper as a PID (Lohmueller, Armel, & Silver, 2012). When internally placed in STFs, leucine zipper domains function at both termini given that the PDZ has to be bound only at the C-terminus. Still all the ligands as PID are added to the 3′ in order to constrain the number of configurations. The presence of a linker is necessary between the DBD and PID. Linker is used to enhance the stability, folding as well as expression of fusion proteins. In between the reporter domain and any other domain, a cleavage domain (CD, e.g. 2A sequence) could be present. In general, at the 5′, the Nuclear Localization Signal (NLS) is preferred, but if there is a CD, then NLS could be located at the 3′. While the effector domain (ED) fused with DBD is responsible for activation or repression of transcription, reporter domain quantitates the scale or activation/repression. To quantify the concentration of STFs in the cell, a reporter domain is sometimes placed at the 5′ end. But the presence of this additional reporter gene impairs the folding of STFs. Placement of a CD before the reporter diminishes such a problem. Presence of CD ensures the cleavage at an appropriate position during translation. As the upstream sequence determines the cleavage efficiency of 2A protein, addition of the Gly-Ser-Gly (GSG linker) before 2A sequence can double the efficiency (Purcell, Peccoud, & Lu, 2014). Presence of multiple fusion domains (exceeding two) complicates the grammar of STF design. The testing of all configurations becomes impractical is case of the availability of a large number of possible configurations. However, it has been possible to implement the grammar of STF design in computer-aided synthetic biology software, GenoCAD (Czar, Cai, & Peccoud, 2009). Besides construction and management of library on biological parts, GenoCAD provides user with customizable workspace.

There is no generalized rule that can depict the relationship of amino acids with the DNA recognition. However, the knowledge about the several modes of DNA recognition of the natural TFs facilitate the design of synthetic DNA binding proteins. *De novo* design of the artificial proteins, generation of miniature model of natural TFs and the construction of chimera of the recognition motifs found in different TFs serves the purpose of recognition of a wide variety of DNA sequences. For the derivation of STF from a naturally occurring TF, to achieve precision DNA binding with minimum alteration, the factors underlying TF-DNA interaction could be taken into account. This is because the DNA binding mode/protein-DNA interface are major targets for alteration. To illustrate further—(1) salt bridge formed between the basic side chains of the amino acids and phosphodiester backbone of DNA confers thermodynamic stability to the TF-DNA complex; (2) hydrogen bonds and bidentate hydrogen bonds formed between the amino acid residue and DNA bases provide specificity to the binding; (3) direct interaction of the amino acids with the base edges at minor groove is crucial for the supply of binding energy; (4) Recognition helix that acting as a scaffold organize the recognition elements in an appropriate orientation to make contacts with the DNA; (5) last but not the least, the phenomenon of dimerization or multimerization with partners to maximize substrate DNA repertoire. For the *de novo* designing of a synthetic DNA binding peptide, both structural as well as thermodynamic parameters need to be considered. The strategies that are usually adopted STF designs are—artificial dimerization of the DNA binding monomeric proteins, small molecule conjugation with DNA reading module, thermodynamic stabilization of DNA binding motif, residue grafting etc.

In 1980s, first attempt was made for the generation of artificial transcription factors by domain swapping (Mapp, 2003). Later, the strategy of modular substitution of the functional domain of an endogenous transcription factor with a synthetic counterpart was adopted (Ansari & Mapp, 2002; Mapp, Ansari, Ptashne, & Dervan, 2000). Another approach is the use of DNA binding molecules, like oligonucleotides capable of triple helix formation; natural compounds like netropsin, distamycin that bind to minor groove; PNA (peptide nucleic acid) that targets specific DNA sequences; various alkylating agents; polyamides with heterocyclic amino acids and many other intercalating agents. The most commonly used strategy is the construction of a chimeric fusion of the DBD from a naturally occurring TF with different regulatory domains (RD) through a covalent or noncovalent tethering linker (Ansari & Mapp, 2002). Length of the linker is crucial for the functional optimization of the regulatory domain (Arora, Ansari, Best, Ptashne, & Dervan, 2002; Ansari, Mapp, Nguyen, Dervan, & Ptashne, 2001).

Accurate functioning of a TF requires precise and perfect interaction between the protein and Transcription Factor Binding Site (TFBS). In a study by Farrel, Murphy, and Guo (2016), the authors discussed a structure-based computational method of TFBS prediction utilizing a new Integrative Energy (IE) function. The primary problem in TFBS projection is the scoring function for determining binding affinity and energy for protein-DNA interaction. There are two primary energy functions: the molecular mechanics force field, which is based on physics and statistical potential, which is knowledge-driven. Here, the authors came up with a new IE function that has three modules—a multibody (MB) potential that is knowledge-based, a function relating to hydrogen bond energy, and an electrostatic potential for π interaction energy. This novel function is shown to predict TFBS more accurately, e.g. in case of homeodomains. This study incorporates the essence of hydrogen-bond and π-interactions towards structure based TFBS prediction.

A case by case analysis of different STF designs including the purpose and strategy of design is presented below. The STF constructs are collectively depicted in Table 1.

5. STF designs involving zinc fingers

The most common DNA binding motif of natural proteins is Zinc finger (ZnF). ZnF is widely used for the engineering of STF for its modularity and versatility (Pabo, Peisach, & Grant, 2001). It is generally a β-β-α structure that folds around a central zinc ion. Individual finger is 30 amino acids long module (C-X_{4-5}-C-X_{12}-H-X_{3-5}-H where C = cysteine, H = histidine, X = any other amino acid). This is composed of an α helix consisting of two histidine residues which coordinates with two cysteines of the antiparallel β sheet (Corbi, Libri, & Passananti, 2004). The ZnF acts as the most potential scaffold to be exploited in the design of novel TFs because of its highly adaptable DNA binding specificities. For the engineering of zinc finger proteins two strategies are generally used. First is the selection through phages from the ZnF libraries displayed on the phage surface, while the second approach relies on the prediction from the amino acid base interaction rules obtained from natural ZnF protein database. As these ZnF proteins do not require a symmetric binding site and each finger is able to recognize a different nucleotide subsite, use of this motif is advantageous for the design and engineering of STFs over others. A new transcription factor for a desired target sequence can be constructed by mixing and matching synthetic ZnFs that are specifically developed for an individual subsite with any existing natural finger with the known subsite specificity (Corbi et al., 2004) and all these sequence variants of ZnF motifs can be evaluated without hindering its 3D structural stability. Some key site randomization followed by the selection of the variant fingers were reported (Choo & Klug, 1995; Jamieson, Kim, & Wells, 1994). Here, the spacing as well as overlapping of the different subsites vary in every cases along with the size of the subsites (Corbi et al., 2004). First of all, the binding site of a

Table 1 List of different DNA binding synthetic constructs, designed based on the major families of naturally occurring transcription factors.

Category of transcription factor	Strategy adopted	Name	Construct design	Purpose	Reference
bZIP	Disulfide dimerization	Adamantane-cyclodextrin -GCN4 bZIP	Non-covalent connector adamantane-cyclodextrin was added to the C-terminal cysteine of the basic region of GCN4 bZIP	Leucine zipper domain mediated dimerization can be substituted by disulfide bridging	Ueno, Murakami, Makino, and Morii (2002)
		pCN (v-jun-C-s-s-v-jun-N)	A cysteine linker was added to the C-terminus and another cysteine was added to the N-terminus of the v-Jun bZIP to promote the disulfide dimerization between the two bZIP monomers	Addition of cysteine linker can promote dimerization by disulfide bond formation	Park, Campbell, and Goddard (1993)
	Metalodimerization	Fe(*II*)complex-bZIP	Octahedral coordination FeII complex attached to the basic region of bZIP	Iron complexes can act as the dimerization domain for bZIPs	Cuenoud and Schepartz (1993)
	Photo-induced dimerization	XAFosW	Azobromoacetyl derivatives were added to the C-terminal cysteine residue of the Fos bZIP to regulate its heterodimerization with Jun	Since azobenzene is a light responsive compound, it helps in regulation of binding of bZIPs in response to light	Zhang, Timm, Arndt, and Woolley (2010)

	Metal chelating	GCN4-(Ida) 4	Metal-coordinating iminodiacetic acid side chain was incorporated into leucine zipper portion of the GCN4 bZIP at positions I and i+2	Metal responsive bZIP construction	Azuma, Imanishi, Yoshimura, Kawabata, and Futaki (2009)
	Lactum bridging	GCN4-Lactum	Three Alanine and one Aspartic acid were added at the C-terminus of the 25 amino acids long analogue of the basic region of GCN4 bZIP. The aspartic acid here forms a lactum bond with lysine present four residues away towards N-terminus	α helix stabilization	Taylor (2002)
	Residue grafting	GCN4-aPP	Solvent exposed residues of the alpha helix of avian pancreatic polypeptide (aPP) were replaced by the residues of GCN4 important for GCN4-DNA interaction	To develop a stable construct with DNA binding specify similar to GCN4	Vázquez, Caamaño, and Mascareñas (2003)

Continued

Table 1 List of different DNA binding synthetic constructs, designed based on the major families of naturally occurring transcription factors.—cont'd

Category of transcription factor	Strategy adopted	Name	Construct design	Purpose	Reference
	Conjugation of small molecules	Distamycin-GCN4bZIP	Basic region of the bZIP protein GCN4 was tethered to a tripyrol antibiotic distamycin-A. Arg245 present in the basic region of GCN4 bZIP was substituted with a Glutamic acid to ensure the connection between side chain of Glu and the nitrogen atom of the N-terminal pyrrol present in distamycin	Delivery of the GCN4 basic region into the major groove with the help of minor groove binder distamycin	Vázquez et al. (2003)
	Fusion with light activated protein	GCN4D25PYP-v2	Prototypical GCN4 bZIP was fused to photoactive yellow protein (PYP) by replacement of the first 25 residues of the N-terminal cap of PYP by the C-terminal leucine zipper domain of GCN4	To develop a construct with light regulated DNA binding affinity	Morgan and Woolley (2010)

HTH	Alteration of interaction strength/specificity by structural modeling—Dimerization	ChD and ReD	ChD has palindromic symmetry as the two Nter are connected at the C terminal region by a synthetic flexible linker. ReD consists of two direct repeats of Nter sequence connected with a longer linker	Dimeric architecture is critically responsible for the tightness and the specificity of the binding	Percipalle et al. (1995)
	Alteration of interaction strength/specificity by structural modeling-modularity	LuxR	The N terminal domain of different LuxR proteins are placed in tandem repeats or they are interchanged, followed by a C terminal DNA binding domain	The effective functionality of a protein can be altered by altering the modularity	da Silva et al. (2015)
	Metal ion induced TF dimerization	Cu-HTH peptide P7	A chimeric Cu-binding peptide is designed on the basis of turn substitution of the prion (PrP) octarepeat into a HTH motif	A chimeric Cu-binding peptide on the basis of turn substitution of the prion (PrP) octarepeat into a HTH motif	Shields and Franklin (2004)

Continued

Table 1 List of different DNA binding synthetic constructs, designed based on the major families of naturally occurring transcription factors.—cont'd

Category of transcription factor	Strategy adopted	Name	Construct design	Purpose	Reference
	Synthesis of hybrid HTH from scratch	Hybrid HTH	Designed hybrid HTH by synthesis of peptide from scratch. The various peptides were designed from β-, γ- and δ- amino acids with carbohydrate side chain and L-ala	Only study available where a hybrid HTH is designed from scratch and its conformation is analyzed by NMR, MD and CD studies	Sharma et al. (2014)
	Light induced TF-DNA interaction	LovTAP	Allosteric switch is designed by joining two distant domains so that they have a continuous helix in between them. Here fusion of the LOV2 domain from *Avena sativa* (AsLOV2) phototropin 1 and Trp repressor (TrpR) domain of *Escherichia coli* is done to design a light regulated DNA	Allowing a signal sensed by one domain to be allosterically propagated to the other domain	Strickland, Moffat, and Sosnick (2008)

	Light induced TF-DNA interaction	EL222	binding protein. ChIP-Seq to identify the naturally binding sites of EL222 in the entire genome, SELEX was used predict the artificial cognate DNA sequences	Identification of natural and artificial cognate DNA sequences	Rivera-Cancel, Motta-Mena, and Gardner (2012)
Homeodomain-ZnF	Fusion of three DNA binding domains	ZFHD1	A transcription factor containing zinc fingers 1 and 2 of Zif268, a four-residue linker, and the Oct-1 homeodomain	Development of structure-based design through the fusion of known DNA binding modules	Pomerantz, Pabo, and Sharp (1995)
ZnF	Attachment of intercalating dye	Cyanidin-GR	Cyanine dye was tethered to the Zn finger peptide of 29 amino acids	Addition of cyanidine imparts extra affinity tc the existing ZnF	Vázquez et al. (2003)
		ZnF—thiazole peptide conjugate	Thiazole orange was attached to the N-terminal of the Zn finger of glucocorticoid receptor	This compound by intercalating into the DNA helps in binding of the ZnF of GR with DNA	Thompson and Woodbury (2000)

Continued

Table 1 List of different DNA binding synthetic constructs, designed based on the major families of naturally occurring transcription factors.—cont'd

Category of transcription factor	Strategy adopted	Name	Construct design	Purpose	Reference
	Conjugation of two zinc fingers	268//NRE	Finger 2, 6 and 8 of the Zif268 was tethered to the three finger domains of NRE via an appropriate linker	To design a fusion construct that can bind to a wide range of desired sequence exploiting the adaptability of zinc finger	Kang and Kim (2000)
		ZFGD1	Fusion of part of the dimerization domain of Gal4 and Zn finger1 and 2 of Zif268	Formation of six finger motif for the recognition of a longer DNA sequence	Wolfe, Grant, and Pabo (2003)

specific ZnF protein is fused with the desired target sequence. Then it is connected to two fixed fingers of a three finger motif, and the randomized third finger positions itself accordingly. One of the original fingers is then substituted by target's 3'end specific finger pools obtained through selection. In the subsequent way novel three finger protein is developed that is able to specifically bind to the desired sequence (Pabo et al., 2001).

5.1 Fusion of DNA binding domain ZnF to different functional domain

The regulation of a wide range of genes can be performed with the use of ZnF proteins by the fusion of the DNA binding ZnFs domains with the various functional domains (Corbi et al., 2004) (Fig. 1). For example, a zinc finger protein (ZFP) can be attached to viral VP16 co-activator domain of HSV for transcriptional activation. And further for the repression of genes, ZnF can be linked to Krüppel Associated Box (KRAB) domain. A ZFP may also be associated with a DNA methyltransferase to be used in DNA modification. For the site-specific cleavage of DNA, nuclease domain of the restriction enzyme FoKI can be fused. Zinc finger nuclease (ZFN) which is designed for editing DNA at specific sites is composed of three

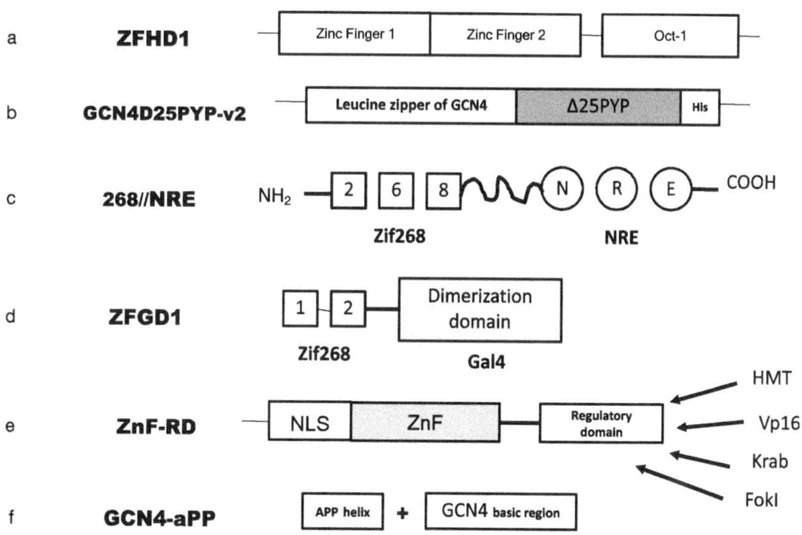

Fig. 1 Schematic representation of different synthetic constructs involving zinc finger as well as basic leucine zipper based designs. Domain organization of the constructs are given.

to six Zn finger motifs which remain connected by a Krüppel type of linker (TGEKP). Design of appropriate linkers enables ZFNs to bind to a wide range of DNA sequences (Anand, Schug, & Wenzel, 2013). In fact the specific helix residues of Rev. peptide of HIV, which are responsible for rev response element binding, is found to be conjugated with the ZnF of Zif268. This inhibits HIV propagation as a result of the competitive inhibition of this engineered protein to that of the Rev. protein (Moore & Ullman, 2003). It can also activate HIV-1 LTR-CAT reporter if attached to the Tat activation domain of HIV-1 (Moore & Ullman, 2003). For the recognition of extended stretch of DNA, 6-finger ZFP was constructed by linking two 3-finger motifs with a canonical linker (Moore & Ullman, 2003). The performance of ZnF based STFs depends on the distance of DBD from the transcription initiation point and many other things (Yaghmai & Cutting, 2002). For example- fusion of ZnF to VP16 results in 6–8 fold up-regulation of erbB-2 gene while the addition of two VP16 activation domain exhibits an activation range of 5–40 folds (Kuznetsova et al., 1999).

5.2 Conjugation of two ZnFs

To design a fusion peptide that can bind to a wide range of desired sequence exploiting the adaptability of the Cys_2His_2 zinc finger, ZFGD1 & ZFGD2 were constructed. In these constructs, the dimerization domain of Gal4 remains attached to ZnF1 and 2 of Zif268 and ZnF2 and 3 of Zif268 respectively. These STFs can bind to a predicted site that is composed of two 6 base-paired ZnF subsites with a 13 base-pair spacer in between with high specificity and affinity (Wolfe et al., 2003). In another six finger fusion protein 268//NRE which act as a genome specific TF (Fig. 1), Finger 2, 6 and 8 of the Zif268 remains tethered to the three finger domains of NRE via an appropriate linker (Kang & Kim, 2000).

5.3 Attachment of DNA binders

Intercalating dye cyanine when tethered to the ZnF motif of glucocorticoid receptor (GR), increased affinity for the substrate DNA or glucocorticoid response element was observed (Vázquez et al., 2003). Addition of another intercalating dye thiazole orange to the N-terminal ZnF of GR also showed enhanced affinity for its native binding sequence (Thompson & Woodbury, 2000).

6. STF designs involving basic leucine zippers

By virtue of simple bi-helical structural arrangement and stability, basic leucine zippers or bZIPs can act as a useful module for the design of STFs. bZIPs are widely distributed in all kingdoms of life. bZIP is characterized by the presence of an N-terminal basic region (BR) and a C-terminal leucine zipper (LZ) connected through a six amino acids long spacer. The LZ is the dimerization domain while the N-terminal basic region is meant for DNA recognition and binding.

The BR derives its name because of the preponderance of basic amino acids. Otherwise unstructured, BR interacts with the major groove of DNA and adopts the α helical structure. The sudden transition from random coil to α helix in BR leads to loss of degrees of freedom. This makes the bZIP-DNA interaction entropically unfavorable from the thermodynamic point of view. Unable to make up for the entropic cost, bZIP monomers fail to produce high-affinity complex upon association with DNA. However, in an enthalpically favorable process, homo/heterodimerization of bZIP monomeric units can form stable protein-protein interaction complex and interacts with DNA as a dimer.

6.1 Disulfide dimerization

Artificial dimerization of the bZIPs can act as an innovative and useful strategy for the designing of synthetic DNA binding proteins. In one study, following LZ replacement, BR regions of GCN4 bZIP were covalently attached together by the creation of disulfide bond (Fig. 2). This architecture ensured high affinity binding to cognate DNA sequence of the natural counterpart (Talanian, Mcknight, & Kim, 1990). This establishes a wonderful example of substitution of LZ and BR dimerization by disulfide bridging. The only difference is that the binding of this synthetic di-sulfur dimer is restricted only to the lower temperature (4 °C). Heterodimers and even trimers could be formed taking the bZIP of v-jun as an extension to this sulfur based dimerization trick. A cysteine linker was added either to the N- or C-terminus of the recognition helix of v-jun bZIP to form v-jun-N or v-jun-C. Then 3-types of dimers—pNN, pCC, and pCN (v-jun-C-s-s-v-jun-N) were formed (Park et al., 1993). Oligomerization of these three type of V-jun bZIP variant was adopted by the Goddard's group for the recognition of an extended DNA sequence (Pazos, Mosquera, Vázquez, & Mascareñas, 2011).

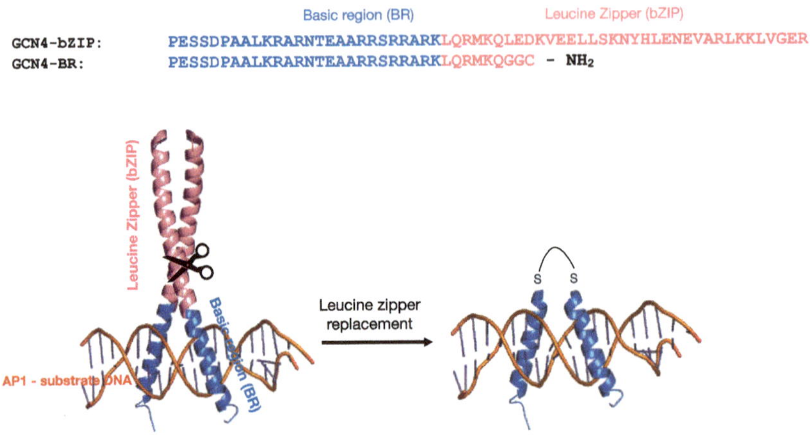

Fig. 2 Strategy of bZIP dimerization by the truncation of leucine zipper portions followed by di-sulfide bond formation. *Adapted from Pazos, E., Mosquera, J., Vázquez, M. E., & Mascareñas, J. L. (2011). DNA recognition by synthetic constructs.* ChemBioChem, 12*(13), 1958–1973. https://doi.org/10.1002/cbic.201100247.*

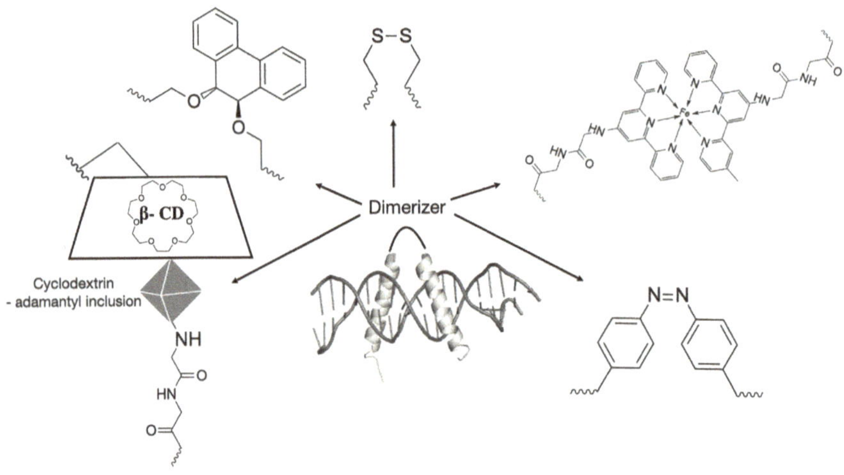

Fig. 3 Synthetic linkers induced bZIP dimerization. Chemical structures of the linkers are depicted here. *Adapted from Pazos, E., Mosquera, J., Vázquez, M. E., & Mascareñas, J. L. (2011). DNA recognition by synthetic constructs.* ChemBioChem, 12*(13), 1958–1973. https://doi.org/10.1002/cbic.201100247.*

Dimerization of the basic regions of GCN4 by a non-covalent connector adamantane-cyclodextrin was also reported (Fig. 3). This construct was formed as a di-sulfur dimer of the cysteine present at its C-terminal with the compound mono-6-deoxy-6-iodo-8-cyclodextrin or *N*-(bromoacetyl)-1-adamantanemethyl-amine (Ueno et al., 2002; Vázquez et al., 2003).

Even when β-cyclodextrine and adamantyl were added to the GCN4 bZIPs, respectively at the N- and C-terminal cysteine residues, it exhibited specificity in DNA recognition as a result of dimerization (Pazos et al., 2011; Sato et al., 2002). The adamantyl-cyclodextrin heterodimer formation between the C/EBP & GCN4 bZIPs resulted in binding to composite non-palindromic regions of DNA (5'-ATGACGCAAT-3′) (Pellegrini & Ebright, 1996). In another case, the C-terminal cysteine residue from GCN4's BR, were linked to activated terpyridine units (Cuenoud & Schepartz, 1993). Presence of C-terminal metallodimer showed higher affinity for the natural ATF/CREB- binding sequence. However, the disulfide dimer formed at the N-terminus showed binding activity towards the inverted sequence with 10 times better efficiency over ATF/CREB site. Such high selectivity of DNA-binding by these dimeric states ensures precision in the DNA targeting from the mixtures of both oligonucleotides. Thus with the use of appropriate external stimuli, the peptide can be targeted to any of the DNA sites. Even reversible "hopping between sites" was also possible in a signal dependent way (Mosquera, Jiménez-Balsa, Dodero, Vázquez, & Mascareñas, 2013).

6.2 Metal induced dimerization

Several iron complexes Fe(II) were also found to be added as the dimerization domain for the basic region of bZIP proteins (Fig. 3) in order to modify the orientation of DNA binding bZIP peptide which is crucial for both the affinity and specificity for DNA (Cuenoud & Schepartz, 1993).

6.3 Light regulated dimerization

Azobromoacetyl derivatives were added to the C-terminal cysteine residue of bZIP for the formation of covalent dimer (Caamaño, Vázquez, Martínez-Costas, Castedo, & Mascareñas, 2000). Basically, DNA binding of azobenzene is light dependent as it undergoes *cis/trans* isomerization (Fig. 4). *Cis* form favors DNA binding due the shorter distance between the atoms of benzylic carbons and more suitable conformation of recognition helix. Activity of all these modified DNA binding bZIP peptides can be altered by external stimuli to achieve DNA binding with desired affinity (Vázquez et al., 2003).

In a similar approach, azobenzene crosslinked FOS subfamily of bZIP protein (XAFosW) was constructed which regulates the formation of the light regulated complex of XAFosW/Jun (Zhang et al., 2010). Here, two cysteine residues, with their side chains located on the same face of the α-helix of the bZIP peptide crosslinks with the azobenzene derivative.

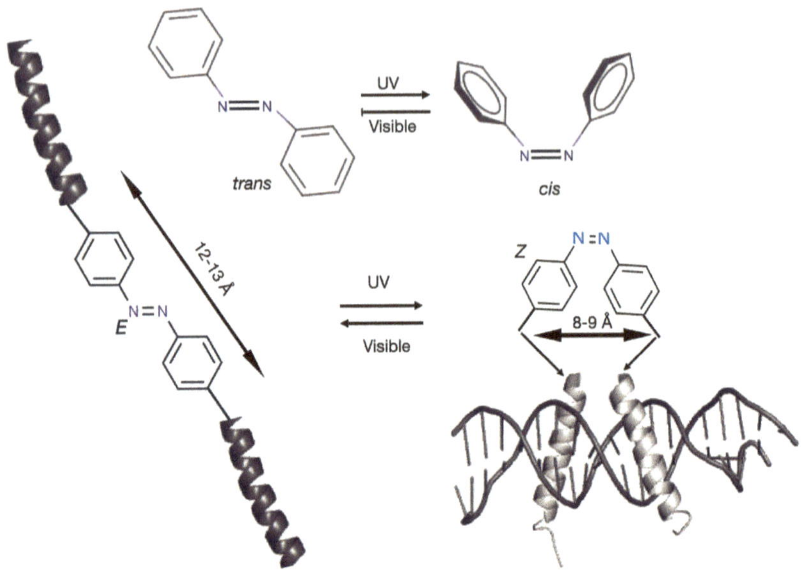

Fig. 4 Light induced conformational change in azobenzene that allows high affinity DNA binding of bZIPs. This is made possible by shortening the distance between the basic regions of two monomeric bZIP units. *Adapted from Vázquez, M. E., Caamaño, A. M., & Mascareñas, J. L. (2003). From transcription factors to designed sequence-specific DNA-binding peptides.* Chemical Society Reviews, 32*(6), 338–349. https://doi.org/10.1039/b206274g.*

Applications of photoreceptor modules, which differentially behaves in dark and light states, are profound in synthetic protein including STF designs for optogenetics (Banerjee & Mitra, 2020). In one such study, chimera of a GCN4 bZIP segment, truncated at the C-terminus, and the PYP photosensor—GCN4(S)D25PYP was constructed (Fig. 1) to promote better DNA binding in dark state. Here, deletion of 8 residues (Leu51-Arg58) from the zipper region of GCN4 bZIP resulted into the suboptimal packing of the zipper region with the PYP core in the dark-adapted state. This further destabilized the helix formation as well as dimerization. The first 25 residues or the N-terminal cap of PYP was replaced by the C-terminal LZ domain of GCN4. The resultant chimera of the prototypical GCN4 bZIP and PYP (GCN4D25PYP-v2) brought two fold increase in the cognate DNA binding upon blue light illumination. In dark, dimerization domain of GCN4 packs against the folded PYP core, whereas upon illumination, the zipper region disintegrates from the core leading to dimerization and DNA binding (Morgan & Woolley, 2010).

6.4 Addition of small molecules to bZIP monomers

Since without dimerization bZIPs cannot stably bind to DNA, conjugation of a bZIP monomer with any natural DNA binding molecule can promote its binding to DNA. Incorporation of the minor groove binding small molecule, distamycin (Fig. 5), into the basic region of GCN4 bZIP via the side chain of glutamic acid gave rise to a hybrid tetrapyrrole peptide. This construct showed a synergistic binding affinity towards the composite site 5'-TCATAAAA-3′, which contains the preferences from both distamycin and GCN4 binding site (Vázquez et al., 2003).

6.5 α-helix stabilization

As the binding of bZIP monomers with DNA is hindered only by the entropic cost of the binding process, imparting stability to the α-helix might solve this problem. Keeping this in mind, two lactum bridges between Lys^{i}–Asp^{i+4} side chains were introduced in the BR analogue of GCN4 to increase the helical content, which further resisted thermal denaturation (Zhang, Wu, Baum, & Taylor, 2000). Another construct was developed by attachment of a fluorescent molecule at the N-terminus and addition of three alanine and one aspartate at the C-terminus to the 25 amino acids taken from the BR of bZIP. The aspartate residue forms a lactum bond with lysine, which

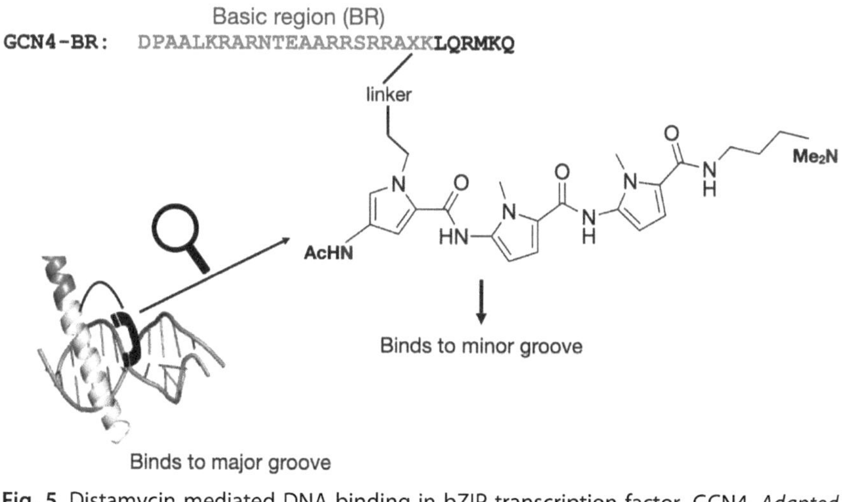

Fig. 5 Distamycin mediated DNA binding in bZIP transcription factor, GCN4. *Adapted from Pazos, E., Mosquera, J., Vázquez, M. E., & Mascareñas, J. L. (2011). DNA recognition by synthetic constructs.* ChemBioChem, 12*(13), 1958–1973. https://doi.org/10.1002/cbic.201100247.*

is present four residues away, towards the N-terminus (Taylor, 2002). This modification also led to an enhanced affinity towards DNA due to the increase in helical content. Substitution of few amino acid residues at the BR of bZIP by alanine is the another strategy adopted to improve α-helical content (Bird & Shin, 2002).

6.6 Residue grafting

Residue grafting is another useful strategy to be employed for the stabilization of bZIP monomer. Following this, the solvent exposed residues of the α-helix of avian pancreatic polypeptide (aPP) were replaced by the residues important for GCN4-DNA interaction (Fig. 1). Hydrophobic interaction with polyproline stabilized aPP but the resulting construct formed by GCN4-residue grafting exhibited extremely high affinity for the specific DNA (Vázquez et al., 2003). Villin, which is F-actin bundling protein, were taken as scaffold for grafting residues from GCN4. This was conducted even though its specific DNA binding affinity was less than that of natural GCN4-DNA (Sato et al., 2002).

6.7 Incorporation of metal chelating residue

Metal chelating iminodiacetic acid (Ida) side chain was incorporated (Fig. 6) into LZ portion of the GCN4 bZIP, precisely at i and i + 2 positions (Azuma et al., 2009). However, the resultant peptide exhibited lower DNA binding affinity when compared to natural GCN4.

7. STF designs involving helix-turn-helix

The helix-turn-helix (HTH) was first discovered in the year 1982 by the pioneer investigators like Matthews, Ohlendorf, Doolittle, Sauer and co-workers. It is a common structural motif of the proteins involved in transcription among all the kingdoms life. The general core structure is comprised of 3α helices where the third helix is generally the recognition helix (Aravind, Anantharaman, Balaji, Babu, & Iyer, 2005). The 2nd and 3rd helix forms the HTH motif. The recognition helix interacts with the major groove of the DNA. HTH can be associated with different types of effector domains like kinase, photo/environmental sensor, etc. to influence diverse biological phenomena. The HTH domain has been classified on the basis of core HTH structure like the bi-helical (e.g., Engralled

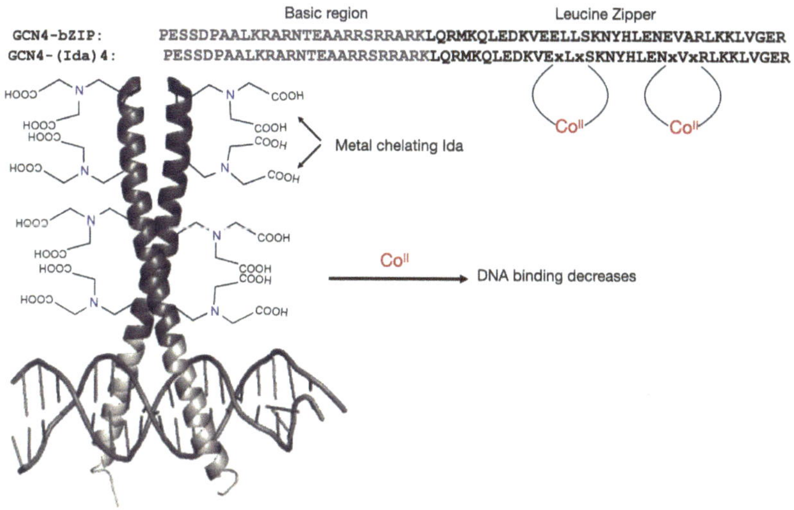

Fig. 6 Metal responsive bZIP protein where the leucine zipper portion of GCN4 bZIP contains metal chelating Ida residues at i and I+4 positions. *Adapted from Pazos, E., Mosquera, J., Vázquez, M. E., & Mascareñas, J. L. (2011). DNA recognition by synthetic constructs.* ChemBioChem, 12*(13), 1958–1973. https://doi.org/10.1002/cbic.201100247.*

Homeodomain), tri-helical (e.g., Myb), tetra-helical and its variations (e.g., LuxR) and winged HTH with its variations (e.g., Ets).

7.1 Alteration of interaction strength/specificity by structural modeling

Dimerization of two different HTH can enhance the affinity and specificity of DNA recognition when compared with respect to the monomeric subunits. In one study, the authors (Percipalle et al., 1995) chose the N-terminal domain (N-ter) of phage 434 repressor, exhibiting low affinity with its cognate DNA. The authors then designed a molecular probe where a covalent linkage mimicked the specificity of the repressor. This resultant molecule, consisting of only one chain was denoted as "single chain repressor." In the first covalent dimer with palindromic symmetry, termed ChD, two N-ters are attached at the C-terminal region by a synthetic flexible linker. The second "single chain repressor," ReD consists of two direct N-ter repeat sequences, bridged via a lengthy linker (Fig. 7). Both of these designed molecules specifically bind with the substrate DNA in presence of a competitor DNA. Moreover, both of them do not show any gel mobility shift in presence of random DNA sequence under the similar conditions of that of cognate DNA. The dissociation constant (K_dS) for ReD was

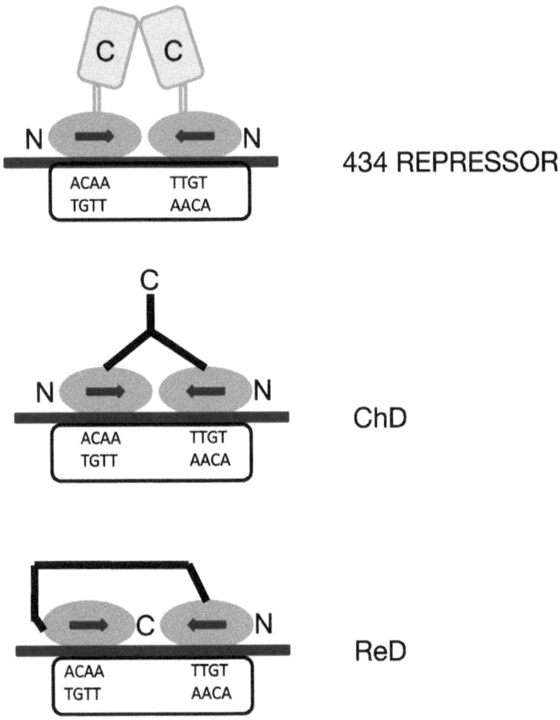

Fig. 7 Diagrammatic representation of binding of the 434 repressor, ChD and ReD to their cognate DNA. The arrows show the N-terminal domain (N to C direction). *Adapted from Percipalle, P., Simoncsits, A., Zakhariev, S., Guarnaccia, C., Sánchez, R., & Pongor, S. (1995). Rationally designed helix-turn-helix proteins and their conformational changes upon DNA binding.* EMBO Journal, 14*(13), 3200–3205. https://doi.org/10.1002/j.1460-2075.1995.tb07322.x.*

calculated to be $5\text{–}6 \times 10^{-9}$ and that of ChD was $2\text{–}3 \times 10^{-8}$ whereas, under same condition the binding of N-ter was not measurable. Another observation is that, there was not much increase in the helical content when N-ter was bound to its cognate DNA. This could mean that the N-ter is supposedly rigid in character. In case of both ChD and ReD, there was a substantial increment in helical content when the cognate DNA is added. This experiment clearly shows, (i) proteins with high DNA-binding affinity, can be successfully designed by dimerizing the domains containing HTH; and (ii) the "single chain repressors" defined in this work could alter their conformation when they interact with random as well as with cognate DNA sites. The results align with the basis of the principle of protein design: dimeric architecture is critically responsible for the strength and specificity of binding. The huge improvement of binding strength in comparison to that of a monomeric N-ter could be attributed partly to an increase in the binding surface and partly to a decrement in K_dS of the bound protein. A possible reason behind high DNA binding affinity is the well-structured and complementary dimer interface of N-ters, which are brought closer to each other by a linker of suitable length.

Modularity of a protein is an intricate property. The effective function ality of a protein can be fine-tuned by altering the modularity. The modular LuxR group of proteins have an important part in quorum sensing. Quorum sensing is an important phenomenon in cellular communication and is widely evident in proteobacteria. LuxR TF is constituted of an autoinducer binding domain at the N-terminal and a HTH at the C-terminal side. In one study, the authors (da Silva et al., 2015) performed a series of specific domain swapping to engineer new synthetic signaling molecules taking into consideration the modularity of LuxR. In this experiment, the authors placed the N-terminal domain of different LuxR proteins in tandem repeats or they are interchanged, followed by a C terminal DNA binding domain. The authors designed hybrid synthetic variants Ola—"PSM-binding domain of OryR + DNA-binding domain of LasR" (Fig. 8A), Lao—AHL-binding domain of LasR + DNA-binding domain of OryR' (Fig. 8B) and Rhlas—"AHL-binding domain of RhlR + DNA-binding domain of LasR" (Fig. 8C). The autoinducer domains of RhlR, OryR and LasR were exchanged for the construction of these synthetic hybrids. They also checked the ability of these variants to bind to the specific promoter region in presence of new specific signals. Two other LuxR synthetic variants, Dahl1 and 2, consisting of two autoinducer domains and one DNA binding domain at the N- and C-terminus respectively, were constructed (Fig. 8D and E). Further experiments confirmed the functionality of hybrid proteins, despite swapping AHL domains from LuxR homologs during design. The authors also wanted to evaluate the functionality of hybrids formed between AHL (N-acyl-L-homoserine lactone) and plant derived signal molecules. The hybrid variants Lao and Ola were tested to evaluate their response to OC12-HSL and rice macerate respectively. In both Lao and Ola variants, it was observed that there was activation of reporter in the presence of specific signals in the N-terminal domains. This result clearly shows that though the activation levels are not as high as compared to the hybrids of two canonical LuxR proteins yet Lao and Ola are functional. The authors next wanted to check the functionality of Dahl1. Here AHL binding domain from RhlR was placed before of the native AHL binding domain of LasR.

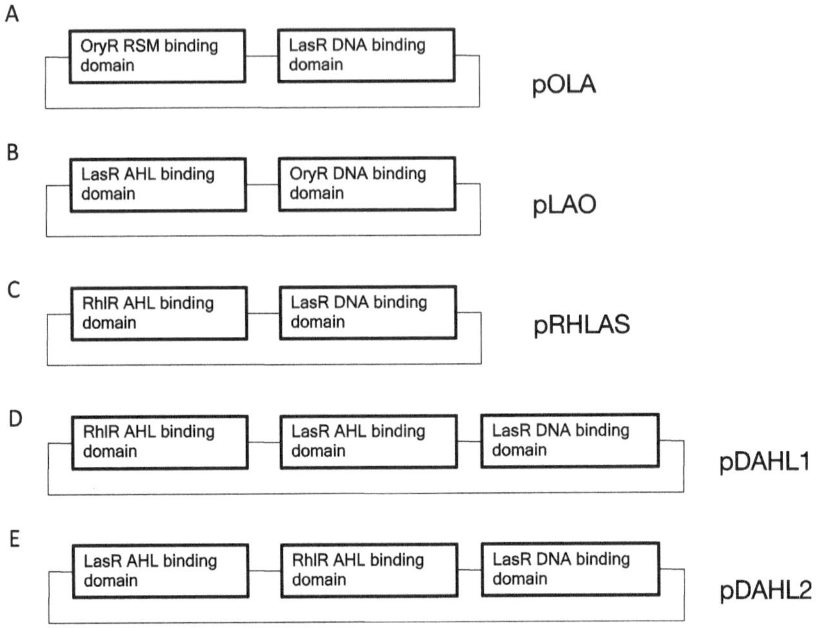

Fig. 8 Domain architecture of the HTH-based hybrid constructs. (A) PSM-binding domain of OryR and DNA-binding domain of LasR, (B) AHL-binding domain of LasR and DNA-binding domain of OryR, (C) AHL-binding domain of RhlR and DNA-binding domain of LasR, (D) RhlR AHL binding domain upstream of the native LasR AHL binding domain, (E) RhlR AHL binding domain downstream of native LasR AHL binding domain. *Adapted from da Silva, D. P., Patel, H. K., González, J. F., Devescovi, G., Meng, X., Covaceuszach, S., Lamba, D., Subramoni, S., & Venturi, V. (2015). Studies on synthetic LuxR solo hybrids.* Frontiers in Cellular and Infection Microbiology, 5*(Jun), 1–10. https://doi.org/10.3389/fcimb.2015.00052.*

In this case, the authors tried different combinations of signals to check the activity: only OC12-HSL, only C4-HSL and a mixture of both. It was seen that the sole presence of C4-HSL was not enough to activate the lasI promoter. On the other hand, only OC12-HSL was capable of reporter expression. Moreover, the combination of two signal molecules kept the promoter activation level comparable to what was seen in case of OC12-HSL alone. This result suggests that addition of RhlR does not affect the activity of the native LasR protein. The Dahl2 was constructed by keeping the RhlR AHL binding domain downstream to that of native LasR AHL binding domain. In this case, the functionality was checked in the similar manner to that of Dahl1. Similar to Dahl1, Dahl2 showed zero activity in presence of

C4-HSL alone. However, there was 1.5 times increment in level of promoter activation when Dahl2 was exposed to only OC12-HSL. This is in contrary to what is observed for Dahl1. Moreover, unlike Dahl1, there was a threefold increment in the level of promoter activation when Dahl2 was exposed to combination of two signal molecules. These results suggest that in contrast to Dahl1, Dahl2 shows activity when both N-terminal autoinducer binding domains had their specific signal molecules. Finally, the whole work shows that domains of various LuxR family proteins responding to divergent signal molecules when swapped, still remain functional—this is indicative of the presence of a common ancestor. Moreover, designing of functional tandem double AHL-binding domain responding to different type of AHLs opens an exciting avenue for future research exploring the modularity of LuxR family proteins.

7.2 Metal ion induced dimerization

De novo protein designing is used as a technique for the bioinorganic chemist where it is used to isolate and explore metal sites in a given protein complex. In one study, the authors (Shields & Franklin, 2004) designed a chimeric Cu-binding peptide based on turn substitution of the prion (PrP) octarepeat into a HTH motif. One octarepeat can bind with one Cu(II), and under certain condition, this metal ion induced dimerization in the N-terminal region facilitates mis-folding of the Prion protein at the C-terminus. The Cu-HTH peptide, P7 is 28-mer and it spectroscopically resembles the PrP octarepeat. This shows Cu(II) specific tryptophan quenching, which is otherwise unaffected in the presence of other divalent ions like Zn^{2+}, Mn^{2+}, Co^{2+}, Ni^{2+} or Ca^{2+}. The Cu-PrP HTH peptide mimics the Cu-binding behavior of a single PrP octarepeat in a novel designed protein (Fig. 9).

7.3 Synthesis of hybrid HTH from scratch

In one study (Sharma et al., 2014), the authors have designed hybrid HTH by synthesis of peptide from scratch. The various peptides were designed from β-, γ- and δ-amino acids with carbohydrate side chain and L-ala. This protein of hybrid HTH contains different types of helices including a classical 11/9-helix nucleating a 13-atom H-bonding turn. This is the only study available where a hybrid HTH is designed from scratch and its conformation is analyzed by NMR, MD and CD studies.

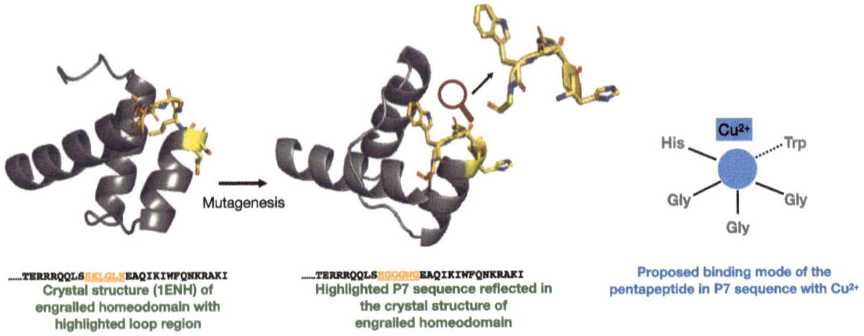

Fig. 9 Left: Crystal structure of 1ENH, loop region is shown distinctively, the sequence of the loop region is underlined. Center: Model structure of P7 with the substituted loop region highlighted and magnified, the sequence of the substituted loop region underlined. Right: Diagrammatic representation of the Cu(II) binding loop consisting of the substituted amino acids in P7. *Adapted from Shields, S. B., & Franklin, S. J. (2004). De novo design of a copper(II)-binding helix-turn-helix chimera: The prion octarepeat motif in a new context.* Biochemistry, *43(51), 16086–16091. https://doi.org/10.1021/bi048555k.*

7.4 Light induced TF-DNA interaction

Allosteric switch can be designed by joining two distant domains so that they have a continuous helix in between them. In a study by Strickland et al. (2008), the authors fused the light-oxygen-voltage (LOV2) photosensor domain from *Avena sativa* (*As*LOV2) phototropin 1 and Trp repressor (TrpR) domain of *Escherichia coli* to design a light regulated DNA binding protein. The core LOV domain comprises of antiparallel β sheets arranged in the topology of 2-1-5-4-3 and several α helices arranged on both sides of the antiparallel β sheets (Möglich, Yang, Ayers, & Moffat, 2010). The *As*LOV2 domain, one of the most sought after photosensor module used in ontogenetic designs (Banerjee & Mitra, 2020), absorbs blue light leading to the adduct formation in between a conserved cysteine residue and a flavin chromophore. This adduct formation is coupled by unfolding of a C terminal Jα helix that transduces the signal to the output domain. TrpR is the effector domain. TrpR and its cofactor L-tryptophan bind with the DNA as a homodimer. Mutations throughout the protein affect the cofactor operator site interaction suggesting the presence of several allosterically important sites. The designed LOV-Trp Activator Protein (LovTAP) in presence of light interacts with the cognate DNA in similar manner as that of classical TrpR. In absence of light, the Jα helix is associated with the LOV domain and the TrpR domain is in inactive confirmation (Fig. 10A). Upon absorption of light, the Jα helix unfolds and the TrpR domain acquires the active confirmation (Fig. 10B). Then the TrpR domain along with its cofactor

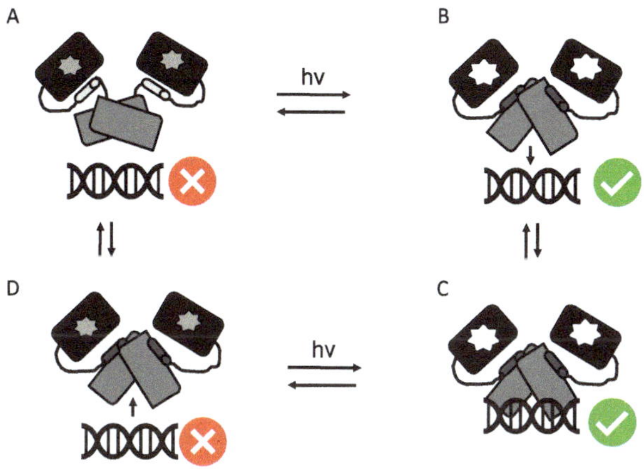

Fig. 10 LOV Domain shown by black box, TrpR domain by gray box, operator dsDNA in black, shared helix shown as white cylinder when contacting LOV domain, dark gray cylinder when contacting the TrpR domain, ground state FMN shown in light gray star, excited state FMN in white star. (A) In the dark DNA-dissociated state, the shared helix contacts the LOV domain, an inactive conformation of the TrpR domain. (B) Photoexcitation breaks contacts between the shared helix and the LOV domain, leading to an active conformation of the TrpR domain. (C) LovTAP binds DNA. (D) The LOV domains return to the dark state upon withdrawal of blue light. LovTAP dissociates from the DNA, the shared helix and the LOV domain contacts are restored, and the system returns to the initial inactive state. *Adapted from Strickland, D., Moffat, K., & Sosnick, T. R. (2008). Light-activated DNA binding in a designed allosteric protein.* Proceedings of the National Academy of Sciences of the United States of America, 105*(31), 10709–10714. https://doi.org/10.1073/pnas.0709610105.*

binds with the DNA (Fig. 10C). Once the light is withdrawn, the DNA is dissociated and the LovTAP regains its initial confirmation (Fig. 10D). Therefore, this work suggests that LovTAP has two confirmations, one inactive form and another active form. Further, this work also suggests that selecting a linker with correct length and composition is an important parameter towards the rational design of photoactive switch.

In synthetic biology, we can engineer the naturally occurring protein as well as the cognate DNA so as to regulate the DNA-protein interaction. In one study (Rivera-Cancel et al., 2012), the authors used a combination of genomic and *in vitro* selection approach to identify the natural as well as artificial substrate DNA of a naturally occurring LOV-HTH, EL222. Light regulated DNA binding is an intrinsic property of EL222. Here, the authors used ChIP-Seq to identify the naturally binding sites of EL222 in the entire genome. Another complementary approach of in vitro selection, SELEX ("Systematic Evolution of Ligands by EXponential enrichment") was also used to predict the artificial cognate DNA sequences. The sequences identified with SELEX were analyzed computationally. Using the motif-based sequence analysis tool MEME ("Multiple Em for Motif Elicitation") algorithm, top 10 scoring DNA were chosen for the gel shift assay. It was found that a particular pool of DNAs could bind 45 times tighter than that of their naturally occurring counterpart. This study definitely opens up new avenues as it investigates binding of a naturally occurring TF with artificially designed cognate DNA.

8. STF designs involving homeodomains

Homeodomain is a common DNA binding structure that is present in several eukaryotic regulatory proteins. Homeodomain proteins are very essential and 143 human loci have been associated with many genetic and genomic disorders (Banerjee-Basu, Ferlanti, Ryan, & Baxevanis, 1999). Several structural studies reveal that homeodomain is made up of three α helices that are compactly folded. Helix1 and helix2 lie parallel to each other and helix3 is present across the two. The third helix is known as the recognition helix as it interacts with the DNA (Banerjee-Basu et al., 1999). This structure of homeodomain is evolutionarily conserved.

Homeodomain is a group of proteins that essays essential roles in protein-DNA and protein-protein interactions. In one study (Pomerantz, Pabo, & Sharp, 1995), the authors described the design of ZFHD1, a TF containing ZnF 1 and 2 of Zif268, a four-residue linker, and the Oct-1 homeodomain (Fig. 1). This chimeric protein has a high affinity and specificity towards a hydrid DNA site 5'-TAATGATGGGCG-3′. The result of this work show that Oct-1 homeodomain actively participates in all protein-protein interactions required to efficiently recruit a transinduction factor and C1 factor into a C1 complex. Moreover, ZFHD1 using a combinatorial approach by fusing different DNA binding modules binds with its hybrid DNA site with high specificity. Importantly, the target site is very much different from the canonical target sites of ZnFs and homeodomains (Pomerantz, Sharp, & Pabo, 1995).

9. Applications of TFs in genetic circuit design

9.1 Logic gates

One of a well known TF is LacI of Lac operon. The lac operon transports and metabolizes lactose in *E.coli* and some other enteric bacteria. It has three structural genes lacZ, lacY and lacA, placed next to each other (Jacob & Monod, 1961). The LacI repressor is a homotetramer and it has two DNA-binding monomeric subunits (Lewis, 2005; Swint-Kruse & Matthews, 2009). That LacI protein binds with the DNA through the N terminal region of a HTH motif (Oehler, Eismann, Kramer, & Muller-Hill, 1990). LacI finds several applications in designing de novo biological gene networks, especially pertaining to gene expression, in various host systems. TetR, also belonging to HTH TF family, is a homodimer that binds with tetO operator DNA (Helbl, Berens, & Hillen, 1995; Hinrichs et al., 1994). Combinations of LacI, TetR and the λ-repressor, cI along with their respective promoters, are used in the construction gene networks. These gene networks show properties of digital logic circuits such as NAND, NOR and NOT IF (Guet, Elowitz, Hsing, & Leibler, 2002). LuxR, another HTH TF involved quorum sensing, binds with AHL through the N- terminal region (Koch et al., 2005). The C-terminal of the TF has a HTH motif and it interacts with the DNA. LuxR homologs function differently—sometimes as activators, sometimes as repressors and sometimes it has dual performances. Such flexibility of LuxR members make them a popular regulator in designing *de novo* gene network controlling a variety of biological processes. Recently, an interacting pair of NAND gates was designed with a variant of lacI and lacO, based on direct protein–protein interactions combined with DNA looping. They are reusable, and with multiple combinations of sequence variants, diverse and complex logic devices can be easily developed (Zhan et al., 2010). In an another experiment, the logical responses of a genetic AND logic gate constructed from LuxI–LuxR bacterial quorum-sensing system, was improved. This LuxR-based AND gate uses directed evolution and a two-step screening process to regulate the activity of LuxR as transcriptional activator. Not only the logical responses are improved but the specificities of these responses are enhanced by 1.5 fold (Sayut, Niu, & Sun, 2011).

9.2 Synthetic oscillators

With recent advancements in the area of biological circuit designs, it has been possible to predict the behavior of synthetic genetic circuit. One such example is the design of "synthetic oscillators." Oscillators are tools for understanding and predicting real cell dynamical properties of genetic oscillatory networks (Singh, 2014). An oscillator can replicate a natural circadian clock in a reliable manner. Both positive and negative feedback is required for designing a perfect oscillator so that it can show the oscillatory behavior in living system. Elowitz and Leibier (2000), designed and characterized a synthetic oscillator in *E. coli*. Their work was based on three genes known as a repressilator because each of the promoters was inhibited by different TFs. These genes were lacI from *E. coli*, TetR from Tn10 transposon and cI from λ phage. In this repressilator system, LacI inhibits transcription of tetR; TetR inhibits transcription of cI, and the cI inhibits transcription of lacI, thus forming a cycle of inhibition. The oscillation in the fluorescence time is checked by the reporter plasmid and around 40% of the cells exhibited oscillations. Even after septation (*process of cell division*) levels of GFP in the two daughter cells often remained same for long periods of time. This experiment showed that the state of the network is passed on to the daughter cells.

9.3 Synchronized oscillators

Hasty and his group designed a synchronized oscillator by using the concept of quorum sensing. They used the quorum sensing mechanism of *Vaucheria fischeri* and *Bacillus thuringiensis*. The luxI (from *V. fischeri*), aiiA (from *B. thuringiensis*), and yemGFP genes were designed to be under the control of three identical copies of luxI promoter. LuxI synthase produces AHL, a small molecule that can pass through the cell membrane and takes part in intercellular coupling. It binds intracellularly and regulates luxR expression constitutively; the luxRAHL complex is an activator of transcription for the luxI promoter (Danino, Mondragón-Palomino, Tsimring, & Hasty, 2010). The AiiA catalyzes the breakdown of AHL through negative regulation of the promoter sequence. In the experimental study, the effective AHL breakdown rate affects the period of the oscillations.

10. Conclusion

Protein folding and design/engineering are opposite sides of a same coin (Khoury, Smadbeck, Kieslich, & Floudas, 2014). The "protein folding problem" remains among the major challenges in structural biochemistry. Research endeavors involving protein design and engineering could be a conclusive test to understand this grandest puzzle. Detailed knowledge on protein structure-function relations is key to generate synthetic proteins including STFs with improved features in terms of size, cost, stability, and distinct functions for potential use in biotechnological and pharmaceutical applications. Levinthal's paradox, which once seemed intractable, poses a lesser degree of challenge today. Newer experimental techniques like hydrogen exchange, φ value methods to identify mutational effects on folding rate/transitions, fast temperature jump methods as well as single molecule experiments have been combined effectively with in silico structure prediction methods (Dill, Ozkan, Shell, & Weikl, 2008). Computational techniques have been indispensable for predicting *de novo* protein structures from amino acid sequences or existing protein structures, which have either been carefully crafted from multiple protein sequences or a single protein sequence/structure. While complete *de novo* designs of STFs could sometimes be computationally expensive, careful observation and analysis of the protein sequence-structure-function correlation is immensely helpful towards structure-based designs of STFs. The field of protein design and engineering, especially protein 3D structure prediction gets a major boost with the emergence of deep neural network. AlphaFold in CASP13 (Senior et al., 2019) or deep neural network, trRosetta (Anishchenko et al., 2020) are consistent in predicting 3D structures for *de novo* amino acid sequences from a mere polypeptide sequence. Their performance even in the absence of information on amino acid residue co-evolution suggests that the fundamental relation between protein sequence and structure is effectively understood by these deep learning methods. The unprecedented ability of these networks to "hallucinate" inhibitors or catalysts, or predicting protein 3D structures from scratch, certainly marks a paradigm shift in the area of protein designs.

References

Alford, R. F., Fleming, P. J., Fleming, K. G., & Gray, J. J. (2020). Protein structure prediction and design in a biologically realistic implicit membrane. *Biophysical Journal, 118*(8), 2042–2055. https://doi.org/10.1016/j.bpj.2020.03.006.

Anand, P., Schug, A., & Wenzel, W. (2013). Structure based design of protein linkers for zinc finger nuclease. *FEBS Letters, 587*(19), 3231–3235. https://doi.org/10.1016/j.febslet.2013.08.015.

Anishchenko, I., Chidyausiku, T. M., Ovchinnikov, S., Pellock, S. J., & Baker, D. (2020). De novo protein design by deep network hallucination. *BioRxiv, 1–19*. https://doi.org/10.1101/2020.07.22.211482.

Ansari, A. Z., & Mapp, A. K. (2002). Modular design of artificial transcription factors. *Current Opinion in Chemical Biology, 6*(6), 765–772. https://doi.org/10.1016/S1367-5931(02)00377-0.

Ansari, A. Z., Mapp, A. K., Nguyen, D. H., Dervan, P. B., & Ptashne, M. (2001). Towards a minimal motif for artificial transcriptional activators. *Chemistry & Biology, 8*(6), 583–592. https://doi.org/10.1016/S1074-5521(01)00037-0.

Arai, R. (2018). Hierarchical design of artificial proteins and complexes toward synthetic structural biology. *Biophysics Reviews, 10*(2), 391–410. https://doi.org/10.1007/s12551-017-0376-1.

Aravind, L., Anantharaman, V., Balaji, S., Babu, M. M., & Iyer, L. M. (2005). The many faces of the helix-turn-helix domain: Transcription regulation and beyond. *FEMS Microbiology Reviews*, *29*(2), 231–262. https://doi.org/10.1016/j.femsre.2004.12.008.

Arora, P. S., Ansari, A. Z., Best, T. P., Ptashne, M., & Dervan, P. B. (2002). Design of artificial transcriptional activators with rigid poly-l-proline linkers. *Journal of the American Chemical Society*, *124*(44), 13067–13071. https://doi.org/10.1021/JA0208355.

Azuma, Y., Imanishi, M., Yoshimura, T., Kawabata, T., & Futaki, S. (2009). Cobalt(II)-responsive DNA binding of a GCN4-bZIP protein containing cysteine residues functionalized with iminodiacetic acid. *Angewandte Chemie International Edition*, *48*(37), 6853–6856. https://doi.org/10.1002/ANIE.200902888.

Banerjee, S., & Mitra, D. (2020). Structural basis of design and engineering for advanced plant optogenetics. *Trends in Plant Science*, *25*(1), 35–65. https://doi.org/10.1016/j.tplants.2019.10.002.

Banerjee-Basu, S., Ferlanti, E. S., Ryan, J. F., & Baxevanis, A. D. (1999). The homeodomain resource: Sequences, structures and genomic information. *Nucleic Acids Research*, *27*(1), 336–337. https://doi.org/10.1093/nar/27.1.336.

Bird, G. H., & Shin, J. A. (2002). MALDI-TOF mass spectrometry characterization of recombinant hydrophobic mutants containing the GCN4 basic region/leucine zipper motif. *Biochimica et Biophysica Acta - Protein Structure and Molecular Enzymology*, *1597*(2), 252–259. https://doi.org/10.1016/S0167-4838(02)00303-5.

Bowers, P. M., Strauss, C. E. M., & Baker, D. (2000). De novo protein structure determination using sparse NMR data. *Journal of Biomolecular NMR*, *18*(4), 311–318. https://doi.org/10.1023/A:1026744431105.

Boyken, S. E., Chen, Z., Groves, B., Langan, R. A., Oberdorfer, G., Ford, A., et al. (2016). De novo design of protein homooligomers with modular hydrogen-bond network-mediated specificity. *Science*, *352*(6288), 69–72. https://doi.org/10.1126/science.aag1318.

Caamaño, A. M., Vázquez, M. E., Martínez-Costas, J., Castedo, L., & Mascareñas, J. L. (2000). A light-modulated sequence-specific DNA-binding peptide. *Angewandte Chemie*, *39*(17), 3104–3107. https://doi.org/10.1002/1521-3773(20000901)39:17<3104::AID-ANIE3104>3.0.CO;2-0.

Choo, Y., & Klug, A. (1995). Designing DNA-binding proteins on the surface of filamentous phage. *Current Opinion in Biotechnology*, *6*(4), 431–436. https://doi.org/10.1016/0958-1669(95)80073-5.

Cooper, S., Khatib, F., Treuille, A., Barbero, J., Lee, J., Beenen, M., et al. (2010). Predicting protein structures with a multiplayer online game. *Nature*, *466*(7307), 756–760. https://doi.org/10.1038/nature09304.

Corbi, N., Libri, V., & Passananti, C. (2004). *Artificial zinc finger peptides: A promising tool in biotechnology and medicine* (pp. 491–507). https://doi.org/10.1007/978-3-642-18932-6_16.

Cuenoud, B., & Schepartz, A. (1993). Altered specificity of DNA-binding proteins with transition metal dimerization domains. *Science*, *259*(5094), 510–513. https://doi.org/10.1126/science.8424173.

Czar, M. J., Cai, Y, & Peccoud, J. (2009). Writing DNA with GenoCAD^{TM}. *Nucleic Acids Research*, *37*, W40–W47. https://doi.org/10.1093/nar/gkp361.

da Silva, D. P., Patel, H. K., González, J. F., Devescovi, G., Meng, X., Covaceuszach, S., et al. (2015). Studies on synthetic LuxR solo hybrids. *Frontiers in Cellular and Infection Microbiology*, *5*(JUN), 1–10. https://doi.org/10.3389/fcimb.2015.00052.

Dahiyat, B. I., & Mayo, S. L. (1997). De novo protein design: Fully automated sequence selection. *Science*, *278*(5335), 82–87. https://doi.org/10.1126/science.278.5335.82.

Danino, T., Mondragón-Palomino, O., Tsimring, L., & Hasty, J. (2010). A synchronized quorum of genetic clocks. *Nature*, *463*(7279), 326–330. https://doi.org/10.1038/nature08753.

Das, R., & Baker, D. (2007). Automated de novo prediction of native-like RNA tertiary structures. *Proceedings of the National Academy of Sciences of the United States of America, 104*(37), 14664–14669. https://doi.org/10.1073/pnas.0703836104.

Dill, K. A., Ozkan, S. B., Shell, M. S., & Weikl, T. R. (2008). The protein folding problem. *Annual Review of Biophysics, 37*, 289–316. https://doi.org/10.1146/annurev.biophys.37.092707.153558.

Elowitz, M, & Leibier, S. (2000). A synthetic oscillatory network of transcriptional regulators. *Nature, 403*, 335–338. https://doi.org/10.1038/35002125.

Farrel, A., Murphy, J., & Guo, J. T. (2016). Structure-based prediction of transcription factor binding specificity using an integrative energy function. *Bioinformatics, 32*(12), i306–i313. https://doi.org/10.1093/bioinformatics/btw264.

Fox, N. K., Brenner, S. E., & Chandonia, J. M. (2014). SCOPe: Structural classification of proteins—Extended, integrating SCOP and ASTRAL data and classification of new structures. *Nucleic Acids Research, 42*(D1), 304–309. https://doi.org/10.1093/nar/gkt1240.

Georgiev, I., & Donald, B. R. (2007). Dead-end elimination with backbone flexibility. *Bioinformatics, 23*(13), 185–194. https://doi.org/10.1093/bioinformatics/btm197.

Guet, C. C., Elowitz, M. B., Hsing, W., & Leibler, S. (2002). Combinatorial synthesis of genetic networks. *Science, 296*(5572), 1466–1470. https://doi.org/10.1126/science.1067407.

Havranek, J. J., Duarte, C. M., & Baker, D. (2004). A simple physical model for the prediction and design of protein-DNA interactions. *Journal of Molecular Biology, 344*(1), 59–70. https://doi.org/10.1016/j.jmb.2004.09.029.

Helbl, V., Berens, C., & Hillen, W. (1995). Proximity probing of Tet repressor to tet operator by dimethylsulfate reveals protected and accessible functions for each recognized base-pair in the major groove. *Journal of Molecular Biology, 245*(5), 538–548. https://doi.org/10.1006/jmbi.1994.0044.

Helling, R., Li, H., Mélin, R., Miller, J., Wingreen, N., Zeng, C., et al. (2001). The designability of protein structures. *Journal of Molecular Graphics and Modelling, 19*(1), 157–167. https://doi.org/10.1016/S1093-3263(00)00137-6.

Hinrichs, W., Kisker, C., Düvel, M., Müller, A., Tovar, K., Hillen, W., et al. (1994). Structure of the tet repressor-tetracycline complex and regulation of antibiotic resistance. *Science, 264*(5157), 418–420. https://doi.org/10.1126/science.8153629.

Huang, P. S., Boyken, S. E., & Baker, D. (2016). The coming of age of de novo protein design. *Nature, 537*(7620), 320–327. https://doi.org/10.1038/nature19946.

Jacob, F., & Monod, J. (1961). Genetic regulatory mechanisms in the synthesis of proteins. *Journal of Molecular Biology, 3*(3), 318–356. https://doi.org/10.1016/S0022-2836(61)80072-7.

Jacobs, T. M., Williams, B., Williams, T., Xu, X., Eletsky, A., Federizon, J. F., et al. (2016). Design of structurally distinct proteins using strategies inspired by evolution. *Science, 352*(6286), 687–690. https://doi.org/10.1126/science.aad8036.

Jamieson, A. C., Kim, S.-H., & Wells, J. A. (1994). In vitro selection of zinc fingers with altered DNA-binding specificity. *Biochemistry, 33*(19), 5689–5695. https://doi.org/10.1021/bi00185a004.

Kang, J. S., & Kim, J. S. (2000). Zinc finger proteins as designer transcription factors. *Journal of Biological Chemistry, 275*(12), 8742–8748. https://doi.org/10.1074/jbc.275.12.8742.

Khoury, G. A., Smadbeck, J, Kieslich, C. A., & Floudas, C. A. (2014). Protein folding and de novo protein design for biotechnological applications. *Trends in Biotechnology, 32*, 99–109. https://doi.org/10.1016/j.tibtech.2013.10.008.

Koch, B., Liljefors, T., Persson, T., Nielsen, J., Kjelleberg, S., & Givskov, M. (2005). The LuxR receptor: The sites of interaction with quorum-sensing signals and inhibitors. *Microbiology, 151*(11), 3589–3602. https://doi.org/10.1099/mic.0.27954-0.

Kryshtafovych, A., Schwede, T., Topf, M., Fidelis, K., & Moult, J. (2019). Critical assessment of methods of protein structure prediction (CASP)—Round XIII. *Proteins: Structure, Function and Bioinformatics*, *87*(12), 1011–1020. https://doi.org/10.1002/prot.25823.

Kuznetsova, S., Ait-Si-Ali, S., Nagibneva, I., Troalen, F., Le Villain, J. P., Harel-Bellan, A., et al. (1999). Gene activation by triplex-forming oligonucleotide coupled to the activating domain of protein VP16. *Nucleic Acids Research*, *27*(20), 3995–4000. https://doi.org/10.1093/nar/27.20.3995.

Lewis, M. (2005). The lac repressor. *Comptes Rendus - Biologies*, *328*(6 Spec. Issue), 521–548. https://doi.org/10.1016/j.crvi.2005.04.004.

Lipsh-Sokolik, R., Listov, D., & Fleishman, S. J. (2021). The AbDesign computational pipeline for modular backbone assembly and design of binders and enzymes. *Protein Science*, *30*(1), 151–159. https://doi.org/10.1002/pro.3970.

Lohmueller, J. J., Armel, T. Z., & Silver, P. A. (2012). A tunable zinc finger-based framework for Boolean logic computation in mammalian cells. *Nucleic Acids Research*, *40*(11), 5180. https://doi.org/10.1093/NAR/GKS142.

Loshbaugh, A. L., & Kortemme, T. (2020). Comparison of Rosetta flexible-backbone computational protein design methods on binding interactions. *Proteins: Structure, Function and Bioinformatics*, *88*(1), 206–226. https://doi.org/10.1002/prot.25790.

Maguire, J. B., Boyken, S. E., Baker, D., & Kuhlman, B. (2018). Rapid sampling of hydrogen bond networks for computational protein design. *Journal of Chemical Theory and Computation*, *14*(5). https://doi.org/10.1021/acs.jctc.8b00033.

Mapp, A. K. (2003). Regulating transcription: A chemical perspective. *Organic and Biomolecular Chemistry*, *1*(13), 2217–2220. https://doi.org/10.1039/b302656f.

Mapp, A. K., Ansari, A. Z., Ptashne, M., & Dervan, P. B. (2000). Activation of gene expression by small molecule transcription factors. *Proceedings of the National Academy of Sciences of the United States of America*, *97*(8), 3930–3935. https://doi.org/10.1073/pnas.97.8.3930.

Marcos, E., Basanta, B., Chidyausiku, T. M., Tang, Y., Oberdorfer, G., Liu, G., et al. (2017). Principles for designing proteins with cavities formed by curved b sheets. *Science*, *355*(6321), 201–206. https://doi.org/10.1126/science.aah7389.

Marcos, E., Chidyausiku, T. M., McShan, A. C., Evangelidis, T., Nerli, S., Carter, L., et al. (2018). De novo design of a non-local β-sheet protein with high stability and accuracy. *Nature Structural and Molecular Biology*, *25*(11), 1028–1034. https://doi.org/10.1038/s41594-018-0141-6.

McCafferty, C. L., Verbeke, E. J., Marcotte, E. M., & Taylor, D. W. (2020). Structural biology in the multi-omics era. *Journal of Chemical Information and Modeling*, *60*(5), 2424–2429. https://doi.org/10.1021/acs.jcim.9b01164.

McDonald, I. K., & Thornton, J. M. (1994). Satisfying hydrogen bonding potential in proteins. *Journal of Molecular Biology*, *238*(5), 777–793. https://doi.org/10.1006/jmbi.1994.1334.

Möglich, A., Yang, X., Ayers, R. A., & Moffat, K. (2010). Structure and function of plant photoreceptors. *Annual Review of Plant Biology*, *61*, 21–47. https://doi.org/10.1146/annurev-arplant-042809-112259.

Moore, M., & Ullman, C. (2003). Recent developments in the engineering of zinc finger proteins. *Briefings in Functional Genomics and Proteomics*, *1*(4), 342–355. https://doi.org/10.1093/bfgp/1.4.342.

Morgan, S. A., & Woolley, G. A. (2010). A photoswitchable DNA-binding protein based on a truncated GCN4-photoactive yellow protein chimera. *Photochemical and Photobiological Sciences*, *9*(10), 1320–1326. https://doi.org/10.1039/c0pp00214c.

Mosquera, J., Jiménez-Balsa, A., Dodero, V. I., Vázquez, M. E., & Mascareñas, J. L. (2013). Stimuli-responsive selection of target DNA sequences by synthetic bZIP peptides. *Nature Communications*, *4*(1), 1–8. https://doi.org/10.1038/ncomms2825.

Oehler, S., Eismann, E. R., Kramer, H., & Muller-Hill, B. (1990). The three operators of the lac operon cooperate in repression. *EMBO Journal, 9*(4), 973–979. https://doi.org/10.1002/j.1460-2075.1990.tb08199.x.

Ollikainen, N., de Jong, R. M., & Kortemme, T. (2015). Coupling protein side-chain and backbone flexibility improves the re-design of protein-ligand specificity. *PLoS Computational Biology, 11*(9), 1–22. https://doi.org/10.1371/journal.pcbi.1004335.

Pabo, C. O., Peisach, E., & Grant, R. A. (2001). Design and selection of novel Cis2His2 zinc finger proteins. *Annual Review of Biochemistry, 70*, 313–340. https://doi.org/10.1146/annurev.biochem.70.1.313.

Pan, X., & Kortemme, T. (2021). Recent advances in de novo protein design: Principles, methods, and applications. *Journal of Biological Chemistry, 296*, 100558. https://doi.org/10.1016/j.jbc.2021.100558.

Pan, X., Thompson, M. C., Zhang, Y., Liu, L., Fraser, J. S., Kelly, M. J. S., et al. (2020). Expanding the space of protein geometries by computational design of de novo fold families. *Science, 369*(6507), 1132–1136. https://doi.org/10.1126/science.abc0881.

Park, C., Campbell, J. L., & Goddard, W. A. (1993). Design superiority of palindromic DNA sites for site-specific recognition of proteins: Tests using protein stitchery. *Proceedings of the National Academy of Sciences of the United States of America, 90*(11), 4892–4896. https://doi.org/10.1073/pnas.90.11.4892.

Pazos, E., Mosquera, J., Vázquez, M. E., & Mascareñas, J. L. (2011). DNA recognition by synthetic constructs. *ChemBioChem, 12*(13), 1958–1973. https://doi.org/10.1002/cbic.201100247.

Pellegrini, M., & Ebright, R. H. (1996). Artificial sequence-specific DNA binding peptides: Branched-chain basic regions. *Journal of the American Chemical Society, 118*(25), 5831–5835. https://doi.org/10.1021/ja9528083.

Percipalle, P., Simoncsits, A., Zakhariev, S., Guarnaccia, C., Sánchez, R., & Pongor, S. (1995). Rationally designed helix-turn-helix proteins and their conformational changes upon DNA binding. *EMBO Journal, 14*(13), 3200–3205. https://doi.org/10.1002/j.1460-2075.1995.tb07322.x.

Pomerantz, J. L., Pabo, C. O., & Sharp, P. A. (1995). Analysis of homeodomain function by structure-based design of a transcription factor. *Proceedings of the National Academy of Sciences of the United States of America, 92*(21), 9752–9756. https://doi.org/10.1073/pnas.92.21.9752.

Pomerantz, J. L., Sharp, P. A., & Pabo, C. O. (1995). Structure-based design of transcription factors. *Science, 267*(5194), 93–96. https://doi.org/10.1126/science.7809612.

Purcell, O., Peccoud, J., & Lu, T. K. (2014). Rule-based design of synthetic transcription factors in eukaryotes. *ACS Synthetic Biology, 3*(10), 737. https://doi.org/10.1021/SB400134K.

Rivera-Cancel, G., Motta-Mena, L. B., & Gardner, K. H. (2012). Identification of natural and artificial DNA substrates for light-activated LOV-HTH transcription factor EL222. *Biochemistry, 51*(50), 10024–10034. https://doi.org/10.1021/bi301306t.

Rocklin, G. J., Chidyausiku, T. M., Goreshnik, I., Ford, A., Houliston, S., Lemak, A., et al. (2017). Global analysis of protein folding using massively parallel design, synthesis, and testing. *Science, 357*(6347), 168–175. https://doi.org/10.1126/science.aan0693.

Sato, S.-i., Honjyo, H., Tsuji, S., Hagihara, M., Makino, K., & Morii, T. (2002). A new strategy to determine the target DNA sequence of a short peptide. *Nucleic Acids Research. Supplement, 2001*(2), 205–206. https://doi.org/10.1093/nass/2.1.205.

Sayut, D. J., Niu, Y., & Sun, L. (2011). Engineering the logical properties of a genetic AND gate. *Methods in Molecular Biology (Clifton, N.J.), 743*(3), 175–184. https://doi.org/10.1007/978-1-61779-132-1_14.

Senior, A. W., Evans, R., Jumper, J., Kirkpatrick, J., Sifre, L., Green, T., et al. (2019). Protein structure prediction using multiple deep neural networks in the 13th critical assessment of protein structure prediction (CASP13). *Proteins*, *87*, 1141–1148. https://doi.org/10.1002/prot.25834.

Sesterhenn, F., Yang, C., Bonet, J., Cramer, J. T., Wen, X., Wang, Y., et al. (2020). De novo protein design enables the precise induction of RSV-neutralizing antibodies. *Science*, *368*(6492), eaay5051. https://doi.org/10.1126/science.aay5051.

Sharma, G. V. M., Thodupunuri, P., Sirisha, K., Basha, S. J., Reddy, P. G., & Sarma, A. V. S. (2014). Design and synthesis of peptides with hybrid helix-turn-helix (HTH) motif and their conformational study. *Journal of Organic Chemistry*, *79*(18), 8614–8628. https://doi.org/10.1021/jo501267k.

Sheffler, W., & Baker, D. (2009). RosettaHoles: Rapid assessment of protein core packing for structure prediction, refinement, design, and validation. *Protein Science*, *18*(1), 229–239. https://doi.org/10.1002/pro.8.

Shields, S. B., & Franklin, S. J. (2004). De novo design of a copper(II)-binding helix-turn-helix chimera: The prion octarepeat motif in a new context. *Biochemistry*, *43*(51), 16086–16091. https://doi.org/10.1021/bi048555k.

Simons, K. T., Kooperberg, C., Huang, E., & Baker, D. (1997). Assembly of protein tertiary structures from fragments with similar local sequences using simulated annealing and Bayesian scoring functions. *Journal of Molecular Biology*, *268*(1), 209–225. https://doi.org/10.1006/jmbi.1997.0959.

Singh, V. (2014). Recent advancements in synthetic biology: Current status and challenges. *Gene*, *535*(1), 1–11. https://doi.org/10.1016/j.gene.2013.11.025.

Strickland, D., Moffat, K., & Sosnick, T. R. (2008). Light-activated DNA binding in a designed allosteric protein. *Proceedings of the National Academy of Sciences of the United States of America*, *105*(31), 10709–10714. https://doi.org/10.1073/pnas.0709610105.

Swint-Kruse, L., & Matthews, K. S. (2009). Allostery in the LacI/GalR family: Variations on a theme. *Current Opinion in Microbiology*, *12*(2), 129–137. https://doi.org/10.1016/j.mib.2009.01.009.

Talanian, R. V., Mcknight, C. J., & Kim, P. S. (1990). Sequence-specific DNA binding by a short peptide dimer. *Science*, *249*(4970), 769–771. https://doi.org/10.1126/SCIENCE.2389142.

Taylor, J. W. (2002). The synthesis and study of side-chain lactam-bridged peptides. *Peptide Science*, *66*(1), 49–75. https://doi.org/10.1002/BIP.10203.

Thompson, M., & Woodbury, N. W. (2000). Fluorescent and photochemical properties of a single zinc finger conjugated to a fluorescent DNA-binding probe. *Biochemistry*, *39*(15), 4327–4338. https://doi.org/10.1021/bi991907g.

Ueno, M., Murakami, A., Makino, K., & Morii, T. (2002). Arranging quaternary structure of peptides by cyclodextrin-guest inclusion complex: Sequence-specific DNA binding by a peptide dimer with artificial dimerization module. *Journal of the American Chemical Society*, *115*(26), 12575–12576. https://doi.org/10.1021/JA00079A043.

Vázquez, M. E., Caamaño, A. M., & Mascareñas, J. L. (2003). From transcription factors to designed sequence-specific DNA-binding peptides. *Chemical Society Reviews*, *32*(6), 338–349. https://doi.org/10.1039/b206274g.

Wolfe, S. A., Grant, R. A., & Pabo, C. O. (2003). Structure of a designed dimeric zinc finger protein bound to DNA. *Biochemistry*, *42*(46), 13401–13409. https://doi.org/10.1021/bi034830b.

Yaghmai, R., & Cutting, G. R. (2002). Optimized regulation of gene expression using artificial transcription factors. *Molecular Therapy*, *5*(6), 685–694. https://doi.org/10.1006/mthe.2002.0610.

Yang, C., Sesterhenn, F., Bonet, J., van Aalen, E. A., Scheller, L., Abriata, L. A., et al. (2021). Bottom-up de novo design of functional proteins with complex structural features. *Nature Chemical Biology*, *17*(4), 492–500. https://doi.org/10.1038/s41589-020-00699-x.

Zhan, J., Ding, B., Ma, R., Ma, X., Su, X., Zhao, Y., et al. (2010). Develop reusable and combinable designs for transcriptional logic gates. *Molecular Systems Biology*, *6*(388), 388. https://doi.org/10.1038/msb.2010.42.
Zhang, F., Timm, K. A., Arndt, K. M., & Woolley, G. A. (2010). Photocontrol of coiled-coil proteins in living cells. *Angewandte Chemie International Edition*, *49*(23), 3943–3946. https://doi.org/10.1002/anie.201000909.
Zhang, M., Wu, B., Baum, J., & Taylor, J. W. (2000). Conformational characterization of a helix-nucleated bicyclic GCN4 decapeptide by proton NMR. *The Journal of Peptide Research*, *55*(5), 398–408. https://doi.org/10.1034/j.1399-3011.2000.00698.x.
Zheng, F., Zhang, J., & Grigoryan, G. (2015). Tertiary structural propensities reveal fundamental sequence/structure relationships. *Structure*, *23*(5), 961–971. https://doi.org/10.1016/j.str.2015.03.015.

CHAPTER NINE

Recent advances in structural studies of 14-3-3 protein complexes

Nikolai N. Sluchanko*
A.N. Bach Institute of Biochemistry, Federal Research Center "Fundamentals of Biotechnology" of the Russian Academy of Sciences, Moscow, Russian Federation
*Corresponding author: e-mail address: nikolai.sluchanko@mail.ru

Contents

Abstract

Being phosphopeptide-binding hubs, 14-3-3 proteins coordinate multiple cellular processes in eukaryotes, including the regulation of apoptosis, cell cycle, ion channels trafficking, transcription, signal transduction, and hormone biosynthesis. Forming constitutive α-helical dimers, 14-3-3 proteins predominantly recognize specifically phosphorylated Ser/Thr sites within their partners; this generally stabilizes phosphotarget conformation and affects its activity, intracellular distribution, dephosphorylation, degradation and interactions with other proteins. Not surprisingly, 14-3-3 complexes are involved in the development of a range of diseases and are considered promising drug targets. The wide interactome of 14-3-3 proteins encompasses hundreds of different phosphoproteins, for many of which the interaction is well-documented *in vitro* and *in vivo* but lack the structural data that would help better understand underlying regulatory mechanisms and develop new drugs. Despite obtaining structural information on 14-3-3 complexes is still lagging behind the research of 14-3-3 interactions on a proteome-wide scale, recent works provided some advances, including methodological improvements and accumulation of new interesting structural data, that are discussed in this review.

Advances in Protein Chemistry and Structural Biology, Volume 130
ISSN 1876-1623
https://doi.org/10.1016/bs.apcsb.2021.12.004

1. Introduction

The normal functioning of a living cell is impossible to imagine without a well-coordinated and dynamic system of protein-protein interactions (PPI). According to estimates, there are at least hundreds of thousands of PPIs (Stumpf et al., 2008; Zhang et al., 2012). The determination of the affinity, selectivity and PPI mechanism is one of the key tasks of modern biochemistry and molecular biology. By addressing these fundamental questions one gains the knowledge particularly necessary for the rational design of new drugs against various pathologies.

A key feature of many PPIs is that they can be regulated. Protein phosphorylation by kinases and dephosphorylation by phosphatases provide for the reversibility of regulation and are used as a means of controlling multiple properties of a regulatory phosphoprotein, including its structure, function, and recognition by other proteins (Brautigan, 1988; Edelman, Blumenthal, & Krebs, 1987; Hardie, 1989; Miller & Turk, 2018; Rubin & Rosen, 1975; Witters, 1990).

Members of the 14-3-3 family are widespread in eukaryotes and are universal regulatory proteins that selectively recognize various protein partners phosphorylated at serine or threonine residues within certain motifs (Aitken, 2006; Muslin, Tanner, Allen, & Shaw, 1996). Through PPIs, 14-3-3 proteins are involved in the regulation of apoptosis, cell cycle, transcription factors, hormone biosynthesis and other processes (Aitken, 2006; Mackintosh, 2004). 14-3-3 complexes are prospective targets for drug development against serious pathologies, including disorders of smooth muscle function, cystic fibrosis, neurodegenerative and oncological diseases (Bartel, Schafer, Stevers, & Ottmann, 2014; Stevers et al., 2018).

The key opportunity for the therapeutic correction of these pathologies might be the determination of the molecular mechanisms governing the formation of complexes involving 14-3-3. Yet, among hundreds of experimentally confirmed, physiologically relevant PPIs involving 14-3-3, there are only a handful of complexes with the established structure, which seriously limits the rational design of small molecule modulators of those interactions. Structural studies are seriously challenged by the prevalence in 14-3-3 partners of intrinsically disordered regions (IDRs) (Bustos & Iglesias, 2006; Sluchanko & Bustos, 2019), and thus the high conformational polymorphism of the corresponding 14-3-3 complexes, and by the need to obtain stoichiometrically and homogenously phosphorylated protein samples. Moreover, many clients of 14-3-3 undergo multisite phosphorylation (e.g., tau protein (Sluchanko, Seit-Nebi, & Gusev, 2009) or CFTR (Stevers et al., 2016)) and have complex domain structures (e.g., LRRK2 kinase of 2500 + residues long (Stevers et al., 2017)). This allows for different 14-3-3 binding modes and stoichiometries, where each can have a distinct physiological effect (Sluchanko, 2020). Notwithstanding the enormous disproportion between the discovered and structurally characterized 14-3-3 PPIs, recent years have evidenced significant progress in structural studies of 14-3-3 complexes, which will be discussed in this review following the description of the main aspects of 14-3-3 biology and nomenclature.

2. Nomenclature, structure, and functioning of 14-3-3

Most eukaryotic organisms have multiple 14-3-3 proteins that are historically called "isoforms," although they are expressed from separate genes (Aitken, 2002). These isoforms of about 30 kDa size are named by Greek letters and form various sets of homo- and heterodimeric assemblies that are considered the main functional phosphopeptide-binding units of 14-3-3. There are seven 14-3-3 isoforms in humans (β, γ, ε, ζ, η, σ, τ/θ), plants can have more than ten different 14-3-3 isoforms, whereas yeast and *Drosophila* possess only two isoforms each (Aitken, 2002).

Because of the long history of research since their discovery in 1967 (Moore & Perez, 1967), 14-3-3 proteins acquired different names. In most databases, members of the 14-3-3 family are named YWHA, which stands for "tyrosine (Y) and tryptophan (W) hydroxylase (H) activators (A)." This abbreviation is normally followed by a capital letter designating the isoform—B (beta), G (gamma), E (epsilon), Z (zeta), Q (eta), and T (theta/tau). Among the human 14-3-3 isoforms, only 14-3-3 sigma (14-3-3σ) is formally called SFN standing for its historic name "stratifin." The most common naming in publications is 14-3-3ζ, 14-3-3σ, 14-3-3ε, etc.

14-3-3 proteins form constitutive α-helical dimers having clamp-like shapes and featuring the so-called amphipathic grooves (AGs) for the phosphopeptide binding, one per subunit. The crystal structure of 14-3-3 dimers has been known since as early as 1995 (Liu et al., 1995; Xiao et al., 1995). In contrast to the variable C-terminal tails, that are typically not resolved in the electron density maps due to disorder (Fig. 1), and the outer surface of the 14-3-3 dimer, the AGs are very conserved among different 14-3-3 isoforms (Fig. 1A). This holds also for the seven human

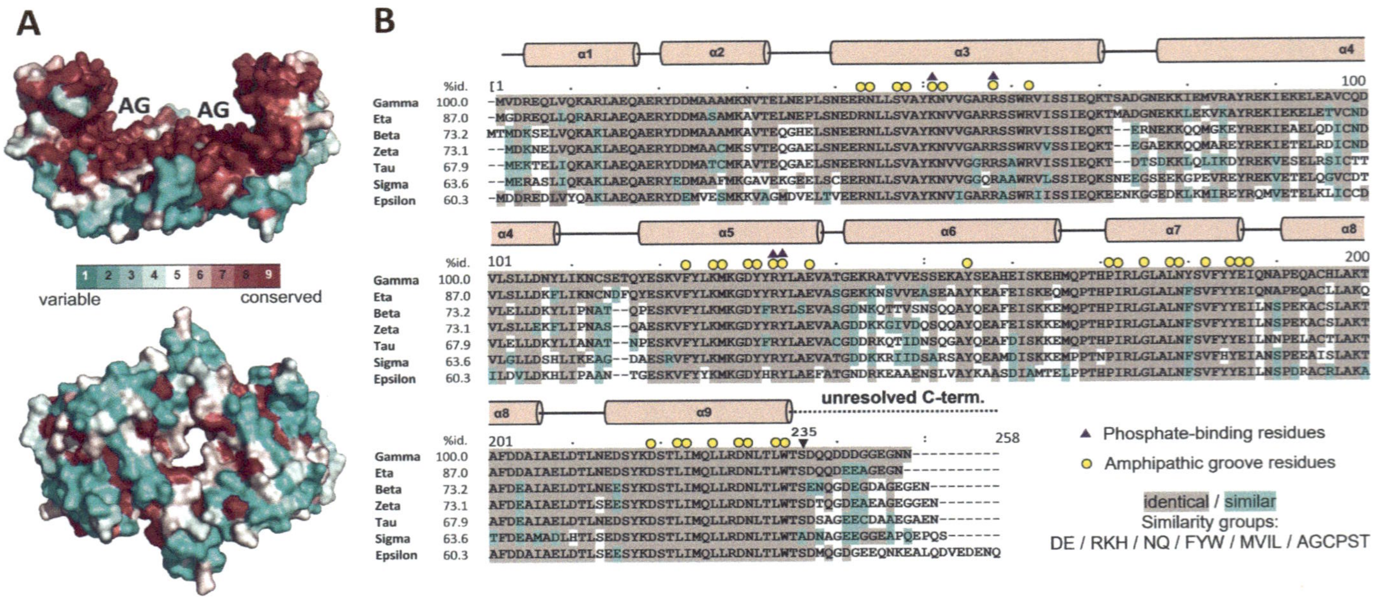

Fig. 1 Evolutionary conservation of 14-3-3 proteins. (A) ConSurf analysis (Ashkenazy et al., 2016) of human 14-3-3 proteins showing conserved and variable regions mapped on the three-dimensional structure according to the scale from 1 (variable) to 9 (conserved). Two orthogonal views are shown. AG, amphipathic groove. (B) Sequence divergence of human 14-3-3 isoforms. Full-length human 14-3-3 isoforms aligned by Clustal Omega (https://www.ebi.ac.uk/Tools/msa/clustalo/) using 14-3-3γ as a reference show the sequence divergence order correlating with the peptide-affinity trends observed for multiple unrelated peptides binding to 14-3-3. Identical residues are highlighted by gray, similar residues are highlighted by cyan (six similarity groups of amino acids are indicated). Residues forming the phosphopeptide-binding amphipathic grooves are marked by yellow circles, residues directly coordinating the phosphate group are marked by blue triangles. Alpha-helices based on the 6A5S structure are shown above the alignment. The variable C-terminal segments are flexible and usually unresolved in crystal structures (dotted line). The last structured residue (S235 in 14-3-3γ) is marked by an arrow for clarity.

14-3-3 isoforms, which display a remarkable order of gradual divergence of their primary structures relative to 14-3-3γ: $\gamma < \eta < \beta \leq \zeta \leq \tau < \sigma < \varepsilon$, where 14-3-3ε is the most divergent from 14-3-3γ (Fig. 1B) (Gogl et al., 2021). At the same time, the amino acid residues lining up the AGs and in particular residues directly involved in coordination of the phosphate groups of partner polypeptides are identical in all human 14-3-3 isoforms (Fig. 1B). Yet, it has recently been shown that, regardless of the binding phosphopeptide sequence, 14-3-3 proteins display a conserved general peptide-affinity trend - γ/η (usually strong phosphopeptide binders), β/ζ/τ (usually intermediate phosphopeptide binders), σ/ε (usually low-affinity phosphopeptide binders) - which is strongly correlated with the sequence divergence order (Fig. 1B) (Gogl et al., 2021). Intriguingly, it is hardly possible to identify a particular subregion of 14-3-3 that would explain the observed affinity difference among the isoforms as the differences are nearly evenly distributed from the N- to the C-terminus (Fig. 1B) (Gogl et al., 2021). It is tempting to speculate that other organisms encoding several 14-3-3 isoforms would also demonstrate similar phosphopeptide affinity trends, however, this requires further investigation.

AGs of 14-3-3 can accommodate not only phosphorylated peptides but also phosphate or sulfate ions (Fig. 2A), and even inosine monophosphate (IMP) and pyridoxal phosphate (PLP) (Ge et al., 2012; Iralde-Lorente et al., 2020), which are bound by the same positively charged pocket of 14-3-3 that coordinates the phosphate moiety of phosphopeptides and includes side chains of Arg56, Arg127, Tyr128 and, in some cases, a very mobile Lys49 (human 14-3-3ζ numbering, Fig. 2B). This pocket is opposed by a cluster of hydrophobic residues of the AG (Fig. 2B), also taking part in ligand binding. Interestingly, crystallographic 14-3-3 subunits can adopt different conformations making the AG more open or more closed (Fig. 2A), and it is tempting to suggest that such motions take part in the process of ligand binding. Such a scenario is strongly supported by Hamiltonian replica exchange molecular dynamics (Nagy, Oostenbrink, & Hritz, 2017).

14-3-3 proteins can themselves be regulated by posttranslational modifications such as lysine acetylation and Ser/Thr phosphorylation (Aitken, 2011). Phosphorylation of a semi-conserved Ser58 residue in human and plant 14-3-3 proteins regulates dimer stability and phosphopeptide binding (Denison, Gökirmak, & Ferl, 2014; Gokirmak, Denison, Laughner, Paul, & Ferl, 2015; Sluchanko et al., 2008; Sluchanko, Sudnitsyna, Chernik, Seit-Nebi, & Gusev, 2011; Woodcock et al., 2003), which is explained by its location in the dimer interface near the phosphate-binding pocket (Fig. 2B).

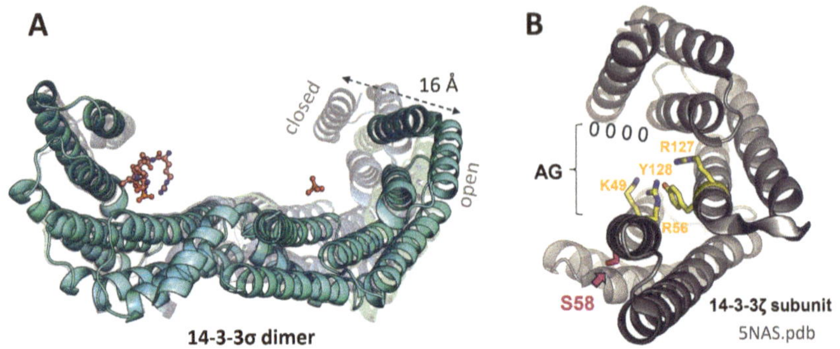

Fig. 2 Plasticity of the 14-3-3 conformation upon ligand binding. (A) The 14-3-3 dimer in which the left subunit is closed upon binding of the HSPB6 phosphopeptide, whereas the right subunit is represented by three different conformations observed in crystal structures (PDB IDs 5OKF, 5OMA) and sampling a ~16 Å amplitude from the most closed to the most open. Of note, in contrast to the phosphopeptide binding, the right open subunit is compatible with phosphate/sulfate ion binding in the amphipathic groove. (B) Close-up view of the 14-3-3 subunit showing the main amino acid residues involved in the coordination of the phosphate moiety of the peptides. AG, amphipathic groove; "0" numbers designate the hydrophobic side of the groove. Note the location of the semi-conserved Ser58 residue, phosphorylatable *in vitro* and *in vivo* (Gu et al., 2006; Powell, Rane, Chen, Singh, & McLeish, 2002; Powell et al., 2003; Sluchanko & Uversky, 2015; Woodcock, Murphy, Stomski, Berndt, & Lopez, 2003), in the vicinity of the phosphopeptide-binding residues of 14-3-3ζ.

While the majority of 14-3-3 ligands are phosphorylated, there are several examples of unphosphorylated proteins and peptides that are specifically bound by 14-3-3. Interestingly, phosphorylated and unphosphorylated peptides are accommodated within the same binding site, AG, but in opposite orientations (Fig. 3A and B). One of the most studied examples of unphosphorylated protein partners of 14-3-3 is the ADP-ribosyltransferase exotoxins S and T (ExoS and ExoT), in which the 14-3-3-anchoring element is represented by C-terminal LLDALDLA motifs (Karlberg et al., 2018; Yasmin et al., 2006). For hijacking the 14-3-3 AGs, they heavily use hydrophobic interactions of the leucines and few polar contacts formed by aspartic acid residues (Karlberg et al., 2018; Ottmann, Yasmin, et al., 2007; Yasmin et al., 2006). Another example includes an artificial high-affinity binding peptide called R18 (-WLDLE-, Fig. 3B), which can inhibit 14-3-3 interactions with both unphosphorylated ExoS and phosphorylated RAF kinase (Petosa et al., 1998). The difopein, a dimeric R18 peptide (Masters & Fu, 2001; Wang et al., 1999), has an exclusively high affinity to 14-3-3 dimers and illustrate that tandem 14-3-3-affinity sites have synergistically increased efficiency

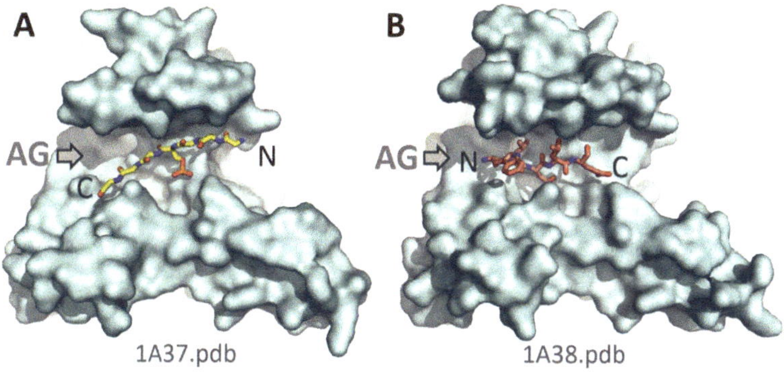

C

Motifs	Sequence
Motif I	RXX[pS/pT]X[P/G]
Motif II	RX[F/Y]X[pS/pT]X[P/G]
Motif III	[pS/pT]X_{0-2}-COOH

Fig. 3 Peptide binding principles of 14-3-3 proteins. (A, B) 14-3-3 subunits with the bound phosphopeptide and unphosphorylated peptide demonstrate opposite orientations of the peptide chains within the amphipathic grooves (AG) of 14-3-3. (C) Typical phosphorylated motifs I–III recognized by 14-3-3 proteins (Coblitz, Wu, Shikano, & Li, 2006; Ganguly et al., 2005; Yaffe et al., 1997). pS/pT designates phosphorylated serine or threonine, X—any amino acid. Motifs I, II are internal, whereas motif III peptides are located at the very C-terminus of 14-3-3 partners (the carboxy group is indicated by -COOH).

of the interaction with both AGs simultaneously (Obsil, Ghirlando, Anderson, Hickman, & Dyda, 2003; Yaffe et al., 1997). Such bidentate binding to 14-3-3 requires at least 15 residues separating the two 14-3-3-binding anchors (Johnson et al., 2010).

The predominant number of 14-3-3 ligands is phosphorylated at Ser or Thr residues which, as a rule, are placed in sequences matching one of three main 14-3-3-binding motifs (Fig. 3C). Those can be either internal (motifs I and II (Yaffe et al., 1997)), or located at the C-terminus of 14-3-3 partners (motif III (Coblitz et al., 2006; Ganguly et al., 2005)) and generally require positively charged Arg (or less often, Lys) residues upstream and either Pro, Gly (usually, at +2 position) or an end of the polypeptide downstream of the phosphosites (Fig. 3C). Based on the refined consensus sequences, several bioinformatic tools have been developed and widely utilized to predict the 14-3-3-binding sites (Madeira et al., 2015; Obenauer, Cantley, & Yaffe, 2003). Still, as evident from the rich repertoire of 14-3-3 complexes studied to date, many 14-3-3-binding sites deviate from the consensus sequences (examples are presented in Table 1). This makes prediction of

Table 1 Examples of noncanonical phosphopeptides forming stable complexes with 14-3-3.

Peptide sequence	Protein	PDB ID	Reference
KSA**pT**TTV	Integrin beta	2V7D	Takala et al. (2008)
KAL**pS**FDNR	PKC epsilon	2WH0	Kostelecky, Saurin, Purkiss, Parker, and McDonald (2009)
KRQYK**pS**IL	Integrin alpha	4HKC	Bonet, Vakonakis, and Campbell (2013)
RRAV**pS**ELD	Tyrosine hydroxylase	4J6S	Skjevik et al. (2014)
KSA**pS**LSN	Chibby	4WRQ	Killoran, Fan, Yang, Shilton, and Choy (2015)
NKV**pS**LAP	CFTR	5D3E	Stevers et al. (2016)
PRI**pS**VIS	CFTR	5D3F	Stevers et al. (2016)
RSA**pS**FS	Gab2	5EWZ	Bier et al. (2016)
KLMFK**pT**EGPDSD	p53	5MHC	Doveston et al. (2017)
KSN**pS**ISVGE	LRRK2	5MYC	Stevers et al. (2017)
QG**pS**FRSQ	Numb	5YQG	Chen et al. (2018)
GVTSS**pS**CPAD	TFEB	6A5Q	Xu et al. (2019)
SYSN**pT**LPVRK	IRSp53	6BCR	Kast and Dominguez (2019)
TENK**pT**LPRSS	IRSp53	6BCY	Kast and Dominguez (2019)
RSR**pS**SLAT	Shroom3	6FCP	Prokop et al. (2018)
RSM**pS**ETGT	CIC	6KZG	Ren et al. (2020)
RSF**pS**LHRM	Na+/H+ antiporter	6QK8	Smidova et al. (2019)
RRAA**pS**MDN	FOXO1	6QZS	Saline et al. (2019)
VEH**pS**LDN	Caspase-2	6GKG	Smidova et al. (2018)
DYDL**pS**LPFP	Caspase-2	6SAD	Kalabova et al. (2020)
SCEH**pS**LD	Caspase-2	6SAD	Kalabova et al. (2020)
RTRTM**pS**VFD	Nth1	5N6N	Alblova et al. (2017)
TRRG**pS**EDD	Nth1	5N6N	Alblova et al. (2017)

the 14-3-3-binding motifs in unknown potential 14-3-3 binders less certain and reliable without experimental verification.

Likewise, the apparent similarity to the consensus sequences I–III is a poor indicator of the 14-3-3-binding affinity (reviewed in Sluchanko, 2018). The situation is complicated by the high probability of *in-cis* interactions within the 14-3-3-binding peptides in their unbound state, which can either decrease or improve the efficiency of complexation with 14-3-3. Indeed, short linear motifs phosphorylatable by basophilic protein kinases usually have a positively charged cluster of residues upstream the phosphosite, the phosphorylated residue itself, and a different number of negatively charged side chains or the C-terminal carboxyl group. This provokes the formation in such peptides of stable charge clamps that restrict the phosphopeptide conformation and complicate their docking in the AG of 14-3-3 and other proteins (Gogl et al., 2018, 2021; Rieloff & Skepö, 2020). Some *in-cis* interactions are revealed by crystal structures of the 14-3-3-bound phosphopeptides (Fig. 4). The underestimated *in-cis* interactions within phosphopeptides can dramatically affect the apparent affinity of

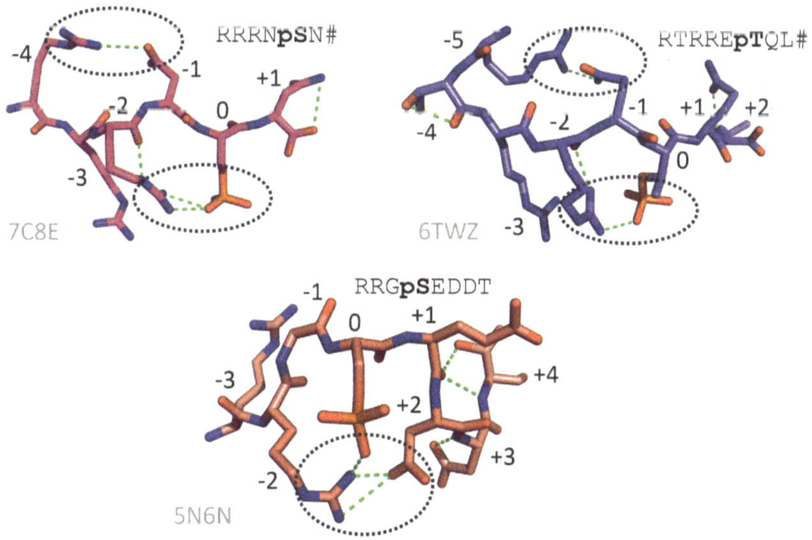

Fig. 4 Conformations of three exemplary 14-3-3-bound phosphopeptides revealing *in-cis* interactions. Sequences of the peptides are shown above, with the phosphosites in bold font and # denoting the C-terminal COOH group. Positions relative to the phosphosites ("0") are numbered. Polar contacts within the peptides are shown by green dashed lines. The charge clamps involving the phosphate moiety, side chains of Arg, and other polar residues are marked by dotted ovals. The image is created using PyMol 2.20.

binding, making it rather difficult to predict solely from peptide sequences. This was recently demonstrated by binding assays and crystallography of 14-3-3 complexes with C-terminal phosphopeptides belonging to different HPV E6 oncoproteins. Rather unexpectedly, quite similar sequences of the 33E6 (SRRREpTAL#), 18E6 (QRRREpTQV#), 16E6 (RTRREp TQL#), and 35E6 (PTRREpTEV#) phosphopeptides, all closely matching the motif III consensus (Fig. 3C), showed significantly different affinities to any selected human 14-3-3 isoform, forming a 33E6 > 18E6 > 16E6 ≫ 35E6 affinity trend, consistent for all seven 14-3-3 isoforms (Gogl et al., 2021). The lowest affinity binder phosphopeptide of 35E6 could be made almost as efficient as the strongest, 33E6, by mutating Thr in position −4 relative to phospho-Thr to Arg, and Glu in position +1 to Ala, which could at least partially be explained by altered charge clamps in the HPV E6 phosphopeptide (Gogl et al., 2021).

An interesting concept was introduced recently concerning the dual-specificity motifs that can be recognized by 14-3-3. It was shown that certain phosphorylatable sequences simultaneously match consensus motifs for the recognition by 14-3-3 proteins (in the phosphorylated state) and by the so-called PDZ domains (in the unphosphorylated state) and that phosphorylation of these motifs may play a role of a switch controlling these two functional interactions (Boon & Banks, 2013; Espejo et al., 2017; Gogl et al., 2020).

While 14-3-3-binding phosphosites tend to be located in the IDRs (Bustos & Iglesias, 2006; Sluchanko & Bustos, 2019), some functional phosphosites crucial for the interaction with 14-3-3 are located in structured protein regions (e.g., in nucleophosmin (Bose et al., 2021) or steroidogenic acute regulatory protein STARD1 (Aghazadeh, Ye, Blonder, & Papadopoulos, 2014; Tugaeva, Titterington, et al., 2020)). This implies that such proteins undergo conformational rearrangements necessary for the 14-3-3 recognition, the so-called conditional binding (Jakob, Kriwacki, & Uversky, 2014; Tugaeva, Titterington, et al., 2020). Given that many partners of 14-3-3 represent multiply phosphorylated proteins and that 14-3-3 binding covers only two sufficiently separated phosphosites simultaneously, binding to different sets of phosphosites may yield distinct functional outputs. In this respect, 14-3-3 proteins recognizing specific phosphosites and their combinations were considered as *logical gates* (Johnson et al., 2010) and readers of the *phosphorylation code* (Sluchanko, 2020).

14-3-3 binding to various phosphotargets may provide for (i) the regulation of their enzymatic activity and protection against dephosphorylation;

(ii) the regulation of their intracellular redistribution via sequestering of phosphorylated protein forms in certain cellular compartments; (iii) the modulation of other protein-protein interactions via shielding of certain recognition motifs and scaffolding activity (Obsil & Obsilova, 2011; van Hemert, Steensma, & van Heusden, 2001). In addition, 14-3-3 proteins display a moonlighting antiaggregation (chaperone-like) activity on a range of protein substrates including those implicated in neurodegenerative and conformational diseases, however, structural data explaining such activity are limited (Goodwin, 2015; Sluchanko & Gusev, 2017; Sluchanko & Uversky, 2015).

3. Approaches to obtain 14-3-3 complexes for structural studies

Structural studies of 14-3-3 complexes require reproducible protocols for obtaining sufficient amounts of protein material, including the challenge of obtaining 14-3-3 partners that must be site-specifically phosphorylated. Importantly, as a rule, 14-3-3 proteins are very selective in binding phosphorylated ligands and do not accept phosphomimicking amino acid substitutions of phosphorylatable residues by Glu or Asp (Chernik, Seit-Nebi, Marston, & Gusev, 2007; Ku, Liao, & Omary, 1998; Sluchanko et al., 2012). In this light, the most straightforward approach has been to obtain recombinant unphosphorylated proteins and then introduce the required phosphosites using *in vitro* protein kinase assays (Kalabova et al., 2017; Obsil et al., 2003; Obsil, Ghirlando, Klein, Ganguly, & Dyda, 2001; Rezabkova et al., 2012). Although proved successful for at least several 14-3-3 complexes, this approach has serious limitations. First of all, it may be rather difficult to obtain stoichiometric phosphorylation of the desired sites and the success largely depends on the nature of the 14-3-3 partner. For example, the small heat shock protein HSPB6 contains the main phosphosite at Ser16, which is a very good substrate for cAMP and cGMP-dependent protein kinases *in vitro* and *in vivo* (Beall et al., 1999, p. 20; Brophy, Dickinson, & Woodrum, 1999, p. 20), and even rather short incubation of HSPB6 with PKA *in vitro* results in almost complete phosphorylation of Ser16, sufficient to form stable assembly with 14-3-3 proteins (Chernik et al., 2007; Sluchanko et al., 2012, 2011). A similar situation of efficient *in vitro* phosphorylation was observed for other protein partners of 14-3-3 (Obsil et al., 2001; Rezabkova et al., 2012). Since the identity of the phosphorylating kinase is not primarily important for obtaining protein phosphorylated at a given position, if the goal is to get a 14-3-3 complex for structural studies, sequences of selected sites can be slightly adjusted to ensure phosphorylation by the available protein kinase (Rezabkova et al., 2012).

Nevertheless, many partners of 14-3-3 contain multiple phosphorylation sites, making the production of the corresponding phosphoforms competent for 14-3-3 binding a big challenge (Tugaeva, Tsvetkov, & Sluchanko, 2017; Zhu, Gafken, Mehl, & Cooley, 2019). This problem often overlaps with difficulties to obtain soluble and stable protein preparations for their phosphorylation *in vitro* (Sahdev, Khattar, & Saini, 2008; Sluchanko, Tugaeva, Faletrov, & Levitsky, 2016). Besides the careful design of 14-3-3 phosphotarget constructs to improve protein yield and characteristics, the use of split intein constructs (Stevens et al., 2016) for separate production of protein fragments and their consequent intein-mediated reunion into one protein entity may be considered. This has been successfully done to obtain BRAF protein kinase in insect cells (Kondo et al., 2019).

Alternatively, coexpression of 14-3-3 targets with protein kinases can be attempted (Flynn et al., 2007; Sluchanko et al., 2017; Sluchanko, Tugaeva, Greive, & Antson, 2017; Stevers et al., 2016; Tugaeva, Kalacheva, Cooley, Strelkov, & Sluchanko, 2019). This approach can provide phosphorylated proteins straight from bacterial cells upon routine purification. As an expansion of this approach, it was proposed to use triple coexpression of a 14-3-3 target, a kinase, and 14-3-3 proteins in *E. coli* (Tugaeva et al., 2017). If designed carefully, this system can in principle lead to robust phosphorylation of a tagged 14-3-3 target and direct purification of its complexes with untagged 14-3-3, expressed in excess (Tugaeva et al., 2017). These three components should be encoded by compatible vectors that are stably maintained in one host bacterial cell (compatible origins of replication and different antibiotic resistances), which can be facilitated either by the duet or polycistronic format (Selleck & Tan, 2008). This can be used also for multisubunit and supramolecular complexes including 14-3-3. In cases when eukaryotic hosts are selected for 14-3-3 target expression, endogenous 14-3-3 proteins can compensate for the necessity of heterologous coexpression. Indeed, recent work on the 14-3-3/BRAF complex showed that it is, in fact, difficult to displace abundant endogenous insect 14-3-3 isoforms by the human counterparts (Park et al., 2019). Due to the high evolutionary conservation of the 14-3-3 structure and phosphopeptide-binding mechanism across species, the endogenous 14-3-3 proteins are fine for preparing complexes for structural characterization, especially when the determination of isoform-specific contacts is not the main goal of the research (Park et al., 2019).

In either *in vitro* phosphorylation or coexpression format, a 14-3-3 target sequence can be engineered to avoid phosphorylation of undesired sites by their mutation to alanine (Park et al., 2019; Veisova et al., 2012), although this approach may not always be optimal. As a very attractive alternative, a growing number of options based on genetic code expansion (GCE) and nonnatural amino acids can be considered (Jin, Park, & Hong, 2019; Manandhar, Chun, & Romesberg, 2021). For example, a recently improved system of engineered *E. coli* cells allows for obtaining strictly site-specifically phosphorylated proteins, including multiply phosphorylated ones, with the help of cotranslational inclusion of a phosphoserine (Schiapparelli et al., 2021; Zhu et al., 2019). This is made possible by mutating the desired phosphorylatable serines to TAG codons, while keeping *bona fide* stop codons either TAA or TGA. As the result, specifically phosphorylated proteins can be obtained even in the absence of protein kinases normally phosphorylating the corresponding residues. The approach has been successfully verified using the physiologically relevant interaction of 14-3-3 with the proapoptotic protein BAD (Zhu et al., 2019). Although very promising on its own, this approach may suffer from partial spontaneous dephosphorylation of proteins and trace misincorporation of undesired amino acids at positions of TAG codons, however, this can be overcome by the future development of similar systems utilizing not phosphoserine itself, but its nonhydrolyzable analogs (Beránek et al., 2018).

4. Crystal structures of 14-3-3 complexes with partner proteins

According to the recent Gene Ontology analysis, 14-3-3 proteins partner with more than 1200 mammalian proteins including nonreceptor Ser/Thr protein kinases (25%), DNA-binding proteins (20%), cell division/cell cycle proteins (12%), RNA-binding proteins (10%), mRNA processing and translation-related proteins (11%), cytoskeletal proteins (4%), ubiquitination system proteins (4%), G-proteins (4%), calcium-sensing (3%), apoptotic (1%) and other proteins (Sluchanko & Bustos, 2019). Many of these proteins represent promising targets for therapeutic intervention, yet there is a gap in structural understanding of their complexation with 14-3-3 proteins (reviewed in Sluchanko, 2018). The great challenge here is reflected in the fact that, despite significant efforts of several laboratories,

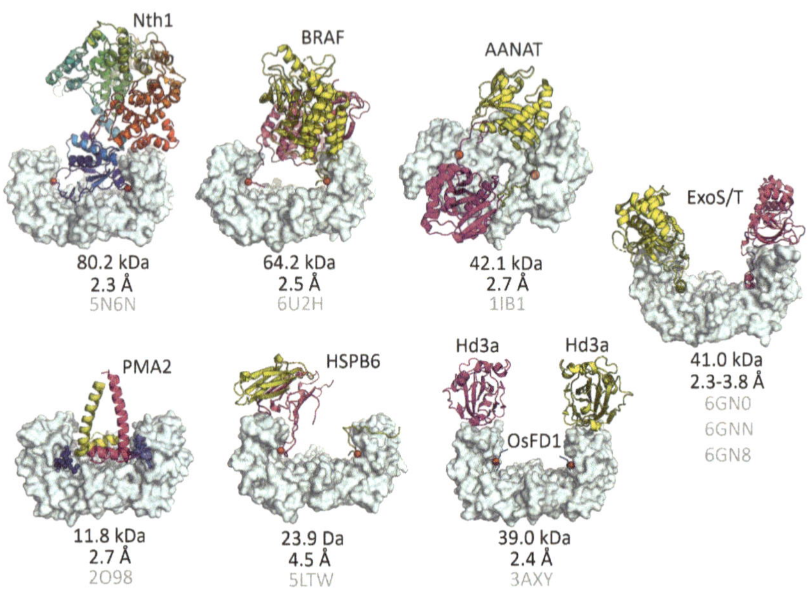

Fig. 5 Crystal structures of 14-3-3 complexes with partner protein fragments larger than 10 kDa. 14-3-3 dimer is shown by semitransparent cyan surface, partner polypeptides are shown by yellow and magenta in all cases except for the largest 14-3-3:pNth1 structure, where the partner protein is shown by a color gradient from blue (N-terminus) to red (C-terminus). Phosphosites occupying the amphipathic grooves of the 14-3-3 dimer are shown by red spheres. Note that ExoS/T and PMA2 represent unphosphorylated 14-3-3 partners and that obtaining of the complex with PMA2 required stabilization by fusicoccin (blue spheres). For all structures, the resolution, PDB IDs, and the molecular weight of the structures excluding the weight of the 14-3-3 dimer are indicated for comparison. The 6XAG structure of the apo BRAF dimer complex with 14-3-3 (Liau, Venkatanarayan, et al., 2020) is very similar to the ligand-bound BRAF-14-3-3 complex (6U2H structure (Liau, Wendorff, et al., 2020), Ca RMSD <0.6 Å), but has a lower resolution (3.3 Å) and is not shown. All figures were drawn using PyMol 2.20.

there are, to the best of my knowledge, only seven nonredundant crystal structures of 14-3-3 complexed with ligands larger than 10 kDa (Fig. 5). Moreover, out of these seven, only three represent crystal structures with human/mammalian proteins—those with AANAT (Obsil et al., 2001), HSPB6 (Sluchanko, Beelen, et al., 2017), and BRAF (Liau, Venkatanarayan, et al., 2020; Liau, Wendorff, et al., 2020). Other successfully determined structures are those with plant Hd3a florigen (Taoka et al., 2011), plant H^+-ATPase PMA2 (Ottmann, Marco, et al., 2007), both not directly linked to human health, and the interspecies complexes of human 14-3-3 proteins with exotoxins S (and T) from *Pseudomonas aeruginosa* causing severe hospital-acquired infections (Karlberg et al., 2018).

Another crystal structure belongs to the yeast protein complex between 14-3-3 and neutral trehalase Nth1 (Alblova et al., 2017), which is also not linked to human health, but represents the biggest crystallographic structure among any 14-3-3 complexes determined to date, also featuring the highest resolution of 2.3 Å (Fig. 5). In addition, this protein partner molecule is unique in that it was trapped by the 14-3-3 dimer in a 1:2 stoichiometry (2:2 in other crystal structures in Fig. 5); 14-3-3 anchoring was achieved via two noncanonical phosphosites (see Table 1) located in the N-terminal IDR of Nth1, which displays several remarkable *in-cis* interactions in the 14-3-3-bound conformation (see an example in Fig. 4) (Alblova et al., 2017). Among the seven highlighted crystal structures, only five represent 14-3-3 complexes with phosphorylated proteins or fragments, whereas the other two (with ExoS/T (Karlberg et al., 2018) and PMA2 (Ottmann, Marco, et al., 2007)) are with unphosphorylated ligands. In all structures with phosphorylated ligands, 14-3-3-binding is facilitated by the IDRs of those ligands, in line with the dominant presence of IDRs in target proteins, especially around the phosphosites (Bustos, 2012; Bustos & Iglesias, 2006; Sluchanko & Bustos, 2019). In six of seven structures, there are secondary interactions of ligands with 14-3-3 outside the AGs of the latter (Fig. 5), yet the overall assemblies are initiated by the primary interactions within the AGs. A possible exception might be the florigen complex, which represents a ternary complex of 14-3-3 with the Hd3a domain and short phosphopeptides belonging to OsFD1 (Fig. 5) (Taoka et al., 2011). Due to the chief importance of the primary interaction in the AGs, the state-of-the-art approach has become to focus on 14-3-3 interactions with representative peptide fragments of partner proteins (see below).

Chronologically, the 14-3-3 structure with the AANAT enzyme was the first achievement, reported in 2001 (Obsil et al., 2001), whereas the second crystal structure of a phosphorylated human protein, HSPB6, bound to 14-3-3 could become possible only 16 years later (Sluchanko, Beelen, et al., 2017). Although 20 years ago it was a quantum leap in understanding of how 14-3-3 proteins associate with protein partners, that first structure was obtained for the monophosphorylated C-terminally trimmed version of ovine AANAT ortholog lacking the second 14-3-3-binding site at the C-terminus, which likely changed the stoichiometry of the interaction from one 14-3-3 dimer per one AANAT to one 14-3-3 dimer per two independent monomers of AANAT (Obsil et al., 2001). This also left the 14-3-3 interaction with the C-terminal fragment of AANAT unaddressed.

Taking into account 20 years that passed since the first structure of a 14-3-3 protein complex was determined, the currently observed disproportion between the known and structurally studied 14-3-3 complexes point to the limited applicability of protein crystallography to solving the problem on a proteome-wide scale and to the urgent need of bringing in alternative techniques.

Many recent studies have reported hybrid structural models of 14-3-3 complexes obtained via a combination of crystallography, small-angle X-ray scattering, electron microscopy, nuclear magnetic resonance, cross-linking and hydrogen/deuterium exchange coupled to mass spectrometry, and other auxiliary techniques, which gained mechanistic understanding of functioning of the corresponding 14-3-3 partners and principles of their interaction with 14-3-3 (Horvath, Petrvalska, Herman, Obsilova, & Obsil, 2021; Hu et al., 2012; Kacirova et al., 2015; Kopecka et al., 2014; Leysen et al., 2021; Petrvalska et al., 2016; Pohl, Joshi, Petrvalska, Obsil, & Obsilova, 2021; Rezabkova et al., 2011). However, such structural models typically have suboptimal resolution and, therefore, can be of limited use for drug discovery in particular.

5. Cryoelectron microscopy structures of 14-3-3 complexes

The largest crystallographic structure of a 14-3-3 complex with Nth1 (Alblova et al., 2017) has a molecular weight of ~132 kDa, however, the average weight of crystallographic structures from Fig. 5 is only ~95 kDa. Many physiological assemblies involving 14-3-3 proteins are much larger, which, given the prevalence of IDRs in 14-3-3 ligands, makes crystallization of the corresponding complexes a very difficult, if not impossible, task. However, these sizes and conformational polymorphism can in theory be tackled by cryoelectron microscopy (cryo-EM). Recent years have witnessed a very promising step forward, as two studies nearly simultaneously have shown the applicability of this powerful technique to solving structures of big signaling complexes of 14-3-3 with kinases of the MAPK cascade (Kondo et al., 2019; Park et al., 2019). This resulted in several nonredundant cryo-EM structures of 14-3-3 complexes with BRAF dimer, BRAF monomer complexed with MEK1 monomer, or BRAF dimer complexed with either one MEK1 or two MEK1 monomers (Fig. 6). The great biomedical importance of these achievements in understanding RAF

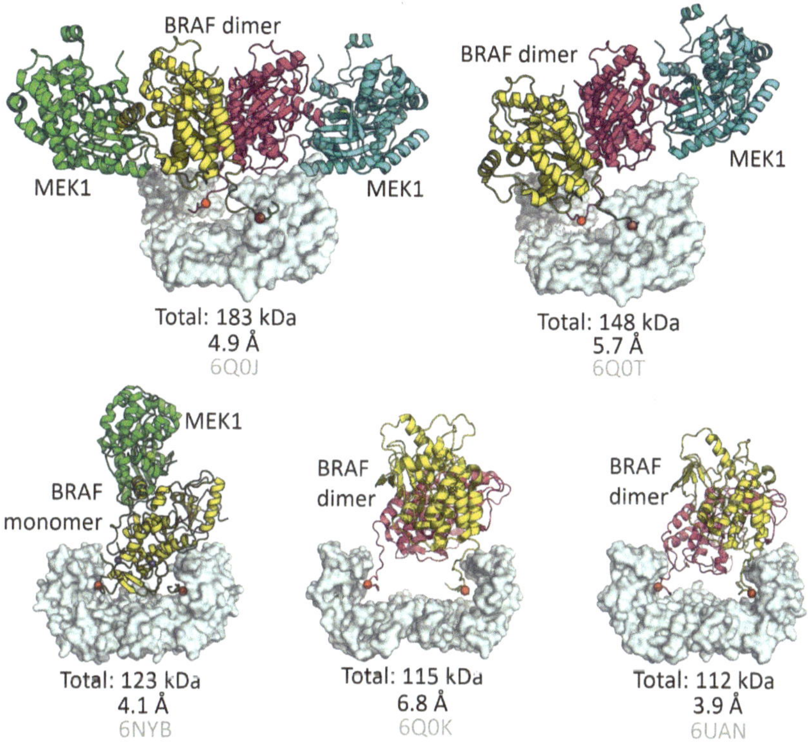

Fig. 6 Cryo-EM structures of 14-3-3 complexes. The 14-3-3 dimer is shown by semitransparent cyan surface, BRAF molecules are indicated by yellow and magenta ribbons, whereas MEK1 molecules are indicated by green and cyan ribbons. Phosphosites occupying the amphipathic grooves of the 14-3-3 dimer are shown by red spheres. The resolution, PDB IDs, and total structure weights (including 14-3-3 and counting only modeled residues with side chains) are indicated. All figures were drawn using PyMol 2.20.

regulation has recently been discussed in the review by Kondo, Paul, Subramaniam, and Kuriyan (2021) and will not be detailed here.

The sizes of all these complexes exceed 100 kDa (Fig. 6). On the one hand, this steps into the zone inaccessible to NMR and less accessible to crystallography, which is promising for many large complexes of 14-3-3 awaiting structural analyses. On the other hand, given that the sizes of biomacromolecules suitable for cryo-EM decrease and resolutions increase as the method progresses (Fan et al., 2019; Nakane et al., 2020), the applicability of cryo-EM to also smaller 14-3-3 complexes already seems very probable. Yet, current resolutions of the 14-3-3 complexes determined by cryo-EM are only 3.9 Å and poorer, which matches the lowest resolution of the crystallographic structures (4.5 Å, for the 14-3-3σ/phospho-HSPB6 complex (Sluchanko, Beelen, et al., 2017), Fig. 5). As a result, only in two cryo-EM structures of around 4 Å resolution, 6UAN and 6NYB, the side chains of the residues could be modeled. This limits the use of this structural information for drug discovery, but the corresponding structures (Kondo et al., 2019; Park et al., 2019) present a very much-needed proof-of-principle. Moreover, given the characteristic shape of the 14-3-3 dimer, it is not impossible to expect that even smaller 14-3-3 complexes could be suitable for cryo-EM in the nearest future.

6. The protein-peptide binding approach in studying 14-3-3 complexes

As noted, in most of the cases, phosphorylated fragments of partner proteins are the prerequisite of their recognition by 14-3-3 proteins; they are located in flexible IDRs facilitating the interaction with much more rigid 14-3-3 dimers (Oldfield et al., 2008; Sluchanko & Bustos, 2019). Given that it is the phosphopeptide recognition that triggers the formation of 14-3-3 complexes with partner proteins, the most prolific structural biology approach has been cocrystallization of 14-3-3 proteins with the corresponding synthetic phosphopeptides. The use of synthetic peptides also allows one to characterize the affinity of their binding to 14-3-3 proteins, which gives a great advantage to this strategy.

As of October 2021, there are 380 redundant X-ray structures of 14-3-3 complexes with various peptides released with the Protein Data Bank, spanning resolutions from 1.05 Å (PDB ID: 3IQU) to 3.7 Å (PDB ID: 6RLZ). Of these, ~360 structures correspond to human 14-3-3 complexes, and the majority are 14-3-3σ complexes (266). These structures cover multiple biomedically important interactions and provide the opportunity to design different small molecules that could affect the stability of the corresponding 14-3-3/peptide complexes (Stevers et al., 2018). One of the pioneers and leaders in this field is the team of Prof. Christian Ottmann, who have developed a very efficient strategy based on cocrystallization of synthetic phosphopeptides and 14-3-3 proteins, and soaking of such crystals with a variety of chemicals that serve as prototypes of PPI modulators and drugs (Andrei, de Vink, et al., 2018; Bartel et al., 2014; Bier et al., 2017; de Vink et al., 2017; Kaplan et al., 2020; Kaplan, Ottmann, & Fournier, 2017; Milroy, Brunsveld, & Ottmann, 2013; Ottmann et al., 2009; Stevers et al., 2018; Wolter et al., 2020). This team has optimized the most productive crystallization conditions that allow for a rapid obtaining of crystals diffracting to a near-atomic resolution (Ballone, Lau, Zweipfenning, & Ottmann, 2020), which is especially helpful for chemical biology aimed at finding efficient inhibition or stabilization of 14-3-3 PPIs.

The 14-3-3/peptide approach has recently been expanded further by the following soaking method (Ballone et al., 2020). It was suggested that even more 14-3-3/phosphopeptide complexes can become amenable to crystal structure determination via a simple two-step procedure. In the first step, 14-3-3σ is cocrystallized with a small excess of a low-affinity cJun-pS227 peptide to ensure trapping of the 14-3-3 dimer in the peptide-bound conformation enabling crystal growth (Fig. 2A). Importantly, this cocrystallization happens in the already known conditions, in a 24-well plate format, and reproducibly gives well-diffracting, robust and stable crystals amenable to ligand soaking (Ballone et al., 2020). The crystal structure does not show good occupancy of the cJun-pS227 peptide in the AG of 14-3-3, where only a few atoms of the peptide are traced in the electron density maps. However, this intermediate step is followed by soaking of the obtained crystals with an excess of a high-affinity peptide of interest, which effectively displaces the low-affinity peptide in a time-dependent manner and already after >24 h provides for the structure determination of the new complex with this high-affinity peptide. This approach proved successful for at least several peptide complexes of 14-3-3 (e.g., peptides of CaMKK2, Caspase-2, MLF1, SOS1, Gab2), including complexes which have not been obtained by the traditional cocrystallization, such as with PLN-pS16 peptide (Ballone et al., 2020). The main advantage of this soaking procedure is that a new complex does not require costly and time-consuming crystal screening, which is otherwise expected for each new object (Ballone et al., 2020).

However, the described methodology is solely based on the use of synthetic peptides which may not be easily afforded by all laboratories. In addition, some peptides of interest are not sufficiently stable or soluble, or their cocrystallization with 14-3-3 may still fail to give diffraction-quality crystals. There are such examples even among high-affinity peptides binding to 14-3-3 (Ballone et al., 2020; Stevers et al., 2017).

7. Chimeric 14-3-3 proteins with phosphorylated peptides

Alternative to the traditional use of synthetic peptides, it was proposed to genetically fuse the corresponding fragments with 14-3-3 proteins (Sluchanko, Tugaeva, et al., 2017). The approach benefits from the proximity of the 14-3-3's C-terminus and the N-terminus of the interacting phosphopeptides in nearly all structures reported to date. The phosphopeptide sequences are tethered to 14-3-3 via a short flexible linker (GSGS or GGGG), which secures binding stoichiometry and full occupancy even for low-affinity phosphopeptides (Sluchanko, Tugaeva, et al., 2017). The 14-3-3 chimeras are coexpressed with protein kinase A in *E. coli* (Tugaeva et al., 2017), which leads to robust phosphorylation of the peptides and their binding in the AGs of 14-3-3 in normal orientation (Sluchanko, Tugaeva, et al., 2017). Such complexes could be efficiently purified using cleavable His-tags present on the N-terminus of 14-3-3 and crystallized, and the entire procedure is compatible with the high-throughput format (Sluchanko, Tugaeva, et al., 2017). Crystallization of 14-3-3 chimeras with phosphopeptide fragments is facilitated by amino acid substitutions within the 14-3-3 component of the chimera, following the so-called surface entropy reduction approach (Goldschmidt, Cooper, Derewenda, & Eisenberg, 2007b). First, the 14-3-3 protein sequence is analyzed by the SERP web server (http://services.mbi.ucla.edu/SER/), which identifies clusters of charged residues that could prevent crystal contacts formation (Fig. 7A). The high-scoring clusters of residues located at the protein surface are mutated to alanine (Fig. 7B) and the resulting mutant versions are purified and crystallized (Gogl et al., 2021; Sluchanko, Tugaeva, et al., 2017; Tugaeva, Remeeva, Gushchin, Cooley, & Sluchanko, 2020; Tugaeva, Titterington, et al., 2020). It was demonstrated that SERP mutations can indeed localize in crystal contacts made by 14-3-3 molecules (Fig. 7C–E), thereby increasing the speed and likelihood of crystallization.

The chimeric approach has been successfully applied to obtain crystal structures of several physiologically relevant 14-3-3/peptide complexes (Table 2 and Fig. 8), providing structural information equivalent to the use of synthetic phosphopeptides (Gogl et al., 2021; Sluchanko, Tugaeva, et al., 2017). Fig. 8 demonstrates the use of the 14-3-3σ chimera with the C-terminal phosphopeptide of human AANAT which was missing from the protein-protein complex reported in 2001 (Obsil et al., 2001). The chimeric protein formed a crystal lattice in which neighboring 14-3-3 dimers were organized in arrays of a Greek key-like assembly stabilized by the phosphopeptide swap (Fig. 8A). The overall positioning of the AANAT phosphopeptide in the 14-3-3 AGs appeared canonical, in line with its correspondence to the motif III consensus recognized by 14-3-3 (Fig. 3C). The structure revealed coordination of the phosphate group by the side chains of Arg56, Arg129, and Tyr130 of 14-3-3σ, H-bonding interactions of the

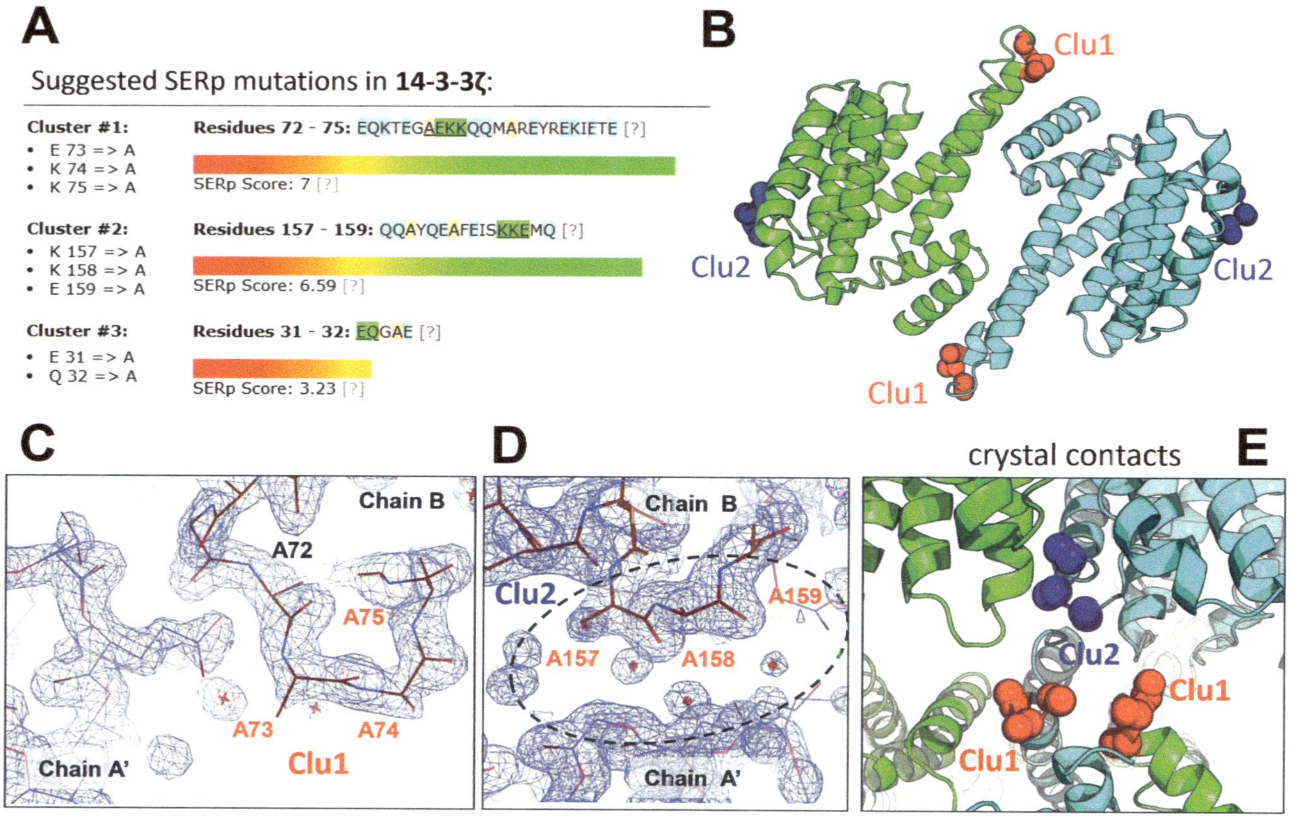

Fig. 7 See figure legend on next page.

AANAT peptide backbone with the side chains of Asn226 and Asn175 of 14-3-3σ, and prominent electrostatic interactions of the Arg203 side chain with the phosphate group of the AANAT peptide *in-cis*, and with the Glu182 side chain of 14-3-3σ *in-trans* (Fig. 8B–D). Nevertheless, two dimers in the asymmetric unit of the crystal revealed at least three subtly different conformations of the AANAT phosphopeptide. The side chain of Asn204 of AANAT formed an H-bond with either the backbone carbonyl of Leu222 of 14-3-3σ (Fig. 8B) or the Asp225 side chain of 14-3-3σ (Fig. 8C). Likewise, the peptide backbone bent differently about the Gly206 amino group, which resulted in its polar contact with the phosphate group of the peptide in one of the conformations (Fig. 8B). In addition, the flexible Lys49 side chain of 14-3-3σ made different polar contacts in all three conformations of the AANAT peptide (Fig. 8B–D). As do many other structures of 14-3-3/peptide complexes, the obtained structure illustrates the plasticity of the AG of 14-3-3, its ability to adapt to different phosphorylated ligands, enabling different placement and chemical bonding of these ligands.

Fig. 9 illustrates that the chimeric approach can be used to obtain not only structural information on the 14-3-3/phosphopeptide interactions but also on ternary complexes with small molecules. This example shows that the well-known small molecule modulator of 14-3-3 PPIs called fusicoccin-A (Kaplan et al., 2020; Marra, Camoni, Visconti, Fiorillo, & Evidente, 2021; Paiardini et al., 2014) binds in the 14-3-3 AG and displaces the C-terminus of the bound phosphopeptide of the HPV 18E6 oncoprotein, destabilizing the 14-3-3/phosphopeptide complex, which was confirmed separately by *in vitro* binding assays (Gogl et al., 2021).

Fig. 7 Identification of clusters of charged surface residues of 14-3-3ζ using surface entropy reduction (SERP) approach (Goldschmidt, Cooper, Derewenda, & Eisenberg, 2007a; Goldschmidt et al., 2007b) and the effect of their mutation to alanines. (A) The prediction yields several groups of charged residues ordered according to the overall score. A patch of residues within a cluster that is predicted to be most successful is highlighted; proposed mutations are shaded green and underlined. (B) The highest scoring clusters 1 and 2 at the periphery of the 14-3-3 dimer structure were chosen for mutation. (C–E) The alanines introduced are often found in crystal contacts formed by 14-3-3 molecules. (C, D) Fragments of the electron density maps (2Fo-Fc, contoured at 1σ) confirming the amino acid substitutions near crystal contacts. (E) A magnified view of the crystal contacts between the three nearest 14-3-3 dimers within the 6ZFD structure (Gogl et al., 2021). Note that the bulky wild-type side chains would prevent such contact from forming.

Table 2 14-3-3 complexes with various peptide fragments crystallized using the chimeric approach (Sluchanko, Tugaeva, et al., 2017).

Partner protein	Fragment boundaries	Native sequence	Resolution (Å)	PDB ID	Reference
HSPB6	12–19	LRRA**pS**^{16}APL	2.3	5OK9	Sluchanko, Tugaeva, et al. (2017)
HSPB6	12–19	LRRA**pS**^{16}APL	3.2	5OKF	Sluchanko, Tugaeva, et al. (2017)
Gli1-3	637–645	RRA**pS**^{640}DPAQA	3.2	5OM0	Sluchanko, Tugaeva, et al. (2017)
AANAT	201–207	LRRN**pS**^{205}GC#	3.0	6T80	—
STARD1	54–62	RRS**pS**^{57}LLGSR	2.0	6T5H	Tugaeva, Titterington, et al. (2020)
STARD1	192–200	RRG**pS**^{195}TCVLA	2.6	6T5F	Tugaeva, Titterington, et al. (2020)
BAD	71–81	SRH**pS**^{74}SYPAGTE	2.4	7Q16	Sluchanko (2021)
HPV18E6	152–158	RRRE**pT**^{156}QV#	1.9	6ZFD, 6ZFG	Gogl et al. (2021)

denote the C-terminal carboxy group.

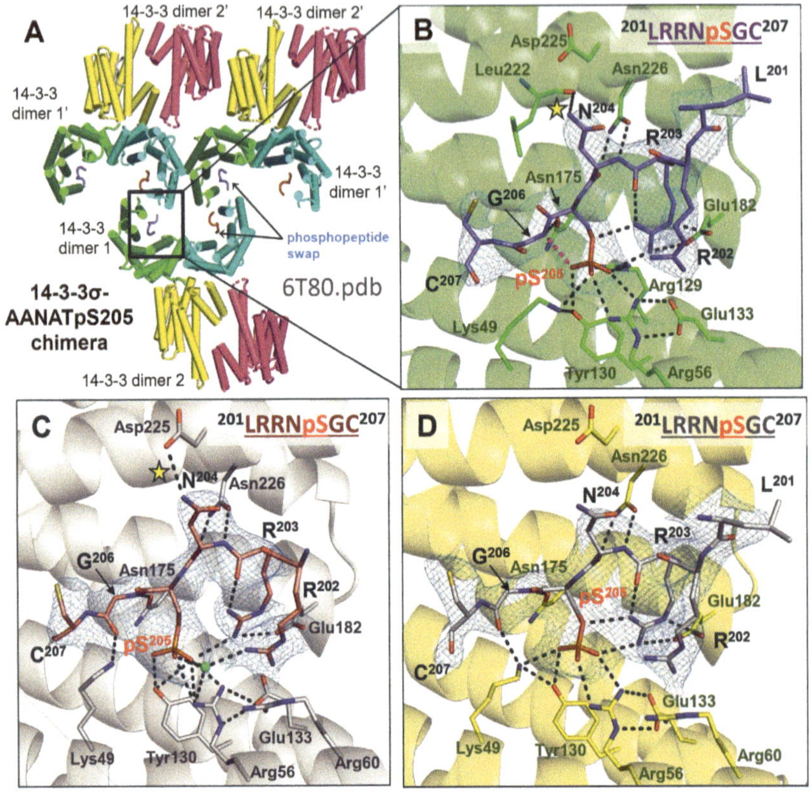

Fig. 8 Crystal structure of the 14-3-3σ chimera with the C-terminal phosphopeptide of human AANAT. (A) Overall crystal packing of the chimera showing the phosphopeptide swap between the nearest subunits belonging to different 14-3-3 dimers. (B–D) AANAT phosphopeptides, shown with the 2Fo-Fc electron density maps contoured at 1σ, display slightly different conformations and chemical contacts with 14-3-3σ within the asymmetric unit of the crystal. The H-bonds formed by Asn204 of AANAT and different atoms of 14-3-3 are shown by yellow stars on (B) and (C). A flexible Gly206 makes a backbone H-bond to the phosphate group in (B) (magenta dashed line). The water molecule mediating contacts between 14-3-3σ residues and the phosphate group is shown by lime green in (C). Note that, depending on the phosphopeptide conformation, Lys49 of 14-3-3σ makes different contacts with the phosphate group and/or the phosphopeptide backbone carbonyl of Gly206.

While the chimeric approach proved useful in all trials (Table 2), its main disadvantage is that, unlike the use of synthetic phosphopeptides, it does not enable the determination of the apparent binding constants of the corresponding peptide fragments to 14-3-3. By contrast, although in many cases the affinity of synthetic phosphopeptides to 14-3-3 can be routinely determined by fluorescence polarization or titration calorimetry (for example,

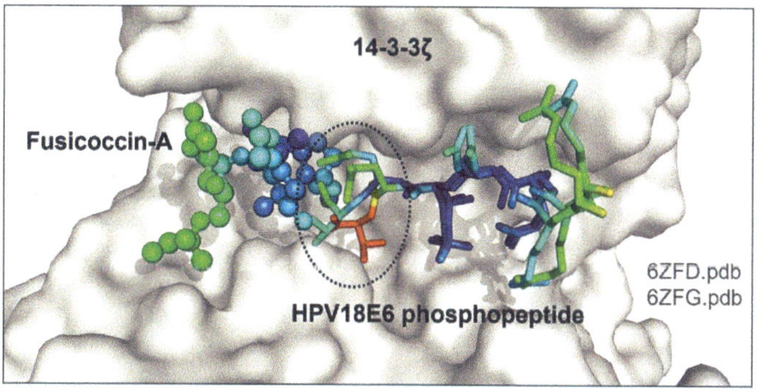

Fig. 9 The effect of fusicoccin-A binding on the conformation of the HPV18E6 phosphopeptide bound to 14-3-3ζ. Fusicoccin-A is shown by spheres, whereas the phosphopeptide is shown by sticks. Both peptide and fusicoccin-A are colored according to the B-factors from low (blue) to high (red). The most pronounced change is marked by the dotted oval. The crystal structure of the 14-3-3ζ-HPV18E6 peptide chimera (PDB: 6ZFD) was superimposed with the structure of the 14-3-3ζ-HPV18E6 peptide chimera with bound fusicoccin-A (PDB: 6ZFG). The image is created in PyMol 2.20.

see Stevers et al., 2016), this traditional approach is more expensive and the phosphopeptides can be poorly soluble and stable on their own.

Depending on the domain structure of a 14-3-3 target protein, the chimeric approach can be expanded to larger fragments interacting with 14-3-3 (Tugaeva et al., 2019), but it is limited by the kinase specificity and availability. So far, all successful attempts dealt with the use of protein kinase A, which is particularly active and soluble upon expression in *E. coli* (Gogl et al., 2021; Sluchanko, Beelen, et al., 2017; Sluchanko, Tugaeva, et al., 2017; Tugaeva et al., 2019; Tugaeva, Remeeva, et al., 2020). Other protein kinases phosphorylating 14-3-3 targets can most likely be used as well, but this requires further investigation. Alternatively, the combination of the chimeric approach and GCE methods allowing for the site-specific cotranslational phosphoserine incorporation (see above) seems very attractive. In addition, it has been proposed to use engineered heterodimeric 14-3-3 proteins where each 14-3-3 subunit is tethered to a distinct phosphopeptide to produce and crystallize ternary complexes—phosphopeptide 1/14-3-3 heterodimer/phosphopeptide 2 (Tugaeva et al., 2021).

8. 14-3-3 studies by nuclear magnetic resonance

While most of the protein-protein complexes of 14-3-3 are too large for NMR, this structural biology method was successfully used for screening and validation of small molecule inhibitors or stabilizers of 14-3-3 PPIs

(Andrei, Meijer, et al., 2018; Iralde-Lorente et al., 2019, p. 02; Valenti et al., 2019). This could be aided by the backbone chemical shift assignment, which has been reported recently for human 14-3-3σ (Neves et al., 2019). In addition, NMR was used to analyze the interaction of 14-3-3 with peptides of human tau protein (Andrei, Meijer, et al., 2018; Joo et al., 2015; Neves et al., 2021), tyrosine hydroxylase (Hritz et al., 2014), Chibby (Killoran et al., 2015), integrin tails (Bonet et al., 2013) and other proteins.

NMR has been very useful to characterize the phosphorylation process of 14-3-3 targets (Danis et al., 2016; Louša, Nedozrálová, Župa, Nováček, & Hritz, 2017; Saline et al., 2019). Nevertheless, NMR cannot probably be considered as an independent structural biology technique for studying 14-3-3/phosphotarget complexes and usually works complementary to other methods.

9. Conclusions

Recent advances in structural studies of 14-3-3 complexes certainly include (i) crystal structure determination of hundreds of 14-3-3/peptide complexes representing primary, druggable interactions of different proteins with 14-3-3, (ii) determination of several rather big crystal structures of 14-3-3 complexes with unphosphorylated and phosphorylated ligands at different stoichiometries (Fig. 5), (iii) the first successful applications of cryo-electron microscopy to large and dynamic signaling complexes of 14-3-3 proteins (Fig. 6), and (iv) methodological improvements and innovations concerning the protein design, sample preparation, phosphorylation, crystallization, and crystal soaking. The accumulation of the wealth of structural data on 14-3-3/peptide interactions (more than 300 crystal structures in the PDB) is expected to help refine the bioinformatic algorithms aimed at predicting new and yet uncharacterized 14-3-3 partners and 14-3-3-binding motifs, as well as the corresponding binding affinities. The rational design of new selective modulators of 14-3-3 PPIs against a range of pathological conditions is also expected to strongly benefit from these big data. The new methods including soaking procedures, coexpression systems, GCE tools for site-specific phosphosite inclusion, chimeric approaches, and others discussed above are believed to facilitate further expansion of these data. Finally, while the successful crystallographic studies have spent titanic efforts to determine a few structures of 14-3-3 complexed with protein fragments larger than 10 kDa (Fig. 5), a new turn is expected to be undertaken with the help of cryo-EM, as this technique is currently mature enough to cope with most of the tasks faced by structural biology of 14-3-3, including small and large, less and more dynamic assemblies involving 14-3-3 proteins, and have all chances to shape the future of the 14-3-3 research.

Acknowledgments

The study was supported by the Ministry of Science and Higher education of the Russian Federation in the framework of an agreement no. 075-15-2021-1354 (07.10.2021). Analysis of the 14-3-3/AANAT structure and properties of 14-3-3 proteins was partially supported by the Russian Science Foundation (grant no. 19-74-10031).

References

Aghazadeh, Y., Ye, X., Blonder, J., & Papadopoulos, V. (2014). Protein modifications regulate the role of 14-3-3gamma adaptor protein in cAMP-induced steroidogenesis in MA-10 Leydig cells. *Journal of Biological Chemistry*, *289*(38), 26542–26553. https://doi.org/10.1074/jbc.M114.569079.

Aitken, A. (2002). Functional specificity in 14-3-3 isoform interactions through dimer formation and phosphorylation. Chromosome location of mammalian isoforms and variants. *Plant Molecular Biology*, *50*(6), 993–1010.

Aitken, A. (2006). 14-3-3 proteins: A historic overview. *Seminars in Cancer Biology*, *16*(3), 162–172.

Aitken, A. (2011). Post translational modification of 14-3-3 isoforms and regulation of cellular function. *Seminars in Cell & Developmental Biology*, *22*(7), 673–680. https://doi.org/10.1016/j.semcdb.2011.08.003.

Alblova, M., Smidova, A., Docekal, V., Vesely, J., Herman, P., Obsilova, V., et al. (2017). Molecular basis of the 14-3-3 protein-dependent activation of yeast neutral trehalase Nth1. *Proceedings of the National Academy of Sciences of the United States of America*, *114*(46), E9811–E9820. https://doi.org/10.1073/pnas.1714491114.

Andrei, S. A., de Vink, P., Sijbesma, E., Han, L., Brunsveld, L., Kato, N., et al. (2018). Rationally designed semisynthetic natural product analogues for stabilization of 14-3-3 protein-protein interactions. *Angewandte Chemie (International Ed. in English)*, *57*(41), 13470–13474. https://doi.org/10.1002/anie.201806584.

Andrei, S. A., Meijer, F. A., Neves, J. F., Brunsveld, L., Landrieu, I., Ottmann, C., et al. (2018). Inhibition of 14-3-3/Tau by hybrid small-molecule peptides operating via two different binding modes. *ACS Chemical Neuroscience*, *9*(11), 2639–2654. https://doi.org/10.1021/acschemneuro.8b00118.

Ashkenazy, H., Abadi, S., Martz, E., Chay, O., Mayrose, I., Pupko, T., et al. (2016). ConSurf 2016: An improved methodology to estimate and visualize evolutionary conservation in macromolecules. *Nucleic Acids Research*, *44*(W1), W344–W350. https://doi.org/10.1093/nar/gkw408.

Ballone, A., Lau, R. A., Zweipfenning, F. P. A., & Ottmann, C. (2020). A new soaking procedure for X-ray crystallographic structural determination of protein–peptide complexes. *Acta Crystallographica Section F: Structural Biology Communications*, *76*(10), 501–507. https://doi.org/10.1107/S2053230X2001122X.

Bartel, M., Schafer, A., Stevers, L. M., & Ottmann, C. (2014). Small molecules, peptides and natural products: Getting a grip on 14-3-3 protein-protein modulation. *Future Medicinal Chemistry*, *6*(8), 903–921. https://doi.org/10.4155/fmc.14.47.

Beall, A., Bagwell, D., Woodrum, D., Stoming, T. A., Kato, K., Suzuki, A., et al. (1999). The small heat shock-related protein, HSP20, is phosphorylated on serine 16 during

cyclic nucleotide-dependent relaxation. *Journal of Biological Chemistry*, *274*(16), 11344–11351.
Beránek, V., Reinkemeier, C. D., Zhang, M. S., Liang, A. D., Kym, G., & Chin, J. W. (2018). Genetically encoded protein phosphorylation in mammalian cells. *Cell Chemical Biology*, *25*(9), 1067–1074.e5. https://doi.org/10.1016/j.chembiol.2018.05.013.
Bier, D., Bartel, M., Sies, K., Halbach, S., Higuchi, Y., Haranosono, Y., et al. (2016). Small-molecule stabilization of the 14-3-3/Gab2 protein-protein interaction (PPI) interface. *ChemMedChem*, *11*(8), 911–918. https://doi.org/10.1002/cmdc.201500484.
Bier, D., Mittal, S., Bravo-Rodriguez, K., Sowislok, A., Guillory, X., Briels, J., et al. (2017). The molecular tweezer CLR01 stabilizes a disordered protein-protein interface. *Journal of the American Chemical Society*, *139*(45), 16256–16263. https://doi.org/10.1021/jacs.7b07939.
Bonet, R., Vakonakis, I., & Campbell, I. D. (2013). Characterization of 14-3-3-zeta interactions with integrin tails. *Journal of Molecular Biology*, *425*(17), 3060–3072. https://doi.org/10.1016/j.jmb.2013.05.024.
Boon, S. S., & Banks, L. (2013). High-risk human papillomavirus E6 oncoproteins interact with 14-3-3zeta in a PDZ binding motif-dependent manner. *Journal of Virology*, *87*(3), 1586–1595. https://doi.org/10.1128/JVI.02074-12.
Bose, A., Modi, K., Dey, S., Dalvi, S., Nadkarni, P., Sudarshan, M., et al. (2021). 14-3-3γ prevents centrosome duplication by inhibiting NPM1 function. *Genes to Cells*, *26*(6), 426–446. https://doi.org/10.1111/gtc.12848.
Brautigan, D. L. (1988). Molecular defects in ion channel regulation in cystic fibrosis predicted from analysis of protein phosphorylation/dephosphorylation. *International Journal of Biochemistry*, *20*(8), 745–752. https://doi.org/10.1016/0020-711X(88)90059-6.
Brophy, C. M., Dickinson, M., & Woodrum, D. (1999). Phosphorylation of the small heat shock-related protein, HSP20, in vascular smooth muscles is associated with changes in the macromolecular associations of HSP20. *Journal of Biological Chemistry*, *274*(10), 6324–6329. https://doi.org/10.1074/jbc.274.10.6324.
Bustos, D. M. (2012). The role of protein disorder in the 14-3-3 interaction network. *Molecular BioSystems*, *8*(1), 178–184. https://doi.org/10.1039/c1mb05216k.
Bustos, D. M., & Iglesias, A. A. (2006). Intrinsic disorder is a key characteristic in partners that bind 14-3-3 proteins. *Proteins*, *63*(1), 35–42. https://doi.org/10.1002/prot.20888.
Chen, X., Liu, Z., Shan, Z., Yao, W., Gu, A., & Wen, W. (2018). Structural determinants controlling 14-3-3 recruitment to the endocytic adaptor Numb and dissociation of the Numb·α-adaptin complex. *Journal of Biological Chemistry*, *293*(11), 4149–4158. https://doi.org/10.1074/jbc.RA117.000897.
Chernik, I., Seit-Nebi, A., Marston, S., & Gusev, N. (2007). Small heat shock protein Hsp20 (HspB6) as a partner of 14-3-3gamma. *Molecular and Cellular Biochemistry*, *295*(1–2), 9–17.
Coblitz, B., Wu, M., Shikano, S., & Li, M. (2006). C-terminal binding: An expanded repertoire and function of 14-3-3 proteins. *FEBS Letters*, *580*(6), 1531–1535.
Danis, C., Despres, C., Bessa, L. M., Malki, I., Merzougui, H., Huvent, I., et al. (2016). Nuclear magnetic resonance spectroscopy for the identification of multiple phosphorylations of intrinsically disordered proteins. *Journal of Visualized Experiments*, *118*, e55001. https://doi.org/10.3791/55001.
Denison, F. C., Gökirmak, T., & Ferl, R. J. (2014). Phosphorylation-related modification at the dimer interface of 14-3-3ω dramatically alters monomer interaction dynamics. *Archives of Biochemistry and Biophysics*, *541*, 1–12.
de Vink, P. J., Briels, J. M., Schrader, T., Milroy, L. G., Brunsveld, L., & Ottmann, C. (2017). A binary bivalent supramolecular assembly platform based on cucurbit[8]uril and dimeric adapter protein 14-3-3. *Angewandte Chemie (International Ed. in English)*, *56*(31), 8998–9002. https://doi.org/10.1002/anie.201701807.

Doveston, R. G., Kuusk, A., Andrei, S. A., Leysen, S., Cao, Q., Castaldi, M. P., et al. (2017). Small-molecule stabilization of the p53—14-3-3 protein-protein interaction. *FEBS Letters*, *591*(16), 2449–2457. https://doi.org/10.1002/1873-3468.12723.

Edelman, A. M., Blumenthal, D. K., & Krebs, E. G. (1987). Protein serine/threonine kinases. *Annual Review of Biochemistry*, *56*(1), 567–613. https://doi.org/10.1146/annurev.bi.56.070187.003031.

Espejo, A. B., Gao, G., Black, K., Gayatri, S., Veland, N., Kim, J., et al. (2017). PRMT5 C-terminal phosphorylation modulates a 14-3-3/PDZ interaction switch. *Journal of Biological Chemistry*, *292*(6), 2255–2265. https://doi.org/10.1074/jbc.M116.760330.

Fan, X., Wang, J., Zhang, X., Yang, Z., Zhang, J.-C., Zhao, L., et al. (2019). Single particle cryo-EM reconstruction of 52 kDa streptavidin at 3.2 Angstrom resolution. *Nature Communications*, *10*(1), 2386. https://doi.org/10.1038/s41467-019-10368-w.

Flynn, C. R., Smoke, C. C., Furnish, E., Komalavilas, P., Thresher, J., Yi, Z., et al. (2007). Phosphorylation and activation of a transducible recombinant form of human HSP20 in Escherichia coli. *Protein Expression and Purification*, *52*(1), 50–58. https://doi.org/10.1016/j.pep.2006.08.015.

Ganguly, S., Weller, J. L., Ho, A., Chemineau, P., Malpaux, B., & Klein, D. C. (2005). Melatonin synthesis: 14-3-3-dependent activation and inhibition of arylalkylamine N-acetyltransferase mediated by phosphoserine-205. *Proceedings of the National Academy of Sciences of the United States of America*, *102*(4), 1222–1227. https://doi.org/10.1073/pnas.0406871102.

Ge, Q., Huang, N., Wynn, R. M., Li, Y., Du, X., Miller, B., et al. (2012). Structural characterization of a unique interface between carbohydrate response element-binding protein (ChREBP) and 14-3-3beta protein. *Journal of Biological Chemistry*, *287*(50), 41914–41921. https://doi.org/10.1074/jbc.M112.418855.

Gogl, G., Biri-Kovacs, B., Poti, A. L., Vadaszi, H., Szeder, B., Bodor, A., et al. (2018). Dynamic control of RSK complexes by phosphoswitch-based regulation. *The FEBS Journal*, *285*(1), 46–71. https://doi.org/10.1111/febs.14311.

Gogl, G., Jane, P., Caillet-Saguy, C., Kostmann, C., Bich, G., Cousido-Siah, A., et al. (2020). Dual specificity PDZ- and 14-3-3-binding motifs: A structural and interactomics study. *Structure*, *28*(7). https://doi.org/10.1016/j.str.2020.03.010. 747–759.e3.

Gogl, G., Tugaeva, K. V., Eberling, P., Kostmann, C., Trave, G., & Sluchanko, N. N. (2021). Hierarchized phosphotarget binding by the seven human 14-3-3 isoforms. *Nature Communications*, *12*(1), 1677. https://doi.org/10.1038/s41467-021-21908-8.

Gokirmak, T., Denison, F. C., Laughner, B. J., Paul, A. L., & Ferl, R. J. (2015). Phosphomimetic mutation of a conserved serine residue in Arabidopsis thaliana 14-3-3omega suggests a regulatory role of phosphorylation in dimerization and target interactions. *Plant Physiology and Biochemistry*, *97*, 296–303. https://doi.org/10.1016/j.plaphy.2015.10.022.

Goldschmidt, L., Cooper, D. R., Derewenda, Z. S., & Eisenberg, D. (2007a). *SERp server.* http://services.mbi.ucla.edu/SER/.

Goldschmidt, L., Cooper, D. R., Derewenda, Z. S., & Eisenberg, D. (2007b). Toward rational protein crystallization: A Web server for the design of crystallizable protein variants. *Protein Science*, *16*(8), 1569–1576. https://doi.org/10.1110/ps.072914007.

Goodwin, K. L. (2015). Localisation of the molecular chaperone site of 14-3-3ζ: An intracellular protein associated with toxic neurological protein aggregates. In *PhD thesis*, School of Physical Sciences, University of Adelaide.

Gu, Y.-M., Jin, Y.-H., Choi, J.-K., Baek, K.-H., Yeo, C.-Y., & Lee, K.-Y. (2006). Protein kinase A phosphorylates and regulates dimerization of 14-3-3 epsilon. *FEBS Letters*, *580*(1), 305–310.

Hardie, D. G. (1989). Protein phosphorylation and dephosphorylation. *Current Opinion in Cell Biology*, *1*(2), 220–226. https://doi.org/10.1016/0955-0674(89)90091-4.

Horvath, M., Petrvalska, O., Herman, P., Obsilova, V., & Obsil, T. (2021). 14-3-3 proteins inactivate DAPK2 by promoting its dimerization and protecting key regulatory phosphosites. *Communications Biology*, *4*(1), 1–14. https://doi.org/10.1038/s42003-021-02518-y.

Hritz, J., Byeon, I. J., Krzysiak, T., Martinez, A., Sklenar, V., & Gronenborn, A. M. (2014). Dissection of binding between a phosphorylated tyrosine hydroxylase peptide and 14-3-3zeta: A complex story elucidated by NMR. *Biophysical Journal*, *107*(9), 2185–2194. https://doi.org/10.1016/j.bpj.2014.08.039.

Hu, S. H., Whitten, A. E., King, G. J., Jones, A., Rowland, A. F., James, D. E., et al. (2012). The weak complex between RhoGAP protein ARHGAP22 and signal regulatory protein 14-3-3 has 1:2 stoichiometry and a single peptide binding mode. *PLoS One*, 7(8), e41731. https://doi.org/10.1371/journal.pone.0041731.

Iralde-Lorente, L., Cau, Y., Clementi, L., Franci, L., Tassone, G., Valensin, D., et al. (2019). Chemically stable inhibitors of 14-3-3 protein–protein interactions derived from BV02. *Journal of Enzyme Inhibition and Medicinal Chemistry*, *34*(1), 657–664. https://doi.org/10.1080/14756366.2019.1574779.

Iralde-Lorente, L., Tassone, G., Clementi, L., Franci, L., Munier, C. C., Cau, Y., et al. (2020). Identification of phosphate-containing compounds as new inhibitors of 14-3-3/c-Abl protein-protein interaction. *ACS Chemical Biology*, *15*(4), 1026–1035. https://doi.org/10.1021/acschembio.0c00039.

Jakob, U., Kriwacki, R., & Uversky, V. N. (2014). Conditionally and transiently disordered proteins: Awakening cryptic disorder to regulate protein function. *Chemical Reviews*, *114*(13), 6779–6805. https://doi.org/10.1021/cr400459c.

Jin, X., Park, O.-J., & Hong, S. H. (2019). Incorporation of non-standard amino acids into proteins: Challenges, recent achievements, and emerging applications. *Applied Microbiology and Biotechnology*, *103*(7), 2947–2958. https://doi.org/10.1007/s00253-019-09690-6.

Johnson, C., Crowther, S., Stafford, M. J., Campbell, D. G., Toth, R., & MacKintosh, C. (2010). Bioinformatic and experimental survey of 14-3-3-binding sites. *Biochemical Journal*, *427*(1), 69–78. https://doi.org/10.1042/BJ20091834.

Joo, Y., Schumacher, B., Landrieu, I., Bartel, M., Smet-Nocca, C., Jang, A., et al. (2015). Involvement of 14-3-3 in tubulin instability and impaired axon development is mediated by Tau. *The FASEB Journal*, *29*(10), 4133–4144. https://doi.org/10.1096/fj.14-265009.

Kacirova, M., Kosek, D., Kadek, A., Man, P., Vecer, J., Herman, P., et al. (2015). Structural characterization of phosducin and its complex with the 14-3-3 protein. *Journal of Biological Chemistry*, *290*(26), 16246–16260. https://doi.org/10.1074/jbc.M115.636563.

Kalabova, D., Filandr, F., Alblova, M., Petrvalska, O., Horvath, M., Man, P., et al. (2020). 14-3-3 protein binding blocks the dimerization interface of caspase-2. *The FEBS Journal*, *287*(16), 3494–3510. https://doi.org/10.1111/febs.15215.

Kalabova, D., Smidova, A., Petrvalska, O., Alblova, M., Kosek, D., Man, P., et al. (2017). Human procaspase-2 phosphorylation at both S139 and S164 is required for 14-3-3 binding. *Biochemical and Biophysical Research Communications*, *493*(2), 940–945. https://doi.org/10.1016/j.bbrc.2017.09.116.

Kaplan, A., Andrei, S. A., van Regteren Altena, A., Simas, T., Banerjee, S. L., Kato, N., et al. (2020). Polypharmacological perturbation of the 14-3-3 adaptor protein interactome stimulates neurite outgrowth. *Cell Chemical Biology*, *27*(6). https://doi.org/10.1016/j.chembiol.2020.02.010. 657–667.e6.

Kaplan, A., Ottmann, C., & Fournier, A. E. (2017). 14-3-3 adaptor protein-protein interactions as therapeutic targets for CNS diseases. *Pharmacological Research*, *125*(Pt. B), 114–121. https://doi.org/10.1016/j.phrs.2017.09.007.

Karlberg, T., Hornyak, P., Pinto, A. F., Milanova, S., Ebrahimi, M., Lindberg, M., et al. (2018). 14-3-3 proteins activate Pseudomonas exotoxins-S and -T by chaperoning a hydrophobic surface. *Nature Communications*, *9*(1), 3785. https://doi.org/10.1038/s41467-018-06194-1.

Kast, D. J., & Dominguez, R. (2019). Mechanism of IRSp53 inhibition by 14-3-3. *Nature Communications*, *10*(1), 483. https://doi.org/10.1038/s41467-019-08317-8.

Killoran, R. C., Fan, J., Yang, D., Shilton, B. H., & Choy, W. Y. (2015). Structural analysis of the 14-3-3zeta/Chibby interaction involved in Wnt/beta-catenin signaling. *PLoS One*, *10*(4), e0123934. https://doi.org/10.1371/journal.pone.0123934.

Kondo, Y., Ognjenovic, J., Banerjee, S., Karandur, D., Merk, A., Kulhanek, K., et al. (2019). Cryo-EM structure of a dimeric B-Raf:14-3-3 complex reveals asymmetry in the active sites of B-Raf kinases. *Science*, *366*(6461), 109–115. https://doi.org/10.1126/science.aay0543.

Kondo, Y., Paul, J. W., Subramaniam, S., & Kuriyan, J. (2021). New insights into Raf regulation from structural analyses. *Current Opinion in Structural Biology*, *71*, 223–231. https://doi.org/10.1016/j.sbi.2021.07.005.

Kopecka, M., Kosek, D., Kukacka, Z., Rezabkova, L., Man, P., Novak, P., et al. (2014). Role of the EF-hand-like motif in the 14-3-3 protein-mediated activation of yeast neutral trehalase Nth1*. *Journal of Biological Chemistry*, *289*(20), 13948–13961. https://doi.org/10.1074/jbc.M113.544551.

Kostelecky, B., Saurin, A., Purkiss, A., Parker, P., & McDonald, N. (2009). Recognition of an intra-chain tandem 14-3-3 binding site within PKCepsilon. *EMBO Reports*, *10*(9), 983–989.

Ku, N. O., Liao, J., & Omary, M. B. (1998). Phosphorylation of human keratin 18 serine 33 regulates binding to 14-3-3 proteins. *The EMBO Journal*, *17*(7), 1892–1906.

Leysen, S., Burnley, R. J., Rodriguez, E., Milroy, L.-G., Soini, L., Adamski, C. J., et al. (2021). A structural study of the cytoplasmic chaperone effect of 14-3-3 proteins on Ataxin-1. *Journal of Molecular Biology*, *433*(19), 167174. https://doi.org/10.1016/j.jmb.2021.167174.

Liau, N. P. D., Venkatanarayan, A., Quinn, J. G., Phung, W., Malek, S., Hymowitz, S. G., et al. (2020). Dimerization induced by C-terminal 14–3–3 binding is sufficient for BRAF kinase activation. *Biochemistry*, *59*(41), 3982–3992. https://doi.org/10.1021/acs.biochem.0c00517.

Liau, N. P. D., Wendorff, T. J., Quinn, J. G., Steffek, M., Phung, W., Liu, P., et al. (2020). Negative regulation of RAF kinase activity by ATP is overcome by 14-3-3-induced dimerization. *Nature Structural & Molecular Biology*, *27*(2), 134–141. https://doi.org/10.1038/s41594-019-0365-0.

Liu, D., Bienkowska, J., Petosa, C., Collier, R. J., Fu, H., & Liddington, R. (1995). Crystal structure of the zeta isoform of the 14-3-3 protein. *Nature*, *376*(6536), 191–194. https://doi.org/10.1038/376191a0.

Louša, P., Nedozrálová, H., Župa, E., Nováček, J., & Hritz, J. (2017). Phosphorylation of the regulatory domain of human tyrosine hydroxylase 1 monitored using non-uniformly sampled NMR. *Biophysical Chemistry*, *223*, 25–29. https://doi.org/10.1016/j.bpc.2017.01.003.

Mackintosh, C. (2004). Dynamic interactions between 14-3-3 proteins and phosphoproteins regulate diverse cellular processes. *Biochemical Journal*, *381*(Pt. 2), 329–342. https://doi.org/10.1042/BJ20031332.

Madeira, F., Tinti, M., Murugesan, G., Berrett, E., Stafford, M., Toth, R., et al. (2015). 14-3-3-Pred: Improved methods to predict 14-3-3-binding phosphopeptides. *Bioinformatics*, *31*(14), 2276–2283. https://doi.org/10.1093/bioinformatics/btv133.

Manandhar, M., Chun, E., & Romesberg, F. E. (2021). Genetic code expansion: Inception, development, commercialization. *Journal of the American Chemical Society*, *143*(13), 4859–4878. https://doi.org/10.1021/jacs.0c11938.

Marra, M., Camoni, L., Visconti, S., Fiorillo, A., & Evidente, A. (2021). The surprising story of fusicoccin: A wilt-inducing phytotoxin, a tool in plant physiology and a 14-3-3-targeted drug. *Biomolecules*, *11*(9), 1393. https://doi.org/10.3390/biom11091393.

Masters, S. C., & Fu, H. (2001). 14-3-3 proteins mediate an essential anti-apoptotic signal. *Journal of Biological Chemistry*, *276*(48), 45193–45200. https://doi.org/10.1074/jbc.M105971200.

Miller, C. J., & Turk, B. E. (2018). Homing in: Mechanisms of substrate targeting by protein kinases. *Trends in Biochemical Sciences*, *43*(5), 380–394. https://doi.org/10.1016/j.tibs.2018.02.009.

Milroy, L. G., Brunsveld, L., & Ottmann, C. (2013). Stabilization and inhibition of protein-protein interactions: The 14-3-3 case study. *ACS Chemical Biology*, *8*(1), 27–35. https://doi.org/10.1021/cb300599t.

Moore, B. W., & Perez, V. J. (1967). In E. Cliffs (Ed.), *Specific acid proteins in the nervous system*.

Muslin, A. J., Tanner, J. W., Allen, P. M., & Shaw, A. S. (1996). Interaction of 14-3-3 with signaling proteins is mediated by the recognition of phosphoserine. *Cell*, *84*(6), 889–897. https://doi.org/10.1016/S0092-8674(00)81067-3.

Nagy, G., Oostenbrink, C., & Hritz, J. (2017). Exploring the binding pathways of the 14-3-3zeta protein: Structural and free-energy profiles revealed by Hamiltonian replica exchange molecular dynamics with distancefield distance restraints. *PLoS One*, *12*(7), e0180633. https://doi.org/10.1371/journal.pone.0180633.

Nakane, T., Kotecha, A., Sente, A., McMullan, G., Masiulis, S., Brown, P. M. G. E., et al. (2020). Single-particle cryo-EM at atomic resolution. *Nature*, *587*(7832), 152–156. https://doi.org/10.1038/s41586-020-2829-0.

Neves, J. F., Landrieu, I., Merzougui, H., Boll, E., Hanoulle, X., & Cantrelle, F.-X. (2019). Backbone chemical shift assignments of human 14-3-3σ. *Biomolecular NMR Assignments*, *13*(1), 103–107. https://doi.org/10.1007/s12104-018-9860-1.

Neves, J. F., Petrvalská, O., Bosica, F., Cantrelle, F.-X., Merzougui, H., O'Mahony, G., et al. (2021). Phosphorylated full-length Tau interacts with 14-3-3 proteins via two short phosphorylated sequences, each occupying a binding groove of 14-3-3 dimer. *The FEBS Journal*, *288*(6), 1918–1934. https://doi.org/10.1111/febs.15574.

Obenauer, J. C., Cantley, L. C., & Yaffe, M. B. (2003). Scansite 2.0: Proteome-wide prediction of cell signaling interactions using short sequence motifs. *Nucleic Acids Research*, *31*(13), 3635–3641.

Obsil, T., Ghirlando, R., Anderson, D. E., Hickman, A. B., & Dyda, F. (2003). Two 14-3-3 binding motifs are required for stable association of Forkhead transcription factor FOXO4 with 14-3-3 proteins and inhibition of DNA binding. *Biochemistry*, *42*(51), 15264–15272. https://doi.org/10.1021/bi0352724.

Obsil, T., Ghirlando, R., Klein, D. C., Ganguly, S., & Dyda, F. (2001). Crystal structure of the 14-3-3zeta:serotonin N-acetyltransferase complex. A role for scaffolding in enzyme regulation. *Cell*, *105*(2), 257–267.

Obsil, T., & Obsilova, V. (2011). Structural basis of 14-3-3 protein functions. *Seminars in Cell & Developmental Biology*, *22*(7), 663–672. https://doi.org/10.1016/j.semcdb.2011.09.001.

Oldfield, C. J., Meng, J., Yang, J. Y., Yang, M. Q., Uversky, V. N., & Dunker, A. K. (2008). Flexible nets: Disorder and induced fit in the associations of p53 and 14-3-3 with their partners. *BMC Genomics*, *9*(Suppl. 1), S1. https://doi.org/10.1186/1471-2164-9-S1-S1.

Ottmann, C., Marco, S., Jaspert, N., Marcon, C., Schauer, N., Weyand, M., et al. (2007). Structure of a 14-3-3 coordinated hexamer of the plant plasma membrane H+ -ATPase by combining X-ray crystallography and electron cryomicroscopy. *Molecular Cell*, *25*(3), 427–440. https://doi.org/10.1016/j.molcel.2006.12.017.

Ottmann, C., Weyand, M., Sassa, T., Inoue, T., Kato, N., Wittinghofer, A., et al. (2009). A structural rationale for selective stabilization of anti-tumor interactions of 14-3-3 proteins

by cotylenin A. *Journal of Molecular Biology*, *386*(4), 913–919. https://doi.org/10.1016/j.jmb.2009.01.005.
Ottmann, C., Yasmin, L., Weyand, M., Veesenmeyer, J., Diaz, M., Palmer, R., et al. (2007). Phosphorylation-independent interaction between 14-3-3 and exoenzyme S: From structure to pathogenesis. *The EMBO Journal*, *26*(3), 902–913.
Paiardini, A., Aducci, P., Cervoni, L., Cutruzzola, F., Di Lucente, C., Janson, G., et al. (2014). The phytotoxin fusicoccin differently regulates 14-3-3 proteins association to mode III targets. *IUBMB Life*, *66*(1), 52–62. https://doi.org/10.1002/iub.1239.
Park, E., Rawson, S., Li, K., Kim, B. W., Ficarro, S. B., Pino, G. G., et al. (2019). Architecture of autoinhibited and active BRAF-MEK1-14-3-3 complexes. *Nature*, *575*(7783), 545–550. https://doi.org/10.1038/s41586-019-1660-y.
Petosa, C., Masters, S. C., Bankston, L. A., Pohl, J., Wang, B., Fu, H., et al. (1998). 14-3-3zeta binds a phosphorylated Raf peptide and an unphosphorylated peptide via its conserved amphipathic groove. *Journal of Biological Chemistry*, *273*(26), 16305–16310.
Petrvalska, O., Kosek, D., Kukacka, Z., Tosner, Z., Man, P., Vecer, J., et al. (2016). Structural insight into the 14-3-3 protein-dependent inhibition of protein kinase ASK1 (apoptosis signal-regulating kinase 1). *Journal of Biological Chemistry*, *291*(39), 20753–20765. https://doi.org/10.1074/jbc.M116.724310.
Pohl, P., Joshi, R., Petrvalska, O., Obsil, T., & Obsilova, V. (2021). 14-3-3-protein regulates Nedd4-2 by modulating interactions between HECT and WW domains. *Communications Biology*, *4*(1), 1–15. https://doi.org/10.1038/s42003-021-02419-0.
Powell, D., Rane, M., Chen, Q., Singh, S., & McLeish, K. (2002). Identification of 14-3-3zeta as a protein kinase B/Akt substrate. *Journal of Biological Chemistry*, *277*(24), 21639–21642.
Powell, D., Rane, M., Joughin, B., Kalmukova, R., Hong, J.-H., Tidor, B., et al. (2003). Proteomic identification of 14-3-3zeta as a mitogen-activated protein kinase-activated protein kinase 2 substrate: Role in dimer formation and ligand binding. *Molecular and Cellular Biology*, *23*(15), 5376–5387.
Prokop, J. W., Yeo, N. C., Ottmann, C., Chhetri, S. B., Florus, K. L., Ross, E. J., et al. (2018). Characterization of coding/noncoding variants for SHROOM3 in patients with CKD. *Journal of the American Society of Nephrology*, *29*(5), 1525–1535. https://doi.org/10.1681/ASN.2017080856.
Ren, Y., Ouyang, Z., Hou, Z., Yan, Y., Zhi, Z., Shi, M., et al. (2020). CIC is a mediator of the ERK1/2-DUSP6 negative feedback loop. *iScience*, *23*(11). https://doi.org/10.1016/j.isci.2020.101635.
Rezabkova, L., Kacirova, M., Sulc, M., Herman, P., Vecer, J., Stepanek, M., et al. (2012). Structural modulation of phosducin by phosphorylation and 14-3-3 protein binding. *Biophysical Journal*, *103*(9), 1960–1969. https://doi.org/10.1016/j.bpj.2012.09.021.
Rezabkova, L., Man, P., Novak, P., Herman, P., Vecer, J., Obsilova, V., et al. (2011). Structural basis for the 14-3-3 protein-dependent inhibition of the regulator of G protein signaling 3 (RGS3) function. *Journal of Biological Chemistry*, *286*(50), 43527–43536. https://doi.org/10.1074/jbc.M111.273573.
Rieloff, E., & Skepö, M. (2020). Phosphorylation of a disordered peptide—Structural effects and force field inconsistencies. *Journal of Chemical Theory and Computation*, *16*(3), 1924–1935. https://doi.org/10.1021/acs.jctc.9b01190.
Rubin, C. S., & Rosen, O. M. (1975). Protein phosphorylation. *Annual Review of Biochemistry*, *44*(1), 831–887. https://doi.org/10.1146/annurev.bi.44.070175.004151.
Sahdev, S., Khattar, S. K., & Saini, K. S. (2008). Production of active eukaryotic proteins through bacterial expression systems: A review of the existing biotechnology strategies. *Molecular and Cellular Biochemistry*, *307*(1), 249–264. https://doi.org/10.1007/s11010-007-9603-6.

Saline, M., Badertscher, L., Wolter, M., Lau, R., Gunnarsson, A., Jacso, T., et al. (2019). AMPK and AKT protein kinases hierarchically phosphorylate the N-terminus of the FOXO1 transcription factor, modulating interactions with 14-3-3 proteins. *Journal of Biological Chemistry*, *294*(35), 13106–13116. https://doi.org/10.1074/jbc.RA119.008649.

Schiapparelli, P., Pirman, N. L., Mohler, K., Miranda-Herrera, P. A., Zarco, N., Kilic, O., et al. (2021). Phosphorylated WNK kinase networks in recoded bacteria recapitulate physiological function. *Cell Reports*, *36*(3). https://doi.org/10.1016/j.celrep.2021.109416.

Selleck, W., & Tan, S. (2008). Recombinant protein complex expression in E. coli. *Current Protocols in Protein Science*, *52*(1). https://doi.org/10.1002/0471140864.ps0521s52. 5.21.1–5.21.21.

Skjevik, A. A., Mileni, M., Baumann, A., Halskau, O., Teigen, K., Stevens, R. C., et al. (2014). The N-terminal sequence of tyrosine hydroxylase is a conformationally versatile motif that binds 14-3-3 proteins and membranes. *Journal of Molecular Biology*, *426*(1), 150–168. https://doi.org/10.1016/j.jmb.2013.09.012.

Sluchanko, Nikolai, et al. (2021). Crystal structure of human 14-3-3ζ complexed with the noncanonical phosphopeptide from proapoptotic BAD. *Biochemical and Biophysical Research Communications*, *583*, 100–105. https://doi.org/10.1016/j.bbrc.2021.10.053.

Sluchanko, N. N. (2018). Association of multiple phosphorylated proteins with the 14-3-3 regulatory hubs: Problems and perspectives. *Journal of Molecular Biology*, *430*(1), 20–26. https://doi.org/10.1016/j.jmb.2017.11.010.

Sluchanko, N. N. (2020). Reading the phosphorylation code: Binding of the 14-3-3 protein to multivalent client phosphoproteins. *Biochemical Journal*, *477*(7), 1219–1225. https://doi.org/10.1042/BCJ20200084.

Sluchanko, N. N., Artemova, N. V., Sudnitsyna, M. V., Safenkova, I. V., Antson, A. A., Levitsky, D. I., et al. (2012). Monomeric 14-3-3zeta has a chaperone-like activity and is stabilized by phosphorylated HspB6. *Biochemistry*, *51*(31), 6127–6138. https://doi.org/10.1021/bi300674e.

Sluchanko, N. N., Beelen, S., Kulikova, A. A., Weeks, S. D., Antson, A. A., Gusev, N. B., et al. (2017). Structural basis for the interaction of a human small heat shock protein with the 14-3-3 universal signaling regulator. *Structure*, *25*(2), 305–316. https://doi.org/10.1016/j.str.2016.12.005.

Sluchanko, N. N., & Bustos, D. M. (2019). Intrinsic disorder associated with 14-3-3 proteins and their partners. *Progress in Molecular Biology and Translational Science*, *166*, 19–61. https://doi.org/10.1016/bs.pmbts.2019.03.007.

Sluchanko, N. N., Chernik, I. S., Seit-Nebi, A. S., Pivovarova, A. V., Levitsky, D. I., & Gusev, N. B. (2008). Effect of mutations mimicking phosphorylation on the structure and properties of human 14-3-3zeta. *Archives of Biochemistry and Biophysics*, *477*(2), 305–312. https://doi.org/10.1016/j.abb.2008.05.020.

Sluchanko, N. N., & Gusev, N. B. (2017). Moonlighting chaperone-like activity of the universal regulatory 14-3-3 proteins. *The FEBS Journal*, *284*(9), 1279–1295. https://doi.org/10.1111/febs.13986.

Sluchanko, N. N., Seit-Nebi, A. S., & Gusev, N. B. (2009). Phosphorylation of more than one site is required for tight interaction of human tau protein with 14-3-3zeta. *FEBS Letters*, *583*(17), 2739–2742. https://doi.org/10.1016/j.febslet.2009.07.043.

Sluchanko, N. N., Sudnitsyna, M. V., Chernik, I. S., Seit-Nebi, A. S., & Gusev, N. B. (2011). Phosphomimicking mutations of human 14-3-3zeta affect its interaction with tau protein and small heat shock protein HspB6. *Archives of Biochemistry and Biophysics*, *506*(1), 24–34. https://doi.org/10.1016/j.abb.2010.11.003.

Sluchanko, N. N., Tugaeva, K. V., Faletrov, Y. V., & Levitsky, D. I. (2016). High-yield soluble expression, purification and characterization of human steroidogenic acute regulatory protein (StAR) fused to a cleavable Maltose-Binding Protein (MBP). *Protein Expression and Purification, 119*, 27–35. https://doi.org/10.1016/j.pep.2015.11.002.

Sluchanko, N. N., Tugaeva, K. V., Greive, S. J., & Antson, A. A. (2017). Chimeric 14-3-3 proteins for unraveling interactions with intrinsically disordered partners. *Scientific Reports*, 7(1), 12014. https://doi.org/10.1038/s41598-017-12214-9.

Sluchanko, N. N., & Uversky, V. N. (2015). Hidden disorder propensity of the N-terminal segment of universal adapter protein 14-3-3 is manifested in its monomeric form: Novel insights into protein dimerization and multifunctionality. *Biochimica et Biophysica Acta, 1854*(5), 492 504. https://doi.org/10.1016/j.bbapap.2015.02.017.

Smidova, A., Alblova, M., Kalabova, D., Psenakova, K., Rosulek, M., Herman, P., et al. (2018). 14-3-3 protein masks the nuclear localization sequence of caspase-2. *The FEBS Journal, 285*(22), 4196–4213. https://doi.org/10.1111/febs.14670.

Smidova, A., Stankova, K., Petrvalska, O., Lazar, J., Sychrova, H., Obsil, T., et al. (2019). The activity of Saccharomyces cerevisiae Na+, K+/H+ antiporter Nha1 is negatively regulated by 14-3-3 protein binding at serine 481. *Biochimica et Biophysica Acta (BBA). Molecular Cell Research, 1866*(12), 118534. https://doi.org/10.1016/j.bbamcr.2019.118534.

Stevens, A. J., Brown, Z. Z., Shah, N. H., Sekar, G., Cowburn, D., & Muir, T. W. (2016). Design of a split intein with exceptional protein splicing activity. *Journal of the American Chemical Society, 138*(7), 2162–2165. https://doi.org/10.1021/jacs.5b13528.

Stevers, L. M., de Vries, R. M., Doveston, R. G., Milroy, L. G., Brunsveld, L., & Ottmann, C. (2017). Structural interface between LRRK2 and 14-3-3 protein. *Biochemical Journal, 474*(7), 1273–1287. https://doi.org/10.1042/BCJ20161078.

Stevers, L. M., Lam, C. V., Leysen, S. F., Meijer, F. A., van Scheppingen, D. S., de Vries, R. M., et al. (2016). Characterization and small-molecule stabilization of the multisite tandem binding between 14-3-3 and the R domain of CFTR. *Proceedings of the National Academy of Sciences of the United States of America, 113*(9), E1152–E1161. https://doi.org/10.1073/pnas.1516631113.

Stevers, L. M., Sijbesma, E., Botta, M., MacKintosh, C., Obsil, T., Landrieu, I., et al. (2018). Modulators of 14-3-3 protein-protein interactions. *Journal of Medicinal Chemistry, 61*(9), 3755–3778. https://doi.org/10.1021/acs.jmedchem.7b00574.

Stumpf, M. P., Thorne, T., de Silva, E., Stewart, R., An, H. J., Lappe, M., et al. (2008). Estimating the size of the human interactome. *Proceedings of the National Academy of Sciences of the United States of America, 105*(19), 6959–6964. https://doi.org/10.1073/pnas.0708078105.

Takala, H., Nurminen, E., Nurmi, S. M., Aatonen, M., Strandin, T., Takatalo, M., et al. (2008). Beta2 integrin phosphorylation on Thr758 acts as a molecular switch to regulate 14-3-3 and filamin binding. *Blood, 112*(5), 1853–1862. https://doi.org/10.1182/blood-2007-12-127795.

Taoka, K., Ohki, I., Tsuji, H., Furuita, K., Hayashi, K., Yanase, T., et al. (2011). 14-3-3 proteins act as intracellular receptors for rice Hd3a florigen. *Nature, 476*(7360), 332–335. https://doi.org/10.1038/nature10272.

Tugaeva, K. V., Kalacheva, D. I., Cooley, R. B., Strelkov, S. V., & Sluchanko, N. N. (2019). Concatenation of 14-3-3 with partner phosphoproteins as a tool to study their interaction. *Scientific Reports, 9*(1), 15007. https://doi.org/10.1038/s41598-019-50941-3.

Tugaeva, K. V., Kalacheva, D. I., Remeeva, A., Gushchin, I., Cooley, R. B., & Sluchanko, N. N. (2021). Devising heterodimeric 14-3-3 chimeras with different phosphopeptides for structural studies of ternary 14-3-3 complexes. *FEBS Open Bio, 11*(S1), 216. P-03.2-11 https://doi.org/10.1002/2211-5463.13205.

Tugaeva, K. V., Remeeva, A., Gushchin, I., Cooley, R. B., & Sluchanko, N. N. (2020). Design, expression, purification and crystallization of human 14-3-3zeta protein chimera with phosphopeptide from proapoptotic protein BAD. *Protein Expression and Purification, 175*, 105707. https://doi.org/10.1016/j.pep.2020.105707.

Tugaeva, K. V., Titterington, J., Sotnikov, D. V., Maksimov, E. G., Antson, A. A., & Sluchanko, N. N. (2020). Molecular basis for the recognition of steroidogenic acute regulatory protein by the 14-3-3 protein family. *The FEBS Journal*, *287*(18), 3944–3966. https://doi.org/10.1111/febs.15474.

Tugaeva, K. V., Tsvetkov, P. O., & Sluchanko, N. N. (2017). Bacterial co-expression of human Tau protein with protein kinase A and 14-3-3 for studies of 14-3-3/phospho-Tau interaction. *PLoS One*, *12*(6), e0178933. https://doi.org/10.1371/journal.pone.0178933.

Valenti, D., Neves, J. F., Cantrelle, F.-X., Hristeva, S., Santo, D. L., Obšil, T., et al. (2019). Set-up and screening of a fragment library targeting the 14-3-3 protein interface. *MedChemComm*, *10*(10), 1796–1802. https://doi.org/10.1039/C9MD00215D.

van Hemert, M. J., Steensma, H. Y., & van Heusden, G. P. (2001). 14-3-3 proteins: Key regulators of cell division, signalling and apoptosis. *BioEssays*, *23*(10), 936–946. https://doi.org/10.1002/bies.1134.

Veisova, D., Macakova, E., Rezabkova, L., Sulc, M., Vacha, P., Sychrova, H., et al. (2012). Role of individual phosphorylation sites for the 14-3-3-protein-dependent activation of yeast neutral trehalase Nth1. *Biochemical Journal*, *443*(3), 663–670. https://doi.org/10.1042/bj20111615.

Wang, B., Yang, H., Liu, Y. C., Jelinek, T., Zhang, L., Ruoslahti, E., et al. (1999). Isolation of high-affinity peptide antagonists of 14-3-3 proteins by phage display. *Biochemistry*, *38*(38), 12499–12504. https://doi.org/10.1021/bi991353h.

Witters, L. A. (1990). Protein phosphorylation and dephosphorylation. *Current Opinion in Cell Biology*, *2*(2), 212–220. https://doi.org/10.1016/0955-0674(90)90009-4.

Wolter, M., de Vink, P., Neves, J. F., Srdanovic, S., Higuchi, Y., Kato, N., et al. (2020). Selectivity via cooperativity: Preferential stabilization of the p65/14-3-3 interaction with semisynthetic natural products. *Journal of the American Chemical Society*, *142*(27), 11772–11783. https://doi.org/10.1021/jacs.0c02151.

Woodcock, J. M., Murphy, J., Stomski, F. C., Berndt, M. C., & Lopez, A. F. (2003). The dimeric versus monomeric status of 14-3-3zeta is controlled by phosphorylation of Ser58 at the dimer interface. *Journal of Biological Chemistry*, *278*(38), 36323–36327. https://doi.org/10.1074/jbc.M304689200.

Xiao, B., Smerdon, S. J., Jones, D. H., Dodson, G. G., Soneji, Y., Aitken, A., et al. (1995). Structure of a 14-3-3 protein and implications for coordination of multiple signalling pathways. *Nature*, *376*(6536), 188–191. https://doi.org/10.1038/376188a0.

Xu, Y., Ren, J., He, X., Chen, H., Wei, T., & Feng, W. (2019). YWHA/14-3-3 proteins recognize phosphorylated TFEB by a noncanonical mode for controlling TFEB cytoplasmic localization. *Autophagy*, *15*(6), 1017–1030. https://doi.org/10.1080/15548627.2019.1569928.

Yaffe, M. B., Rittinger, K., Volinia, S., Caron, P. R., Aitken, A., Leffers, H., et al. (1997). The structural basis for 14-3-3:phosphopeptide binding specificity. *Cell*, *91*(7), 961–971. https://doi.org/10.1016/s0092-8674(00)80487-0.

Yasmin, L., Jansson, A., Panahandeh, T., Palmer, R., Francis, M., & Hallberg, B. (2006). Delineation of exoenzyme S residues that mediate the interaction with 14-3-3 and its biological activity. *The FEBS Journal*, *273*(3), 638–646.

Zhang, Q. C., Petrey, D., Deng, L., Qiang, L., Shi, Y., Thu, C. A., et al. (2012). Structure-based prediction of protein-protein interactions on a genome-wide scale. *Nature*, *490*(7421), 556–560. https://doi.org/10.1038/nature11503.

Zhu, P., Gafken, P. R., Mehl, R. A., & Cooley, R. B. (2019). A highly versatile expression system for the production of multiply phosphorylated proteins. *ACS Chemical Biology*, *14*(7), 1564–1572. https://doi.org/10.1021/acschembio.9b00307.

CHAPTER TEN

Immunosafe(r)-by-design nanoparticles: Molecular targets and cell signaling pathways in a next-generation model *proxy for* humans

Annalisa Pinsino[a,b,*] **and Maria Di Bernardo**[b]
[a]Consiglio Nazionale delle Ricerche, Istituto di Farmacologia Traslazionale (IFT), Palermo, Italy
[b]Consiglio Nazionale delle Ricerche, Istituto per la Ricerca e l'Innovazione Biomedica (IRIB), Palermo, Italy
*Corresponding author: e-mail address: annalisa.pinsino@cnr.it

Contents

Abstract

Nanotechnology research covers a wide field of studies pointing to design and shape complex matter in a scale between 1 and 100nm, with unique size-depending properties and applications. The value and potential of engineered nanoparticles in human diagnostics and therapies essentially relay on their safety and biocompatibility. Entering a cell, in fact, these particles take complex interactions with the surrounding biological environment, dramatically changing their own identity. The formation of a custom-made protein corona is the first signal of their interplay with the cell defensive mechanisms, and a major issue in their application in medicine. Preliminary in-depth studies in model organisms have been developed to assess immunological safety and competence in facing the host immune system and its defensive response. New affordable animal models are emerging in pilot nano-response and safety studies.

Advances in Protein Chemistry and Structural Biology, Volume 130
ISSN 1876-1623
https://doi.org/10.1016/bs.apcsb.2022.01.002

Sea urchins, benthic marine Echinoderms, have a wide and very efficient immune system working with innate defense mechanisms and are widely used in immune studies. Nano-safety studies have been showing that the sea urchin *Paracentrotus lividus* displays an excellent sensing system and high defensive capability, joined to the availability of easily accessible immune cells. As in mammals, nanoparticle recognition and interaction activate specific signaling pathways, metabolic rewiring and homeostasis maintenance. In this chapter, we point to the value of planning new research and developing nano-immune studies using an easy nonmammalian next-generation model, able to unravel new specific response mechanisms to nanoparticles.

Abbreviations

AMPs	antimicrobial peptides
Ca^{2+}	calcium ions
CCM	coelomocyte culture medium
DNA	desoxy- (o deoxy-) ribonucleic acid
EPAA	European partnership for alternative approaches to animal testing
GRNs	gene regulatory networks
Mg^{2+}	magnesium ions
$MSP\text{-}1_{19}$	C-terminal 19 kDa fragment of merozoite surface protein 1
NLRs	NOD/NALP-like cytoplasmic recognition proteins
nm	nanometer
PANDORA	probing safety of nano-objects by defining immune responses of environmental organisms
PDA	polydopamine
PEG	polyethyleneglycol
pI	isoelectric point
PRRs	pattern recognition receptors
PVA	polyvinylalcohol
PVP	polyvinylpyrrolidone
siRNA	short-interfering ribonucleic acid
SpRAG1L/*SpRAG2L*	recombination activating genes 1/2 like in *S. purpuratus*
***Sp*Transformer proteins**	*S. purpuratus* transformer proteins
SRCR	scavenger receptor cysteine-rich
3R principle	reducing, refining, and replacing
TLRs	toll-like receptors
V(D)J	variability, diversity, and joining
μm	micrometer

1. Safe(r)-by-design concept for nanoparticle development: Immunological point of view

Nanotechnology or nanotech is a field of research and innovation devoted to designing and producing a wide-ranging class of materials that include ultrafine particles with at least one nanometric scale dimension (from 1 to 100 nm) (e.g., nanoparticles). These new materials that are obtained by manipulating matter at the scale of atoms and molecules, acquire peculiar characteristics and produce effects not always easily predicted from current knowledge. Nanoparticles (nano-objects with three external nanoscale dimensions) behave very differently if compared to the same materials made at a large scale (bulk materials), because of their exceptional physical, chemical, electrical, magnetic, optical and biological properties related to the nano form (Asha & Narain, 2020). As the diversity of nanoparticles manufacture and usage has increased, the need to assess the level of safety becomes an important issue for the beneficial handling of these new materials. The concept of nanosafe-by-design concerns predicting the risk and limiting the impact of the nanotechnology-based innovation on human health and environmental safety (Bottero et al., 2017). Within the near-limitless variety of these new materials, some of them may be toxic to biological systems, others relatively harmless, while others may promote health and welfare. Based on the nanosafe-by-design concept, engineered nanoparticles are expected to neither distress human health nor negatively affect the environment along their life cycle. This is why a rational science-based approach is needed to minimize the risk for health, focusing on continuing study strategy and appropriate industrial development, adopting sustainable health policies and encompassing actions at individual and community levels. In our era, health promotion and environmental protection are necessary and can be addressed by adopting a holistic approach, promoting knowledge and learning, empowering society to foster control for human and environmental safety, pushing for socio-political and economic change, and investing in fundamental research on nanoproducts. We are the responsible actors in the circle of life, guardians of our planet and defenders of its health, as one entity of circular nature (Capua, 2020). A proactive approach is needed to minimize the gap between the speed of innovation and the speed of developing nano-risks, risks that may not be excluded in absolute terms. Rather than talking about the "safe-by-design" concept we have to focus on a "safer-by-design" concept in nanotechnology, based on three major challenges as follows: (i) Safe(r) products by design, to identify and develop less hazardous nanomaterials; (ii) Safe(r) use of products, to evaluate the risks during their use; (iii) Safe(r) industrial production, to control the industrial processes along the production chain (e.g., minimizing waste) (Fig. 1). Thus, the Safe(r)-by-Design aims at modifying nanoparticle design to minimize the risk at all stages of their lifecycle, from its manufacture to its end of life, through the consumer use and its effects on the exposed environment.

Fig. 1 Toward safety-by-design: representation of the main checkpoints.

What does safe(r) mean? How is it possible to determine if one nanoparticle is harmless for living organisms? A very important issue to consider is whether or not there exist tools to approach the "safety assessment" from a comparative perspective. The response is affirmative, now we know that the safety or the harmless of nanoparticles can be approached by a comparative analysis of their impact on immunity across living organisms, as defined within the EU Horizon 2020 project PANDORA (Pinsino et al., 2020). The immune system is one of the main biological systems committed to the recognition, interaction and response to foreign matters, including nanoparticles, and the innate immune system represents the best ground for understanding the shared or even different mechanisms of defense from plants to humans (Boraschi et al., 2020). This is why the high evolutionary maintenance of innate immunity has produced several types of potent defensive reactions that are conserved (with slight changes) across evolution, and that co-occur with adaptive immunity in vertebrates. Antibodies are a critical part of the adaptive immune system that involves both humoral and cell-mediated immune responses to protect against foreign matters (e.g., viruses, bacteria, particles). Invertebrates lack adaptive immune responses, yet some of them show specific genetic mechanisms known to increase the diversity and specificity of the innate immune responses, including the diversification mechanisms of Ig superfamily genes in snail (Zhang, Adema, Kepler, & Loker, 2004), and the gene somatic diversification of the *Sp*Transformer proteins (formerly Sp185/333) in sea urchin (Oren et al., 2019). Plants do not possess specialized immune cells but all cells can sense a threat, and the signal emitted by a cell (e.g., phytocytochine realize) is capable to alert adjacent cells (Gust, Pruitt, & Nürnberger, 2017).

Based on the notion that the immune system works to meet the organism's changing needs for responding to a foreign input, nanoparticles always stimulate an immune reaction to maintain metabolic homeostasis and self-integrity of cell/tissue/organ functions (Pinsino et al., 2020). The immune system can be recruited at different levels, directly (e.g., innate immune cells) or indirectly (e.g., microbiota), resulting in either its suppression or overstimulation. We can consider nanoparticles immunologically harmful only in the case they damage physiological health-preserving defensive responses and cause pathological immune reactions. All new foreign matters demand adequate inquiries into their interactions with the immune system before their use in industry, and medical care (Food and Drug Administration, 2012). The question is whether it could be possible to predict the effects of nanoparticles on the innate response relying on how the immune system copes with natural challenges (e.g., infections). Nanoparticles may (or not) trigger an immune response, interact with cells, cellular compartments and macromolecules (e.g., receptors, proteins, metabolites) and drive a signaling cascade enhancing or attenuating immune response, depending on their properties, especially their surface properties. Design particles with selective physicochemical properties of which we can predict the interaction with immune cells mark a turning point in biomedicine, because it allows us to attain desirable immunomodulation for a specific therapeutic use, and avoid dangerous pathological reactions.

2. Bio-corona: An important issue in nano-immune interactions

The relationship between inorganic nanoparticles and the immune system is strongly determined by the absorption of organic matters or biomacromolecules (lipids, sugars, nucleic acids, metabolites, and particularly proteins) existing in a biological medium. The particle-biomolecule interaction gives rise to a dynamic bio-molecular coating or "bio-corona," which is the molecular pattern sensed by immune cells during the nanoparticle-organism meet. In some cases, the dynamic bio-corona evolves from transient to permanent defining the nanoparticle surface characteristics, and increasing biocompatibility (Lee, Choi, Webster, Kim, & Khang, 2014). Among the extensive range of biomacromolecules, proteins are considered the most important, due to their high amount and variety in the biological media (Docter et al., 2015). Proteins mask the nanoparticle properties and create a new biological identity that controls the interface with the biological environment, changing the behavior and fate of the particle themselves. At the start, proteins and particles are governed by motion forces (e.g., diffusion, convection) that help them to enter in contact with each other. Later, proteins will interact and bind with the particles based on the different bonding forces governing molecule interactions (e.g., electrostatic, π-effects, van der Waals forces, and hydrophobic effects) (García-Álvarez & Vallet-Regí, 2021). The structure of the bio-corona can be differed in hard and soft corona, depending on the protein affinity on the particle surface (García-Álvarez & Vallet-Regí, 2021). Nanoparticle size and shape, such as surface curvature and some non-electrostatic residue-specific interactions, are critical factors in determining the affinity of protein binding. However, the main force determining this binding is dependent on the possibility to maintain the 3D conformation of proteins themselves on the particle surface (Haynes & Norde, 1994). Particles surface, pH of the media, and protein charge are the main players on which depends the formation of the bio-corona (Barbero et al., 2017). For example, it is known that nanoparticles of the same surface charge and curvature but different sizes absorb proteins with different affinity, or that charge can denature proteins. As well, it is recognized that positively charged nanoparticles absorb proteins with an isoelectric point (pI) lower than 5.5, whereas those negatively charged absorb proteins with pI higher than 5.5 (Barbero et al., 2017). Nevertheless, after an early layer of proteins is adsorbed on the particle surface, the particle charge can be altered to neutral or differently charged, opening the way to other secondary interactions with neutrally, positively, and negatively charged proteins.

On the one hand, proteins may form a monolayer on the particle surface, retaining their native structure and behaving such as the free ones (e.g., interact with receptors, promote biological processes). On the other hand, they can be partially denatured when absorbed into the nanoparticles, leading to protein-protein aggregation, multi-layered coatings, and nanoparticle crosslinking, in addition to the expression of new epitopes. Indeed, a hydrophobic and cationic surface induces a strong interaction with the unfolded proteins leading to particle-protein aggregate formation, especially in small nanoparticles with a spherical shape, in which the increased surface curvature is a further consequence of the particle size decrease (Dennison, Zupancic, Lin, Dwyer, & Murphy, 2017; Dominguez-Medina et al., 2016; Wang et al., 2013).

The stability of a colloidal suspension of naked nanoparticles is based on the balance between the electrostatic repulsion and the universal attractive van der Waals forces (DLVO Theory). When nanoparticles are introduced into a complex medium, different competing scenarios are likely to occur, balanced by respective kinetics. As well known, some ions of opposite charge adsorb onto the nanoparticle surface forming the Stern layer, while the high content in organic matters alters the stability of the nanoparticle and changes surface chemical traits (Guerrini, Alvarez-Puebla, & Pazos-Perez, 2018; Noventa, Rowe, & Galloway, 2018). Moreover, divalent ions, such as Ca^{2+} and Mg^{2+}, can act as an effective bridge to adsorb negatively charged proteins onto the negatively charged surface of such nanoparticles, including titanium dioxide nanoparticles. In such conditions, divalent cations, can neutralize particle electrostatic stabilization that is enhanced by macromolecules, and induce aggregation through cation bridging and charge neutralization mechanisms (Philippe & Schaumann, 2014). Thus, salt concentrations of the majority of biological fluids tend to screen electrostatic repulsions between head-to-head nanoparticles, favoring homo-aggregation (particle-particle complexes); while the abundance of biomacromolecules suspended in such fluids are expected to help hetero-aggregation (particle-protein complexes). In agreement, when nanoparticles are introduced into a complex biological medium, the balance between homo- and hetero-aggregation scenarios are driven by the relative collision frequencies of particles together and with biomacromolecules, respectively (Slomberg et al., 2019). These chemical and physical nanoparticle changes cause an increase in the particle size, and in turn, lead the immune-particle recognition and interaction. Therefore, the bio-corona results in a dynamic multi-layered structure, where not only the protein-particle interaction plays a central role in the particle behavior, but also the protein-protein interaction is critical (Zanganeh, Spitler, & Mahmoudi, 2016). The dynamism of the structure is well defined by the hard and soft bonding proteins, that drive the formation of a more (hard) or less (soft) stable protein-particle complex over time (García-Álvarez & Vallet-Regí, 2021). A depiction of the protein corona formation is presented in Fig. 2.

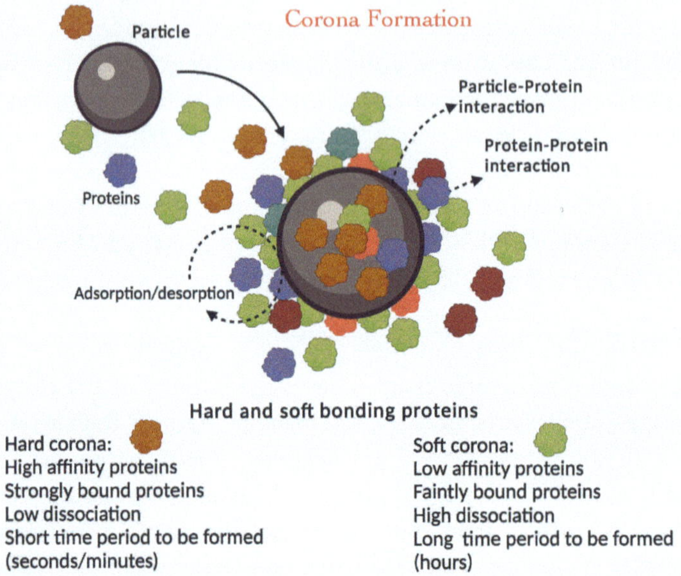

Fig. 2 Schematic representation of the protein corona formation. *Created with BioRender.com.*

Thus, if the nanoparticle-protein affinity is high enough, and there is no alteration in the protein conformation, the resulting nanoparticle-protein complex is rather soluble and tolerated by the immune system. On the other hand, if the nanoparticle-protein affinity is low, the effect will be the formation of large aggregates that could annoy the immune system. For example, particle complexes with a prevailing average diameter larger than 10 μm (polydisperse micro aggregates) may stimulate frustrated phagocytosis leading to chronic inflammation, permanent tissue damage and cancer, eventually (Baranov, Kumar, Sacanna, Thutupalli, & Van Den Bogaart, 2021; Murphy, Schinwald, Poland, & Donaldson, 2012). In summary, the immune system tolerates, removes or attacks the particles depending on what it will feel. Nanoparticle surface passivation or functionalization, which refers to a material becoming inert or designed to elicit the desired response, become essential to increase particle stability/biocompatibility before exposing it to the working biological media.

3. Nanoparticle surface functionalization for biomedical use

Since nanoparticles have the same scale dimension of biomolecules and cellular structures, they can head over to intra-cellular structures miming a pathogen. The capacity of nanoparticles to be internalized and retained within cells opened a new frontier in biomedicine: the development of therapies based on nanoscale interactions. The uptake of the extracellular matter is necessary for the basic cellular functions, including immune functions and tissue remodeling. This physiological event is required for the clearance of pathogens, foreign matters and cells (e.g., senescent, apoptotic, and cancer cells) by specialized immune cells (phagocytes) bringing surface receptors sensing the extracellular signals (Flannagan, Jaumouille, & Grinstein, 2012).

Functionalization of the particle surface by using ligands with high affinity and specificity for such receptors and other molecules, including secreted proteins, allows us to manipulate the properties and characteristics of nanoparticles through surface modification, increasing the applicability and functionality of the particles in hitting the cell/tissue/organ target with high precision (Sanità, Carrese, & Lamberti, 2020). Concordantly, the (bio) functionalization makes nanoparticles capable of specifically interacting with such selective cellular and molecular processes, and of manipulating biological states, structures and functions based on a specific biomedical application (Schutz et al., 2013). Functionalizing the nanoparticles means changing chemically the particle surface structure under controlled conditions, incorporating on their surface a specific targeting ligand. The target specificity and selectivity for nanoparticle applications in diagnostics, prognostics, and therapeutics are made by conjugation of chemicals or biomolecules like folic acid, biotin molecules, oligonucleotides, peptides, antibodies, polymers, drugs, etc., onto the particle surface (Yoo, Park, Yi, Lee, & Koo, 2019). A high number of methods have been carried out to enhance the nanoparticle properties and hit the target with high precision (Mourdikoudis, Pallares, & Thanh, 2018). Thus, a specific monolayer structure can increase the ability of a particle to be internalized within cells, decrease toxicity, and raise the binding capability necessary for efficient intracellular delivery; as well, surface functionality can ensure the exclusive recognition and interaction/non-interaction with the target against which it was designed, and prevent unwanted immune responses.

Several methods are available for introducing chemical functionality on a particle surface and promoting strong binding affinity. For example, albuminization leads to stable permanent biocompatible bio-corona on top of the desired nanoparticles, or polyethyleneglycol(*PEG*)ylation limits particle aggregation, opsonization and particle-protein crosslinking, enhancing stabilization; other techniques include surface coating with polyvinylpyrrolidone (PVP), humic acid, hyaluronic acid, serum (Mout, Moyano, Rana, & Rotello, 2012; Vismara et al., 2017). Nanoparticles covered by PEG or other polymers create a hydrophilic environment, thereby escaping from immune recognition. Hydrophilic polymers bind water molecules and produce a shield onto the nanoparticle surface, reducing the interaction of particles with opsonin and the recognition by macrophage receptors. This strategy should enhance the time of the particle circulation and reduce the time of clearance. Based on specific parameters of nanoparticle design, it is possible to predict the biological interaction of particles with the immune system, and functionalize them to induce immune and inflammatory signaling that may both support or suppress inflammation, depending on the therapeutic needs (Jiao, Li, Mu, & Zhang, 2014).

For example, gold nanoparticles with different surface modifications showed different immunogenicity. Among many nano-candidates, these particles have been distinguished for their use in vaccine development, because of their capability to tune and regulate immune responses. Gold nanoparticles can be good candidates as next-generation vaccines when functionalized with C-terminal 19 kDa fragment of merozoite surface protein 1 (MSP-1_{19}) which confers poor immunogenicity (Parween, Gupta, & Chauhan, 2011), but also, they can be employed as delivery systems and adjuvants to vaccines to stimulate the immune system.

Beyond the easy modification of the gold nanoparticle surface for attaching any targeting molecule, the exceptional biocompatibility and colloidal stability, especially in theragnostic applications (bio-imaging + therapy), make gold nanoparticles an excellent tool for biomedical applications (Bansal, Kumar, Karimi, Singha, & Kumar, 2020), although some issues need to be focused to appreciate their full potential. For example, the low rate of clearance of gold nanoparticles from circulation and tissues can lead to health problems, thus specific cells/tissues targeting must be achieved using chemical/biomolecule functionalized particles (Zhang, 2015). New solutions have attracted considerable attention, particularly those based on hybrid nanostructures, including iron–gold-based nanoparticles that incorporate potentiated magnetic response, light absorption, and thermal effect properties in one, providing an additional therapeutic opportunity for disease treatment (e.g., cancer cure) due to their combined exceptional optical and magnetic properties (Tarkistani, Komalla, & Kayser, 2021). Iron nanoparticles can be stabilized by shells of inorganic or organic materials increasing particle stability, including gold, PEG, polydopamine (PDA), chitosan, polyvinylalcohol (PVA), PVP, and silica. Iron–gold-based nanoparticles simplify the attachment of different functional groups, including thiols, amines (high gold nanoparticle affinity), catechols, silanes, diol carboxylic groups (high Iron-oxide nanoparticle affinity), acquiring multiple functionalities. Some applications include: hyperthermia, protein biomarkers and gene expression product detection, DNA targeted delivery, siRNA drug solubility improvement, imaging, and targeted drug delivery (Reviewed by Tarkistani et al., 2021). However, despite progress in the design and synthesis of functionalized nanoparticles, only a few of them are employed in real-world clinical settings. The main barrier on the way is to predict the nature of the bio-corona formation and its ensuing in vivo effects in humans based on accurate in vitro/ex vivo studies on appropriate models.

Knowing and understanding bio-corona formation in an organism, including mechanisms involving the immune system, ensure a detailed control and predictable behavior of the protein corona formation, preserving the function for which nanoparticles have been designed and functionalized (Nienhausa, Wang, & Nienhaus, 2020).

To cope with the nanoparticle-protein interactions within a living system is necessary to reduce protein corona formation by functionalizing particle surfaces with zwitterionic polymers, hydrophilic polymers, or carbohydrates (Yu, Zhao, Guo, Yan, & Su, 2020). Protein adsorption from particle surfaces cannot be avoided, but it can be controlled in amount, type and orientation of the entrapped proteins, in such a way that their selective attachment becomes an advantage to drive the desired biological response. For example, particle surface pre-coating with opsonins enhances the phagocytic uptake, whereas pre-coating with pre-apolipoproteins and -albumin inhibits particle uptake, and make particles "invisible" (Ritz et al., 2015; Schöttler et al., 2016). The potential applications of the functionalized nanoparticles in medicine and biology are summarized in Fig. 3.

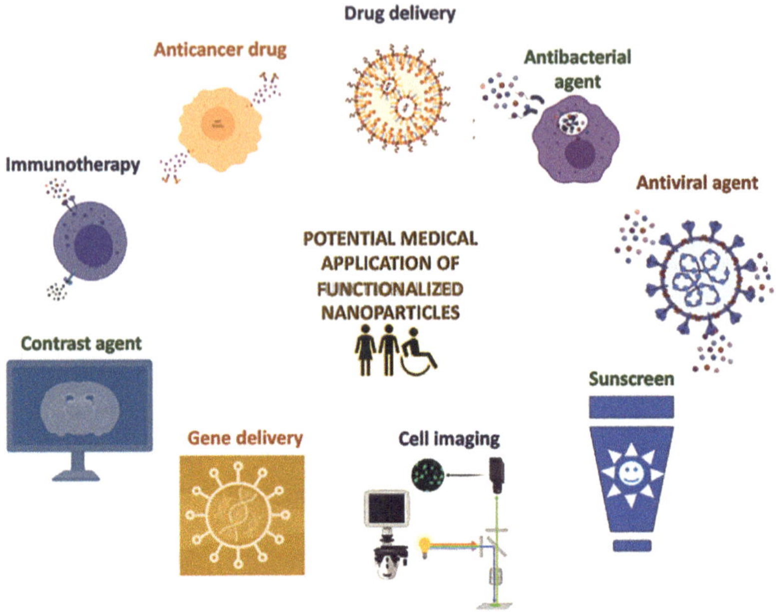

Fig. 3 Potential application of the (bio)functionalized nanoparticles in biomedicine.

4. The innate immunity of the sea urchin from the human perspective

Sea urchin (phylum Echinodermata) is an ancient and very successful marine invertebrate that appeared more than 500 years ago, before the Cambrian bang (Bottjer, Davidson, Peterson, & Cameron, 2006). About 950 species live in the sea, widely distributed across all the seas (from the Mediterranean Sea to the Pacific Ocean), all climates (from tropical to polar environments), and inhabiting all marine benthic environments (from rocky shores to hadopelagic zones). These echinoderms play a central role in the safeguarding of the marine ecosystem integrity and they are incessantly exposed to environmental and anthropogenic pressure, including predation, climate changes, pathogens, pollutants (e.g., chemicals, nanomaterials, plastics) (Smith et al., 2018). In the sea urchin, environmental and anthropogenic pressures favored phenotypic plasticity both during the embryonic and adult life-cycle stages, conferring a notable ability to generate multiple phenotypic shapes and responses in an always-changing world (Hamdoun & Epel, 2007). Concordantly, sea urchins are capable of adjusting to environmental changes due to their effective innate immune system: an integrated, highly complex system with discriminatory properties and multi-defense responses against environmental challenges that constitute a persistent, long-standing evolutionary "arms race" (Hibino et al., 2006). Sea urchin is small and spiny, it has no eyes and it is mainly a grazer, voracious of kelp and algae. Its genome uncovers a remarkable connection to humans, sharing with them 7077 genes, despite the evident morphological differences; 4–5% of the genes are involved directly in innate immunity. Detailed annotation of the sea urchin genome reveals a large-scale duplication and divergence of multigene families encoding effector and receptor molecules (PRRs, Pattern Recognition Receptors) of the innate immune system, including Toll-like receptors (TLRs) and the directly associated signaling adaptors, NOD/NALP-like cytoplasmic recognition proteins (NLRs), and scavenger receptor cysteine-rich (SRCR) (Rast, Smith, Loza-Coll, Hibino, & Litman, 2006). It has been hypothesized that this exceptional diversity and expansion could be related to an increase in the potential microbial recognition ability and host defense of these animals. Although adaptive immunity is absent, sea urchins possess functional equivalents of the acquired responses of vertebrates (e.g., *Sp*Transformer proteins), whose expression is upregulated in the circulating immune cells responding to immune challenge (Smith et al., 2018). Furthermore, the broad repertoire of the sea urchin immunity gene toolkit includes a gene pair, *SpRAG1L* and *SpRAG2L* (Recombination Activating Genes 1/2 like in *S. purpuratus*). The two genes produce full-length proteins, strikingly similar to the RAG1/2 enzymes that mediate specific V(*D*)J recombination, at the basis of immunological memory in differentiating lymphocytes of vertebrates. Although their function is currently unknown, the two genes are co-expressed both during development and in immune cells.

All these elements argue in favor of the existence of a sharp specificity and memory within the echinoid immune system and indicate that sea urchins have developed all the basic components required for a functional vertebrate-like recombination mechanism (Oren et al., 2019; Yakovenko, Agronin, Smith, & Oren, 2021).

Overall, the expansion in the innate immune repertoire reveals a different solution for diversifying the sea urchin immune catalog, in the absence of an immunoglobulin-based adaptive immune system. It is evident a high-level diversity in gene family expansion to generate PRR heterogeneity in sea urchin, although with the distinctive trait of an amazingly huge number of multigene families encoding for molecules of the vertebrate innate immune system. On the other hand, sea urchin presents some

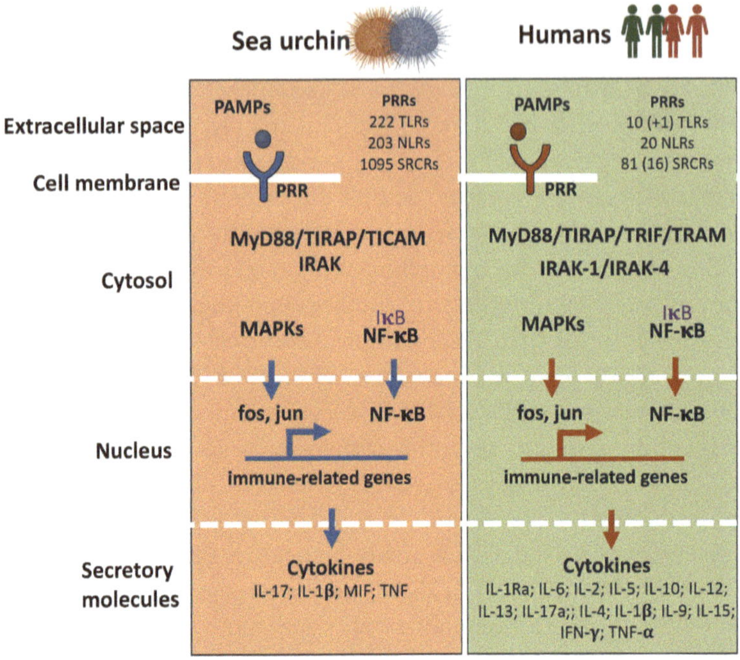

Fig. 4 Innate immune molecules and signaling: comparison between sea urchin and humans.

immune-related genes that are sea urchin-specific and invertebrate-specific, suggesting their loss in humans and other vertebrates. Fig. 4 represents the comparison between the main innate immune molecules and signaling occurring in sea urchins and humans.

For over a century, sea urchin has served as a model organism in developmental biology, contributing to understanding broad biological principles of development, differentiation, cellular and molecular biology (Ernst, 2011; McClay, 2011). The annotation of the sea urchin genome opened a new opportunity for research: the use of the sea urchin immune cells as a nonmammalian next-generation model for in vitro and in vivo basic and translational immunology. The reason for the great interest in sequencing the sea urchin genome was because, as a Deuterostome (second mouth, the mouth develops from a second opening in the embryo), it shares the last common ancestor with humans dating back to 540 million years ago "only" (Sodergren, Weinstock, Davidson, Cameron, et al., 2006). In fact, from an evolutionary perspective, sea urchins are closer to vertebrates than other more widely studied animal models, such as fruit fly and worms (Protostomes).

A heterogeneous free-circulating population of blood like-cells present in the coelomic fluid (blood equivalent) mediates immunity in adult sea urchins (Smith et al., 2018). The phagocytes are the most abundant immune cell types and have functions similar to those of human macrophages, including phagocytosis, rejection, cytotoxic activity and destruction of pathogens (Pinsino & Matranga, 2015). They can initiate inflammation by releasing functional biomolecules (the sea urchin coelomic fluid is a big source of macromolecules), and inducing an immune signaling cascade (Alijagic, Benada, Kofroňová, Cigna, & Pinsino, 2019; Alijagic & Pinsino, 2017). When exposed to a range of materials or pathogens, sea urchin immune cells show selective changes in specific pathways similar to those known in humans (Smith et al., 2018). Thus, sea urchin immunity may hold the key to uncovering basic molecular and regulatory mechanisms of the immune response, immune disease and immuno-toxicity common to all organisms including humans, and may contribute to highlighting the key protective molecules for preventing and curing several human diseases.

5. The sea urchin *Paracentrotus lividus*: A next-generation model to study the immunosafety of nanoparticles

Widespread application of nanomaterials emphasizes the necessity to assess their immunological safety by examining the nanomaterial ability to initiate inflammation and/or to interfere with the ongoing defensive response. Within this framework, the sea urchin *P. lividus* immune system holds an excellent sensing and defensive capability and, along with the easily accessible immune cells, features large potential in the development of a next-generation model to study immunosafety of nanoparticles both in vivo and in vitro/ex vivo. In particular, this model strictly adheres to the 3R principle of reducing, refining, and replacing animal experimentation (Herrmann, Pistollato, & Stephens, 2019), and our recent works addressed the question based on the responsible use of animals in nanobiomedical research, focusing on a new methodology for in vitro/ex vivo nanostudies based on *P. lividus* primary immune cell cultures (Pinsino & Alijagic, 2019). The sea urchin immune cell-based cultures preserve high viability and phagocytic capability, and in a simple way reproduce different phases of the physiological immune response. Primary sea urchin immune cell cultures function as a simplified model of a complex cellular network in an intact organism, allowing us to learn about the role of extracellular protein/peptide molecules/metabolites/other signals involved in cellular communication and potential particle functionalization (Pinsino & Alijagic, 2019). Sea urchin immune cell system does not need to be supplied by a medium with all essential nutrients (e.g., vitamins, amino acids, lipids, nucleic acid precursors, carbohydrates, trace elements), but simply exploits the nature of the coelomic fluid, in which immune cells reside and move (Pinsino & Matranga, 2015). Thus, harvesting and culture maintenance is based on the Coelomocyte Culture Medium (CCM), a high-salt medium that presents density and salinity values close to those of the sea urchin coelomic fluid, able to preserve the morphological structures of sea urchin immune cells, and block calcium-dependent cell coagulation. A gradual replacement with sterile artificial seawater ensures the maintenance of cell survival and in vitro-ex vivo physiological homeostasis over 2 weeks (Pinsino & Alijagic, 2019).

The methodology facilitates the unraveling of complex cellular phenomena and is a pivotal element in establishing the sea urchin invitrome, as already tested for the understanding of the immune compatibility of those nanoparticles potentially useful for the development in nanomedicine, including gold and titanium nanoparticles (Alijagic et al., 2020, Alijagic, Barbero, et al., 2021, Alijagic, Bonura, et al., 2021) (Fig. 5).

Sea urchin immune cells possess an extraordinary responsiveness capability, which enables us to learn about the molecular signaling pathways involved in the nano-immune interaction. Thus, a good understanding of gene function and developmental Gene Regulatory Networks (GRNs), combined with a close genetic relationship to humans, make sea urchin an attractive and emerging probing model to study the nanomaterial safety/toxicity at the cellular and molecular levels. Based on these considerations and robust experimental data, the sea urchin *Paracentrotus lividus* has been selected for inclusion in the list of alternative animal models presented by the EPAA (European Partnership for Alternative Approaches to Animal Testing). The EPAA, which involves the European Commission and European industry sectors, is committed to pooling knowledge and resources aiming to accelerate the development, validation and acceptance of alternative approaches (the 3R principle) of animal use in regulatory testing. Hence, immune studies in emerging nonmammalian models, including the sea urchin *P. lividus*, should be encouraged, helping to solve successfully the apparent problems to translate nano-results to human immunity.

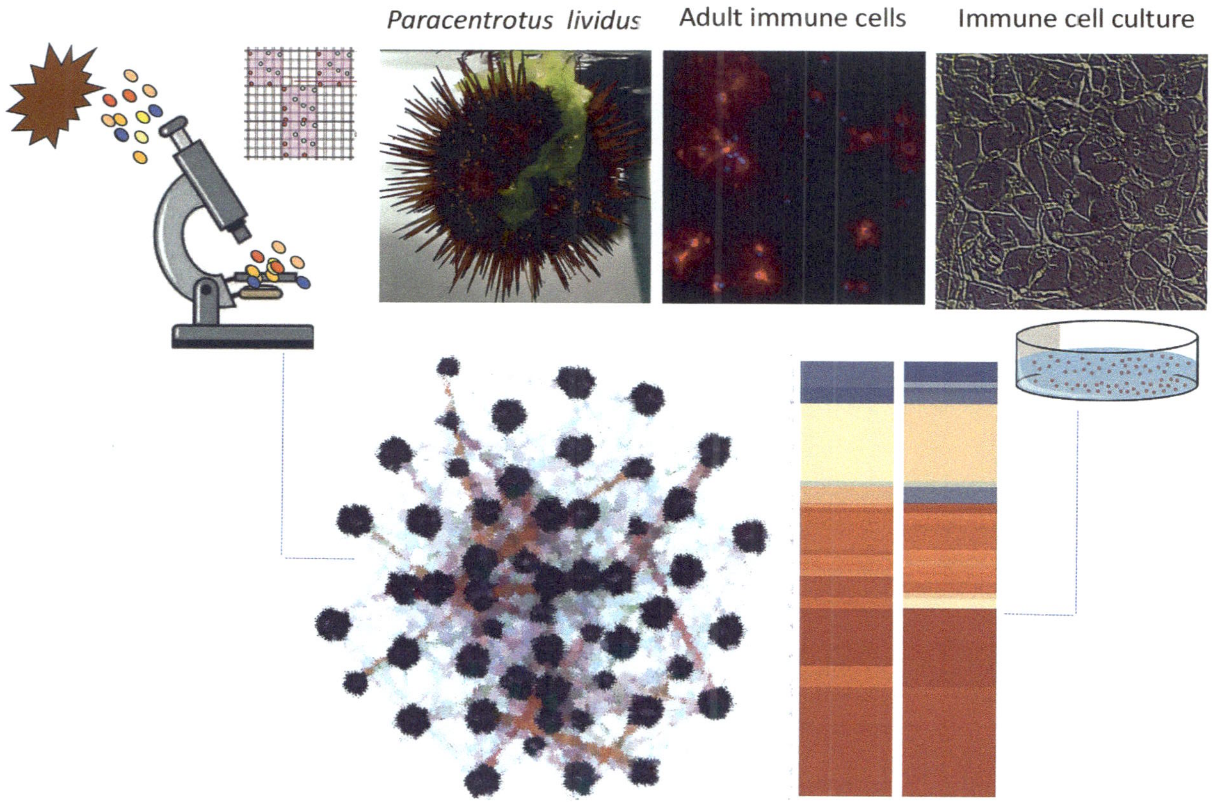

Fig. 5 *Paracentrotus lividus* immune cells in culture: a simplified model of a complex cellular network to study the immunosafety of nanoparticles.

6. Insights into the molecular targets and cell signaling pathways involved in immune-nanoparticle recognition/interaction of the sea urchin *Paracentrotus lividus*

The characterization of macromolecules bound to the surface of the nanoparticles is the first point to explore the accurate nature of nanoparticles-mediated biological effects. All cells produce extracellular molecules to control and maintain cell growth and metabolism, cell cycle progression, cell signaling, proliferation and nutrient uptake (Lloyd, 2013). *Paracentrotus lividus* blood fluid is rich in essential trophic and activating factors (bioactive molecules) produced by cells to maintain homeostasis and intercellular crosstalk, and through which cells sense the environment and communicate all biochemical changes (Alijagic et al., 2019; D'Alessio, Buckley, Kraev, Hayes, & Lange, 2021). When in contact with the blood fluid of sea urchins, nanoparticles undergo a transformation and adhere to various organic molecules that shape a complex protein corona before contacting cells and tissues (Alijagic et al., 2019, Alijagic, Barbero, et al., 2021). Several *P. lividus* extracellular proteins interact with nanoparticles and bind swiftly to the particle surface. The existence of a *P. lividus* bio-corona was detected at the nanoparticle surface upon their entering in contact with the sea urchin blood fluid in vitro, indicating that the nanoparticles interact with the components of the biological medium and influence the sea urchin immune cells behavior (Alijagic et al., 2019, Alijagic, Barbero, et al., 2021). The principal components shaping the protein corona from media conditioned by *P. lividus* immune cells are a subset of adhesion and cytoskeletal proteins, including galectin, nectin, toposome, actin, and tubulin. This is in agreement with the most recent knowledge concerning a broad spectrum of proteins that form human nanoparticle-protein complexes (e.g., immunoglobulins, lipoproteins, complement components, coagulation factors) (Lee et al., 2014).

Notably, protein corona modifies the size and surface composition of the nanoparticles, giving them a new identity, which provides a discriminating affinity for some proteins and a non-selective affinity for others. For example, toposome, and actin were detected in the corona of titanium dioxide, iron oxide and PVP-coated gold nanoparticles. Galectin-8 and tubulin are found in the corona of titanium dioxide and iron oxide nanoparticles only; Nectin uniquely in the titanium dioxide and gold nanoparticles coated with PVP (Alijagic et al., 2019, Alijagic, Barbero, et al., 2021). The interaction between PVP-coated gold nanoparticles and soluble proteins of the sea urchin blood fluid results in the formation of a protein "corona" made of only three proteins, probably because PVP, acts as a stabilizer agent and minimizes protein adsorption compared to that of an uncoated particle surface (Alijagic, Barbero, et al., 2021). This data endorses the possibility to modulate the configuration of the protein corona which is crucial for the design of ideal and safe nanoparticles, for guiding their design, synthesis, and application in nanomedicine.

Moreover, the development of sea urchin-based nanoparticles may be an opportunity to produce nanoparticles with selective biological activities (e.g., antimicrobial, anti- and pro-inflammatory, antitumoral) based on an alternative, safe, economical and eco-friendly approach (green-synthesis approach). For example, a few recent reports on sea urchin immune cells exposed to titanium dioxide and gold coated with polyvinylpyrrolidone nanoparticles underlined the capability of particles to activate a transient immune response (Pinsino et al., 2015; Alijagic et al., 2020, Alijagic, Barbero, et al., 2021). The biological regulatory mechanism in signal transduction underlying the effects of titanium dioxide nanoparticles on *P. lividus* immune cell behavior in vivo, demonstrated that phagocytes interact with these particles eliciting a receptor-mediated phagocytic mechanism involving Toll-like receptor 4 (TLR4)/ 38 mitogen-activated protein kinase (p38 MAPK) signaling pathway coupled to an anti-inflammatory response (Pinsino et al., 2015). Similarly, titanium dioxide nanoparticle activates suppressive mechanisms by down-regulating the expression of genes encoding immune-related and apoptotic proteins (e.g., NF-κB, FGFR2, JUN, p38MAPK). This prompts a metabolic rewiring by boosting the immune cell antioxidant activity (e.g., pentose phosphate, cysteine-methionine, glycine-serine pathways), and restores homeostasis by keeping at physiological levels some key immune-related proteins (e.g., nectin, TLR4, P-p38MAPK, HSP70, P-ERK) in vitro (Alijagic et al., 2020). On the other hand, PVP-coated gold nanoparticles drive inflammatory signaling by the phagocytes in vitro through a transient involvement of the TLR4/ p42/44 MAP Kinase (ERK) signaling pathway based on increased protein levels (e.g., TLR4, NF-kB, nectin, IL-6) followed by a rewiring to control inflammatory responses (Alijagic, Barbero, et al., 2021).

Overall, results highlight titanium dioxide nanoparticles as an effective immunosuppressant, and gold functionalized with polyvinylpyrrolidone nanoparticles as a pro-inflammatory substance. Thus, knowing the specific configuration of the protein corona on the particle surface, it is possible to predict downstream immune signaling that may elicit both anti- and pro-inflammation, thus making nanoparticles useful for a specific therapeutic plan. Table 1 summarizes the reported examples of nanoparticles–sea urchin immune cell interaction.

Table 1 Molecular targets and biological mechanisms activated by nanoparticles.

Nanoparticles	*P. lividus* proteins bonding particles in vitro	Immunological activity	Receptor-mediated mechanism	Metabolic activity
Titanium dioxide (TiO_2)	Toposome Actin Galectin-8 Tubulin Nectin Other not identify (Alijagic et al., 2019)	Anti-inflammatory (Pinsino et al., 2015) (Alijagic et al., 2020)	Toll-like receptor 4/38 mitogen-activated protein kinase (Pinsino et al., 2015) (Alijagic et al., 2020)	Antioxidant activity (Alijagic et al., 2020)
Iron oxide (Fe_3O_4)	Toposome Actin Galectin-8 Tubulin (Alijagic et al., 2019)	Transient pro-inflammatory and apoptotic activation (Personal communication)	Toll-like receptor 4/unknown kinase cascade (Personal communication)	–
PVP-coated gold (PVP–Au)	Toposome Actin Nectin (Alijagic, Barbero, et al., 2021)	Transient pro-inflammatory (Alijagic, Barbero, et al., 2021)	Toll-like receptor 4/p42/44 MAP Kinase (Alijagic, Barbero, et al., 2021)	Metabolic shift to an inflammatory resolution (Alijagic, Barbero, et al., 2021)

The widespread consumption of antibiotics has led to the emergence of multidrug-resistant bacterial strains. Finding alternatives to antibiotics is one of the greatest challenges facing the community of immunologists, and nanoparticles seem to be promising antimicrobial tools since main antibiotic resistance mechanisms do not concern nanoparticles, as they act directly against the bacterial surface. For example, a recent work investigating the role of PVP-coated gold nanoparticles in the sea urchin innate immune response against an acute *Vibrio* (*Vibrio anguillarum*) infection in vitro shows that these particles can stimulate a subset of rare cells (defined Group 3) that express on their surface CD45 and CD14 antigens, which are known to be involved in immune cell maturation and macrophage activation in humans (Alijagic, Bonura, et al., 2021). Data represent an important starting point for planning new research work on the topic. Notably, the sea urchin phagocytes secrete a subset of antimicrobial peptides (AMPs), including strongylocin, centrocin, paracentrin, β-thymosin, showing antibacterial activity against both Gram-positive and Gram-negative bacteria (Smith et al., 2018). Considering the heat-resistant nature, chemical and proteolytic stability, AMPs should be used for the rational design of nanoparticles (e.g., gold nanoparticles) to enhance their microbial killing activities.

7. Concluding remarks

Nanotechnology promises innovation in many strategic areas, including biomedicine, energy, agriculture, consumer products and manufacturing. Researching, developing, and utilizing engineered materials with novel sizes, shapes, and physicochemical features represents a big opportunity for new and improved uses but make the risk assessment a timeless challenge at all stages of their lifecycle. The speed of innovation obliges us to study and readily understand the relationship between the properties of the new nanoproducts and their biological effects, which must act as a guide to design safety materials and underline those that require further structural changes and evaluations. Safe(r)-by-design concept is one of the terms used to refer to reducing risks of materials, constructions and products at the design stage from accidents to the intrinsic toxicity hazards of nanoparticles. Understanding whether and how immune cells of many model organisms respond to nanoparticles represents a valid approach to assess the safety of the innovative materials throughout the innovation process. On the other hand, the nanoimmunity-by-design concept is now making its way also in biomedicine as the particle immunological properties (e.g., immunosuppression, immune activation, inflammation) may be explored for specific diagnostics, prognostics, and therapeutic use. The suitable design of new nanoparticles for biomedical applications requires a deep characterization in their immunological mechanism of action using a wide-ranging approach, including omics, both at the pristine and (bio)functionalized form. Living cells secrete extracellular molecules, including proteins and other macromolecules, that form a protein corona or biomolecular corona around the nanoparticles giving them a new biological identity. For a successful design of nanoparticles for biomedical use, it is mandatory to characterize and predict the formation of this added "natural" layer and its related effects. Medicine is continuously looking for new efficient and cost-effective treatments for diseases. Society asks for new tools for cures that can target only the illness and not harm other healthy parts of the body. Nanoparticles try to answer this question.

The sea urchin blood fluid is a rich resource of bioactive molecules including anticancer, anticoagulant/antithrombotic, antimicrobial and antioxidant molecules that has to be explored to find applications in biomedical and pharmaceutical applications (Sibiya et al., 2021). Based on our current studies, sea urchin extracellular proteins form a corona on the particle surface, eliciting a selective immunological response; the development of functionalized nanoparticles with these molecules can be viewed as a concrete opportunity for new and green therapeutic applications. Notably, sea urchin does not show a reduction in metabolic activity and fertility over time, maintains a strong antioxidant response and negligible oxidative cell damage during all life; no cases of cancer, immune and age-related diseases have been detected (Bodnar, 2009, 2013; Du, Anderson, Lortie, Parsons, & Bodnar, 2013).

References

Alijagic, A., Barbero, F., Gaglio, D., Napodano, E., Benada, O., Kofroňová, O., et al. (2021). Gold nanoparticles coated with polyvinylpyrrolidone and sea urchin extracellular molecules induce transient immune activation. *Journal of Hazardous Materials*, *402*, 123793. https://doi.org/10.1016/j.jhazmat.2020.123793.

Alijagic, A., Benada, O., Kofroňová, O., Cigna, D., & Pinsino, A. (2019). Sea urchin extracellular proteins design a complex protein corona on titanium dioxide nanoparticle surface influencing immune cell behavior. *Frontiers in Immunology*, *10*, 2261. https://doi.org/10.3389/fimmu.2019.02261.

Alijagic, A., Bonura, A., Barbero, F., Puntes, V. F., Gervasi, F., & Pinsino, A. (2021). Immunomodulatory function of polyvinylpyrrolidone (PVP)-functionalized gold nanoparticles in vibrio-stimulated sea urchin immune cells. *Nanomaterials*, *11*, 2646. https://doi.org/10.3390/nano11102646.

Alijagic, A., Gaglio, D., Napodano, E., Russo, R., Costa, C., Benada, O., et al. (2020). Titanium dioxide nanoparticles temporarily influence the sea urchin immunological state suppressing inflammatory-relate gene transcription and boosting antioxidant metabolic activity. *Journal of Hazardous Materials*, *384*, 121389. https://doi.org/10.1016/j.jhazmat.2019.121389.

Alijagic, A., & Pinsino, A. (2017). Probing safety of nanoparticles by outlining sea urchin sensing and signaling cascades. *Ecotoxicology and Environmental Safety*, *144*, 416–421. https://doi.org/10.1016/j.ecoenv.2017.06.060.

Asha, A. B., & Narain, R. (2020). Nanomaterials properties. In R. Narain (Ed.), *Polymer science and nanotechnology* (pp. 343–359). Elsevier. https://doi.org/10.1016/B978-0-12-816806-6.00015-7.

Bansal, S. A., Kumar, V., Karimi, J., Singha, A. P., & Kumar, S. (2020). Role of gold nanoparticles in advanced biomedical applications. *Nanoscale Advances*, *2*, 3764–3787. https://doi.org/10.1039/D0NA00472C.

Baranov, M. V., Kumar, M., Sacanna, S., Thutupalli, S., & Van Den Bogaart, G. (2021). Modulation of immune responses by particle size and shape. *Frontiers in Immunology*, *11*, 607945. https://doi.org/10.3389/fimmu.2020.607945.

Barbero, F., Russo, L., Vitali, M., Piella, J., Salvo, I., Borrajo, M. L., et al. (2017). Formation of the protein corona: The interface between nanoparticles and the immune system. *Seminars in Immunology*, *34*, 52–60. https://doi.org/10.1016/j.smim.2017.10.001.

Bodnar, A. G. (2009). Marine invertebrates as models for aging research. *Experimental Gerontology*, *44*, 477–484. https://doi.org/10.1016/j.exger.2009.05.001.

Bodnar, A. G. (2013). Proteomic profiles reveal age-related changes in coelomic fluid of sea urchin species with different life spans. *Experimental Gerontology*, *48*, 525–530. https://doi.org/10.1016/j.exger.2013.01.014.

Boraschi, D., Alijagic, A., Auguste, M., Barbero, F., Ferrari, E., Hernadi, S., et al. (2020). Addressing nanomaterial immunosafety by evaluating innate immunity across living species. *Small*, *2000598*. https://doi.org/10.1002/smll.202000598.

Bottero, J. Y., Rose, J., de Garidel, C., Masion, A., Deutsch, T., et al. (2017). Serenade: Safer and ecodesign research and education applied to nanomaterial development, the new generation of materials safer by design. *Environmental Science: Nano*, *4*, 526–538. https://doi.org/10.1039/C6EN00282J.

Bottjer, D. J., Davidson, E. H., Peterson, K. J., & Cameron, R. A. (2006). Paleogenomics of echinoderms. *Science*, *314*, 956–960. https://doi.org/10.1126/science.1132310.

Capua, I. (2020). *Circular health: Empowering the one health revolution*. EGEA Spa - Bocconi University Press.

D'Alessio, S., Buckley, K. M., Kraev, I., Hayes, P., & Lange, S. (2021). Extracellular vesicle signatures and post-translational protein deimination in Purple Sea urchin (Strongylocentrotus purpuratus) Coelomic fluid—Novel insights into echinodermata biology. *Biology*, *10*, 866. https://doi.org/10.3390/biology10090866.

Dennison, J. M., Zupancic, J. M., Lin, W., Dwyer, J. H., & Murphy, C. J. (2017). Protein adsorption to charged gold Nanospheres as a function of protein deformability. *Langmuir*, *33*, 7751. https://doi.org/10.1021/acs.langmuir.7b01909.

Docter, D., Westmeier, D., Markiewicz, M., Stolte, S., Knauer, S. K., & Stauber, R. H. (2015). The nanoparticle biomolecule corona: Lessons learned - challenge accepted? *Chemical Society Reviews*, *44*, 6094–6121. https://doi.org/10.3390/nano11040888.

Dominguez-Medina, S., Kisley, L., Tauzin, L. J., Hoggard, A., Shuang, B., Indrasekara, A. S., et al. (2016). *ACS Nano*, *10*, 2103. https://doi.org/10.1021/acsnano.5b06439.

Du, C., Anderson, A., Lortie, M., Parsons, R., & Bodnar, A. (2013). Oxidative damage and cellular defense mechanisms in sea urchin models of aging. *Free Radical Biology & Medicine*, *63*, 254–263.

Ernst, S. G. (2011). Offerings from an urchin. *Developmental Biology, 358*, 285–294.

Flannagan, R. S., Jaumouille, V., & Grinstein, S. (2012). The cell biology of phagocytosis. *The Annual Review of Pathology: Mechanisms of Disease*, 7, 61–98. https://doi.org/10.1146/annurev-pathol-011811-132445.

Food and Drug Administration. (2012). International Conference on Harmonisation; addendum to International Conference on Harmonisation Guidance on S6 Preclinical Safety Evaluation of Biotechnology-Derived Pharmaceuticals; availability. *Federal Register*, 77, 29665–29666. Notices.

García-Álvarez, R., & Vallet-Regí, M. (2021). Hard and soft protein Corona of nanomaterials: Analysis and relevance. *Nanomaterials (Basel), 11*, 888. https://doi.org/10.3390/nano11040888.

Guerrini, L., Alvarez-Puebla, R. A., & Pazos-Perez, N. (2018). Surface modifications of nanoparticles for stability in biological fluids. *Materials (Basel), 11*, E1154. https://doi.org/10.3390/ma11071154.

Gust, A. A., Pruitt, R., & Nürnberger, T. (2017). Sensing danger: Key to activating plant immunity. *Trends in Plant Science, 22*, 779–791. https://doi.org/10.1016/j.tplants.2017.07.005.

Hamdoun, A., & Epel, D. (2007). Embryo stability and vulnerability in an always changing world. *Proceedings of the National Academy of Sciences of the United States of America, 104*, 1745–1750. https://doi.org/10.1073/pnas.0610108104.

Haynes, C. A., & Norde, W. (1994). Globular proteins at solid/liquid interfaces. *Colloids and Surfaces B: Biointerfaces, 2*, 517–566. https://doi.org/10.1016/0927-7765(94)80066-9.

Herrmann, K., Pistollato, F., & Stephens, M. L. (2019). Beyond the 3Rs: Expanding the use of human-relevant replacement methods in biomedical research. *ALTEX-Alternatives to Animal Experimentation, 36*, 343–352. https://doi.org/10.14573/altex.1907031.

Hibino, T., Loza-Coll, M., Messier, C., Majeske, A. J., Cohen, A. H., Terwilliger, D. P., et al. (2006). The immune gene repertoire encoded in the purple sea urchin genome. *Developmental Biology, 300*, 349–365. https://doi.org/10.1016/j.ydbio.2006.08.065.

Jiao, Q., Li, L., Mu, Q., & Zhang, Q. (2014). Immunomodulation of nanoparticles in nanomedicine applications. *BioMed Research International, 2014*. https://doi.org/10.1155/2014/426028.

Lee, Y. K., Choi, E., Webster, T., Kim, S., & Khang, D. (2014). Effect of the protein corona on nanoparticles for modulating cytotoxicity and immunotoxicity. *International Journal of Nanomedicine, 10*, 97–113. https://doi.org/10.2147/IJN.S72998.

Lloyd, A. C. (2013). The regulation of cell size. *Cell, 154*, 1194–1205. https://doi.org/10.1016/j.cell.2013.08.053.

McClay, D. R. (2011). Evolutionary crossroads in developmental biology: Sea urchins. *Development, 138*, 2639–2648. https://doi.org/10.1242/dev.048967.

Mourdikoudis, S., Pallares, R. M., & Thanh, N. T. K. (2018). Characterization techniques for nanoparticles: Comparison and complementarity upon studying nanoparticle properties. *Nanoscale, 10*, 12871–12934. https://doi.org/10.1039/C8NR02278J.

Mout, R., Moyano, D. F., Rana, S., & Rotello, V. M. (2012). Surface functionalization of nanoparticles for nanomedicine. *Chemical Society Reviews, 41*, 2539–2544. https://doi.org/10.1039/c2cs15294k.

Murphy, F. A., Schinwald, A., Poland, C. A., & Donaldson, K. (2012). The mechanism of pleural inflammation by long carbon nanotubes: Interaction of long fibres with macrophages stimulates them to amplify pro-inflammatory responses in mesothelial cells. *Particle and Fibre Toxicology, 9*, 8. https://doi.org/10.1186/1743-8977-9-8.

Nienhausa, K., Wang, H., & Nienhaus, U. (2020). Nanoparticles for biomedical applications: Exploring and exploiting molecular interactions at the nano-bio interface. *Materials Today Advances, 5*, 100036. https://doi.org/10.1016/j.mtadv.2019.100036.

Noventa, S., Rowe, D., & Galloway, T. (2018). Mitigating effect of organic matter on the in vivo toxicity of metal oxide nanoparticles in the marine environment. *Environmental Science: Nano*, *5*, 1764–1777. https://doi.org/10.1039/C8EN00175H.

Oren, M., Rosental, B., Hawley, T. S., Kim, G.-Y., Agronin, J., Reynolds, C. R., et al. (2019). Individual sea urchin coelomocytes undergo somatic immune gene diversification. *Frontiers in Immunology*, *10*, 1298. https://doi.org/10.3389/fimmu.2019.01298.

Parween, S., Gupta, P. K., & Chauhan, V. S. (2011). Induction of humoral immune response against PfMSP-119 and PvMSP-119 using gold nanoparticles along with alum. *Vaccine*, *29*(2451–2460), 2011. https://doi.org/10.1016/j.vaccine.2011.01.014.

Philippe, A., & Schaumann, G. E. (2014). Interactions of dissolved organic matter with natural and engineered inorganic colloids: A review. *Environmental Science & Technology*, *48*, 8946–8962. https://doi.org/10.1021/es502342r.

Pinsino, A., & Alijagic, A. (2019). Sea urchin *Paracentrotus lividus* immune cells in culture: Formulation of the appropriate harvesting and culture media and maintenance conditions. *Biology Open*, *8*(3). https://doi.org/10.1242/bio.039289. bio039289.

Pinsino, A., Bastús, N. G., Busquets-Fité, M., Canesi, L., Cesaroni, P., Drobne, D., et al. (2020). Probing the immune responses to nanoparticles across environmental species. A perspective of the EU horizon 2020 project PANDORA. *Environmental Science: Nano*, 7, 3216–3232. https://doi.org/10.1039/D0EN00732C.

Pinsino, A., & Matranga, V. (2015). Sea urchin immune cells as sentinels of environmental stress. *Developmental & Comparative Immunology*, *49*, 198–205. https://doi.org/10.1016/j.dci.2014. 11.013.

Pinsino, A., Russo, R., Bonaventura, R., Brunelli, A., Marcomini, A., & Matranga, V. (2015). Titanium dioxide nanoparticles stimulate sea urchin immune cell phagocytic activity involving TLR/p38 MAPK-mediated signalling pathway. *Scientific Reports*, *5*, 14492. https://doi.org/10.1038/srep14492.

Rast, J. P., Smith, L. C., Loza-Coll, M., Hibino, T., & Litman, G. W. (2006). Genomic insights into the immune system of the sea urchin. *Science*, *314*, 952–956. https://doi.org/10.1126/science.1134301.

Ritz, S., Schöttler, S., Kotman, N., Baier, G., Musyanovych, A., Kuharev, J., et al. (2015). Protein corona of nanoparticles: Distinct proteins regulate the cellular up- take. *Biomacromolecules*, *16*, 1311–1321. https://doi.org/10.1021/acs.biomac.5b00108.

Sanità, G., Carrese, B., & Lamberti, A. (2020). Nanoparticle surface functionalization: How to improve biocompatibility and cellular internalization. *Frontiers in Molecular Biosciences*, 7, 587012. https://doi.org/10.3389/fmolb.2020.587012.

Schöttler, S., Becker, G., Winzen, S., Steinbach, T., Mohr, K., Landfester, K., et al. (2016). Protein adsorption is required for stealth effect of poly (ethylene glycol)-and poly (phosphoester)-coated nanocarriers. *Nature Nanotechnology*, *11*, 372377. https://doi.org/10.1038/NNANO.2015.330.

Schutz, C. A., Juillerat-Jeanneret, L., Mueller, H., Lynch, I., Riediker, M., & NanoImpactNet, C. (2013). Therapeutic nanoparticles in clinics and under clinical evaluation. *Nanomedicine (London, England)*, *8*, 449–467. https://doi.org/10.2217/nnm.13.8.

Sibiya, A., Jeyavani, J., Sivakamavalli, J., Ravi, C., Divya, M., & Vaseeharan, B. (2021). Bioactive compounds from various types of sea urchin and their therapeutic effects — A review. *Regional Studies in Marine Science*, *44*, 101760. https://doi.org/10.1016/j.rsma.2021.101760.

Slomberg, D. L., Ollivier, P., Miche, H., Angeletti, B., Bruchet, A., Philibert, M., et al. (2019). Nanoparticle stability in lake water shaped by natural organic matter properties and presence of particulate matter. *Science of the Total Environment*, *656*, 338. https://doi.org/10.1016/j.scitotenv.2018.11.279.

Smith, L. C., Arizza, V., Hudgell, M. A. B., Barone, G., Bodnar, A. G., Buckley, K. M., et al. (2018). Echinodermata: The complex immune system in echinoderms. In E. L. Cooper (Ed.), *Advances in comparative immunology* (pp. 409–501). Springer International Publishing AG.

Sodergren, E., Weinstock, G. M., Davidson, E. H., Cameron, R. A., et al. (2006). The genome of the sea urchin *Strongylocentrotus purpuratus*. *Science*, *314*, 941. https://doi.org/10.1126/science.1133609.

Tarkistani, M. A. M., Komalla, V., & Kayser, V. (2021). Recent advances in the use of iron–gold hybrid nanoparticles for biomedical applications. *Nanomaterials*, *11*, 1227. https://doi.org/10.3390/nano11051227.

Vismara, E., Bongio, C., Coletti, A., Edelman, R., Serafini, A., Mauri, M., et al. (2017). Albumin and hyaluronic acid-coated superparamagnetic Iron oxide nanoparticles loaded with paclitaxel for biomedical applications. *Molecules*, *22*, 1030. https://doi.org/10.3390/molecules22071030.

Wang, L., Li, J., Pan, J., Jiang, X., Ji, Y., Li, Y., et al. (2013). Revealing the binding structure of the protein Corona on gold Nanorods using synchrotron radiation-based techniques: Understanding the reduced damage in cell membranes. *Journal of the American Chemical Society*, *135*, 17359. https://doi.org/10.1021/ja406924v.

Yakovenko, I., Agronin, J., Smith, L. C., & Oren, M. (2021). Guardian of the genome: An alternative RAG/Transib co-evolution hypothesis for the origin of V(D)J recombination. *Frontiers in Immunology*, *12*, 709165. https://doi.org/10.3389/fimmu.2021.709165.

Yoo, J., Park, C., Yi, G., Lee, D., & Koo, H. (2019). Active targeting strategies using biological ligands for nanoparticle drug delivery systems. *Cancers*, *11*, 640. https://doi.org/10.3390/cancers11050640.

Yu, Q., Zhao, L., Guo, C., Yan, B., & Su, G. (2020). Regulating protein Corona formation and dynamic protein exchange by controlling nanoparticle hydrophobicity. *Frontiers in Bioengineering and Biotechnology*, *8*, 210. https://doi.org/10.3389/fbioe.2020.00210.

Zanganeh, S., Spitler, R., & Mahmoudi, M. E. M. (2016). Protein corona. Opportunities and challenges. *International Journal of Biochemistry & Cell Biology*, *75*, 5. https://doi.org/10.1016/j.biocel.2016.01.005.

Zhang, X. (2015). Gold nanoparticles: Recent advances in the biomedical applications. *Cell Biochemistry and Biophysics*, *72*, 771–775. https://doi.org/10.1007/s12013-015-0529-4.

Zhang, S. M., Adema, C. M., Kepler, T. B., & Loker, E. S. (2004). Diversification of Ig superfamily genes in an invertebrate. *Science*, *305*, 251–254. https://doi.org/10.1126/science.1088069.

CHAPTER ELEVEN

Mutations in *G6PC2* gene with increased risk for development of type 2 diabetes: Understanding via computational approach

S. Udhaya Kumar[a], Balu Kamaraj[b], Rinku Polachirakkal Varghese[a], V. Anu Preethi[c], R. Bithia[a], and C. George Priya Doss[a,*]

[a]Laboratory of Integrative Genomics, Department of Integrative Biology, School of Bio Sciences and Technology, Vellore Institute of Technology, Vellore, Tamil Nadu, India
[b]Department of Neuroscience Technology, College of Applied Medical Sciences, Imam Abdulrahman Bin Faisal University, Jubail, Saudi Arabia
[c]School of Computer Science and Engineering, Vellore Institute of Technology, Vellore, Tamil Nadu, India
*Corresponding author: e-mail address: georgepriyadoss@vit.ac.in

Contents

Advances in Protein Chemistry and Structural Biology, Volume 130
ISSN 1876-1623
https://doi.org/10.1016/bs.apcsb.2022.02.005

Abstract

An increase in the fast blood glucose (FBG) levels has been linked to an increased risk of developing a chronic condition, type 2 diabetes (T2D). The mutation in the *G6PC2* gene was identified to have a lead role in the modulation of FBG levels. The abnormal regulation of this enzyme influences glucose-stimulated insulin secretion (GSIS), which controls the insulin levels corresponding to the system's glucose level. This study focuses on the mutations at the *G6PC2* gene, which cause the variation from normal expression levels and increase the risk of T2D. We examined the non-synonymous single nucleotide polymorphisms (nsSNPs) present in the G6PC2 and subjected them to pathogenicity, stability, residue conservation, and membrane simulation. The individual representation of surrounding amino acids in the mutant (I63T) model showed the loss of hydrophobic interactions compared to the native G6PC2. In addition, the trajectory results from the membrane simulation exhibited reduced stability, and the least compactness was identified for the I63T mutant model.

Our study shed light on the structural and conformational changes at the transmembrane region due to the I63T mutation in G6PC2. Additionally, the Gibbs free energy landscape analysis against the two principal components showed structural differences and decreased the conformational stability of the I63T mutant model compared to the native. Like those presented in this study, dynamical simulations may indeed be crucial to comprehending the structural insights of G6PC2 mutations in cardiovascular-associated mortality and T2D.

Abbreviations

FBG fast blood glucose
FEL Gibbs free energy landscape
GWAS genome-wide association studies
HGMD human gene mutation database
MSA multiple sequence alignment
nsSNPs non-synonymous single nucleotide polymorphisms
PC principal component
RMSD root mean square deviation
RMSF root mean square fluctuation
T2D type 2 diabetes

1. Introduction

Diabetes is a prevalent chronic condition characterized by an unusual rise in high blood glucose levels caused by a malfunction in insulin action or secretion affecting more than 425 million people worldwide, with the estimates rising to 693 million by 2045 (Cho et al., 2018; Selvin & Juraschek, 2020). Diabetes is traditionally categorized into two types: Type 1 diabetes

(T1D, early-onset autoimmune diabetes) and Type 2 diabetes (T2D, late-onset, non-autoimmune diabetes). T1D is defined by the American Diabetes Association (ADA) as the destruction of autoimmune β-cell resulting in insulin deficiency, and T2D is the escalating loss of β-cell insulin secretion that often occurs in insulin resistance (American Diabetes Association, 2018). T2D accounts for more than 90% of all diabetes cases (Bruno et al., 2005; Holman, Young, & Gadsby, 2015). The key drivers of the global pandemic of T2D are obesity, lifestyle choices, genetic influence, and aging (Chatterjee, Khunti, & Davies, 2017; Olokoba, Obateru, & Olokoba, 2012). T2D has also increased in tandem with obesity making it one of the primary drivers of the global diabetes epidemic; by 2025, it is estimated that more than 300 million individuals globally will suffer from T2D as a result of obesity (Chait & den Hartigh, 2020; NCD Risk Factor Collaboration (NCD-RisC), 2016). Genetic mutations and environmental factors influence T2D; when exposed to an environment characterized by sedentary behavior and excessive intake of calories, the genetic variables get influenced (Galicia-Garcia et al., 2020). While genetic factors certainly have a role in the development of diabetes, the specific genetic variants involved in these inherited risks were unclear until the advent of current genetic technologies. In the 1980s, the advances in human genetics made it possible to uncover genetic loci that underpin this hereditary component (Rother, 2007).

Despite this, little is known about the significance of genetic variations in T2D development. This study will focus on G6PC2 and its association with susceptibility to T2D. G6PC2 encodes an enzyme of the glucose-6-phosphatase catalytic subunit family, which that is engaged in the final stage of the glycogenolytic and gluconeogenic pathways, enabling glucose to enter the bloodstream (Hutton & O'Brien, 2009). G6PC2 has four exons, mapped to human chromosome 2q31.1, and is generally expressed in the pancreatic cells (Arden et al., 1999). The *G6PC2* gene codes for a 355 amino acid length protein, a glucose negative regulator of insulin secretion stimulated by glucose-6-phosphate, assisting in hydrolyzing the glucose-6-phosphate, thus counteracting the glucose sensor's activity (Li et al., 2020; O'Brien, 2013). The deletion of the *G6PC2* gene in the pancreatic islet beta cells was reported to reduce blood glucose levels during fasting (Bosma et al., 2020). Studies conducted in the EndoC-*β* H1 human cell lines revealed that knocking down the *G6PC2* gene boosted the immune secretion at subliminal glucose levels, implying that boosting glycolytic flux at stimulatory glucose concentrations helps improve basal glucose sensitivity (Ng et al., 2019). Several Genome-wide association studies (GWAS)

(Kanai et al., 2018; O'Brien, 2013; Wheeler, Marenne, & Barroso, 2017), exome regions studies (Mahajan et al., 2015; Ng et al., 2019; Spracklen et al., 2018; Wessel et al., 2015) and Whole-genome sequencing (WGS) studies (Flannick et al., 2017; Fuchsberger et al., 2016) have revealed that multiple genetic variants (nsSNPs) of the *G6PC2* gene are related to the fasting blood glucose levels (FBG). As per the study conducted by Mahajan et al. (2015), when three genetic variations were introduced into cell lines of human embryonic kidneys and pancreatic cells of mice, the levels of G6PC2 proteins were reduced, and the loss of this function resulted in a reduction in fasting blood glucose levels (FBG) (Mahajan et al., 2015).

The evidence from germline (Pound et al., 2013; Wang et al., 2007) and β cell-selective deletion (Westermeier et al., 2019) of G6PC2 revealed decreased levels of glucose levels in mice, making G6PC2 a potential target in T2D treatment. A genetic association study conducted in three populations revealed the nsSNP rs560887 found in the *G6PC2* gene to be more closely linked to pancreatic β cell functioning and FBG but not associated with susceptibility of T2D (Bouatia-Naji et al., 2008). A French population cohort study revealed the rs560887, rs13431652, and rs2232316 causative mutations that contributed to the association signaling between the *G6PC2* gene and FBG (Baerenwald et al., 2013). A large-scale exome chip-wide analysis identified rs138726309, rs2232323, rs146779637, and rs2232326 of the *G6PC2* gene to be strongly linked FBG, increasing the susceptibility risks of T2D (Wessel et al., 2015). Moreover, various research groups have widely utilized computational techniques to understand the significance of the mutations, functional characterization, evolutionary analysis, and structural insights towards personalized medicines on G6PC2 protein, cancer, and single-gene disorders (Al-Daghri et al., 2017; Kumar et al., 2021; Mahajan et al., 2015; Overway et al., 2022). In addition, our previous study explored the protein interactions and pathway mechanisms from candidate genes that are responsible for obesity and diabetes (Udhaya Kumar et al., 2020). The present study investigates the effect of the I63T (rs201561079) mutation of the G6PC2 gene on susceptibility to T2D using various algorithms and a membrane simulation approach.

2. Materials and methods

2.1 Collection of data

The nsSNPs/missense mutations occurring in the *G6PC2* gene were collected from the dbSNP database. Further, mutations were also retrieved from other sources like UniProt, Human Gene Mutation Database (HGMD), along with the data published in previous research (Stenson et al., 2014; UniProt Consortium, 2019). The common mutations found among these databases were combined and listed as one. In FASTA format, the protein sequence data were obtained from UniProt (Accession No.: Q9NQR9).

2.2 Determination of pathogenicity

The mutations were evaluated for their pathogenicity. This was done by using the PredictSNP web-server, where it showed the consequence of the mutation either leading to a disease condition or remaining neutral (Bendl et al., 2014). We collected combined data from tools like PredictSNP, MAPP, SIFT, PhD-SNP, PolyPhen1&2, and SNAP. We obtained either diseases or neutral for each mutation site under all of these tools and noted corresponding values. The mutation sites which show disease conditions in all the tools were shortlisted for the stability prediction.

2.3 Stability of protein upon mutation

Using the iStable web-server, we computed the stability of the protein upon amino acid changes (Chen, Lin, & Chu, 2013). The FASTA sequence of G6PC2 and the mutations with positions were given as input. The iStable suite gives us the combined results from different predictors such as i-Mutant2.0, MUpro, and iStable. We obtained results from three predictors such as i-Mutant2.0, MUpro, and iStable, along with the corresponding values indicating the stability. The mutation, which shows a decrease in the protein stability in all three algorithms, was shortlisted for further process.

2.4 Biophysiochemical prediction

Here we check the change in functionality of the mutated protein for a given mutation. Align-GVGD is a web server that combines MSA (multiple sequence alignment) and biophysical predictions of amino acids where the mutation is positioned (Mathe et al., 2006). The FASTA sequence of G6PC2 was given as an input along with the shortlisted mutant positions. The results were obtained in Class 0 to 65, corresponding to enriched neutral and deleterious, respectively. While C0 shows that it is least likely to interfere with the function, C65 shows that it is most likely to interrupt function. The data was collected with the GD and GV values to calculate in the prediction process, and results were obtained in classes where the mutations belong. The mutation showed that C65 was considered for the amino acid interaction and simulation studies.

2.5 Evolutionarily conserved amino acids

The deleterious mutants with decreased stability and altered biophysical characteristics were considered for the evolutionary conservation study. Finally, evolutionary conservation is studied between homologous sequences with phylogenetic relations with the help of the ConSurf web-server, and that examines whether the amino acid positions are conserved in a protein (Glaser et al., 2003). The ConSurf result is obtained as a color-coded sequence ranging from dark green (grade1) to white (grade 5) and then to dark purple (grade 9), indicating varied, average, and conserved amino acid positions, respectively.

2.6 Homology modeling

We modeled the G6PC2 protein using the I-Tasser webserver (Roy, Kucukural, & Zhang, 2010). The structure prediction was performed with template-based fragment assembly simulations, resulting in the top 5 predictive protein models. The models were checked for integrity, C-score, normalized *Z*-score, estimated TM-score, and RMSD (Yang et al., 2015). Furthermore, the selected protein model was validated using SAVES v6.0 (https://saves.mbi.ucla.edu/) (Colovos & Yeates, 1993; Laskowski, MacArthur, Moss, & Thornton, 1993). Here, we check the overall structure quality factors, allowed and disallowed regions. The depiction of membrane topology was drawn using the PROTTER v1.0 (https://wlab.ethz.ch/protter/start/) (Omasits, Ahrens, Müller, & Wollscheid, 2014).

2.7 Mutant protein modeling and interactions

DynaMut is a web-server that can examine the influence of the mutations on the protein. The mutation was induced with the help of a SwissPDB viewer for the selected model. The native and mutant structures of G6PC2 were viewed through PyMol, and DynaMut provided the interactive amino acid surrounding regions (Rodrigues, Pires, & Ascher, 2018). The surrounding interactions of the mutant amino acid were analyzed by comparing them with the interactions present in the native model.

2.8 Molecular dynamics simulations

2.8.1 Membrane system building

The modeled native and mutant (I63T) G6PC2 protein-membrane was built individually using the CHARMM-GUI Membrane builder (Lee et al., 2016). In steps 1 and 2, the native and mutant G6PC2 model protein structures were pre-orientated to the normal membrane (z-axis) and translated to the XY center. Then, along the z-axis, the protein cross-sectional area profile was estimated. An asymmetric bilayer was created in steps 3 and 4 by replacing each pseudo atom with a di-palmitoyl-phosphatidyl-choline (DPPC) lipid molecule using the replacement approach (Woolf & Roux, 1994, 1996). The system size across the Z-axis was calculated by the addition of 22.5 Å to the protein's edges, providing a $\sim 72 \times 72 \times 133\,\text{Å}^3$ initial system size. In this study, the DPPC membranes are asymmetric with ratios of inner and outer leaflets, 81 and 80 respectively, used in both native and mutant G6PC2 protein molecules. The TIP3P water model and a 0.15 M KCl solution were also used (Durell, Brooks, & Ben-Naim, 1994; Jorgensen, Chandrasekhar, Madura, Impey, & Klein, 1983). The total number of atoms is 62,174 (16,216 water molecules, 43 K^+, and 48 Cl^-). In step 5, the individual components, bulk water, including bilayer, and ions, were formed, assembled, and generated the input for the GROMACS simulation.

2.8.2 Membrane simulation

We selected the CHARMM 36m force field (Huang et al., 2017). The minimization was performed using the van der Waals interactions cut off over 10–12 Å by force switching function (Steinbach & Brooks, 1994) and the steepest descent algorithm. Using the parameters offered by CHARMM-GUI (Lee et al., 2016), we employed GROMACS for equilibration and MD production with the LINCS algorithm (Hess, Bekker, Berendsen, & Fraaije, 1997). To maintain the temperature (323.15 K), a Nose-Hoover temperature coupling method (Hoover, 1985; Nosé, 1984) with a tau-t of 1 ps was used. In addition, a semi-isotropic Parrinello-Rahman method for pressure coupling (1 bar) (Nosé, 1984; Parrinello & Rahman, 1981) with a tau-p of 5 ps and compressibility of $4.5 \times 10\text{–}5\,\text{bar}^{-1}$ was used. NVT dynamics was first applied during the equilibration run with a 1 fs time step for 250 ps. Subsequently, the NPT ensemble was applied with a 2-fs time step (for 2 ns). Because of the system size, this is the only modification we applied from the input data produced by CHARMM-GUI for extended equilibration. Dihedral restraint and various positional potentials were executed during the equilibration, and their force constants were gradually reduced. No restraint potential was used for production. For both native and mutant G6PC2, all equilibration processes were performed. Finally, the MD production run was performed individually for both native and mutant G6PC2 proteins with a 2-fs time-step for 100 ns. The results are further analyzed by following GROMACS tools.

2.8.3 MD analysis

The GROMACS inbuilt plugins, namely gmx rms, covar, rmsf, anaeig, sham, and gyrate, were used for the trajectory analysis to identify the structural changes that occurred in the protein. The ggplot2 plugin from R software was used for plotting trajectory graphs (Wickham, 2016). UCSF chimera and Maestro from Schrödinger were used for structural analysis and visualization (Maestro, 2021; Meng, Pettersen, Couch, Huang, & Ferrin, 2006).

3. Results and discussion

3.1 Data retrieval, pathogenicity, and stability testing

For this study, we collected the nsSNPs/missense mutations present in the G6PC2 protein from HGMD, UniProt, dbSNP databases, and a literature survey. In total, we obtained 32 mutations by combining the data collected from all the databases. PredictSNP and iStable web-servers evaluated the pathogenicity and stability effect. The G6PC2 protein sequence was provided input in the web servers to obtain the results. Here, the PredictSNP server integrated with the other computational algorithms such as MAPP, SIFT, PhD-SNP, PolyPhen-1, SNAP, and PolyPhen-2. The server predicted whether the mutation exhibited neutral or deleterious from the given position and sequence of G6PC2 protein. We found that the seven mutations such as G8E (rs368382511), Q16H (CM093014), I63T (rs201561079), R79Q (rs144254880), G114R (rs149663725), Y207S (rs2232323), and R293W (rs374055555) exhibited deleterious among all integrated computational pathogenicity algorithms (Supplementary Table 1 in the online version at https://doi.org/10.1016/bs.apcsb.2022.02.005).

Further, we intended to identify how these mutations affect the G6PC2 protein's stability by utilizing an iStable web server. iStable integrated with three predictors such as MUpro, iStable, and i-Mutant2.0 results in decreasing or increasing the protein's stability. As a result, we found that I63T (rs201561079), Y207S (rs2232323), and R293W (rs374055555) have shown a decrease in stability upon mutation in G6PC2 (Supplementary Table 2 in the online version at https://doi.org/10.1016/bs.apcsb.2022.02.005). Therefore, three mutations, I63T (rs201561079), Y207S (rs2232323), and R293W (rs374055555), have been theoretically shown to impact the G6PC2 structure and stability.

3.2 Biophysicochemical and conservation analysis

Align-GVGD is a web server that combines MSA and biophysical predictions of amino acids where the mutations are positioned. The biophysicochemical prediction was performed for the seven mutations to identify the biochemical/physical features upon mutation. The results were obtained in classes that are calculated from the GD values.

According to the class, the mutations had more chance to cause a change in function (Class 65) or significantly less chance to cause any functional changes of the protein (Class 0). As a result, the mutations such as G8E, I63T, G114R, Y207S, and R293W exhibited a score of class C65, whereas the Q16H and R79Q exhibited class C15 and C35, respectively (Table 1). Therefore, these mutation have the most chance of causing the G6PC2 protein function to be altered.

Furthermore, we utilized the ConSurf web-server to find the amino acids that are conserved or varied, which eventually shows how the residue in G6PC2 protein has been the same/varied over the period. It signifies whether a particular residue is functionally effective and conserved among various species. The protein FASTA sequence was provided as an input; as a

Table 1 Prediction of biophysiochemical and evolutionary conservation of amino acids.

Accession number	Mutation	Align GV/GD prediction	ConSurf	Domain
rs368382511	G8E	Class C65	9	Topological domain
CM093014	Q16H	Class C15	9	Topological domain
rs201561079	**I63T**	**Class C65**	**5**	**Transmembrane domain**
rs144254880	R79Q	Class C35	9	Topological domain
rs149663725	G114R	Class C65	9	Topological domain
rs2232323	Y207S	Class C65	4	Topological domain
rs374055555	R293W	Class C65	5	Topological domain

Bold one present transmembrane domain.

result, we obtained the residue conservation scores along with the functionally and structurally important sites (Supplementary Fig. 1 in the online version at https://doi.org/10.1016/bs.apcsb.2022.02.005). The color-coded system shows variable and conserved residues in the whole sequence. The color dark blue shows the position is varied, and purple shows that the position is conserved. The mutations such as G8E, Q16H, R79Q, and G114R showed a conservation score of 9, whereas the I63T and R293W exhibited a score of 5, and Y207S showed a score of 4 (Table 1). Among the seven shortlisted mutations, only I63T is located in the transmembrane domain, and other mutations are located in the topological domain. The hydrophobic change to hydrophilic amino acid (I63T) in membrane protein could result in protein malfunction compared to the other soluble protein (Partridge, Therien, & Deber, 2004). For more structural insights of G6PC2 protein, we considered the I63T mutation for membrane simulation. The pictorial representation of the G6PC2 membrane protein is shown in Fig. 1A with I63T mutation as a red circle. Overall, nine transmembrane domains have been labeled mutations (yellow diamond) and post-translational modification sites (dark green square). The location of G6PC2 transmembrane regions was retrieved from the UniProt database.

3.3 Homology modeling, mutant modeling, and interactions

Since there was no structure available for G6PC2, we used I-TASSER for the homology model. The G6PC2 protein sequence was provided as an input in the I-TASSER server, and several models were predicted. Among them, we considered the model with a better C-score (confidence) and estimated TM-score. The selected G6PC2 model exhibited -1.16 of C-score and 0.57 ± 0.15 of TM-score and represented in cartoon model (Fig. 1B). The higher C-score and the TM-score > 0.5 indicate a predicted G6PC2 model with a high degree of confidence and better topology (Roy et al., 2010; Zhang & Skolnick, 2004). The selected model was checked for accuracy and quality using the SAVES v6.0 web server. As a result, the model exhibited an overall quality factor of 96.532, and the error value has been calculated. The ERRAT2 (Colovos & Yeates, 1993) program provided the error value as Y-axis with residue number as X-axis. Two lines were drawn to indicate the regions that can be rejected if the error value is exceeded (Fig. 1C). The high-resolution models generally exhibit 95% or higher values, and our selected model exhibited 96.532 overall quality factors (Fig. 1C). Then, the PROCHECK was utilized to check the selected protein model's stereochemical integrity based on the phi/psi

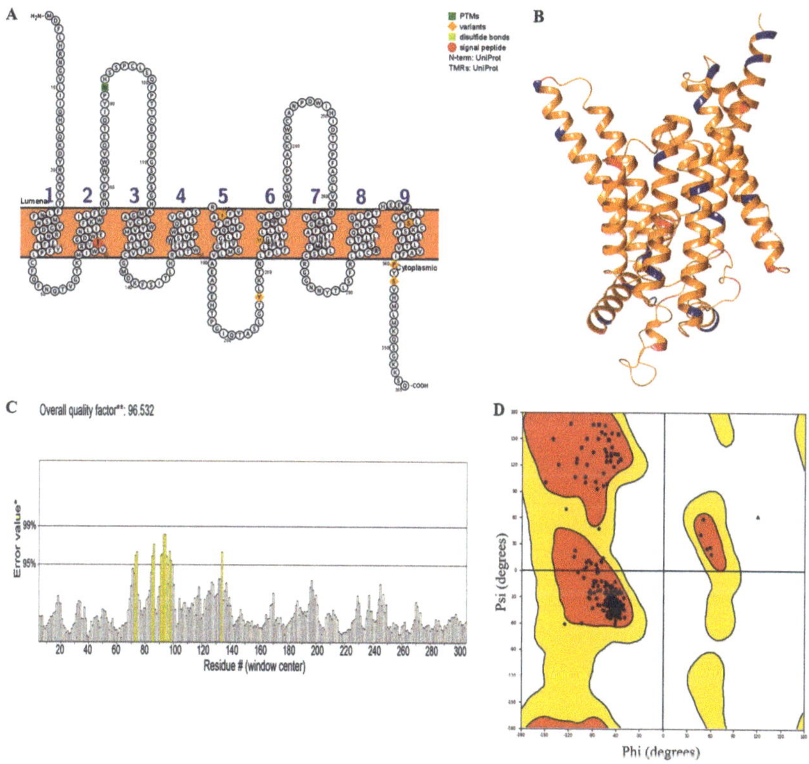

Fig. 1 (A) The topological structure of G6PC2 transmembrane protein predicted using the PROTTER v1.0. The mutation site is represented in the red circle; other mutations in diamond yellow; transmembrane domains are depicted in number (blue). (B) A cartoon-style representation of G6PC2 protein modeled using I-TASSER with a C-score of −1.16 and TM score of 0.57 ± 0.15. (C) The ERRAT value/plot demonstrates the overall quality factor of 96.532% using the SAVES server. (D) The Ramachandran plot for the predicted G6PC2 protein model using the Maestro program: 96.4% residues in favorable regions, 5.4% residues in additionally allowed regions, and 0% in disallowed regions.

angle pattern (Laskowski et al., 1993), and the Ramachandran plot was depicted to check the residues are located at the allowed, favored, and not allowed regions (Fig. 1D). Here, 96.4% of the residues were present in the most favored regions, 5.4% present in the additional allowed regions, and no residues present in the disallowed regions (0%) (Fig. 1D). In general, high-quality protein models are assumed to have more than 90% of the amino acid residues located at the most favored regions (Laskowski, Rullmannn, MacArthur, Kaptein, & Thornton, 1996). As a result, it is considered a high-quality protein model.

Furthermore, the structural effects of G6PC2 protein upon I63T mutation were studied with the help of DynaMut. Here, the protein along with the mutation was given as an input. The results from the DynaMut web-server was analyzed individually to identify the interaction of native (I63) and mutant (I63T) with its surrounding amino acids. Fig. 2 depicts native (I63) and mutant (I63T) interactions with surrounding amino acids. The native showed a hydrogen bond interaction with D65 and W59, weak hydrogen bond with W59, and hydrophobic interaction with V189, L67, L214, W66, V60, and F193. The electrostatic interaction between W66 and L214 mediated through I63 was found (Fig. 2A). In mutant I63T, the hydrogen bond was present as in native. The hydrophobic interaction such as F193 and N68 was not found in the mutant type. Many bonds remained the same except for the hydrophobic and carbonyl contact in I63T (Fig. 2B). The change of isoleucine to threonine is present at the helical and transmembrane region of the G6PC2 protein. The mutation eventually changes charges since the threonine is smaller and slightly polar than the isoleucine (Betts & Russell, 2003). In general, threonine amino acid is involved in the phosphorylation of intracellular proteins; the mutation (I63T) at this region might act as a phosphorylation site, which is not required for the G6PC2 protein (Betts & Russell, 2003). Further, the G6PC2 protein packed with the DPPC membrane with the help of CHARMM 36m force field was used as an input to perform the membrane simulation using GROMACS. The G6PC2 protein along with the membrane is depicted in Fig. 3.

3.4 Membrane simulation

The molecular dynamics simulation of native and mutant G6PC2 proteins was employed to study the structural and functional variations caused by the I63T mutation. Determining the disease's structural mechanism using experimental methods requires a lot of time and resources. In order to comprehend and classify the mutations more precisely and efficiently, computational methods are essential due to the high expense of laboratory procedures. The RMSD is a widely utilized quantitative approach to assess the conformational stability perturbations within the protein backbone during MD simulations on various nanosecond scales. The native and mutant RMSD values were graphically measured to investigate the structural stability indicated by the divergence in time scale simulation. The RMSD values for the native and I63T structures showed slight convergence

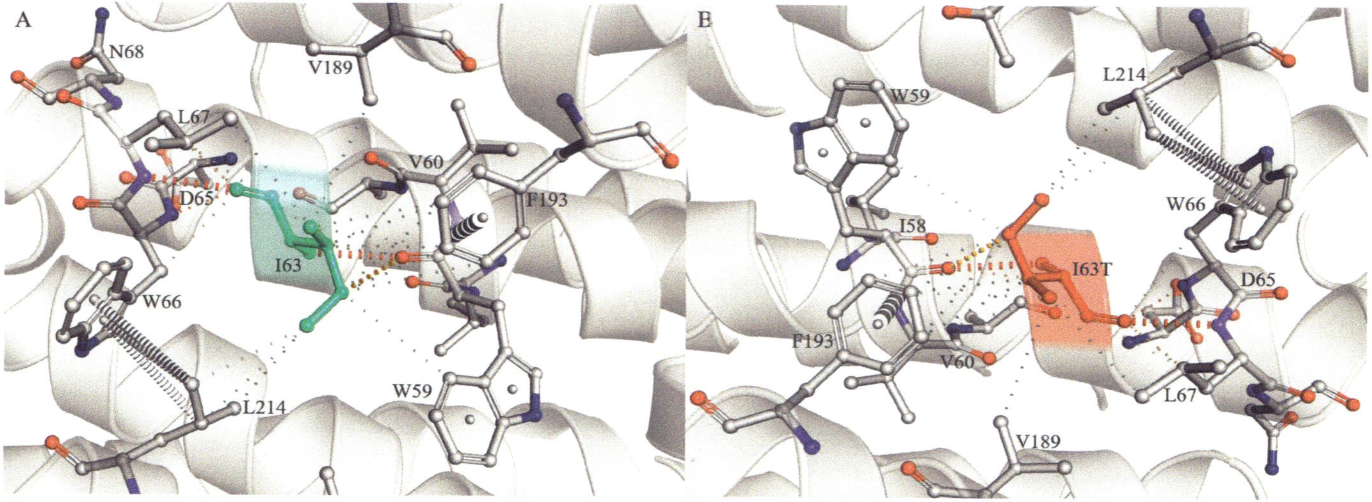

Fig. 2 The representation of native G6PC2 protein residues and mutated residues interacting at positions I58, D65, and F193. The interatomic interactions of the residues of G6PC2 and mutant (p.I63T) are depicted. (A) The native turquoise color, (B) mutant, is red, whereas the surrounding interacting residues are gray in color. The color scheme: H-bond interactions are red dotted lines, ionic interactions are represented in yellow dotted lines.

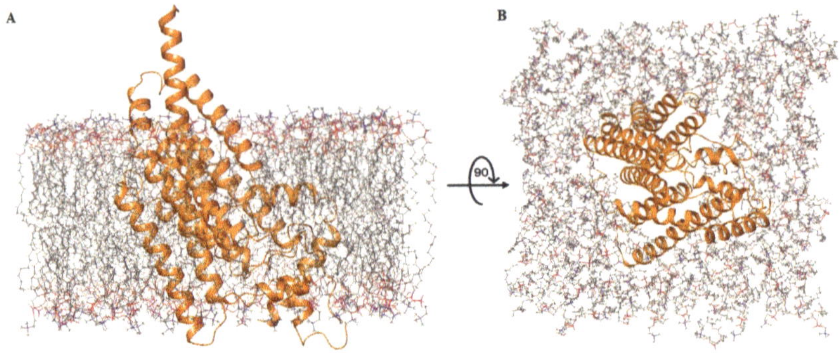

Fig. 3 The 3D depiction of G6PC2 protein with DPPC membrane visualized using Maestro software. (A) G6PC2 complex with membrane 90° orientation (side view); (B) G6PC2 complex with membrane top rotated orientation (top view).

from 53 to 81 ns with an approximate value of 0.27 nm; after 82 ns, an increase in deviations of the mutant structure was observed with an approximate value of 0.33 nm (Fig. 4A). The flexibility of a protein's structure determine its dynamics (Shoichet, Baase, Kuroki, & Matthews, 1995). To anticipate whether the mutation disrupts the dynamic behavior of residues, the RMSF was computed. The RMSF plot helps to elucidate the fluctuations per residue in a protein structure. The native and mutant structures showed mean RMSFs in 0.20–0.92 nm. A significant fluctuation was observed between the residues 230–260 (~0.30 nm) of the mutant structure compared with the native structure exhibiting a change in structural flexibility of protein due to I63T mutation (Fig. 4B). In addition, 230–260 residues are located in the lumenal region (topological domain) of the G6PC2 protein, where it might undergo intra-lumenal interaction with other proteins. Rg is a crucial parameter determining the compactness of proteins in solution (Grantcharova, Riddle, Santiago, & Baker, 1998; Lobanov, Bogatyreva, & Galzitskaya, 2008). The radius of gyration (Rg) represents the compactness of a protein structure and can be related to stability; the Rg was computed to describe the compactness of native and mutant G6PC2 structures. Higher deviations were seen in the mutant structure (~2.33 nm), depicting the least compactness than the native G6PC2 structure, which showed comparatively lesser deviations, making it more compact (Fig. 4C). The analysis of both the trajectory files is well-associated with the pathogenicity and stability computation results (Supplementary Tables 1 and 2 in the online version at https://doi.org/10.1016/bs.apcsb.2022.02.005).

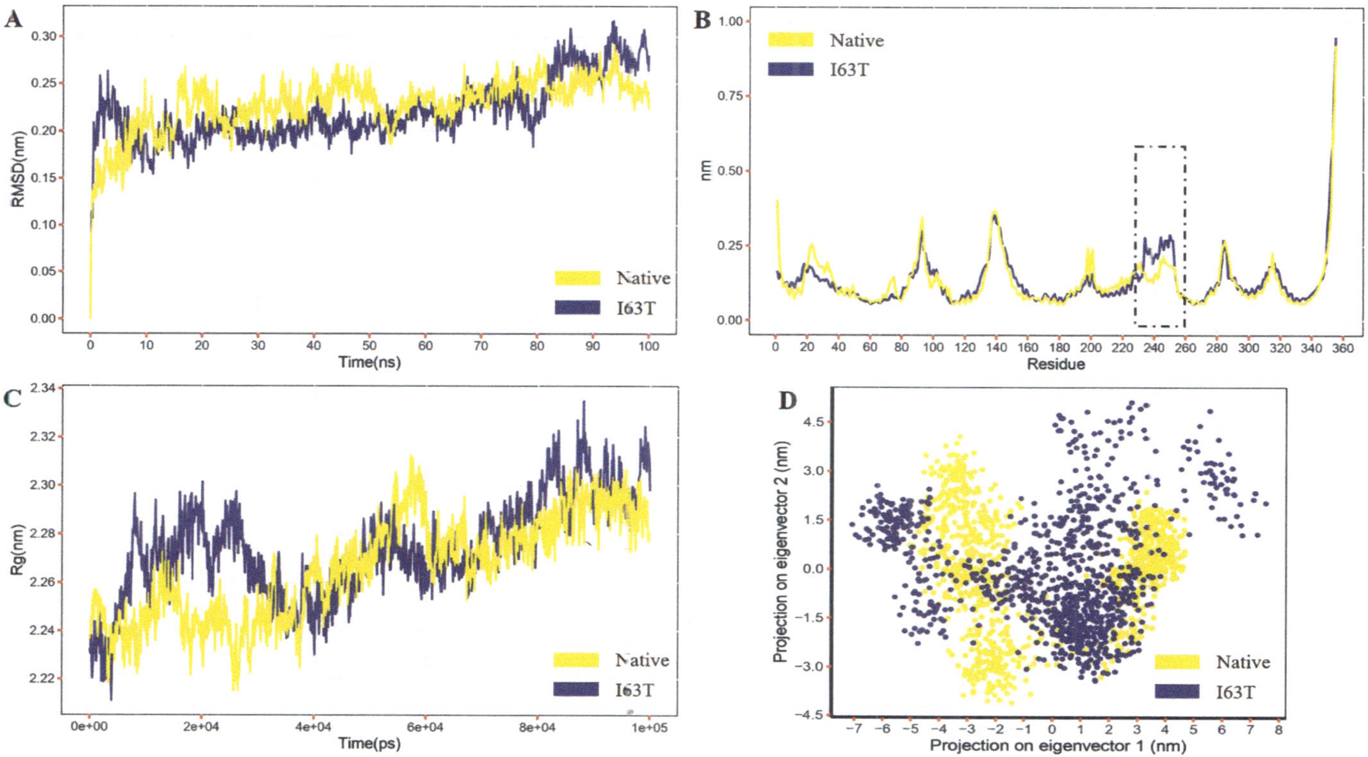

Fig. 4 (A) Graphical depiction of RMSD plot for the G6PC2 protein and mutant (I63T) at 100 ns, the x-axis indicates time (ns), and the y-axis indicates the RMSD (nm). (B) Graphical representation of the RMSF plot for the G6PC2 protein and mutant (I63T) at 100 ns. The x-axis specifies residues in number, and the y-axis indicates the RMSF values (nm). (C) Graphical depiction of Rg plot for the G6PC2 protein and mutant (I63T) at 100 ns. The x-axis shows time (ps), and the y-axis specifies the Rg values (nm). (D) PCA scatter plot and the pair of first two principal components, PC1 and PC2, depict the variations between the native and mutant systems. The color-coding schemes for the plots are as follows: Native—yellow; Mutant—blue.

Furthermore, our study examined the essential dynamics of G6PC2 protein from the simulation trajectory. Proteins, in general, perform their specific functions through a series of atomic motions. As a result, the aggregate or basic motions of a particular protein can be utilized as a metric to determine the protein's stability (Bloom, Wilke, Arnold, & Adami, 2004; Bryngelson, Onuchic, Socci, & Wolynes, 1995). PCA can be typically used to break down a protein's collective motions into a few principal motions that can be defined by eigenvectors and eigenvalues (Studer, Dessailly, & Orengo, 2013). Our previous study exhibited the protein's conformational motion changes and dynamic behavior upon mutation (Kumar et al., 2021; Udhaya Kumar et al., 2020). The native protein structure of G6PC2 showed changes in atomic motions ranging from −3.3 to 5 nm in Y-axis and X-axis from −7 to 7.8 nm.

In contrast, the mutant G6PC2 structure displayed changes in atomic motions ranging from − 4.9 to 4.3 nm in Y-axis and X-axis it ranges from −4.7 to 4.8 nm. The results show that the I63T mutant covers a wide range of phase spaces than the native structure. The significant increase in the total motion of I63T could be the primary reason for the protein functional impairment. The conformational states of native and mutant G6PC2 were computed using the Gibbs Free Energy Landscape (FEL); the first two principal components were considered as the reaction coordinates (Fig. 5). The FEL analysis against the two principal components, PC1, and PC2, revealed ΔG values ranged from 0 to 7.87, 0 to 8.17 kJ/mol for native and mutant G6PC2 structures, respectively (Fig. 5B and D). The stability of the native and mutant structures is depicted by the size and shape of the minimal energy area represented in blue color (Fig. 5A and C). The native G6PC2 showed the presence of global free energy minima in one basin.

A study by Boortz, Syring, Dai, et al. (2016) and Boortz, Syring, Pound, et al. (2016) showed that R79Q, S324P, G114R mutations alter the activity of mouse G6PC1, whereas the mutations such as G8E, S324P, H177Y, R293W, P340L, Y207S, and P313L resulted in the reduced expression of G6PC2 protein (Boortz, Syring, Dai, et al., 2016; Boortz, Syring, Pound, et al., 2016). In contrast, the mutant structure showed more wide basins showing evidence that I63T mutation leads to a decrease in the conformational stability of G6PC2. The G6PC2 expressed as a membrane-bound enzyme within the endoplasmic reticulum in the pancreatic islet beta cells function to hydrolyze the glucose-6-phosphate and counter the action of glucokinase to release the glucose back into the system (O'Brien, 2013). The mutation at G6PC2 has a lead role in disrupting the FBG levels, and

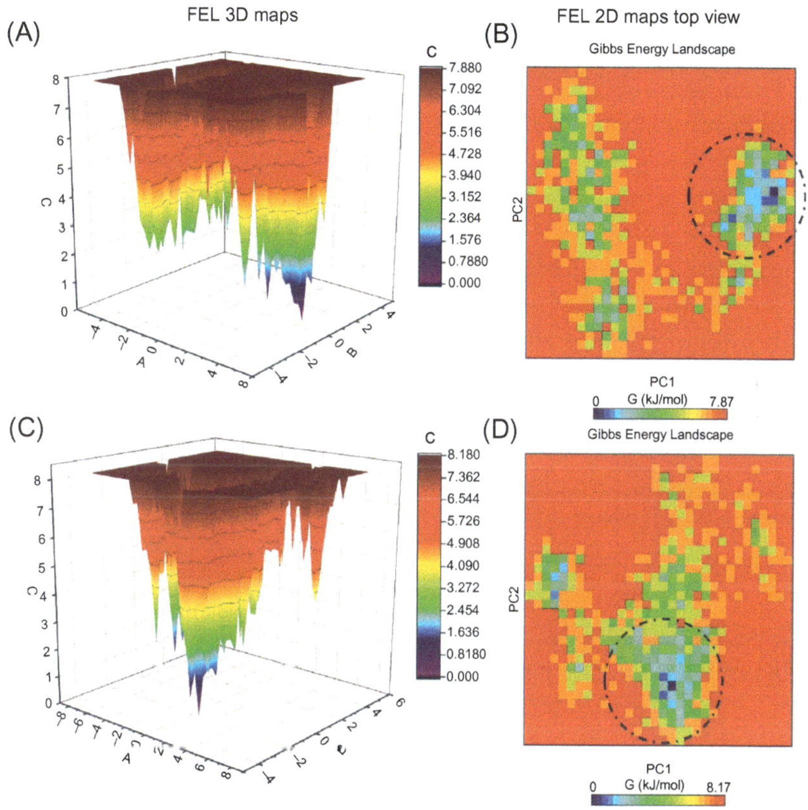

Fig. 5 Plots representing the projections of Gibbs free energy landscape (FEL) of native and mutant G6PC2 conformational space onto PC1 and PC2 for (A, B) Native type and (C, D) Mutant where dark blue color indicates the lowest energy configuration and red color specifies the highest energy conformation.

previous molecular studies resulted in elevated downstream signaling owing to the mutation that might modulate the FBG levels (Boortz, Syring, Pound, et al., 2016; Wang et al., 2007). Our study is the first to computationally analyze the functional role of I63T at the transmembrane region of G6PC2 protein. Taken together, our study exhibited that the I63T mutation disrupts the G6PC2 protein structurally and is also associated with the disturbed FBG and HbA1c levels that lead to T2D in humans.

4. Conclusion

Disease-causing mutations alter the structure of proteins, resulting in negative consequences and disrupting the molecular signaling cascades.

Our study explored the clinical mutations present in the *G6PC2* gene and have been reported to reduce the function of the G6PC2. We selected I63T mutation from our stability, pathogenicity, and simulation studies located at the transmembrane region of the G6PC2 protein. We found that the I63T was shown to be highly deleterious and disrupt the structural behavior of G6PC2 protein. Furthermore, the simulation results exhibited structural effects such as reduced stability, increased fluctuation, and disturbed surrounding amino acid interactions upon I63T mutation. Consequently, the mutations at the transmembrane region of G6PC2 protein likely to modify the FBG and might exacerbate the risk of T2D in patients. Hence, our computational pipelines could better comprehend the mutations linked to FBG and T2D and aid in the development of combinatorial diabetic drugs.

Conflict of interest

The authors have declared that no conflicts of interest exist.

Author contributions

S.U.K. and C.G.P.D. were involved in the design of the study. S.U.K., B.K., R.P.V., V.A.P., and R.B. were involved in the data collection and experimentation. S.U.K., B.K., R.P.V., V.A.P., R.B., and C.G.P.D. were involved in the acquiring, analyzing, and interpreting results. S.U.K. and B.K. drafted the manuscript. C.G.P.D. supervised the entire study and was involved in study design, the acquisition, analysis, understanding of the data, and critically reviewing the manuscript. All authors edited and approved the submitted version of the article.

Funding

No funding agency was involved in the present study.

Acknowledgments

The authors would like to take this opportunity to thank the management of Vellore Institute of Technology (VIT), Vellore, India, and Bioinformatics Lab, College of Applied Medical Sciences in Jubail, Imam Abdulrahman Bin Faisal University, Jubail, 35816, Saudi Arabia, for providing the necessary facilities and encouragement to carry out this work.

References

Al-Daghri, N. M., Pontremoli, C., Cagliani, R., Forni, D., Alokail, M. S., Al-Attas, O. S., et al. (2017). Susceptibility to type 2 diabetes may be modulated by haplotypes in G6PC2, a target of positive selection. *BMC Evolutionary Biology*, *17*(1), 43. https://doi.org/10.1186/s12862-017-0897-z.

American Diabetes Association. (2018). 2. Classification and diagnosis of diabetes: Standards of medical care in diabetes—2019. *Diabetes Care*, *42*(Suppl. 1), S13–S28. https://doi.org/10.2337/dc19-S002.

Arden, S. D., Zahn, T., Steegers, S., Webb, S., Bergman, B., O'Brien, R. M., et al. (1999). Molecular cloning of a pancreatic islet-specific glucose-6-phosphatase catalytic subunit-related protein. *Diabetes*, *48*(3), 531–542. https://doi.org/10.2337/diabetes.48.3.531.

Baerenwald, D. A., Bonnefond, A., Bouatia-Naji, N., Flemming, B. P., Umunakwe, O. C., Oeser, J. K., et al. (2013). Multiple functional polymorphisms in the G6PC2 gene contribute to the association with higher fasting plasma glucose levels. *Diabetologia*, *56*(6), 1306–1316. https://doi.org/10.1007/s00125-013-2875-3.

Bendl, J., Stourac, J., Salanda, O., Pavelka, A., Wieben, E. D., Zendulka, J., et al. (2014). PredictSNP: Robust and accurate consensus classifier for prediction of disease-related mutations. *PLoS Computational Biology*, *10*(1), e1003440. https://doi.org/10.1371/journal.pcbi.1003440.

Betts, M. J., & Russell, R. B. (2003). Amino acid properties and consequences of substitutions. In *Bioinformatics for geneticists* (pp. 289–316). John Wiley & Sons, Ltd. https://doi.org/10.1002/0470867302.ch14.

Bloom, J. D., Wilke, C. O., Arnold, F. H., & Adami, C. (2004). Stability and the evolvability of function in a model protein. *Biophysical Journal*, *86*(5), 2758–2764. https://doi.org/10.1016/S0006-3495(04)74329-5.

Boortz, K. A., Syring, K. E., Dai, C., Pound, L. D., Oeser, J. K., Jacobson, D. A., et al. (2016). G6PC2 modulates fasting blood glucose in male mice in response to stress. *Endocrinology*, *157*(8), 3002–3008. https://doi.org/10.1210/en.2016-1245.

Boortz, K. A., Syring, K. E., Pound, L. D., Wang, Y., Oeser, J. K., & O'Brien, R. M. (2016). Functional analysis of mouse G6pc1 mutations using a novel in situ assay for glucose-6-phosphatase activity and the effect of mutations in conserved human G6PC1/G6PC2 amino acids on G6PC2 protein expression. *PLoS One*, *11*(9), e0162439. https://doi.org/10.1371/journal.pone.0162439.

Bosma, K. J., Rahim, M., Singh, K., Goleva, S. B., Wall, M. L., Xia, J., et al. (2020). Pancreatic islet beta cell-specific deletion of G6pc2 reduces fasting blood glucose. *Journal of Molecular Endocrinology*, *64*(4), 235–248. https://doi.org/10.1530/JME-20-0031.

Bouatia-Naji, N., Rocheleau, G., Van Lommel, L., Lemaire, K., Schuit, F., Cavalcanti-Proença, C., et al. (2008). A polymorphism within the G6PC2 gene is associated with fasting plasma glucose levels. *Science (New York, N.Y.)*, *320*(5879), 1085–1088. https://doi.org/10.1126/science.1156849.

Bruno, G., Runzo, C., Cavallo-Perin, P., Merletti, F., Rivetti, M., Pinach, S., et al. (2005). Incidence of type 1 and type 2 diabetes in adults aged 30-49 years: The population-based registry in the province of Turin, Italy. *Diabetes Care*, *28*(11), 2613–2619. https://doi.org/10.2337/diacare.28.11.2613.

Bryngelson, J. D., Onuchic, J. N., Socci, N. D., & Wolynes, P. G. (1995). Funnels, pathways, and the energy landscape of protein folding: A synthesis. *Proteins*, *21*(3), 167–195. https://doi.org/10.1002/prot.340210302.

Chait, A., & den Hartigh, L. J. (2020). Adipose tissue distribution, inflammation and its metabolic consequences, including diabetes and cardiovascular disease. *Frontiers in Cardiovascular Medicine*, 7, 22. https://doi.org/10.3389/fcvm.2020.00022.

Chatterjee, S., Khunti, K., & Davies, M. J. (2017). Type 2 diabetes. *Lancet (London, England)*, *389*(10085), 2239–2251. https://doi.org/10.1016/S0140-6736(17)30058-2.

Chen, C.-W., Lin, J., & Chu, Y.-W. (2013). iStable: Off-the-shelf predictor integration for predicting protein stability changes. *BMC Bioinformatics*, *14*(2), S5. https://doi.org/10.1186/1471-2105-14-S2-S5.

Cho, N. H., Shaw, J. E., Karuranga, S., Huang, Y., da Rocha Fernandes, J. D., Ohlrogge, A. W., et al. (2018). IDF diabetes atlas: Global estimates of diabetes prevalence for 2017 and projections for 2045. *Diabetes Research and Clinical Practice*, *138*, 271–281. https://doi.org/10.1016/j.diabres.2018.02.023.

Colovos, C., & Yeates, T. O. (1993). Verification of protein structures: Patterns of non-bonded atomic interactions. *Protein Science: A Publication of the Protein Society*, *2*(9), 1511–1519. https://doi.org/10.1002/pro.5560020916.

Durell, S. R., Brooks, B. R., & Ben-Naim, A. (1994). Solvent-induced forces between two hydrophilic groups. *The Journal of Physical Chemistry*, *98*(8), 2198–2202. https://doi.org/10.1021/j100059a038.

Flannick, J., Fuchsberger, C., Mahajan, A., Teslovich, T. M., Agarwala, V., Gaulton, K. J., et al. (2017). Sequence data and association statistics from 12,940 type 2 diabetes cases and controls. *Scientific Data*, *4*, 170179. https://doi.org/10.1038/sdata.2017.179.

Fuchsberger, C., Flannick, J., Teslovich, T. M., Mahajan, A., Agarwala, V., Gaulton, K. J., et al. (2016). The genetic architecture of type 2 diabetes. *Nature*, *536*(7614), 41–47. https://doi.org/10.1038/nature18642.

Galicia-Garcia, U., Benito-Vicente, A., Jebari, S., Larrea-Sebal, A., Siddiqi, H., Uribe, K. B., et al. (2020). Pathophysiology of type 2 diabetes mellitus. *International Journal of Molecular Sciences*, *21*(17), E6275. https://doi.org/10.3390/ijms21176275.

Glaser, F., Pupko, T., Paz, I., Bell, R. E., Bechor-Shental, D., Martz, E., et al. (2003). ConSurf: Identification of functional regions in proteins by surface-mapping of phylogenetic information. *Bioinformatics*, *19*(1), 163–164. https://doi.org/10.1093/bioinformatics/19.1.163.

Grantcharova, V. P., Riddle, D. S., Santiago, J. V., & Baker, D. (1998). Important role of hydrogen bonds in the structurally polarized transition state for folding of the src SH3 domain. *Nature Structural Biology*, *5*(8), 714–720. https://doi.org/10.1038/1412.

Hess, B., Bekker, H., Berendsen, H. J. C., & Fraaije, J. G. E. M. (1997). LINCS: A linear constraint solver for molecular simulations. *Journal of Computational Chemistry*, *18*(12), 1463–1472. https://doi.org/10.1002/(SICI)1096-987X(199709)18:12<1463::AID-JCC4>3.0.CO;2-H.

Holman, N., Young, B., & Gadsby, R. (2015). Current prevalence of type 1 and type 2 diabetes in adults and children in the UK. *Diabetic Medicine: A Journal of the British Diabetic Association*, *32*(9), 1119–1120. https://doi.org/10.1111/dme.12791.

Hoover, W. G. (1985). Canonical dynamics: Equilibrium phase-space distributions. *Physical Review. A, General Physics*, *31*(3), 1695–1697. https://doi.org/10.1103/physreva.31.1695.

Huang, J., Rauscher, S., Nawrocki, G., Ran, T., Feig, M., de Groot, B. L., et al. (2017). CHARMM36m: An improved force field for folded and intrinsically disordered proteins. *Nature Methods*, *14*(1), 71–73. https://doi.org/10.1038/nmeth.4067.

Hutton, J. C., & O'Brien, R. M. (2009). Glucose-6-phosphatase catalytic subunit gene family. *The Journal of Biological Chemistry*, *284*(43), 29241–29245. https://doi.org/10.1074/jbc.R109.025544.

Jorgensen, W. L., Chandrasekhar, J., Madura, J. D., Impey, R. W., & Klein, M. L. (1983). Comparison of simple potential functions for simulating liquid water. *The Journal of Chemical Physics*, *79*(2), 926–935. https://doi.org/10.1063/1.445869.

Kanai, M., Akiyama, M., Takahashi, A., Matoba, N., Momozawa, Y., Ikeda, M., et al. (2018). Genetic analysis of quantitative traits in the Japanese population links cell types to complex human diseases. *Nature Genetics*, *50*(3), 390–400. https://doi.org/10.1038/s41588-018-0047-6.

Kumar, S. U., Rajan, B., Kumar, D. T., Cathryn, R. H., Das, S., Zayed, H., et al. (2021). Comparison of potential inhibitors and targeting fat mass and obesity-associated protein causing diabesity through docking and molecular dynamics strategies. *Journal of Cellular Biochemistry*. https://doi.org/10.1002/jcb.30109.

Kumar, S. U., Sankar, S., Kumar, D. T., Younes, S., Younes, N., Siva, R., et al. (2021). Molecular dynamics, residue network analysis, and cross-correlation matrix to characterize the deleterious missense mutations in GALE causing galactosemia III. *Cell Biochemistry and Biophysics*. https://doi.org/10.1007/s12013-020-00960-z.

Laskowski, R. A., MacArthur, M. W., Moss, D. S., & Thornton, J. M. (1993). PROCHECK: A program to check the stereochemical quality of protein structures. *Journal of Applied Crystallography*, *26*(2), 283–291. https://doi.org/10.1107/S0021889892009944.

Laskowski, R. A., Rullmannn, J. A., MacArthur, M. W., Kaptein, R., & Thornton, J. M. (1996). AQUA and PROCHECK-NMR: Programs for checking the quality of protein structures solved by NMR. *Journal of Biomolecular NMR*, *8*(4), 477–486. https://doi.org/10.1007/BF00228148.

Lee, J., Cheng, X., Swails, J. M., Yeom, M. S., Eastman, P. K., Lemkul, J. A., et al. (2016). CHARMM-GUI input generator for NAMD, GROMACS, AMBER, OpenMM, and CHARMM/OpenMM simulations using the CHARMM36 additive force field. *Journal of Chemical Theory and Computation*, *12*(1), 405–413. https://doi.org/10.1021/acs.jctc.5b00935.

Li, C., Yang, Y., Liu, X., Li, Z., Liu, H., & Tan, Q. (2020). Glucose metabolism-related gene polymorphisms as the risk predictors of type 2 diabetes. *Diabetology & Metabolic Syndrome*, *12*(1), 97. https://doi.org/10.1186/s13098-020-00604-5.

Lobanov, M. Y., Bogatyreva, N. S., & Galzitskaya, O. V. (2008). Radius of gyration as an indicator of protein structure compactness. *Molecular Biology*, *42*(4), 623–628. https://doi.org/10.1134/S0026893308040195.

Maestro. (2021). *Maestro (Version 2021-4)*. Maestro, Schrödinger (English; LLC, New York, NY).

Mahajan, A., Sim, X., Ng, H. J., Manning, A., Rivas, M. A., Highland, H. M., et al. (2015). Identification and functional characterization of G6PC2 coding variants influencing glycemic traits define an effector transcript at the G6PC2-ABCB11 locus. *PLoS Genetics*, *11*(1), e1004876. https://doi.org/10.1371/journal.pgen.1004876.

Mathe, E., Olivier, M., Kato, S., Ishioka, C., Hainaut, P., & Tavtigian, S. V. (2006). Computational approaches for predicting the biological effect of p53 missense mutations: A comparison of three sequence analysis based methods. *Nucleic Acids Research*, *34*(5), 1317–1325. https://doi.org/10.1093/nar/gkj518.

Meng, E. C., Pettersen, E. F., Couch, G. S., Huang, C. C., & Ferrin, T. E. (2006). Tools for integrated sequence-structure analysis with UCSF Chimera. *BMC Bioinformatics*, 7(1), 339. https://doi.org/10.1186/1471-2105-7-339.

NCD Risk Factor Collaboration (NCD-RisC). (2016). Worldwide trends in diabetes since 1980: A pooled analysis of 751 population-based studies with 4.4 million participants. *Lancet (London, England)*, *387*(10027), 1513–1530. https://doi.org/10.1016/S0140-6736(16)00618-8.

Ng, N. H. J., Willems, S. M., Fernandez, J., Fine, R. S., Wheeler, E., Wessel, J., et al. (2019). Tissue-specific alteration of metabolic pathways influences glycemic regulation. *bioRxiv*, 790618. https://doi.org/10.1101/790618.

Nosé, S. (1984). A molecular dynamics method for simulations in the canonical ensemble. *Molecular Physics*, *52*(2), 255–268. https://doi.org/10.1080/00268978400101201.

O'Brien, R. M. (2013). Moving on from GWAS: Functional studies on the G6PC2 gene implicated in the regulation of fasting blood glucose. *Current Diabetes Reports*, *13*(6), 768–777. https://doi.org/10.1007/s11892-013-0422-8.

Olokoba, A. B., Obateru, O. A., & Olokoba, L. B. (2012). Type 2 diabetes mellitus: A review of current trends. *Oman Medical Journal*, *27*(4), 269–273. https://doi.org/10.5001/omj.2012.68.

Omasits, U., Ahrens, C. H., Müller, S., & Wollscheid, B. (2014). Protter: Interactive protein feature visualization and integration with experimental proteomic data. *Bioinformatics (Oxford, England)*, *30*(6), 884–886. https://doi.org/10.1093/bioinformatics/btt607.

Overway, E. M., Bosma, K. J., Claxton, D. P., Oeser, J. K., Singh, K., Breidenbach, L. B., et al. (2022). Nonsynonymous single-nucleotide polymorphisms in the G6PC2 gene affect protein expression, enzyme activity, and fasting blood glucose. *Journal of Biological Chemistry*, *298*(2), 101534. https://doi.org/10.1016/j.jbc.2021.101534.

Parrinello, M., & Rahman, A. (1981). Polymorphic transitions in single crystals: A new molecular dynamics method. *Journal of Applied Physics*, *52*(12), 7182–7190. https://doi.org/10.1063/1.328693.

Partridge, A. W., Therien, A. G., & Deber, C. M. (2004). Missense mutations in transmembrane domains of proteins: Phenotypic propensity of polar residues for human disease. *Proteins*, *54*, 648–656. https://doi.org/10.1002/prot.10611.

Pound, L. D., Oeser, J. K., O'Brien, T. P., Wang, Y., Faulman, C. J., Dadi, P. K., et al. (2013). G6PC2: A negative regulator of basal glucose-stimulated insulin secretion. *Diabetes*, *62*(5), 1547–1556. https://doi.org/10.2337/db12-1067.

Rodrigues, C. H., Pires, D. E., & Ascher, D. B. (2018). DynaMut: Predicting the impact of mutations on protein conformation, flexibility and stability. *Nucleic Acids Research*, *46*(W1), W350–W355. https://doi.org/10.1093/nar/gky300.

Rother, K. I. (2007). Diabetes treatment—Bridging the divide. *The New England Journal of Medicine*, *356*(15), 1499–1501. https://doi.org/10.1056/NEJMp078030.

Roy, A., Kucukural, A., & Zhang, Y. (2010). I-TASSER: A unified platform for automated protein structure and function prediction. *Nature Protocols*, *5*(4), 725–738. https://doi.org/10.1038/nprot.2010.5.

Selvin, E., & Juraschek, S. P. (2020). Diabetes epidemiology in the COVID-19 pandemic. *Diabetes Care*, *43*(8), 1690–1694. https://doi.org/10.2337/dc20-1295.

Shoichet, B. K., Baase, W. A., Kuroki, R., & Matthews, B. W. (1995). A relationship between protein stability and protein function. *Proceedings of the National Academy of Sciences of the United States of America*, *92*(2), 452–456. https://doi.org/10.1073/pnas.92.2.452.

Spracklen, C. N., Shi, J., Vadlamudi, S., Wu, Y., Zou, M., Raulerson, C. K., et al. (2018). Identification and functional analysis of glycemic trait loci in the China Health and Nutrition Survey. *PLoS Genetics*, *14*(4), e1007275. https://doi.org/10.1371/journal.pgen.1007275.

Steinbach, P. J., & Brooks, B. R. (1994). New spherical-cutoff methods for long-range forces in macromolecular simulation. *Journal of Computational Chemistry*, *15*(7), 667–683. https://doi.org/10.1002/jcc.540150702.

Stenson, P. D., Mort, M., Ball, E. V., Shaw, K., Phillips, A. D., & Cooper, D. N. (2014). The human gene mutation database: Building a comprehensive mutation repository for clinical and molecular genetics, diagnostic testing and personalized genomic medicine. *Human Genetics*, *133*(1), 1–9. https://doi.org/10.1007/s00439-013-1358-4.

Studer, R. A., Dessailly, B. H., & Orengo, C. A. (2013). Residue mutations and their impact on protein structure and function: Detecting beneficial and pathogenic changes. *Biochemical Journal*, *449*(3), 581–594. https://doi.org/10.1042/BJ20121221.

Udhaya Kumar, S., Rajan, B., Thirumal Kumar, D., Anu Preethi, V., Abunada, T., Younes, S., et al. (2020). Involvement of essential signaling cascades and analysis of gene networks in diabesity. *Genes*, *11*(11), 1256. https://doi.org/10.3390/genes11111256.

Udhaya Kumar, S., Thirumal Kumar, D., Mandal, P. D., Sankar, S., Haldar, R., Kamaraj, B., et al. (2020). Chapter Eight—Comprehensive in silico screening and molecular dynamics studies of missense mutations in Sjogren-Larsson syndrome associated with the ALDH3A2 gene. In R. Donev (Ed.), *Vol. 120. Advances in protein chemistry and structural biology* (pp. 349–377). Academic Press. https://doi.org/10.1016/bs.apcsb.2019.11.004.

UniProt Consortium. (2019). UniProt: A worldwide hub of protein knowledge. *Nucleic Acids Research*, *47*(D1), D506–D515. https://doi.org/10.1093/nar/gky1049.

Wang, Y., Martin, C. C., Oeser, J. K., Sarkar, S., McGuinness, O. P., Hutton, J. C., et al. (2007). Deletion of the gene encoding the islet-specific glucose-6-phosphatase catalytic subunit-related protein autoantigen results in a mild metabolic phenotype. *Diabetologia*, *50*(4), 774–778. https://doi.org/10.1007/s00125-006-0564-1.

Wessel, J., Chu, A. Y., Willems, S. M., Wang, S., Yaghootkar, H., Brody, J. A., et al. (2015). Low-frequency and rare exome chip variants associate with fasting glucose and type 2 diabetes susceptibility. *Nature Communications*, *6*(1), 5897. https://doi.org/10.1038/ncomms6897.

Westermeier, F., Holyoak, T., Asenjo, J. L., Gatica, R., Nualart, F., Burbulis, I., et al. (2019). Gluconeogenic enzymes in β-cells: Pharmacological targets for improving insulin secretion. *Trends in Endocrinology and Metabolism: TEM*, *30*(8), 520–531. https://doi.org/10.1016/j.tem.2019.05.004.

Wheeler, E., Marenne, G., & Barroso, I. (2017). Genetic aetiology of glycaemic traits: Approaches and insights. *Human Molecular Genetics*, *26*(R2), R172–R184. https://doi.org/10.1093/hmg/ddx293.

Wickham, H. (2016). *ggplot2: Elegant graphics for data analysis* (2nd ed.). Springer International Publishing. https://doi.org/10.1007/978-3-319-24277-4.

Woolf, T. B., & Roux, B. (1994). Molecular dynamics simulation of the gramicidin channel in a phospholipid bilayer. *Proceedings of the National Academy of Sciences of the United States of America*, *91*(24), 11631–11635. https://doi.org/10.1073/pnas.91.24.11631.

Woolf, T. B., & Roux, B. (1996). Structure, energetics, and dynamics of lipid-protein interactions: A molecular dynamics study of the gramicidin A channel in a DMPC bilayer. *Proteins*, *24*(1), 92–114. https://doi.org/10.1002/(SICI)1097-0134(199601)24:1<92::AID-PROT7>3.0.CO;2-Q.

Yang, J., Yan, R., Roy, A., Xu, D., Poisson, J., & Zhang, Y. (2015). The I-TASSER Suite: Protein structure and function prediction. *Nature Methods*, *12*(1), 7–8. https://doi.org/10.1038/nmeth.3213.

Zhang, Y., & Skolnick, J. (2004). Scoring function for automated assessment of protein structure template quality. *Proteins*, *57*(4), 702–710. https://doi.org/10.1002/prot.20264.

CHAPTER TWELVE

Computational structural assessment of BReast CAncer type 1 susceptibility protein (BRCA1) and BRCA1-Associated Ring Domain protein 1 (BARD1) mutations on the protein-protein interface

D. Thirumal Kumar[a,d], S. Udhaya Kumar[a], Nikita Jain[a], Baviri Sowmya[a], Kamakshi Balsekar[a], R. Siva[a], Balu Kamaraj[b], Mariem Sidenna[c], C. George Priya Doss[a], and Hatem Zayed[c,*]

[a]Laboratory of Integrative Genomics, Department of Integrative Biology, School of Bio Sciences and Technology, Vellore Institute of Technology, Vellore, Tamil Nadu, India
[b]Department of Neuroscience Technology, College of Applied Medical Sciences, Imam Abdulrahman Bin Faisal University, Jubail, Saudi Arabia
[c]Department of Biomedical Sciences, College of Health and Sciences, QU Health, Qatar University, Doha, Qatar
[d]Meenakshi Academy of Higher Education and Research (Deemed to be University), Chennai, Tamil Nadu, India
*Correspondence author: e-mail address: hatem.zayed@qu.edu.qa

Contents

Advances in Protein Chemistry and Structural Biology, Volume 130
ISSN 1876-1623
https://doi.org/10.1016/bs.apcsb.2022.02.003

Abstract

Breast cancer type 1 susceptibility protein (BRCA1) is closely related to the BRCA2 (breast cancer type 2 susceptibility protein) and BARD1 (BRCA1-associated RING domain-1) proteins. The homodimers were formed through their RING fingers; however they form more compact heterodimers preferentially, influencing BRCA1 residues 1–109 and BARD1 residues 26–119. We implemented an integrative computational pipeline to screen all the mutations in BRCA1 and identify the most significant mutations influencing the Protein-Protein Interactions (PPI) in the BRCA1-BARD1 protein complex. The amino acids involved in the PPI regions were identified from the PDBsum database with the PDB ID: 1JM7. We screened 2118 missense mutations in BRCA1 and none in BARD1 for pathogenicity and stability and analyzed the amino acid sequences for conserved residues. We identified the most significant mutations from these screenings as V11G, M18K, L22S, and T97R positioned in the PPI regions of the BRCA1-BARD1 protein complex. We further performed protein-protein docking using the ZDOCK server. The native protein-protein complex showed the highest binding score of 2118.613, and the V11G mutant protein complex showed the least binding score of 1992.949. The other three mutation protein complexes had binding scores between the native and V11G protein complexes. Finally, a molecular dynamics simulation study using GROMACS was performed to comprehend changes in the BRCA1-BARD1 complex's binding pattern due to the mutation. From the analysis, we observed the highest deviation with lowest compactness and a decrease in the intramolecular h-bonds in the BRCA1-BARD1 protein complex with the V11G mutation compared to the native complex or the complexes with other mutations.

Abbreviations

BRCA1 breast cancer type 1 susceptibility protein
BARD1 BRCA1 associated RING domain 1
PPI protein-protein interaction

1. Introduction

Our cells contain an innate ability, in the form of tumor suppressor genes, to control unwarranted cell divisions, repair errors in DNA, and instruct damaged cells to perish (Henderson, 2012). When the function of tumor suppressor genes goes awry, cells grow uncontrollably, triggering cancer. BRCA1, which codes for breast cancer type 1 susceptibility protein, is one of the most essential human tumor suppressor genes and is implicated in numerous cases of breast cancer and rumored to be involved in other cancer types, including ovarian and prostate cancers (Duncan, Reeves, & Cooke, 1998; Kotsopoulos, 2018; O'Donovan & Livingston, 2010; Weigelt et al., 2008). BRCA1 is involved in fixing chromosomal damage by playing a prominent role in the DNA double-strand break repair. The DNA double-strand break repair ensures an error-free genome by either correcting the strands or destroying the cells beyond repair (Friedenson, 2007). BRCA1 is at the hub of a large multi-unit protein complex called the BASC (BRCA1-associated genome surveillance complex). BRCA1's interacting partners include DNA damage repair proteins and tumor suppressors such as ATM, MSH6, MSH2, BLM, and MLH1 and the RAD50-MRE11-NBS1 protein complex (Wang et al., 2000). Thus, BRCA1 is essential in sustaining genome integrity and keeping cell division from becoming aberrant.

The stability of the BRCA1 protein depends upon its interaction with the BARD1 protein (BRCA1-associated RING domain 1). BARD1 has sequence homology with two of the most conserved domains in BRCA1: the C-terminal BRCT domain and the N-terminal RING motif, which contains a cysteine-rich sequence found in various tumors suppressor genes and proteins that govern cell growth (Fox et al., 2008). The affiliation of BARD1 with BRCA1 alters its conformation and facilitates the efficient binding of BRCA1 with BASC for rapid damage control. Hence, the interaction of BRCA1 with BARD1 is quintessential for DNA repair (Baer, 2013; Irminger-Finger & Jefford, 2006; Li & Yu, 2013). It has been suspected that, for most breast cancer cases, the disruption of the BRCA1-BARD1 interface due to mutations is the contributing factor (De Brakeleer et al., 2010; Jakubowska et al., 2008; Thai et al., 1998). The mechanisms by which these mutations alter the interface and actuate the disease remain elusive. To develop an efficient therapeutic arsenal against this dreadful malady, it is highly exigent to understand the metrics by which the mutations aberrate the interaction between BRCA1 and BARD1. Increased efficacy of computational resources to analyze biological data has reduced the reliance on wet bench research. With elevated levels of genomic and proteomic data in the various databases, the burden of identifying, characterizing, and prioritizing mutations and asserting their impacts has been alleviated. Further, advanced computational algorithms have been developed to model mutations in a protein's 3D structure and ascertain their impact on other protein interactions (Cheng, Li, Zhou, Wang, & Bryant, 2012; Mirza, Salmas, Fatmi, & Durdagi, 2016; Ripphausen, Nisius, Peltason, & Bajorath, 2010).

Thus, in this study, we have computationally elucidated the most significant mutations in BRCA1-BARD1 using various computational parameters that are promising in supporting the mutations in PPI in several of the previous reports (Hossain, 2020; Thirumal Kumar et al., 2017, 2019). Further, we have also examined how these mutations alter the BRCA1-BARD1 interface to cause breast cancer. Our analysis will help illuminate and understand the pathological role of tumorigenic amino acid variations.

2. Materials and methods

2.1 Collection of data

The mutations for BRCA1 and BARD1 were obtained from three different databases, ClinVar (https://www.ncbi.nlm.nih.gov/clinvar/), UniProt (https://www.UniProt.org/), and NCBI (https://www.ncbi.nlm.nih.gov/snp/), for our computational analysis (Bateman et al., 2017; Landrum et al., 2014; Sherry et al., 2001). The BRCA1-BARD1 interacting residues were obtained from PDBsum using the protein structure of PDB ID: 1JM7 (Laskowski, Jabłońska, Pravda, Vařeková, & Thornton, 2018).

2.2 Pathogenicity, stability, and conservation prediction

The pathogenicity and stability of the retrieved mutations were analyzed using computational algorithmic procedures. For the pathogenicity prediction, PredictSNP (https://loschmidt.chemi.muni.cz/predictsnp1/) was used (Bendl et al., 2014). This consists of eight different tools, namely, PolyPhen-1, PhD-SNP, PredictSNP, MAPP, PolyPhen-2, PANTHER, SNAP, and SIFT. The obtained results from these tools increases the credence for analysis. To predict and determine the stability of the mutants, the iStable tool (http://predictor.nchu.edu.tw/istable/) was used (Chen, Lin, & Chu, 2013). iStable consists of 3 integrated tools, MUpro, iMutant, and iStable. These tools help us envisage the stability of the retrieved mutants. The web-server ConSurf was utilized to study the conservation level of the most deleterious and destabilizing mutations. It gives a value ranging from 1 to 9, where 1 depicts that amino acid is positioned in the least conserved region and 9 denotes that the amino acid is positioned in the highly conserved region (http://consurf.tau.ac.il/) (Ashkenazy et al., 2016).

2.3 Selection of the mutations

The mutations were selected based on the most pathogenic, destabilizing, and positioning of the conserved amino acids in the PPI region. In silico mutation modeling on the native protein (PDB ID: 1JM7) was then performed with the help of Swiss PDB Viewer (Guex & Peitsch, 1997).

Protein-protein docking was further executed using the ZDOCK server with the default parameters (Pierce et al., 2014).

2.4 Molecular dynamics simulation

The protein structure of native and the four shortlisted mutants of BRCA1 were subjected to GROMACS package to undergo molecular dynamics simulations (Abraham et al., 2015). The OPLS force field was used to generate the essential topology records (Kaminski, Friesner, Tirado-Rives, & Jorgensen, 2001). The protein-protein complex (PDB ID: 1JM7) was converted from the PDB format to the GROMACS format with the help of the pdb2gmx command. The removal of hydrogen atoms from the complex was carried out using the command -ignh. The treated complex was kept in a cubical box and filled with SPC/E water molecules to solvate the system. In order to facilitate neutrality, the protein complex system was exposed to the necessary amount of sodium ions. The system was then set to an equilibrium state using NPT and NVT ensembles, which were carried out over 50,000 steps. After the equilibrium state was attained, the molecular simulation was carried out at 50 ns, and the obtained trajectories were applied to generate the Root-Mean-Square Deviation (RMSD), intramolecular

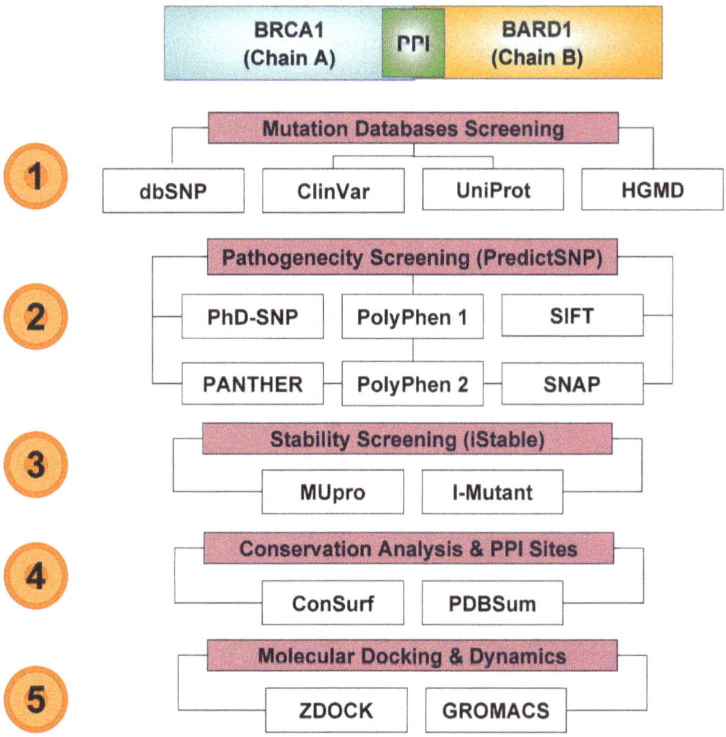

Fig. 1 The detailed methodology flowchart of screening the BRCA1-BARD1 mutations.

hydrogen bonds (h-bond) plots, and Radius of Gyration (Rg). The plots were finally analyzed using grace software. The detailed methodology flowchart is shown in Fig. 1.

3. Results

3.1 Dataset

A total of 2118 BRCA1 mutations were retrieved from the ClinVar, NCBI, and UniProt databases. No mutation reports were identified in BARD1 from these databases till date. Since the proteins are composed of chain A and B chains, mutations in the PPI sites were considered to understand the influence of mutations. From the PPI study using the PDBsum server, we found that 26 amino acids from the A chain (BRCA1) and 23 amino acids from the B chain (BARD1) were in the PPI region of BRCA1-BARD1 complex (Table 1).

3.2 Pathogenicity, stability, and conservation predictions

PredictSNP was implemented to determine the pathogenicity of the mutants retrieved. An overall of 222 mutations showed a deleterious effect when analyzed with all 8 integrated tools, i.e., PolyPhen-1, PhD-SNP, PredictSNP, MAPP, PolyPhen-2, PANTHER, SNAP, and SIFT. As a result, 973, 972, 678, 797, 1242, 1254, and 768 mutations were predicted to be pathogenic by PredictSNP, PhD-SNP, PANTHER, PolyPhen-1, SIFT, PolyPhen-2, and SNAP tools, respectively. Since the MAPP tool could not predict pathogenicity for a significant number of mutations, the results from this tool was ignored in the study (Table S1 in the online version at https://doi.org/10.1016/bs.apcsb.2022.02.003 and Fig. S1 in the online version at https://doi.org/10.1016/bs.apcsb.2022.02.003). To determine the stability of the obtained mutations, iStable was used. The results were obtained from the three tools, MuPro, iMutant, and iStable. Overall, 1255 mutations were predicted to be destabilizing by all three tools. MUpro, iMutant, and iStable predicted 1511, 1619, and 1519 mutations to have destabilizing effect, respectively (Table S2 in the online version at https://doi.org/10.1016/bs.apcsb.2022.02.003 and Fig. S2 in the online version at https://doi.org/10.1016/bs.apcsb.2022.02.003). Overall, with the comparison of the pathogenicity and stability prediction tools, 124 mutations were predicted to be most pathogenic and destabilizing by all the tools. These 124 mutations were further under ConSurf algorithm to

Table 1 List of amino acids involved in the BRCA1-BARD1 interaction.

S. No.	BRCA1	BARD1
1	96ASP	36HIS
2	92ALA	98ASN
3	40ASP	62CYS
4	78ARG	69ILE
5	88LYS	40ALA
6	89ILE	65GLY
7	76SER	66CYS
8	77THR	67GLU
9	93PHE	91TRP
10	38LYS	34TRP
11	97THR	118ASN
12	8VAL	117ASP
13	3LEU	114LEU
14	6LEU	43ARG
15	7ARG	103SER
16	11VAL	47LEU
17	10GLU	101LEU
18	85GLU	100GLN
19	81GLU	104MET
20	82LEU	111LEU
21	79PHE	44LEU
22	21ILE	107LEU
23	22LEU	110LYS
24	18MET	
25	17ALA	
26	14VAL	

Residues retrieved from the PDBSum server with PDB ID 1JM7.

Table 2 Conservation analysis of most pathogenic and destabilizing mutations of BRCA1.

S. No.	Variants	Consurf score
1	V11G	9
2	M18K	9
3	L22S	9
4	C24R	8
5	P25T	8
6	I26N	9
7	C27S	7
8	L28P	9
9	P34L	9
10	V35F	9
11	V35G	9
12	T37K	9
13	T37R	9
14	C39G	9
15	C39R	9
16	F43S	9
17	C44G	9
18	C44S	9
19	C47G	9
20	L49R	8
21	L52P	8
22	C61G	8
23	P62R	8
24	C64G	8
25	C64R	8
26	I68R	8
28	T97R	9
29	G98C	8
30	G129V	8
31	Y130C	7

Table 2 Conservation analysis of most pathogenic and destabilizing mutations of BRCA1.—cont'd

S. No.	Variants	Consurf score
32	Y130N	7
33	L165P	4
34	V178D	8
35	Y179C	9
36	S308C	9
37	E326Q	4
38	R332W	7
39	P334H	2
40	S378N	9
41	I379M	9
42	Q380P	7
43	W385C	8
44	R388T	7
45	S434C	5
46	I460R	9
47	F461L	9
48	G462R	8
49	L471P	2
50	R507I	8
51	L512F	9
52	E515G	8
53	K519Q	9
54	L523W	3
55	N550H	5
56	D749H	4
57	L771S	6
58	S819C	7
59	F861C	8
60	S864A	9
61	R866H	9
62	C882G	3

Continued

Table 2 Conservation analysis of most pathogenic and destabilizing mutations of BRCA1.—cont'd

S. No.	Variants	Consurf score
63	S956C	6
64	I1031F	5
65	D1123Y	6
66	V1145F	4
67	D1151A	7
68	E1172G	9
69	V1176D	8
70	E1210G	8
71	S1321C	8
72	L1404R	9
73	L1407H	6
74	L1517F	5
75	V1665E	7
76	T1685A	9
77	D1692N	8
78	C1697S	8
79	C1697Y	8
80	R1699P	9
81	R1699Q	9
82	T1700A	9
83	Y1703C	8
84	Y1703D	8
85	Y1703H	8
86	Y1703S	8
87	L1705P	8
88	G1706E	8
89	I1707S	9
90	V1714G	9
91	S1715C	9
92	W1718C	8
93	W1718L	8

Table 2 Conservation analysis of most pathogenic and destabilizing mutations of BRCA1.—cont'd

S. No.	Variants	Consurf score
94	W1718S	8
95	I1723N	7
96	D1739Y	9
97	G1748C	9
98	G1748D	9
99	F1761S	1
100	L1764P	8
101	I1766N	8
102	I1766S	8
103	I1766V	8
104	C1767S	8
105	C1768G	9
106	C1768Y	9
107	G1770A	8
108	M1775K	9
109	M1775R	9
110	L1780P	9
111	L1780R	9
112	L1786P	6
113	G1788D	9
114	V1808E	8
115	V1809F	7
116	A1814D	9
117	W1837C	8
118	W1837G	8
119	W1837R	8
120	V1838E	9
121	V1838G	9
122	A1843E	9
123	Y1845C	8
124	Y1853C	9

The conservation analysis was carried out using ConSurf server.

study the conservation nature of the native amino acids based on the position. From the ConSurf analysis, 51 mutations were found to be positioned in the highly conserved region with a ConSurf score of 9 (Table 2).

3.3 Selection of mutants

From the above-mentioned series of in silico analysis, four mutants, namely, V11G, M18K, L22S, and T97R were selected for the structural analysis. These four mutants were observed in the conserved PPI region with high pathogenic and destabilizing effects. The pathogenicity, stability, and conservation predictions of mutations V11G, M18K, L22S, and T97R are listed in Table 3. Protein-protein docking was carried out for the native BRCA1-BARD1 protein and the four selected mutant proteins using the ZDOCK server to understand the influence of these mutations in the PPI. The ZDOCK score for the native protein complex was high (2118.613), and the V11G mutant complex was low (1992.949). The ZDOCK score for mutant complexes M18K, L22S, and T97R were 2114.28, 2086.252, and

Table 3 Compiled in silico tools results of most pathogenic, destabilizing and conserved mutations in the BRCA1-BARD1 interacting region.

Mutations	V11G	M18K	L22S	T97R
PredictSNP	Deleterious	Deleterious	Deleterious	Deleterious
PhD-SNP	Deleterious	Deleterious	Deleterious	Deleterious
PolyPhen-1	Deleterious	Deleterious	Deleterious	Deleterious
PolyPhen-2	Deleterious	Deleterious	Deleterious	Deleterious
SIFT	Deleterious	Deleterious	Deleterious	Deleterious
SNAP	Deleterious	Deleterious	Deleterious	Deleterious
PANTHER	Deleterious	Deleterious	Deleterious	Deleterious
i-Mutant2.0	Decrease	Decrease	Decrease	Decrease
DDG	−2.35	−1.43	−1.99	−0.55
MUpro	Decrease	Decrease	Decrease	Decrease
Conf. Score	−1	−1	−1	−0.29800313
iStable	Decrease	Decrease	Decrease	Decrease
Conf. Score	0.90026	0.853717	0.881919	0.854195
ConSurf Score	9	9	9	9

Table 4 Comparison of docking scores between native and mutant BRCA1-BARD1 complex complexes.

Protein	ZDOCK score
Native	2118.613
V11G	1992.949
M18K	2114.28
L22S	2086.252
T97R	2114.279

The protein-protein docking score was obtained using ZDOCK server.

2114.279, respectively, and these scores were between those for the native and V11G mutant complexes (Table 4).

3.4 Molecular dynamics simulation

Molecular Dynamics Simulation for 50 ns was implemented with the help of the GROMACS package. The trajectories obtained for 50 ns were used to generate plots related to RMSD, h-bonding, and radius of gyration, and the plots were examined with the help of GRACE software. From the RMSD plot, the V11G mutant complex showed the highest deviation, and the native protein complex expressed the lowest deviation. The other three mutant complexes (M18K, L22S, and T97R) showed a deviation between native and V11G mutant complexes (Fig. 2). However, all the plots expressed convergence by the end of 50 ns. A block-based RMSD graph was plotted for the last 10 ns (40–50 ns) to increase the confidence in the convergence. This graph showed a more appropriate deviation range (~0.35 nm to ~0.65 nm) and provided confidence for performing further analyses (Fig. 3). To obtain an idea about the protein's compactness, the plot for the Rg was examined. From the graph obtained, the V11G mutant complex showed the highest Rg value and hence possessed the least compactness in the protein structure, whereas native complex showed the least Rg value and therefore designated as the most compact protein complex structure. Similar to the RMSD plot, the Rg showed deviations between the native and the other three mutant complexes (M18K, L22S, and T97R). Thus, these three mutant complexes possess compactness between those of native and V11G mutant complexes (Fig. 4). Finally, the alterations

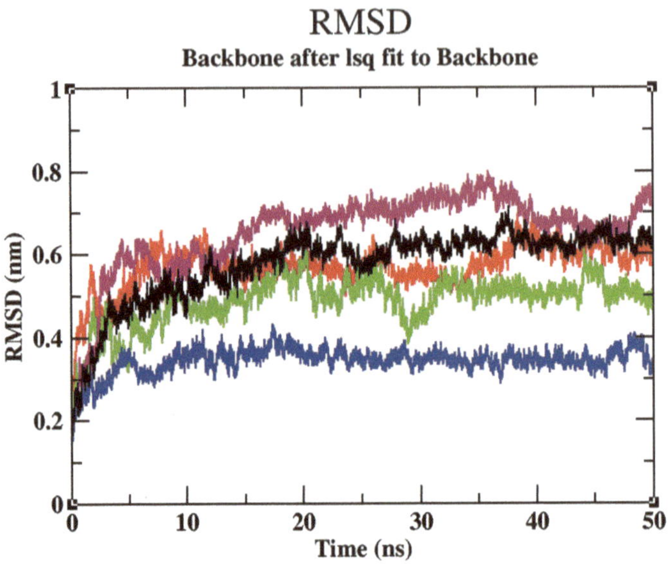

Fig. 2 Root mean square deviation of the BRCA1-BARD1 native and mutant complexes for 50 ns. Color scheme: Native—Blue, M18K—Green, T97R—Red, L22S—Black, V11G—Magenta.

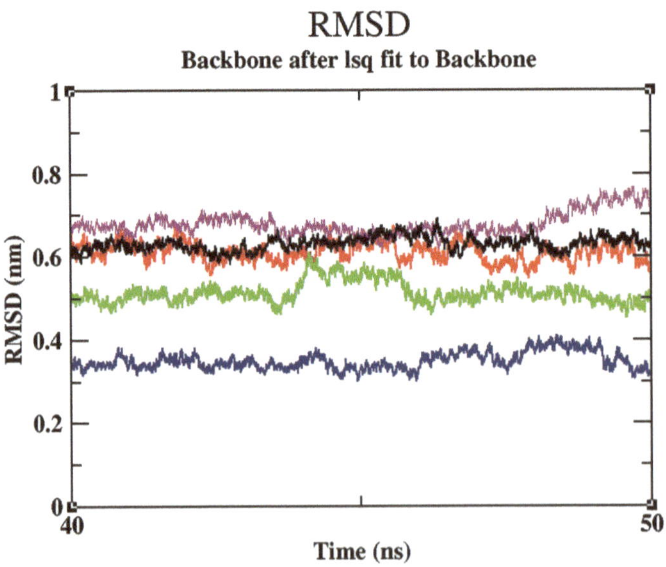

Fig. 3 Block-based root mean square deviation of the BRCA1-BARD1 native and mutant complexes between 40 and 50 ns. Color scheme: Native—Blue, M18K—Green, T97R—Red, L22S—Black, V11G—Magenta.

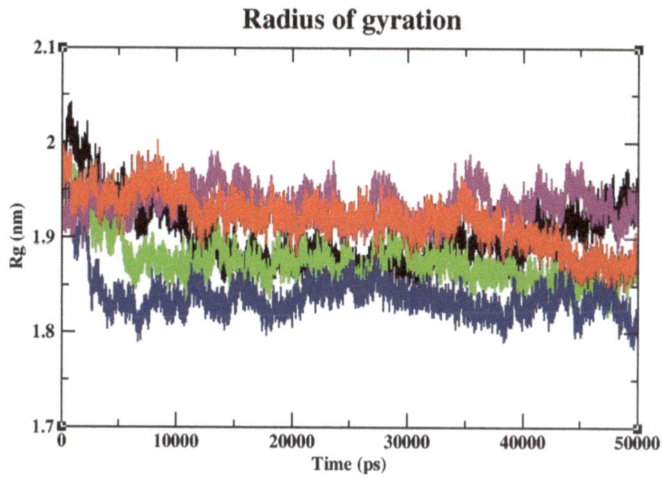

Fig. 4 The Radius of Gyration of the BRCA1-BARD1 native and mutant complexes for 50ns. Color scheme: Native—Blue, M18K—Green, T97R—Red, L22S—Black, V11G—Magenta.

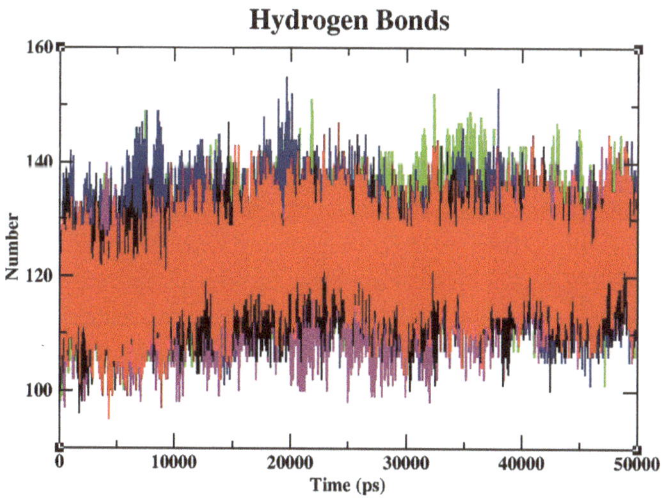

Fig. 5 Number of intramolecular hydrogen bonds in the BRCA1-BARD1 native and mutant complexes for 50ns. Color scheme: Native—Blue, M18K—Green, T97R—Red, L22S—Black, V11G—Magenta.

in a number of intramolecular h-bonds were examined to address the change in deviation and compactness of the protein. The plot for intramolecular h-bonds was analyzed, and the graph obtained depicted that the protein with the V11G mutant complex possessed least number of intramolecular h-bonds, whereas native protein complex possessed most significant number of intramolecular h-bonds. The mutant complexes M18K, L22S, and T97R possessed an amount of intramolecular h-bonds concerning those for the native and V11G mutant complexes (Fig. 5).

4. Discussion

Breast cancer is the 2nd most frequently occurring cancer among women after skin cancer (Breast Cancer, 2018).Understanding mutations that cause breast cancer is an essential task as many proteins are associated with breast cancer. Several proteins, such as BRCA1 (Henneman et al., 2015; Yan et al., 2015), BRCA2 (Daniels et al., 2004; Wooster et al., 1994), CDH1 (Fiolka et al., 2013; Mahmud, Ford, Longacre, Parent, & Norton, 2015), PIK3CA (Kumar & Doss, 2016; Sudhakar et al., 2016), AKT1 (Luo, Manning, & Cantley, 2003; Thirumal Kumar et al., 2019), are most commonly known to be the causative proteins for breast cancer; however, mutations in BRCA1 is the first to cause breast cancer (Abdel-Razeq, Al-Omari, Zahran, & Arun, 2018; Kuchenbaecker et al., 2017; Mincey, 2003). BRCA1 interacts with several protein molecules, including suppressors of tumors such as p53, RB, BRCA2, and ATM (Deng & Brodie, 2000). In addition to these tumor suppressors, BRCA1 also communicates with mismatch repair proteins such as MSH3, MLH, MSH6, and MSH2 (Tavtigian, Byrnes, Goldgar, & Thomas, 2008). Bearing complex mechanisms and a lack of complete 3D structure for the BRCA proteins, the interactions of BRCA proteins with other proteins are still unknown (Begg et al., 2008). However, the details for understanding the complex mechanism in the interaction of BRCA1 with the BARD1 protein (Baer & Ludwig, 2002; Irminger-Finger et al., 2001). Mutations in the interacting region of these two proteins are common causes of breast cancer (Birrane, Varma, Soni, & Ladias, 2007).

In this study, we screened all mutations of the BRCA1-BARD1 complex using various computational algorithms, and we studied the structural changes in the proteins due to the deleterious interface mutations. The computational studies have been beneficially used in earlier studies to prove the genotype and phenotype correlations (Thirumal Kumar, Udhaya Kumar, Magesh, & George Priya Doss, 2021; Zaki et al., 2017; Zhang et al., 2019). A list of 2118 mutations related to BRCA1 was taken from the databases such as ClinVar, UniProt, and NCBI. No mutational reports were found for BARD1 protein in the UniProt, NCBI, and ClinVar databases. Since these proteins are comprised of two chains, chain A and chain B, mutations in the PPI sites were considered in the present study. The list of amino acids in the protein-protein interface was obtained from the PDBsum database (Table 1). The 3D protein model of the BRCA1-BARD1 complex was acquired from the RCSB protein data bank with PDB ID: 1JM7. The entire list of missense mutations retrieved from the public databases was computationally screened for pathogenic and destabilizing properties using the PredictSNP and iStable servers, respectively. From these screenings, 222 and 1255 mutations were predicted to have deleterious effect and destabilizing by PredictSNP and iStable predictors (Tables S1 and S2 in the online version at https://doi.org/10.1016/bs.apcsb.2022.02.003). Overall, 124 missense mutations were found to have both deleterious effect and responsible for destabilizing the protein structure. To further filter and identify the most significant mutations, the conservation position of each mutation was analyzed using the ConSurf server. The mutations positioned at the conserved regions of the proteins expressed higher structural changes when compared to the mutations positioned at the variable region of the proteins (Shastry, 2007). From the screening of 124 mutations against a conservation analysis, 51 deleterious and destabilizing mutations were found to be positioned in the most conserved sites of these proteins with a ConSurf score of 9 (Table 2). The mutations occurring in the interface of the PPI led to higher distortions and structurally affected the protein complex (Brender & Zhang, 2015; Jubb et al., 2017). On this selection basis, four mutations at PPI, V11G, M18K, L22S, and T97R, were pathogenic and destabilizing by all the computational algorithms and were located in a highly conserved position; therefore, these mutations were chosen for the molecular docking and followed by dynamics studies. The results of the chosen mutations are tabulated in Table 3. The distortion due to the PPI was evaluated using protein-protein docking with the ZDOCK server. From the docking analysis, we distinguished the highest interactions with the native BRCA1-BARD1 protein complex, whereas the mutant V11G complex led to the highest distortion. This result was followed in order of L22S, T97R, and M18K mutant complexes for the lowest to highest distortion (Table 4). These protein complexes were further subjected to molecular dynamics simulations to explore the behavior of protein complexes over a simulation time of 50 ns. In this line the RMSD, is used to study the deviation and differentiation of molecules over the time period. The RMSD plot manifested that the native protein complex had a minimal deviation and that the complex with V11G mutation had the highest deviation. However, the other mutant complexes showed significant deviations that were between those of the native and V11G protein complexes (Fig. 2). To increase the confidence in the convergence, a block-based RMSD was plotted for the last 10 ns (40–50 ns). The overall deviation of molecules showing deviations between ~0.3 and ~0.7 nm showed that the molecules attained convergence at the completion of the simulation period (Fig. 3). The radius of gyration helps in explaining the compactness of the protein structure (George Priya Doss & Zayed, 2017; Kumar et al., 2020) and amount of h-bonds in the protein helps in perpetuating the stability of the protein structure (Thirumal Kumar et al., 2021; Zaki et al., 2017). Further, from Rg and intramolecular h-bonds, we spotted that the protein with the V11G mutation was the least compact of these proteins and had a minimal amount of h-bonds when compared to the other mutant complexes (M18K, L22S, and T97R). The native protein complex, exhibited the highest amount of h-bonds and the best compact nature (Figs. 4 and 5). The previous reports by Starita et al. (2018), Machackova, Damborsky, Valik, and Foretova (2001), and Sweet, Senter, Pilarski, Wei, and Toland (2010) have similarly found that the mutations V11G, M18K, and L22S are deleterious, and thus, their studies support our findings (Machackova et al., 2001; Starita et al., 2018; Sweet et al., 2010).

5. Conclusion

The structural studies on the BRCA proteins are minimal due to the unavailability of the 3D crystal structures and the complexity in the PPIs with the other proteins. The BRCA1 protein complexed with the BARD1 protein plays a crucial role in causing breast cancer. In the prevailing study, we computationally screened the entire set of BRCA1 mutations based on their properties of pathogenicity, stability, and conserved amino acids. We also considered the phenomenon that mutations in the PPI region significantly distort the structure. With these considerations, we identified the V11G, M18K, L22S, and T97R mutations, which are positioned in the conserved PPI region, to be more pathogenic and destabilizing. These mutations were considered in the interaction of protein-protein and molecular dynamics simulations studies. From the overall molecular docking and dynamics analyses, the V11G mutation was found to possess the least binding affinity, a higher deviation, and a lower compactness and to possess the least amount of H-bonds when compared to the native protein. Thus, illustrating this mutation could be the most significant in comparison to the other three mutations (M18K, L22S, and T97R) in the protein-protein interaction region. We believe that this study will help in understanding the overall significance of missense mutations as well as help in understanding the structural impact of mutations in the PPI region.

Acknowledgment

The authors take this opportunity to thank the management of VIT and BRAF @ CDAC for providing the facilities and encouragement to carry out this work.

Funding

This work was supported by Qatar University Grant# QUST-2-CHS-2019-3.

Author contributions

T.K.D., U.K.S., N.J., B.S., and K.B. involved in substantial contributions to conception and design, acquisition of data, or analysis and interpretation of data. D.T.K., U.K.S., N.J., B.S., B.K., K.B., M.S., and M.S. involved in drafting the article or revising it critically for important intellectual content. H.Z., S.R., and G.P.D.C. involved in final approval of the version to be published.

Conflict of interest

The authors have declared that no conflicts of interest exist.

References

Abdel-Razeq, H., Al-Omari, A., Zahran, F., & Arun, B. (2018). Germline BRCA1/BRCA2 mutations among high risk breast cancer patients in Jordan. *BMC Cancer, 18*(1), 152. https://doi.org/10.1186/s12885-018-4079-1.

Abraham, M. J., Murtola, T., Schulz, R., Páll, S., Smith, J. C., Hess, B., et al. (2015). GROMACS: High performance molecular simulations through multi-level parallelism from laptops to supercomputers. *SoftwareX, 1–2*, 19–25. https://doi.org/10.1016/j.softx.2015.06.001.

Ashkenazy, H., Abadi, S., Martz, E., Chay, O., Mayrose, I., Pupko, T., et al. (2016). ConSurf 2016: An improved methodology to estimate and visualize evolutionary conservation in macromolecules. *Nucleic Acids Research, 44*(W1), W344–W350. https://doi.org/10.1093/nar/gkw408.

Baer, R. (2013). Luring BRCA1 to the scene of the crime. *Cancer Cell, 23*(5), 565–567. https://doi.org/10.1016/j.ccr.2013.04.013.

Baer, R., & Ludwig, T. (2002). The BRCA1/BARD1 heterodimer, a tumor suppressor complex with ubiquitin E3 ligase activity. *Current Opinion in Genetics & Development, 12*(1), 86–91. https://doi.org/10.1016/s0959-437x(01)00269-6.

Bateman, A., Martin, M. J., O'Donovan, C., Magrane, M., Alpi, E., Antunes, R., et al. (2017). UniProt: The universal protein knowledgebase. *Nucleic Acids Research, 45*(D1), D158–D169. https://doi.org/10.1093/nar/gkw1099.

Begg, C. B., Haile, R. W., Borg, A., Malone, K. E., Concannon, P., Thomas, D. C., et al. (2008). Variation of breast cancer risk among BRCA1/2 carriers. *JAMA, 299*(2), 194–201. https://doi.org/10.1001/jama.2007.55-a.

Bendl, J., Stourac, J., Salanda, O., Pavelka, A., Wieben, E. D., Zendulka, J., et al. (2014). PredictSNP: Robust and accurate consensus classifier for prediction of disease-related mutations. *PLoS Computational Biology*, *10*(1), e1003440. https://doi.org/10.1371/journal.pcbi.1003440.

Birrane, G., Varma, A. K., Soni, A., & Ladias, J. A. A. (2007). Crystal structure of the BARD1 BRCT domains. *Biochemistry*, *46*(26), 7706–7712. https://doi.org/10.1021/bi700323t.

Breast Cancer: Causes, Stage, Diagnosis & Treatment. (2018). *Cleveland clinic*. https://my.clevelandclinic.org/health/diseases/3986-breast-cancer.

Brender, J. R., & Zhang, Y. (2015). Predicting the effect of mutations on protein-protein binding interactions through structure-based Interface profiles. *PLoS Computational Biology*, *11*(10), e1004494. https://doi.org/10.1371/journal.pcbi.1004494.

Chen, C.-W., Lin, J., & Chu, Y.-W. (2013). iStable: Off-the-shelf predictor integration for predicting protein stability changes. *BMC Bioinformatics*, *14*(2), S5. https://doi.org/10.1186/1471-2105-14-S2-S5.

Cheng, T., Li, Q., Zhou, Z., Wang, Y., & Bryant, S. H. (2012). Structure-based virtual screening for drug discovery: A problem-centric review. *The AAPS Journal*, *14*(1), 133–141. https://doi.org/10.1208/s12248-012-9322-0.

Daniels, D. S., Woo, T. T., Luu, K. X., Noll, D. M., Clarke, N. D., Pegg, A. E., et al. (2004). DNA binding and nucleotide flipping by the human DNA repair protein AGT. *Nature Structural & Molecular Biology*, *11*(8), 714–720. https://doi.org/10.1038/nsmb791.

De Brakeleer, S., De Grève, J., Loris, R., Janin, N., Lissens, W., Sermijn, E., et al. (2010). Cancer predisposing missense and protein truncating BARD1 mutations in non-BRCA1 or BRCA2 breast cancer families. *Human Mutation*, *31*(3), E1175–E1185. https://doi.org/10.1002/humu.21200.

Deng, C. X., & Brodie, S. G. (2000). Roles of BRCA1 and its interacting proteins. *BioEssays: News and Reviews in Molecular, Cellular and Developmental Biology*, *22*(8), 728–737. https://doi.org/10.1002/1521-1878(200008)22:8<728::AID-BIES6>3.0.CO;2-B.

Duncan, J. A., Reeves, J. R., & Cooke, T. G. (1998). BRCA1 and BRCA2 proteins: Roles in health and disease. *Molecular Pathology*, *51*(5), 237–247.

Fiolka, R., Zubor, P., Janusicova, V., Visnovsky, J., Mendelova, A., Kajo, K., et al. (2013). Promoter hypermethylation of the tumor-suppressor genes RASSF1A, GSTP1 and CDH1 in endometrial cancer. *Oncology Reports*, *30*(6), 2878–2886. https://doi.org/10.3892/or.2013.2752.

Fox, D., Le Trong, I., Rajagopal, P., Brzovic, P. S., Stenkamp, R. E., & Klevit, R. E. (2008). Crystal structure of the BARD1 ankyrin repeat domain and its functional consequences. *The Journal of Biological Chemistry*, *283*(30), 21179–21186. https://doi.org/10.1074/jbc.M802333200.

Friedenson, B. (2007). The BRCA1/2 pathway prevents hematologic cancers in addition to breast and ovarian cancers. *BMC Cancer*, 7, 152. https://doi.org/10.1186/1471-2407-7-152.

George Priya Doss, C., & Zayed, H. (2017). Comparative computational assessment of the pathogenicity of mutations in the Aspartoacylase enzyme. *Metabolic Brain Disease*, *32*(6), 2105–2118. https://doi.org/10.1007/s11011-017-0090-5.

Guex, N., & Peitsch, M. C. (1997). SWISS-MODEL and the Swiss-Pdb viewer: An environment for comparative protein modeling. *Electrophoresis*, *18*(15), 2714–2723. https://doi.org/10.1002/elps.1150181505.

Henderson, B. R. (2012). The BRCA1 breast Cancer suppressor: Regulation of transport, dynamics, and function at multiple subcellular locations. *Scientifica*, *2012*, 796808. https://doi.org/10.6064/2012/796808.

Henneman, L., van Miltenburg, M. H., Michalak, E. M., Braumuller, T. M., Jaspers, J. E., Drenth, A. P., et al. (2015). Selective resistance to the PARP inhibitor olaparib in a mouse model for BRCA1-deficient metaplastic breast cancer. *Proceedings of the*

National Academy of Sciences of the United States of America, 112(27), 8409–8414. https://doi.org/10.1073/pnas.1500223112.
Hossain, M. S. (2020). In silico analysis predicting effects of deleterious SNPs of human RASSF5 gene on its structure and functions. *Scientific Reports, 3*(10), 14542. https://doi.org/10.1038/s41598-020-71457-1.
Irminger-Finger, I., & Jefford, C. E. (2006). Is there more to BARD1 than BRCA1? *Nature Reviews. Cancer, 6*(5), 382–391. https://doi.org/10.1038/nrc1878.
Irminger-Finger, I., Leung, W. C., Li, J., Dubois-Dauphin, M., Harb, J., Feki, A., et al. (2001). Identification of BARD1 as mediator between proapoptotic stress and p53-dependent apoptosis. *Molecular Cell, 8*(6), 1255–1266. https://doi.org/10.1016/s1097-2765(01)00406-3.
Jakubowska, A., Cybulski, C., Szymańska, A., Huzarski, T., Byrski, T., Gronwald, J., et al. (2008). BARD1 and breast cancer in Poland. *Breast Cancer Research and Treatment, 107*(1), 119–122. https://doi.org/10.1007/s10549-007-9537-4.
Jubb, H. C., Pandurangan, A. P., Turner, M. A., Ochoa-Montaño, B., Blundell, T. L., & Ascher, D. B. (2017). Mutations at protein-protein interfaces: Small changes over big surfaces have large impacts on human health. *Progress in Biophysics and Molecular Biology, 128*, 3–13. https://doi.org/10.1016/j.pbiomolbio.2016.10.002.
Kaminski, G. A., Friesner, R. A., Tirado-Rives, J., & Jorgensen, W. L. (2001). Evaluation and Reparametrization of the OPLS-AA force field for proteins via comparison with accurate quantum chemical calculations on peptides. *The Journal of Physical Chemistry B, 105*(28), 6474–6487. https://doi.org/10.1021/jp003919d.
Kotsopoulos, J. (2018). BRCA mutations and breast cancer prevention. *Cancers, 10*(12), E524. https://doi.org/10.3390/cancers10120524.
Kuchenbaecker, K. B., Hopper, J. L., Barnes, D. R., Phillips, K.-A., Mooij, T. M., Roos-Blom, M.-J., et al. (2017). Risks of breast, ovarian, and contralateral breast cancer for BRCA1 and BRCA2 mutation carriers. *JAMA, 317*(23), 2402–2416. https://doi.org/10.1001/jama.2017.7112.
Kumar, D. T., & Doss, C. G. P. (2016). Investigating the inhibitory effect of Wortmannin in the hotspot mutation at codon 1047 of PIK3CA kinase domain: A molecular docking and molecular dynamics approach. *Advances in Protein Chemistry and Structural Biology, 102*, 267–297. https://doi.org/10.1016/bs.apcsb.2015.09.008.
Kumar, S. U., Kumar, D. T. R. S., Doss, C. G. P., & Zayed, H. (2020). An extensive computational approach to analyze and characterize the functional mutations in the galactose-1-phosphate uridyl transferase (GALT) protein responsible for classical galactosemia. *Computers in Biology and Medicine, 117*, 103583. https://doi.org/10.1016/j.compbiomed.2019.103583.
Landrum, M. J., Lee, J. M., Riley, G. R., Jang, W., Rubinstein, W. S., Church, D. M., et al. (2014). ClinVar: Public archive of relationships among sequence variation and human phenotype. *Nucleic Acids Research, 42*(D1), D980–D985. https://doi.org/10.1093/nar/gkt1113.
Laskowski, R. A., Jabłońska, J., Pravda, L., Vařeková, R. S., & Thornton, J. M. (2018). PDBsum: Structural summaries of PDB entries. *Protein Science, 27*(1), 129–134. https://doi.org/10.1002/pro.3289.
Li, M., & Yu, X. (2013). Function of BRCA1 in the DNA damage response is mediated by ADP-ribosylation. *Cancer Cell, 23*(5), 693–704. https://doi.org/10.1016/j.ccr.2013.03.025.
Luo, J., Manning, B. D., & Cantley, L. C. (2003). Targeting the PI3K-Akt pathway in human cancer: Rationale and promise. *Cancer Cell, 4*(4), 257–262. https://doi.org/10.1016/s1535-6108(03)00248-4.
Machackova, E., Damborsky, J., Valik, D., & Foretova, L. (2001). Novel germline BRCA1 and BRCA2 mutations in breast and breast/ovarian cancer families from the Czech Republic. *Human Mutation, 18*(6), 545. https://doi.org/10.1002/humu.1232.

Mahmud, N., Ford, J. M., Longacre, T. A., Parent, R., & Norton, J. A. (2015). Metastatic lobular breast carcinoma mimicking primary signet ring adenocarcinoma in a patient with a suspected CDH1 mutation. *Journal of Clinical Oncology*, *33*(4), e19–e21. https://doi.org/10.1200/JCO.2013.49.1159.

Mincey, B. A. (2003). Genetics and the management of women at high risk for breast cancer. *The Oncologist*, *8*(5), 466–473. https://doi.org/10.1634/theoncologist.8-5-466.

Mirza, S. B., Salmas, R. E., Fatmi, M. Q., & Durdagi, S. (2016). Virtual screening of eighteen million compounds against dengue virus: Combined molecular docking and molecular dynamics simulations study. *Journal of Molecular Graphics & Modelling*, *66*, 99–107. https://doi.org/10.1016/j.jmgm.2016.03.008.

O'Donovan, P. J., & Livingston, D. M. (2010). BRCA1 and BRCA2: Breast/ovarian cancer susceptibility gene products and participants in DNA double-strand break repair. *Carcinogenesis*, *31*(6), 961–967. https://doi.org/10.1093/carcin/bgq069.

Pierce, B. G., Wiehe, K., Hwang, H., Kim, B.-H., Vreven, T., & Weng, Z. (2014). ZDOCK server: Interactive docking prediction of protein–protein complexes and symmetric multimers. *Bioinformatics*, *30*(12), 1771–1773. https://doi.org/10.1093/bioinformatics/btu097.

Ripphausen, P., Nisius, B., Peltason, L., & Bajorath, J. (2010). Quo vadis, virtual screening? A comprehensive survey of prospective applications. *Journal of Medicinal Chemistry*, *53*(24), 8461–8467. https://doi.org/10.1021/jm101020z.

Shastry, B. S. (2007). SNPs in disease gene mapping, medicinal drug development and evolution. *Journal of Human Genetics*, *52*(11), 871–880. https://doi.org/10.1007/s10038-007-0200-z.

Sherry, S. T., Ward, M.-H., Kholodov, M., Baker, J., Phan, L., Smigielski, E. M., et al. (2001). dbSNP: The NCBI database of genetic variation. *Nucleic Acids Research*, *29*(1), 308–311. https://doi.org/10.1093/nar/29.1.308.

Starita, L. M., Islam, M. M., Banerjee, T., Adamovich, A. I., Gullingsrud, J., Fields, S., et al. (2018). A multiplex homology-directed DNA repair assay reveals the impact of more than 1,000 BRCA1 missense substitution variants on protein function. *American Journal of Human Genetics*, *103*(4), 498–508. https://doi.org/10.1016/j.ajhg.2018.07.016.

Sudhakar, N., Doss, C. G. P., Kumar, D. T., Chakraborty, C., Anand, K., & Suresh, M. (2016). Deciphering the impact of somatic mutations in exon 20 and exon 9 of PIK3CA gene in breast tumors among Indian women through molecular dynamics approach. *Journal of Biomolecular Structure and Dynamics*, *34*(1), 29–41. https://doi.org/10.1080/07391102.2015.1007483.

Sweet, K., Senter, L., Pilarski, R., Wei, L., & Toland, A. E. (2010). Characterization of BRCA1 ring finger variants of uncertain significance. *Breast Cancer Research and Treatment*, *119*(3), 737–743. https://doi.org/10.1007/s10549-009-0438-6.

Tavtigian, S. V., Byrnes, G. B., Goldgar, D. E., & Thomas, A. (2008). Classification of rare missense substitutions, using risk surfaces, with genetic- and molecular-epidemiology applications. *Human Mutation*, *29*(11), 1342–1354. https://doi.org/10.1002/humu.20896.

Thai, T. H., Du, F., Tsan, J. T., Jin, Y., Phung, A., Spillman, M. A., et al. (1998). Mutations in the BRCA1-associated RING domain (BARD1) gene in primary breast, ovarian and uterine cancers. *Human Molecular Genetics*, 7(2), 195–202. https://doi.org/10.1093/hmg/7.2.195.

Thirumal Kumar, D., George Priya Doss, C., Sneha, P., Tayubi, I. A., Siva, R., Chakraborty, C., et al. (2017). Influence of V54M mutation in giant muscle protein titin: a computational screening and molecular dynamics approach. *Journal of Biomolecular Structure and Dynamics*, *35*(5). https://doi.org/10.1080/07391102.2016.1166456.

Thirumal Kumar, D., Jain, N., Evangeline, J., Kamaraj, B., Siva, R., Zayed, H., et al. (2019). A computational approach for investigating the mutational landscape of

RAC-alpha serine/threonine-protein kinase (AKT1) and screening inhibitors against the oncogenic E17K mutation causing breast cancer. *Computers in Biology and Medicine, 115*, 103513. https://doi.org/10.1016/j.compbiomed.2019.103513.

Thirumal Kumar, D., Susmita, B., Judith, E., Priyadharshini Christy, J., George Priya Doss, C., & Zayed, H. (2019). Elucidating the role of interacting residues of the MSH2-MSH6 complex in DNA repair mechanism: A computational approach. *Advances in Protein Chemistry and Structural Biology, 115*. https://doi.org/10.1016/bs.apcsb.2018.11.005.

Thirumal Kumar, D., Udhaya Kumar, S., Magesh, R., & George Priya Doss, C. (2021). Chapter Four—Investigating mutations at the hotspot position of the ERBB2 and screening for the novel lead compound to treat breast cancer—A computational approach. In R. Donev (Ed.), *Vol. 123. Advances in protein chemistry and structural biology* (pp. 49–71). Academic Press. https://doi.org/10.1016/bs.apcsb.2020.10.001.

Wang, Y., Cortez, D., Yazdi, P., Neff, N., Elledge, S. J., & Qin, J. (2000). BASC, a super complex of BRCA1-associated proteins involved in the recognition and repair of aberrant DNA structures. *Genes & Development, 14*(8), 927–939.

Weigelt, B., Horlings, H. M., Kreike, B., Hayes, M. M., Hauptmann, M., Wessels, L. F. A., et al. (2008). Refinement of breast cancer classification by molecular characterization of histological special types. *The Journal of Pathology, 216*(2), 141–150. https://doi.org/10.1002/path.2407.

Wooster, R., Neuhausen, S. L., Mangion, J., Quirk, Y., Ford, D., Collins, N., et al. (1994). Localization of a breast cancer susceptibility gene, BRCA2, to chromosome 13q12-13. *Science (New York, N.Y.), 265*(5181), 2088–2090. https://doi.org/10.1126/science.8091231.

Yan, M., Shield-Artin, K., Byrne, D., Deb, S., Waddell, N., kConFab Investigators, et al. (2015). Comparative microRNA profiling of sporadic and BRCA1 associated basal-like breast cancers. *BMC Cancer, 15*, 506. https://doi.org/10.1186/s12885-015-1522-4.

Zaki, O. K., Priya Doss, C. G., Ali, S. A., Murad, G. G., Elashi, S. A., Ebnou, M. S. A., et al. (2017). Genotype–phenotype correlation in patients with isovaleric acidaemia: Comparative structural modelling and computational analysis of novel variants. *Human Molecular Genetics, 26*(16), 3105–3115. https://doi.org/10.1093/hmg/ddx195.

Zhang, W., Zhang, H., Yang, H., Li, M., Xie, Z., & Li, W. (2019). Computational resources associating diseases with genotypes, phenotypes and exposures. *Briefings in Bioinformatics, 20*(6). https://doi.org/10.1093/bib/bby071.

CHAPTER THIRTEEN

Evolutionary aspect of Miltefosine transporter proteins in *Leishmania major*

Ritika Kabra and Shailza Singh*

National Centre for Cell Science, NCCS Complex, SP Pune University Campus, Pune, India
*Corresponding author: e-mail address: singhs@nccs.res.in

Contents

Abstract

Transporter proteins, P-glycoprotein (P-gp) and P4ATPase-CDC50, are responsible for the transport of Miltefosine drug across cell membrane of a protozoan parasite *Leishmania major*. Mutations or change in activity of these proteins may lead to emergence of resistance in the parasite. Owing to the structural and functional importance of these transporter proteins, we have tried to decipher the evolutionary divergence of these Miltefosine transporter proteins across different forms of life including Protists, Fungi, Plants and Animals. We retrieved 96, 207, and 189 sequences of P-gp, P4ATPase and CDC50 proteins respectively, across diverse variety of organisms for the conserved analysis. Phylogenetic trees were constructed for these three transporter proteins based on Bayesian posterior probability inference. The evolutionary analysis concluded that these proteins remain highly conserved throughout the species diversity but still substantial differences in the proteins for host (*Homo sapiens*) and

Advances in Protein Chemistry and Structural Biology, Volume 130
ISSN 1876-1623
https://doi.org/10.1016/bs.apcsb.2022.01.005

parasite (*L. major*) were observed which have led in targeting these Miltefosine transporter proteins in a parasite specific manner. The functional and structural components observed in terms of pattern resulting from the variability in the phylogenetic tree are outlined.

Abbreviations

ABC ATP binding cassette
ATP adenosine tri phosphate
CFTR cystic fibrosis transmembrane conductance regulator
gGAPDH glycosomal glyceraldehyde phosphate dehydrogenase
HMM hidden Markov model
MalK maltose/maltodextrin import ATP-binding protein
MCMCMC Metropolis Coupled Markov Chain Monte Carlo
MSA multiple sequence alignment
NBD nucleotide binding domain
NCBI National Centre for Biotechnology Information
P-gp P-glycoprotein
SDP positions determining specificity
SSU rRNA small subunit ribosomal RNA
TMD transmembrane domain
TMH transmembrane helices
TMS transmembrane segments

1. Introduction

Leishmaniasis, a neglected tropical disease, is the second largest parasitic killer in the world only after malaria. It is prevalent in more than 90 countries all over world with an estimate of 2 million new cases every year (Herwaldt, 1999; India National vector Borne Diseases Control Programme, 2017). As for treatment regimen, various drugs available are antimonials, Miltefosine, sitamaquine, Paromomycin, amphotericin B, etc. Of these, Miltefosine is the only one to be administered orally, hence widely used. The irregular and extensive use of Miltefosine in the past few years have resulted in development of resistance against this drug (Ghorbani & Farhoudi, 2018; Singh, Kumar, & Singh, 2012). Miltefosine is being transported inside parasite through P4ATPase-CDC50 protein complex while the efflux is carried out by an ATP binding cassette (ABC) transporter protein transporting phospholipid, i.e., P-glycoprotein (P-gp). Structurally, P4ATPase consists of 4 domains, a transmembrane domain (TMD) having 10 transmembrane helices (TMH), an actuator domain (A-domain), a phosphorylating domain (P-domain) and a nucleotide binding domain (NBD). CDC50 has 2 TMH with a large exocytoplasmic loop. Structural composition of P-gp includes 2 TMDs each having 6 TMH and 2 NBDs (Dawson & Locher, 2006; López-Marqués, Poulsen, Bailly, et al., 2015). Alteration in the functional activity of these two proteins have resulted in the development of resistance by the Leishmanial parasite against Miltefosine (Croft, Sundar, & Fairlamb, 2006; Singh & Mandlik, 2015).

The transporters proteins play a pivotal role in the survival of intracellular pathogens inside hosts. They mediate the entry of essential nutrients inside cell, uptake and efflux of ions, maintain homeostasis by regulating concentration of metabolites, and particularly in survival of pathogens by extrusion of drugs, toxins and xenobiotics (Lam, Lee, Silverio, et al., 2011). As transporter proteins play such an important role in survival of organisms, hence a lot of studies have been focused on them, especially on the drug transporters. Since most of the transporter proteins are membrane proteins, hence a large fraction of reports have been on the evolution of the transmembrane helices among various superfamilies. Lam et al. have proposed a pathway of evolution of transmembrane helices (TMH) in drug—metabolite transporter proteins starting from having 2 Transmembrane Segments (TMS) to 4-5TMS and reaching up to 10TMS (Lam et al., 2011). In another study by MH Saier, a pathway of the evolution of transporter proteins in all organisms was proposed which starts from a simple peptide channel comprising of 1–3 helices to evolving to much complex secondary carriers, primary active transporters as well group translocators comprising of 12–14 helices (Saier, 2016).

ATP binding cassette (ABC) transporter proteins constitute one of the largest superfamily of the transporter proteins and are present in all forms of life where gene duplication has played an important role in the evolution of new genes (Jones & George, 2004). Gene duplication events have resulted in the evolution of many ABC genes in vertebrates and most of them are associated with various diseases as well as development of resistance (Moitra & Dean, 2011). Saurin et al. retrieved 197 sequences of ABC transporters homologs to Maltose/maltodextrin import ATP-binding protein (MalK protein), a bacterial transporter. The phylogenetic analysis of these transporter proteins showed a clear divergence in the evolution of ABC importers and ABC exporters and subsequently they proposed a model for studying the generation and diversification of ABC systems (Saurin, Hofnung, & Dassa, 1999). Multigene families like those of ABC transporters having ATP-binding cassette, are subjected to point mutations or duplication events resulting in either few members to be nonfunctional or evolution of new members suggesting that almost all the members have emerged from a common ancestor. Based on this ideology, 68 members of ABC family have been identified from previously known 6 genes in *Dictyostelium discoideum* and other representatives of each family of ABC superfamily. This number is quite more than those present in either yeast or human which are among the best characterized genomes (Anjard & Loomis, 2002). Evolutionary analysis have helped in establishing the connection of Cystic fibrosis transmembrane conductance regulator (CFTR) protein being a member of ABC-transporter superfamily in spite of being an ATP-dependent chloride channel. Although having the functional divergence, CFTR belongs to ABCC subfamily based on the structural similarity and domain rearrangement (Jordan, Kota, Cui, et al., 2008). Igarashi et al. have studied 126 prokaryotic ABC transporters on the basis of nucleotide binding domain and classified them into 11 clusters. This have facilitated in understanding functional divergence and identification of new putative functions of ABC genes and how they are related in both eukaryotes and prokaryotes (Igarashi, Aoki, Mamitsuka, et al., 2004). Another large superfamily of membrane transport proteins utilizing ATP-hydrolysis energy is P-type ATPases. Axelsen and Palmgren analyzed 159 P-type ATPase sequences and classified them in 5 major classes based on their substrate similarity instead of sequence similarity (Axelsen & Palmgren, 1998). Moller et al. studied the evolutionary divergence of P5 subfamily of P-type ATPases in eukaryotes which acts as secretory pumps in Endoplasmic reticulum (Møller, Asp, Holm, & Palmgren, 2008). Similarly substrate specificities based evolutionary analysis was done for maltose transporters in *Thermotoga maritima* (Nanavati, Nguyen, & Noll, 2005). A few research groups have pointed the evolutionary relationship between different subfamilies of the superfamily on the basis of their structure as well as function i.e., their substrate specificity.

Here, we are providing insight on the sequential conservedness of a particular member of a transporter proteins superfamily and their presence in various organisms. To the best of our knowledge, this is the first report of its kind where evolutionary divergence of Miltefosine transporter proteins in *Leishmanial* species is being discussed. Since, Miltefosine is the only orally administered drug available for the treatment of Leishmaniasis, and the protozoan parasite is developing resistance with an alarming rate against this drug. We aim to study the evolutionary relationship of these transporter proteins to find any possible solutions for combating drug resistance. In one of our previous published dataset, we did the conservedness analysis and compared these Miltefosine transporter proteins between host (*Homo sapiens*) and parasite (*L. major*) to identify parasite specific motifs against which peptides were designed in order to allosterically modulate activity of transporter proteins to reverse the resistance effect (Kabra, Ingale, & Singh, 2020). Using the same data set, we did reanalysis aiming at the evolutionary divergence of these 3 transporter proteins across different kingdoms of life.

2. Methodology

2.1 Sequence retrieval

Sequences for P-gp, P4ATPase and CDC50 were extracted manually from National Centre for Biotechnology Information (NCBI, https://www.ncbi.nlm.nih.gov/) in FASTA format. All the annotated, complete and fully characterized sequences available in the database (as of December 2015) ranging from all five kingdoms of life (Monera, Protista, Fungi, Animalia and Plantae) were retrieved. The sequences of P-gp, P4ATPase and CDC50 proteins of L. *major* (parasite) and Human (host) serves as the basis to identify homologous sequences in other organisms via BLASTp and then apparently their retrieval from NCBI following the accession ID. As far as, multiple copy or paralogs are concerned we looked for those variants whose substrates were phospholipids since Miltefosine is a phosphatidylcholine analog and this substrate information was available only in annotated sequences. The sequences ranged from 181 to 1503 amino acids in P-gp; 776 residues to 2061 residues for P4ATPase and from 245 residues to 1537 residues in CDC50 protein. Multiple Sequence Alignment (MSA) for each protein was done using Clustal Omega and the output was saved in Nexus format (Sievers, Wilm, Dineen, et al., 2011). Clustal omega is a tool, available online at https://www.ebi.ac.uk/Tools/msa/clustalo/, for aligning three or more nucleotide or protein sequences. This tool is provided and maintained by EMBL-EBI. It uses either seeded guide trees or Hidden Markov Model (HMM) profile-profile technique to generate MSA.

2.2 Phylogenetic tree construction

Phylogenetic Tree was constructed through MrBayes v3.2.6 (Ronquist, Teslenko, Van Der Mark, et al., 2012) using Metropolis Coupled Markov Chain Monte Carlo (MCMCMC) method. MrBayes is open source software available at https://nbisweden.github.io/MrBayes/download.html.

It is based on Bayesian interference and uses Markov Chain Monte Carlo (MCMC) method to estimate posterior distribution of model parameters. MrBayes accepts multiple sequences aligned in Nexus format as an input file. The amino acid model parameter is set as 'mixed' and default MCMCMC method was used for tree construction considering two runs each having 4 chains. The analysis was terminated when the average standard deviation of split frequency reached below 0.01. It took 2,000,000, 1,150,000 and 2,900,000 generations respectively for P-gp, P4ATPase and CDC50 sequences to get a converged tree with the average standard deviation of split frequency below 0.01. The constructed trees were visualized using FigTree v1.4.2 available at http://tree.bio.ed.ac.uk/software/figtree/.

2.3 Identification of specificity determining positions

JDet software was used to analyze aligned sequences and identify the positions determining specificity through Xdet method (Muth, García-martín, Rausell, et al., 2012). In the graphical user interface, we used predefined Xdet method where the command ".\conf\Maxhom_McLachlan.metric" was run. For the Xdet analysis, the threshold cutoffs considered were 0.6 for Xdet and 2.5 for entropy. Using JDet interface, the sequences were represented in logo format.

3. Results

Fully annotated sequences of P-glycoprotein (P-gp), P4ATPase and CDC50 proteins were retrieved from NCBI (as available till 2015) from archaea to mammalia as well as plantae. The sequences homologs to *Leishmania major* were considered. The sequences of these proteins retrieved for various organisms were classified and represented in Fig. 1. A total of 96, 207 and 189 sequences were retrieved for P-gp, P4ATPase and CDC50 proteins. As can be seen, P4ATPase and CDC50 have been very well characterized in Fungi followed by Chordata, Protozoa and Nemathelminthes. In order to study the evolutionary relationship and sequential conservedness among different organisms, phylogenetic tree was constructed for these three proteins, viz. P-gp, P4ATPase and CDC50.

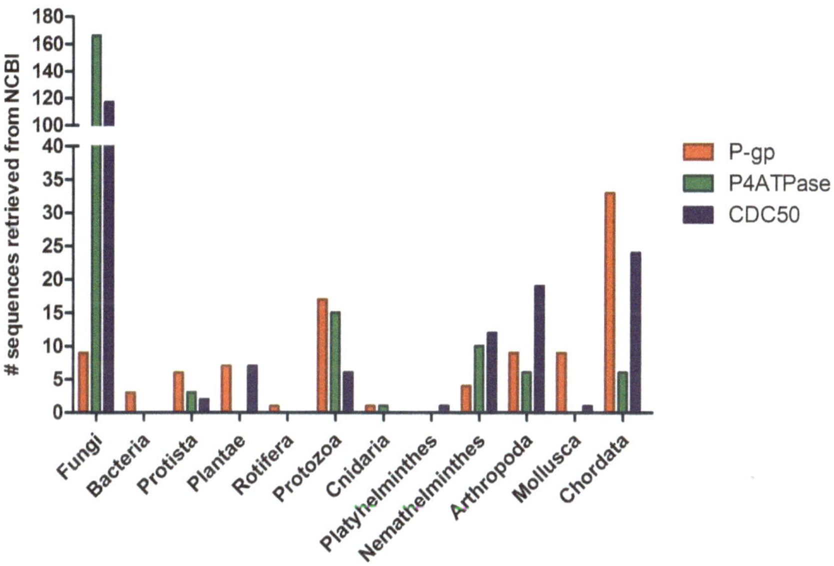

Fig. 1 Classification of P-gp, P4ATPase and CDC50 sequences retrieved from NCBI.

3.1 P-gp

Fully annotated sequences of P-glycoprotein (P-gp) were retrieved from NCBI (as available till 2015) for 96 organisms ranging from archaea to mammalia as well as plantae. The retrieved organisms were classified as highlighted in the phylogenetic tree shown in Fig. 2. The Protozoans are shown in red color and includes 17 organisms, 1 organism (*Hydra vulgaris*) belongs to Cnidaria phylum, 4 Nematodes (Orange), 9 Arthropods (Cyan), 6 Mollusks (Pink), 33 Chordates (Blue), 7 Plants (Green), 6 Protists (Purple), 3 Bacteria (Yellow) and 9 Fungi (Gray) were included. A wide variety in the sequence length was observed, with *Loa loa*, a nematode, having the minimum length of 181 residues while the organism having longest P-gp protein was *Trypanosoma brucei* with 1503 amino acid residues.

Although P-gp is sequentially highly conserved, but it was seen to be more conserved among chordates and plants when compared to other organisms (Figs. S1 and S3 in the online version at https://doi.org/10.1016/bs.apcsb.2022.01.005). Among chordates, P-gp in human was observed to be closely related with African monkey (*Chlorocebus aethiops*) and Rhesus Monkey (*Macaca mulatta*). As expected the P-gp protein in Mammals is closely related to Aves which have diverged from an Amphibian, *Xenopus laevis*. The fishes are represented in purple color in

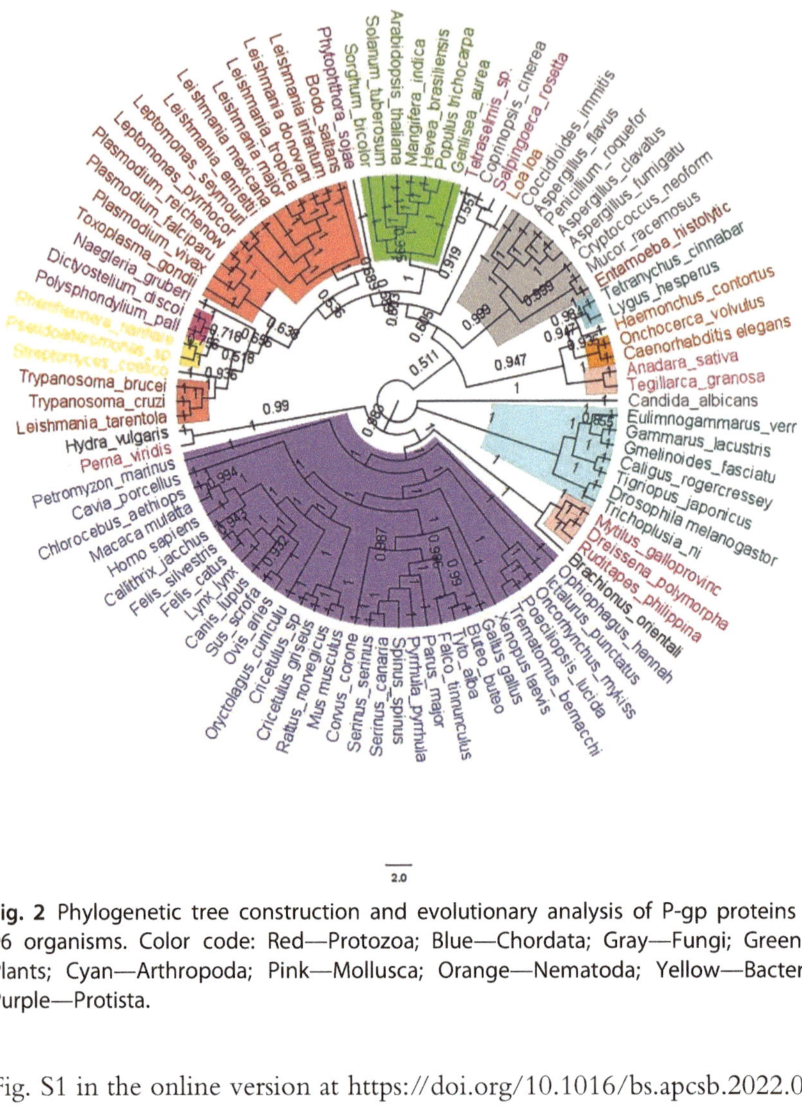

Fig. 2 Phylogenetic tree construction and evolutionary analysis of P-gp proteins of 96 organisms. Color code: Red—Protozoa; Blue—Chordata; Gray—Fungi; Green—Plants; Cyan—Arthropoda; Pink—Mollusca; Orange—Nematoda; Yellow—Bacteria; Purple—Protista.

Fig. S1 in the online version at https://doi.org/10.1016/bs.apcsb.2022.01.005, and *Petromyzon marinus* is found to be diverged from the other 4 organisms representing fish's class. Although, among chordates, *Ophiophagus Hannah,* a reptile, acts as outgroup. Among the lower degree of animals, comparatively less conservedness was observed. As represented in Fig. S2 in the online version at https://doi.org/10.1016/bs.apcsb.2022.01.005, protozoans were seen clustered in three different clades. One cluster comprising of 6 Leishmania species, 2 Leptomonas species and *Bodo saltans*, was seen to share similarity with an algae *Phytophthora sojae* suggesting the transfer of genes horizontally. The P-gp protein of *Leishmania tarentolae* have diverged itself from other species of same genus and was found to share homology with Trypanosomatids. This sister taxa of this clade is a bacteria, *Streptomyces coelicolor*. The third cluster of protozoans consist of plasmodium and toxoplasma species. Even belonging to same order *Trypanosomatidae*, species belonging to genus *Leishmania* and *Leptomonas* share more similarity than *Trypanosomes*. Trypanosomes are more homologs to bacteria and protists and further to other protozoans like *Toxoplasma* and *Plasmodiums*. Similar kind of observations were also made by Hamilton et al. in 2004, where they studied evolutionary relationship between *Trypanosomes* and *Leishmania* on the basis of few genes like small subunit ribosomal RNA (SSU rRNA), glycosomal glyceraldehyde phosphate dehydrogenase (gGAPDH), spliced leader (SL) RNA gene etc. They observed that although belonging to the same order *Trypanosomatidae*, *Trypanosomes* and *Leishmania* have diverged on the genomic level (Hamilton, Stevens, Gaunt, et al., 2004; Maslov, Lukes, Jirku, & Simpson, 1996). P-gp also seems to be one of the genetic marker in differentiating these two. In a detailed review by Lukes and group, they considered genome sequence analysis over morphological features in establishing evolutionary relationship between various members of Kinetoplastidae family (Lukeš, Skalický, Týč, et al., 2014). Seven species of angiosperms representing the Plantae Kingdom is seen to be conserved with *Sorghum bicolor*, a monocot, being diverged from Eudicots, *Genlisea aurea* being an exception (Fig. S3 in the online version at https://doi.org/10.1016/bs.apcsb.2022.01.005). In the evolutionary tree, mollusks are closely related to chordates exception being *Anadara sativa* and *Tegillarca granosa* which are comparatively closer to nematodes. Although nematodes display more homology with arthropods belonging to insecta (*Lygus hesperus*) and arachnida (*Tetranychus cinnabarinus*) class with *Loa loa* being an exception. The phylogenetic tree viewed as a whole for the 96 organisms' displays *Candida albicans*, a fungus, as outgroup. The tree is further subdivided in two clades, one having arthropods (mainly crustaceans and insects) and other clade comprising of rest of the organisms.

For identifying the positions determining specificity (SDPs), JDet was used where *C. albicans* served as the master sequence. Within that, Xdet method was used with a cut off of 0.6 and entropy cutoff as 2.5. From the various position specific residues obtained, we were particularly interested in those residues which directly interacts with P-gp substrate, i.e., Miltefosine. The Miltefosine interacting residues for *L. major* P-gp, which have been previously determined by us (Kabra et al., 2020), are S359,

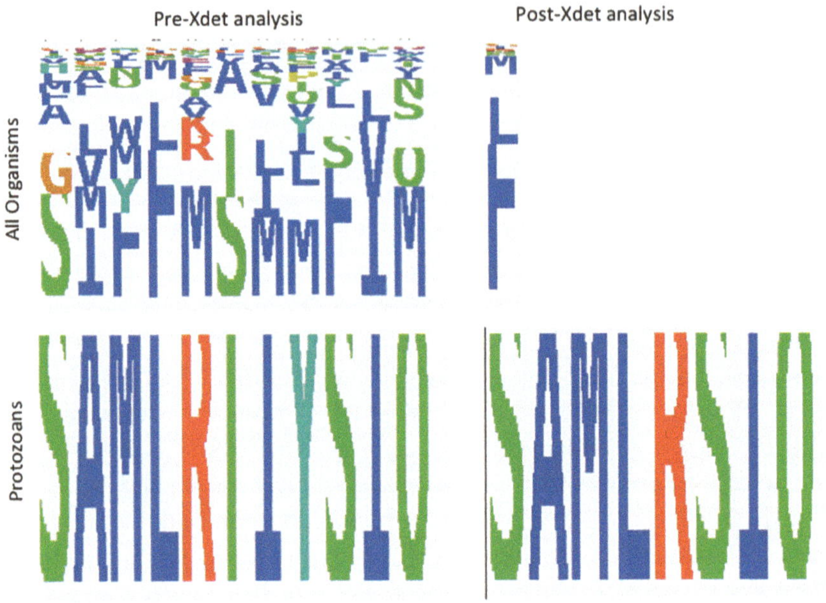

Fig. 3 P-gp residues interacting with Miltefosine in sequence logo format. The upper panel shows sequence logo of P-gp from all 96 organisms while lower panel includes 17 protozoans. The left panel shows all the 11 residues interacting with Miltefosine. The right panel shows only those Miltefosine interacting residues that are also specificity-determining positions obtained through Xdet analysis with xdet cutoff as 0.6 and entropy cutoff as 2.5.

A360, M364, L397, R941, T1008, I1011, Y1012, S1041, I1044 and Q1049. Of these 11 residues, only one residue, L397, was obtained post Xdet analysis. But, in case of protozoan species only, 8 residues (S359, A360, M364, L397, R941, S1041, I1044 and Q1049) were identified as SDPs. The Miltefosine interacting residues for Pgp (all organisms and only protozoans) pre and post Xdet analysis are represented in Fig. 3.

3.2 P4ATPase

P4ATPase is a flippase belonging to type 4 sub-family of P-type ATPases. It mainly transports phospholipids from outer leaflet of membrane toward inner leaflet (cytosol). Fully annotated sequences for P4ATPase of 207 organisms were retrieved from NCBI, as of 2015. Of these, 80% are fungi and the rest 20% comprises of protista, protozoa, cnidarian, nematodes, arthropods and chordates (Fig. 4). The minimum and maximum length of P4ATPase protein was observed in fungi with a *Saccharomycetes*, *Sugiyamaella lignohabitans*, having 776 residue length and *Taleromyces cellulolyticus*, a *Eurotiomycetes*, having the protein length of 2061 residues. It's evident that this subfamily of P-type ATPases have been well characterized in Fungi. As demonstrated in Fig. 4, *Trichomonas vaginalis*, a protozoan parasite, was seen as outgroup. The protozoans were clustered together and similar to that of P-gp proteins, *Leishmania* species and *Leptomonas* species were closely related. Among, *Leishmnaials*, *L. braziliensis* and *L. panamensis* have diverged from other species with L. *amazonensis* and *L. mexicana* being sister taxa and *L. major* and *L. infantum* being sister taxas (Fig. S4 in the online version at https://doi.org/10.1016/bs.apcsb.2022.01.005). Although belonging to Trypanosomatidae family, trypanosomes were observed to be deviated from Leishmania genus as well as Leptomonas genus. Our observation from P-gp and P4ATPase protein backs up the evolutionary relationship established from the small and large subunit rRNAs of different Trypanosomatids as studied by Maslov and group (Maslov et al., 1996). This also validates the consideration of genes/proteins over morphological features while talking about the evolution of these protozoan parasites. In an elaborated review by Lukes and group, they pointed toward more possibility of gene duplication events in trypanosomes having approximately 11,000 protein coding genes when compared with Leptomonas and leishmanials having approximately 8200 protein coding genes (Lukeš, Butenko, Hashimi, et al., 2018). The three protists in which fully annotated P4ATPase sequence have been characterized and retrieved, *Naegleria gruberi* was found to share more similarity with protozoans compared to the other two protists, *Acanthamoeba castellanii* and *Dictyostelium lacteum*. The Cnidarians and arthropods along with a few chordates (*Bos taurus* and *Macaca mulatta*) were observed to be diverged from a common ancestor with *Trichusris trichiura* (a nematode) being an exception (Fig. S5 in the online version at https://doi.org/10.1016/bs.apcsb.2022.01.005). The rest of the chordates follow an independent lineage with P4ATPase in humans sharing similarity with rodents. As foresaid, Fungi P4ATPases have been well characterized and out of sequences belonging to 166 fungal species being retrieved; 117 belongs to Ascomycota division, 39 to Basidiomycota, 3 each to Zygomycota and Chytridiomycota and 1 each to Entomophthoromycota, Mucoromycota, Glomeromycota and Deuteromycota (Fig. S6 in the online version at https://doi.org/10.1016/bs.apcsb.2022.01.005). The evolutionary divergence suggests Ascomycots and Basidiomycots have diverged from a common ancestor and are closely

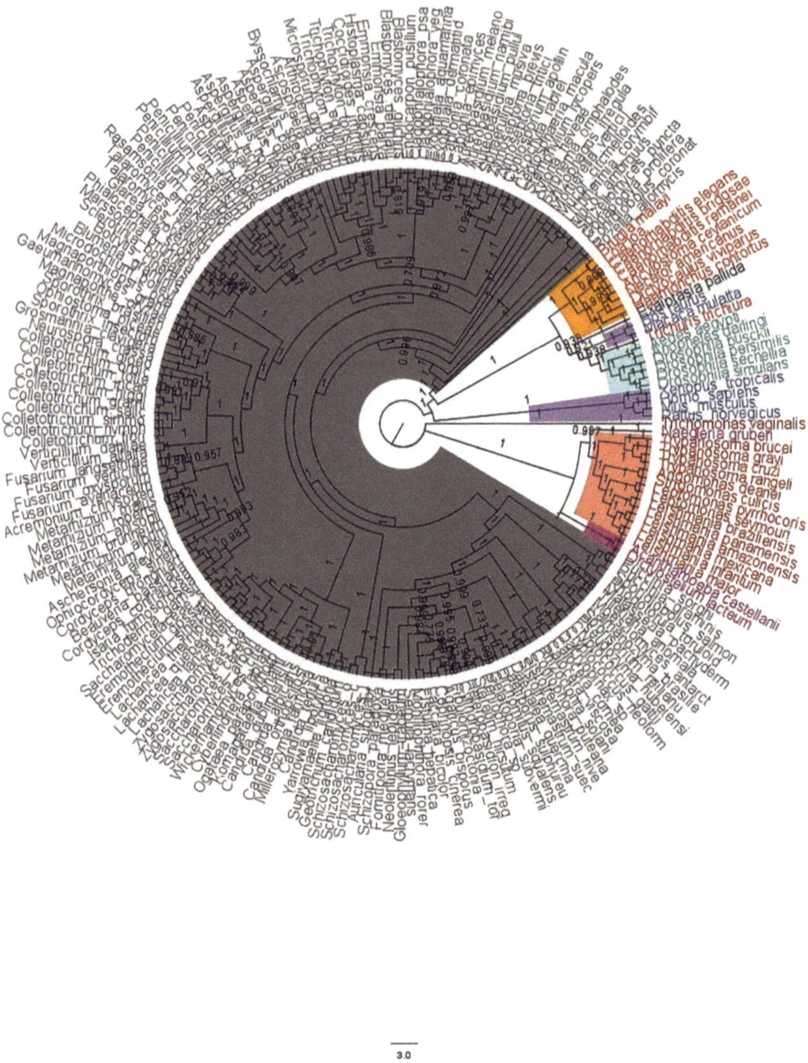

Fig. 4 Phylogenetic tree construction and evolutionary analysis of P4ATPase proteins of 207 organisms. Color code: Red—Protozoa; Blue—Chordata; Gray—Fungi; Cyan—Arthropoda; Orange—Nematoda; Purple—Protista.

related to Glomeromycots. Among these fungal species, *Rozella allomycis*, a Chytridiomycot, was believed to have been evolved first in comparison to other members of same division like *Spizellomyces punctatus* and *Gonapodya prolifera*.

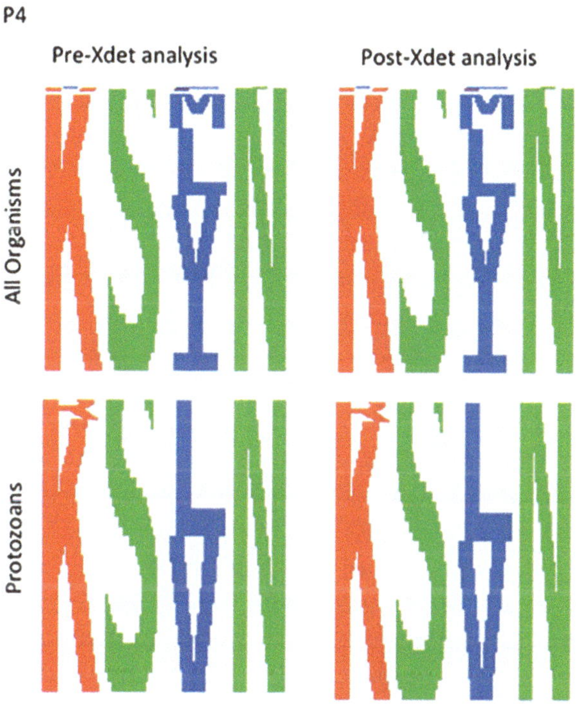

Fig. 5 P4ATPase residues interacting with Miltefosine in sequence logo format. The *upper panel* shows sequence logo of P4ATPase from all 207 organisms while lower panel includes 15 protozoans. The left panel shows all the 4 residues interacting with Miltefosine. The right panel shows only those Miltefosine interacting residues that are also specificity-determining positions obtained through Xdet analysis with xdet cutoff as 0.6 and entropy cutoff as 2.5 (in this case all 4 residues are identified as SDPs).

For the SDPs in P4ATPase protein, *Trichomonas vaginalis* served as the reference sequence. K121, S365, V372 and N870 residues of *L. major* P4ATPase are interacting with Miltefosine (Kabra et al., 2020). All the four residues were identified as SDPs post Xdet analysis whether all 207 organisms or just 15 protozoans were considered. These residues in sequence logo format are shown in Fig. 5. It was observed that residues "K," "S" and "N" were highly conserved in all organisms while "V" shared place with "M" and "L" (majorly in protozoans). This conservedness of Miltefosine interacting residues was expected since P4ATPase proteins are specifically lipid transporters and Miltefosine is an analog of phosphatidylcholine.

3.3 CDC50

CDC50 acts as β-subunit required for the functioning as well as membrane localization of P4ATPase (López-Marqués et al., 2015). The heterodimeric complex of P4ATPase-CDC50 proteins together is responsible for the influx of their substrates toward the cytosol. As of 2015, fully annotated sequences of CDC50 protein belonging to 189 organisms were retrieved from NCBI. Similar to P4ATPase, CDC50 proteins are also very well characterized in Fungi with fully annotated sequences retrieved from 117 organisms comprising of 62% of the total organisms considered for phylogenetic analysis (Fig. 6). Unlike P-gp and P4ATPase, in terms of sequence length CDC50 protein was observed to be much conserved with a range from 245 residues (*Drosophila sechellia*, an arthropod) to 612 residues (*Trichinella sp. T9*, a nematode) with *Amazona aestiva,* a chordate, being an exception with protein length of 1537 residues. As observed in P4ATPase, in case of CDC50 proteins also a protozoan was found to be acting as outgroup, i.e., *Trypanosoma grayi*. The rest of the protozoans, specifically kinetoplastids, have evolved together with CDC50 protein of L. *major* being diverged from other leishmanial species (Fig. S7 in the online version at https://doi.org/10.1016/bs.apcsb.2022.01.005). Arthropods, nematodes and mollusks were observed to have diverged from a point with chordates being the other branch of bifurcation which further seems to have diverged from platyhelminth, *Schistosoma japonicum*. *Stegodyphus mimosarum*, an African spider, shares more sequence similarity with a mollusk, *Crassostrea gigas*, rather than with the members of same phylum and believed to have evolved earlier than other arthropods and nematodes (*Trichinella* species). Another arthropod, *Artemia sinica*, deviates from other members by being homologous to *Trichinella* species (Fig. S8 in the online version at https://doi.org/10.1016/bs.apcsb.2022.01.005). Of the two protists for which annotated CDC50 sequences were available, *Capsaspora owczarzaki*, have evolved much earlier and diverged from the group of taxas which branch into platyhelminthes, nematodes, arthropods and chordates; while all these groups have been diverged from the other protist, *Thecamonas trahens* (Fig. S9 in the online version at https://doi.org/10.1016/bs.apcsb.2022.01.005). Coming to chordates, sequences of CDC50 proteins from 24 chordates were retrieved and their evolutionary relationship is shown in Fig. S10 in the online version at https://doi.org/10.1016/bs.apcsb.2022.01.005. The Mammals and Aves form the sister clades. Unlike, P-gp and P4ATPase, CDC50 protein of humans shares similarity with a Sumatran orangutan, *Pongo abelii*, which is also a critically endangered species.

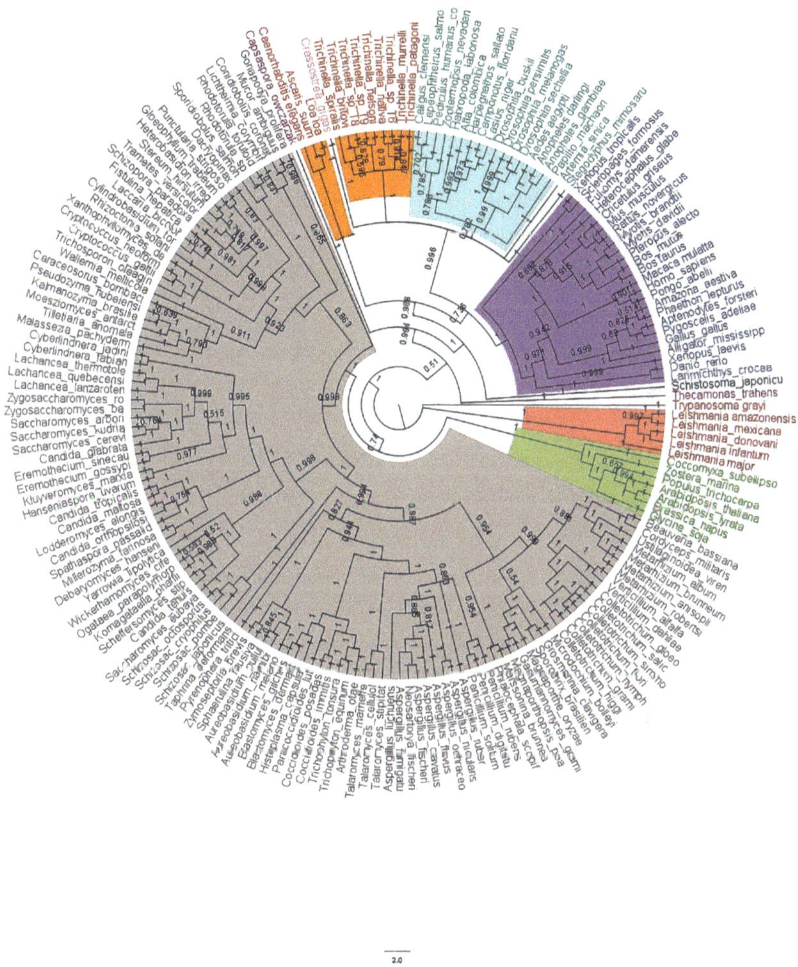

Fig. 6 Phylogenetic tree construction and evolutionary analysis of CDC50 proteins of 189 organisms. Color code: Red—Protozoa; Blue—Chordata; Gray—Fungi; Cyan—Arthropoda; Orange—Nematoda; Purple—Protista; Green—Plants; Dark Green—Platyhelminthes and Pink: Mollusk.

Although belonging to same genus, *Xenopus tropicalis* and *Xenopus laevis,* two different species of aquatic frog, an amphibian, have diverged a lot with the former sharing similarity with *Scleropages formosus*, a fish, and the later acting as an outgroup to other chordates (Fig. S10 in the online version at https://doi.org/10.1016/bs.apcsb.2022.01.005). Of the 117 fungal species, 88 belongs to Ascomycota division, 25 to Basidiomycota, 2 Zygomycota and 1 each to Chytridiomycota and Entomophthoromycota (Fig. S11 in the online version at https://doi.org/10.1016/bs.apcsb.2022.01.005). Among them, *Gonapodya prolifera*, a Chytridiomycot, was perceived to be outgroup. Of the two zygomycots, *Mucor ambiguous*, shares more resemblance with entomophthoromycot, *Conidiobolus coronatus* rather than its fellow member i.e., *Lichtheimia corymbifera*. The sister clade of these three comprises of basidiomycots.

In case of *L. major*'s CDC50 protein, F151, Q159, M235 and W236 interacts directly with Miltefosine (Kabra et al., 2020). When all organisms were considered, none of these residues fit in the criteria of Xdet analysis, but when only protozoans were considered, all these miltefosine interacting residues acted as SDPs post Xdet analysis (Fig. 7). Unlike P4ATPase, in CDC50 proteins, Miltefosine interacting residues are far diverged when all organisms were considered and hence no residues lied in Xdet analysis. In contrast, these residues showed their dominance in protozoans.

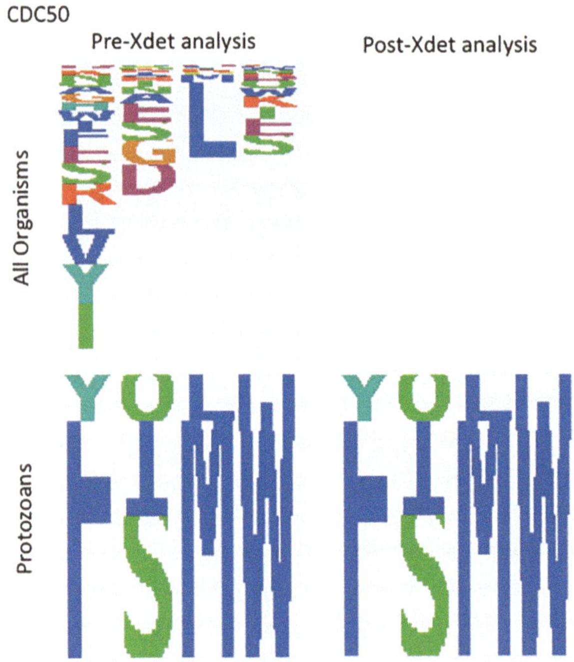

Fig. 7 CDC50 residues interacting with Miltefosine in sequence logo format. The upper panel shows sequence logo of CDC50 from all 189 organisms while lower panel includes 6 protozoans. The left panel shows all the 4 residues interacting with Miltefosine. The right panel shows only those miltefosine interacting residues that are also specificity-determining positions obtained through Xdet analysis with xdet cutoff as 0.6 and entropy cutoff as 2.5.

4. Discussion

The transporter proteins holds utmost position in any cell or organisms as they not only help in the uptake of nutrients, small molecules or ions but are also responsible for the extrusion of toxic substances, xenobiotics, drugs etc. Since different transporters are specific for their substrates hence, these proteins have remained highly conserved in the due course of evolution. Here in this project, we have tried to study the evolutionary relationship of Miltefosine transporter proteins in *Leishmania major* with the homologous proteins present in other organisms. P4ATPase and CDC50 proteins are responsible for the uptake of Miltefosine drug while P-gp is responsible for the extrusion of drug outside parasite. On analyzing the completely annotated sequences of these proteins among four kingdoms of life, i.e., Protista, Fungi, Animalia and Plantae; we found that these proteins have shown quite a high degree of conservedness across different classes of organisms. Most of the organisms belonging to a phylum or division were found to be clustered together with very exceptional cases of horizontal gene transfer. Through these analysis it was realized that Type IV—P-type ATPase have been studied and characterized more than the ABC transporter, P-glycoprotein (P-gp) especially among fungal species. The identification of SDPs through Xdet analysis confirmed that drug binding residues were more conserved in P4ATPase rather than in its accessory protein CDC50 or in P-gp. But when a particular class of organisms was considered, like in protozoans, the three Miltefosine transporter proteins, i.e., P-gp, P4ATPase and CDC50; were found to exhibit more conservedness. Similar observation was observed through phylogenetic analysis too. In spite of being kinetoplastids, Trypanosomes and Leishmanias were observed to have diverged from each other sequentially. But they were found to possess same substrate specificity as various Trypanosome strains like *Trigonoscuta cruzi*, *T. copemani*, etc. were found to be susceptible of Miltefosine (Botero, Keatley, Peacock, & Thompson, 2017; Luna, Hernández, Rueda, et al., 2009; Saraiva, Gibaldi, Previato, et al., 2002). In nutshell, these phylogenetic analysis as well as finding of SDPs not only helped in understanding the evolutionary pattern of a particular protein in an organism but can also help in targeting a protein in an organism specific manner to either increase or decrease their functional activity. Using this concept, the sequential and positional conservedness of P-gp, P4ATPase and CDC50 proteins between Human (host) and

Leishmania major (parasite) have helped in designing parasite specific peptides that allosterically modulate activity of Miltefosine transporter proteins resulting in reversal of Miltefosine resistant parasites to the drug sensitive form (Kabra et al., 2020). Similar strategy can be implied for other proteins in same or different organisms. Thus, the study of evolutionary relationship of a probable target protein across diverse organisms may provide probable and long-lasting solutions in curing a disease or fighting against resistance.

Acknowledgment

Ritika Kabra would like to acknowledge her SRF (Senior Research Fellowship) from Department of Biotechnology (DBT/JRF/14/AL/22/3544), Ministry of Science and Technology, Government of India. Authors would also like to thank the Director, National Centre for Cell Science (NCCS) Pune, for supporting the Bioinformatics and High Performance Computing Facility (BHPCF) at NCCS Pune, India.

Authors contribution

All authors contributed to study design and conception. Ritika Kabra: data curation, methodology implementation and writing the original draft. Shailza Singh: Initialization and conceptualization, data curation, supervision, writing and editing the original draft. All authors agree with the submission of the final version of the manuscript.

References

Anjard, C., & Loomis, W. F. (2002). Evolutionary analyses of ABC transporters of Dictyostelium discoideum. *Eukaryotic Cell*. https://doi.org/10.1128/EC.1.4.643-652.2002.

Axelsen, K. B., & Palmgren, M. G. (1998). Evolution of substrate specificities in the P-type ATPase superfamily. *Journal of Molecular Evolution*. https://doi.org/10.1007/PL00006286.

Botero, A., Keatley, S., Peacock, C., & Thompson, R. C. A. (2017). In vitro drug susceptibility of two strains of the wildlife trypanosome, Trypanosoma copemani: A comparison with Trypanosoma cruzi. *International Journal for Parasitology: Drugs and Drug Resistance*. https://doi.org/10.1016/j.ijpddr.2016.12.004.

Croft, S. L., Sundar, S., & Fairlamb, A. H. (2006). Drug resistance in leishmaniasis. *Clinical Microbiology Reviews*, *19*, 111–126. https://doi.org/10.1128/CMR.19.1.111-126.2006.

Dawson, R. J. P., & Locher, K. P. (2006). Structure of a bacterial multidrug ABC transporter. *Nature*, *443*, 180–185. https://doi.org/10.1038/nature05155.

Ghorbani, M., & Farhoudi, R. (2018). Leishmaniasis in humans: Drug or vaccine therapy? *Drug Design, Development and Therapy*, *12*, 25–40. https://doi.org/10.2147/DDDT.S146521.

Hamilton, P. B., Stevens, J. R., Gaunt, M. W., et al. (2004). Trypanosomes are monophyletic: Evidence from genes for glyceraldehyde phosphate dehydrogenase and small subunit ribosomal RNA. *International Journal for Parasitology*. https://doi.org/10.1016/j.ijpara.2004.08.011.

Herwaldt, B. L. (1999). Leishmaniasis. *Lancet*, *354*, 1191–1199. https://doi.org/10.1016/S0140-6736(98)10178-2.

Igarashi, Y., Aoki, K. F., Mamitsuka, H., et al. (2004). The evolutionary repertoires of the eukaryotic-type ABC transporters in terms of the phylogeny of ATP-binding domains in eukaryotes and prokaryotes. *Molecular Biology and Evolution*, *21*, 2149–2160. https://doi.org/10.1093/molbev/msh226.

India National vector Borne Diseases Control Programme. (2017). *Accelerated plan for Kala-Azar elimination 2017* (pp. 1–94). WHO.

Jones, P. M., & George, A. M. (2004). The ABC transporter structure and mechanism: Perspectives on recent research. *Cellular and Molecular Life Sciences*, *61*, 682–699. https://doi.org/10.1007/s00018-003-3336-9.

Jordan, I. K., Kota, K. C., Cui, G., et al. (2008). Evolutionary and functional divergence between the cystic fibrosis transmembrane conductance regulator and related ATP-binding cassette transporters. *Proceedings of the National Academy of Sciences of the United States of America*. https://doi.org/10.1073/pnas.0806306105.

Kabra, R., Ingale, P., & Singh, S. (2020). Computationally designed synthetic peptides for transporter proteins imparts allostericity in Miltefosine resistant L. major. *The Biochemical Journal*, *477*, 2007–2026. https://doi.org/10.1042/BCJ20200176.

Lam, V. H., Lee, J.-H., Silverio, A., et al. (2011). Pathways of transport protein evolution: Recent advances. *Biological Chemistry*, *392*, 5–12. https://doi.org/10.1515/BC.2011.018.

López-Marqués, R. L., Poulsen, L. R., Bailly, A., et al. (2015). Structure and mechanism of ATP-dependent phospholipid transporters. *Biochimica et Biophysica Acta, General Subjects*, *1850*, 461–475. https://doi.org/10.1016/j.bbagen.2014.04.008.

Lukeš, J., Butenko, A., Hashimi, H., et al. (2018). Trypanosomatids are much more than just trypanosomes: Clues from the expanded family tree. *Trends in Parasitology*, *34*, 466–480. https://doi.org/10.1016/j.pt.2018.03.002.

Lukeš, J., Skalický, T., Týč, J., et al. (2014). Evolution of parasitism in kinetoplastid flagellates. *Molecular and Biochemical Parasitology*, *195*, 115–122.

Luna, K. P., Hernández, I. P., Rueda, C. M., et al. (2009). In vitro susceptibility of Trypanosoma cruzi strains from Santander, Colombia, to hexadecylphosphocholine (miltefosine), nifurtimox and benznidazole. *Biomédica*. https://doi.org/10.7705/biomedica.v29i3.15.

Maslov, D. A., Lukes, J., Jirku, M., & Simpson, L. (1996). Phylogeny of trypanosomes as inferred from the small and large subunit rRNAs: Implications for the evolution of parasitism in the trypanosomatid protozoa. *Molecular and Biochemical Parasitology*. https://doi.org/10.1016/0166-6851(95)02526-X.

Moitra, K., & Dean, M. (2011). Evolution of ABC transporters by gene duplication and their role in human disease. *Biological Chemistry*, *392*, 29–37.

Møller, A. B., Asp, T., Holm, P. B., & Palmgren, M. G. (2008). Phylogenetic analysis of P5 P-type ATPases, a eukaryotic lineage of secretory pathway pumps. *Molecular Phylogenetics and Evolution*, *46*, 619–634. https://doi.org/10.1016/j.ympev.2007.10.023.

Muth, T., García-martín, J. A., Rausell, A., et al. (2012). JDet: Interactive calculation and visualization of function-related conservation patterns in multiple sequence alignments and structures. *Bioinformatics*, *28*, 584–586. https://doi.org/10.1093/bioinformatics/btr688.

Nanavati, D. M., Nguyen, T. N., & Noll, K. M. (2005). Substrate specificities and expression patterns reflect the evolutionary divergence of maltose ABC transporters in Thermotoga maritima. *Journal of Bacteriology*. https://doi.org/10.1128/JB.187.6.2002-2009.2005.

Ronquist, F., Teslenko, M., Van Der Mark, P., et al. (2012). Mrbayes 3.2: Efficient bayesian phylogenetic inference and model choice across a large model space. *Systematic Biology*, *61*, 539–542. https://doi.org/10.1093/sysbio/sys029.

Saier, M. H. (2016). Transport protein evolution deduced from analysis of sequence, topology and structure. *Current Opinion in Structural Biology*, *38*, 9–17. https://doi.org/10.1016/j.sbi.2016.05.001.

Saraiva, V. B., Gibaldi, D., Previato, J. O., et al. (2002). Proinflammatory and cytotoxic effects of hexadecylphosphocholine (miltefosine) against drug-resistant strains of Trypanosoma cruzi. *Antimicrobial Agents and Chemotherapy*. https://doi.org/10.1128/AAC.46.11.3472-3477.2002.

Saurin, W., Hofnung, M., & Dassa, E. (1999). Getting in or out: Early segregation between importers and exporters in the evolution of ATP-binding cassette (ABC) transporters. *Journal of Molecular Evolution*. https://doi.org/10.1007/PL00006442.

Sievers, F., Wilm, A., Dineen, D., et al. (2011). Fast, scalable generation of high-quality protein multiple sequence alignments using Clustal omega. *Molecular Systems Biology*, *7*, 539. https://doi.org/10.1038/msb.2011.75.

Singh, N., Kumar, M., & Singh, R. K. (2012). Leishmaniasis: Current status of available drugs and new potential drug targets. *Asian Pacific Journal of Tropical Medicine*, *5*, 485–497. https://doi.org/10.1016/S1995-7645(12)60084-4.

Singh, S., & Mandlik, V. (2015). Structure based investigation on the binding interaction of transport proteins in leishmaniasis: Insights from molecular simulation. *Molecular BioSystems*, *11*, 1251–1259. https://doi.org/10.1039/C4MB00713A.

FSC
www.fsc.org
MIX
Papier | Fördert
gute Waldnutzung
FSC® C083411

Kolibri 360 GmbH
Ferdinand-Jühlke-Straße 7
99095 Erfurt, Deutschland
produktsicherheit@kolibri360.de